IFIP Advances in Information and Communication Technology 776

IFIP Advances in Information and Communication Technology

The IFIP AICT series publishes state-of-the-art results in the sciences and technologies of information and communication. The scope of the series includes: foundations of computer science; software theory and practice; education; computer applications in technology; communication systems; systems modeling and optimization; information systems; ICT and society; computer systems technology; security and protection in information processing systems; artificial intelligence; and human-computer interaction.

Edited volumes and proceedings of refereed international conferences in computer science and interdisciplinary fields are featured. These results often precede journal publication and represent the most current research.

The principal aim of the IFIP AICT series is to encourage education and the dissemination and exchange of information about all aspects of computing.

More information about this series at https://link.springer.com/bookseries/6102

Maciej Pondel · Iwona Chomiak-Orsa ·
Eunika Mercier-Laurent · Petr Berka ·
Łukasz Przysucha
Editors

Artificial Intelligence for Knowledge Acquisition and Management

11th IFIP WG 12.6 International Workshop, AI4KAM 2025
Held at IJCAI 2025, Montréal, QC, Canada, August 16–17, 2025
Revised Selected Papers

Springer

Editors
Maciej Pondel
Wroclaw University of Economics and Business
Wrocław, Poland

Iwona Chomiak-Orsa
Wroclaw University of Economics and Business
Wrocław, Poland

Eunika Mercier-Laurent
University of Reims Champagne-Ardenne
Reims, France

Petr Berka
Prague University of Economics and Business
Prague, Czech Republic

Łukasz Przysucha
Wroclaw University of Economics and Business
Wrocław, Poland

ISSN 1868-4238 ISSN 1868-422X (electronic)
IFIP Advances in Information and Communication Technology
ISBN 978-3-032-18919-6 ISBN 978-3-032-18920-2 (eBook)
https://doi.org/10.1007/978-3-032-18920-2

This Springer imprint is published by the registered company Springer Nature Switzerland AG
The registered company address is: Gewerbestrasse 11, 6330 Cham, Switzerland

Preface

The IJCAI 2025 Workshop on Artificial Intelligence for Knowledge Acquisition & Management (AI4KAM 2025) was held on August 17, 2025, as part of the 34th International Joint Conference on Artificial Intelligence in Montreal. Continuing a tradition that dates back to 1992, the AI4KAM series provides an international forum for presenting and discussing advances in AI-driven knowledge management. Over the years, the workshop has evolved into a vibrant space where researchers and practitioners examine how artificial intelligence supports knowledge acquisition, representation, and discovery, and organizational decision-making. Its thematic scope includes AI-based business process automation, knowledge representation and reasoning, machine learning for knowledge discovery, natural language processing, AI in health care, smart city solutions, and AI-supported educational environments.

The 2025 edition preserved the classical structure of contributed, peer-reviewed presentations. The workshop program encompassed thematic blocks reflecting the diversity of AI applications in knowledge management: AI for organizational processes, relationship management, smart city solutions, education, business intelligence, multi criteria decision-making, and medical knowledge management. The program also featured an invited talk by Joel Alleyne on "Embedded Intelligence — Implementing Intelligent IT Solutions — Promise or Peril?", providing valuable insights from long-term experience in AI-supported knowledge-intensive organizations. A total of 73 submissions were sent for peer review. Following the open peer-review process, each submission received at least two reviews, involving both Program Committee members and external reviewers. Ultimately, 30 full papers and 17 short papers were accepted for presentation and inclusion in the proceedings. No invited papers were included in the reviewed and accepted scientific content. Contributions co-authored by Program Committee members were handled independently to ensure objectivity and fairness.

The accepted papers collectively illustrate the breadth of current research on AI-supported knowledge management. They range from methodological advances in intelligent agents, natural language processing, large language models, and multi-criteria decision analysis, to practical implementations in business, education, public administration, logistics, smart cities, and medicine. This diversity reflects both the rapid technological progress in artificial intelligence and the expanding need to manage knowledge effectively across domains.

We are deeply grateful to the Program Committee members and external reviewers for their dedicated work in assessing submissions and ensuring the scientific quality of the workshop. We also thank the organizing team and our colleagues who supported the logistics and dissemination of the event. Finally, we express our appreciation to all authors for their high-quality contributions and to all participants for their active engagement during discussions.

We hope that this volume will serve as a valuable reference for researchers and practitioners interested in how AI technologies contribute to knowledge-intensive processes, and that it will inspire further exploration of intelligent methods in organizational and societal contexts.

October 2025

Iwona Chomiak-Orsa
Maciej Pondel

Organization

General Chairs

Iwona Chomiak-Orsa	Wrocław University of Economics and Business, Poland
Maciej Pondel	Wrocław University of Economics and Business, Poland

Program Committee Chairs

Petr Berka	Prague University of Economics and Business, Czech Republic
Agnieszka Siennicka	Wrocław Medical University, Poland
Eunika Mercier-Laurent	University of Reims Champagne-Ardenne, France
Łukasz Przysucha	Wrocław University of Economics and Business, Poland
Mieczysław Owoc	Wrocław University of Economics and Business, Poland

Program Committee

Frederic Andres	National Institute of Informatics, Japan
Yevgeniy Bodyanskiy	Kharkiv National University of Radio Electronics, Ukraine
David Chudán	Prague University of Economics and Business, Czech Republic
Krzysztof Hauke	Wrocław University of Economics and Business, Poland
Marcin Hernes	Wrocław University of Economics and Business, Poland
Artur Rot	Wrocław University of Economics and Business, Poland
Jan Vanthienen	Katholieke Universiteit Leuven, Belgium
Tomáš Kliegr	Prague University of Economics and Business, Czech Republic
Krzysztof Kluza	AGH University of Science and Technology, Poland

Antoni Ligęza	AGH University of Science and Technology, Poland
Agnieszka Matera-Witkiewicz	Wrocław Medical University, Poland
Kazimierz Perechuda	Wrocław University of Economics and Business, Poland
Jeanne Schreurs	Hasselt University, Belgium
Pradeep Singh	KIET Group of Institutions, India
Yashwant Singh	Jaypee University of Information Technology, India
Małgorzata Sobińska	Wrocław University of Economics and Business, Poland
Michael Stankosky	University of Scranton, USA
Sudeep Tanwar	Nirma University, India
Sudhanshu Tyagi	Thapar Institute of Engineering & Technology, India
Julian Vasiliev	University of Economics – Varna, Bulgaria

Additional Reviewers

Elwira Gross Gołacka	University of Warsaw, Poland
Paula Bajdor	Warsaw University of Technology, Poland
Kesra Nermend	University of Szczecin, Poland
Dorota Jelonek	Częstochowa University of Technology, Poland
Artur Karczmarczyk	West Pomeranian University of Technology in Szczecin, Poland
Jaroslaw Watrobski	University of Szczecin, Poland
Ewa Ziemba	University of Economics in Katowice, Poland
Jaroslaw Was	AGH University of Science and Technology, Poland
Kamila Łuczak	Wrocław University of Economics and Business, Poland
Wojciech Sałabun	National Institute of Telecommunications, Poland
Andrii Shekhovtsov	West Pomeranian University of Technology in Szczecin, Poland
Mateusz Piwowarski	University of Szczecin, Institute of Management, Poland
Marta Starostka-Patyk	Częstochowa University of Technology, Poland
Paweł Ziemba	Maritime University of Szczecin, Poland
Joanna Kołodziejczyk	National Institute of Telecommunications, Poland
Michał Jan Petri	Wrocław University of Economics and Business, Poland

Contents

AI for Smart Cities, Public Sector, and Government Innovation

AI in Business Processes, Logistics, Industry, and Market Behavior

AI in Education, Learning, and Academic Environments

Using AI as a New Ethical Challenge in Higher Education Management. An Example of a European Union Country

Anna Brzozowska[1], Elwira Gross-Gołacka[2], and Aleksander Pabian[1(✉)]

[1] Częstochowa University of Technology, Ul. Dąbrowskiego 69, 42-201 Częstochowa, Poland
aleksander.pabian@hotmail.com

[2] University of Warsaw, Ul. Krakowskie Przedmieście 26/28, 00-927 Warszawa, Poland

Abstract. Nowadays, artificial intelligence tools are increasingly used in the processes of knowledge transferring, learning and in conducting scientific research. It can be assumed that in the near future their usage will become common, which will affect the change in the process of transformation of science and time management by students. It seems that AI offers enormous development opportunities but, at the same time, carries certain threats, mainly of an ethical nature. Therefore, one of the new challenges of higher education management is the need to regulate the principles of using AI tools in the teaching and learning process in general as well as when writing theses. Many universities around the world have noticed this new challenge and have already introduced appropriate internal regulations and guidelines, others are only introducing or modifying existing ones. However, regulations cannot be created ad hoc and, above all, their content should be a response to real challenges and threats. Among other things, research must be conducted in the field of social sciences and related disciplines in terms of the scope and methods of using AI tools by the academic community and the degree of acceptance for the introduced solutions and regulations as well as ethical issues. The article meets this need. Taking into account conclusions drawn from the authors' team research being presented in this article will contribute to the better usage of the potential of AI in education. Combining AI technology with social sciences, including behavioural sciences can also contribute to preventing potential abuses.

Keywords: Artificial intelligence · IT · Education · Management · Ethical issues

1 Introduction

The origins of artificial intelligence (AI) date back to the 1950s, when Alan Mathison Turing published the groundbreaking article, "Computer machinery and intelligence" in the Oxford journal Mind [Kaja, Wesołowska & Tomaszewska, 2024]. Today, artificial intelligence (AI) is finding more and more wide application in various areas of life - "it will be either the best thing that has happened to us, or the worst. If we will not be careful, it could be the last thing for us" - as the famous astrophysicist Stephen Hawking

Published by Springer Nature Switzerland AG 2026
M. Pondel et al. (Eds.): AI4KAM 2025, IFIP AICT 776, pp. 3–14, 2026.
https://doi.org/10.1007/978-3-032-18920-2_1

once said about artificial intelligence [Guzik, Flisowski, 2025]. The current value of the global AI market is already valued at USD 279.22 billion (2024 data) and is expected to grow at a CAGR average annual rate of 35.9% between 2025 and 2030 [Grand View Research, 2023].

In recent years, AI technologies have revolutionized many sectors of the economy, offering innovative solutions that enhance efficiency, service personalization, and customer satisfaction [Maurer, 2021, p. 1]. Artificial intelligence tools are also increasingly used in education, including by lecturers in the teaching and research processes and by students of higher education institutions in the learning processes and writing theses. Consequently, scholars and practitioners are expanding their knowledge on the potential benefits of AI applications in education and seeking effective ways to incorporate AI tools. The results of the analysis of the frequency of the expression "artificial intelligence" in various contexts show that in connection with the word "education" (for example "artificial intelligence in education") it appears much more often than in many other contexts, e.g. with "pharmaceutical industry" or "e-commerce" [Raport PARP, 2023]. Currently, applications such as Microsoft Bing Chat, ChatGPT, Google Bard, Claude, Microsoft Copilot, Google Gemini, DEEPL are particularly popular in teaching. As their creators emphasize, they improve the quality of teaching and learning experiences by personalizing learning and providing real-time feedback. In this way, AI is gradually becoming an integral component of academic practice, influencing both the efficiency of teaching and the quality of learning outcomes, and at the same time raising new questions about ethics, responsibility, and the boundaries of its proper use in higher education.

The analysis of AI-based technologies in education allows to identify its main components: learning platforms and virtual facilitators, intelligent tutoring systems (ITS), smart content, fraud & risk management systems and so on [Raport PARP, 2023]. Their usage provides huge development opportunities in terms of speed of data analysis, detection of research gaps and new areas of research, application of natural language processing models or use of large-scale language models. These tools also create opportunities for new models of collaboration between students and educators, offering the potential to enrich academic dialogue and foster innovation in the didactic process. Currently, the artificial intelligence market in education is dominated by educational platforms and virtual moderators - their share in revenue amounted to over 45.0% [Grand View Research, 2023].

Importantly, in addition to educational support, AI can be also used as a tool to support students' mental health (we mean here e.g. intelligent chat) [Śluzek and Suppan, 2024]. Psychological support using chatbots is already used today to reduce anxiety and symptoms of depression [Lattie et al., 2019]. This tool can act as an interlocutor to whom students can turn to express their concerns, fears or frustrations in an anonymous and safe way. In addition, intelligent chat can provide information about healthy lifestyle strategies, provide advice on relaxation techniques, time management and offer educational resources on mental health. Although it will not replace professional help, it provides easily accessible support in everyday stressful situations and emotional challenges. This is a very important social function of AI, especially in the context of the results of the latest research conducted by Cigna US Loneliness Index, which shows that 73% of representatives of Generation Z feel lonely. Similarly, research from Harmony Healthcare IT reveals that 42% of representatives of Generation Z have diagnosed mental health problems and 85% feel anxious about their future [Kotler et al., 2024].

On the other hand, the rapid expansion of these tools stimulates broader discussions about how academic institutions should adapt curricula, teaching strategies, and evaluation methods to the realities of an AI- supported learning environment. Their widespread availability makes it increasingly difficult to separate traditional forms of student work from those supported by digital assistants, which further intensifies the need for clear guidelines and transparent policies. Artificial intelligence carries certain risks, mainly in the area of authorship appropriation, i.e. plagiarism. This fact cannot be ignored in the process of managing a university.

Artificial intelligence should only support creative processes and doesn't replace thinking and creativity. The University of Oxford website emphasizes, among other things, that artificial intelligence tools cannot replace human critical thinking or the development of scientific arguments based on evidence and subject knowledge, which are the basis of an university education. They can only provide support in the learning processes [Oxford Students]. Therefore, one of the new challenges of higher education management is the need to regulate the principles of using AI tools in the teaching and learning process in general as well as when writing theses. The aim is to develop a coherent body of knowledge about good practices in the usage of AI in the process of scientific research and education at a higher education institution and to unify the relevant legal regulations in this respect. This also applies to the aspect of proper time management by students themselves [Raszyd et al., 2024], which is also important in the context of the increasing activity of young people in cyberspace towards narrowing the sphere of life in the real world [Pabian and Pabian, 2025].

However, developing guidelines and regulations in higher education institutions is demanding. It should not be done quickly and spontaneously – it must be carefully thought out. Above all, such regulations have a major impact on the functioning of the entire university, including the quality of education, academic processes and student development. The regulations must also be consistent with the applicable guidelines of the superior law and take into account the specificity of a given university and its needs. Finally, they should take into account the opinions of various stakeholders, such as lecturers, students and administration, in order to be realistic and effective. Only then can the adopted rules be respected and successfully implemented in everyday practice. Properly prepared regulations not only ensure legal compliance and organizational order, but also build trust within the academic community, minimize potential conflicts, and create a stable framework for the responsible use of new technologies such as artificial intelligence.

In order for the process of formulating guidelines to run smoothly and to the benefit of the academic community, research is needed, among other things in the field of social sciences and related disciplines, in terms of the scope and methods of using AI tools by the academic community and the degree of acceptance of the introduced solutions and legal regulations as well as ethical issues. The article addresses this need. Its aim is to diagnose the scope and methods of using artificial intelligence by students in the learning/writing processes and to check whether they have no hesitation in accepting the guidelines and regulations adopted by universities. The results of the study show that students, while recognizing the potential of AI, simultaneously accept legal regulations. However, another question arises here, which the authors decided to answer – does this acceptance result from the fact that students attach great importance to ethical conduct or are they rather afraid of potential negative consequences related to the inappropriate use of AI tools? The authors of the article decided to find answers to this question, among others, using the results of research conducted on a group of students from one of the large universities from an European Union country.

Obtained results, when implemented in practice, will contribute to the better usage of the potential of AI in education and will improve the management of higher education in this area. The results will be also an inspiration for other researchers to undertake further, interdisciplinary and multidirectional research in the field of the use of AI in education.

2 Ethical Aspects of Using AI by Students – Literature Review

As mentioned earlier, artificial intelligence is one of the most important trends in contemporary social reality. This is a term reserved to describe technologies that fundamentally transform our lives, that is, technologies such as the steam engine and the Internet causing significant and widespread impact on society and the economy [Przegalińska, 2022]. AI is also the future of education. It has already begun to revolutionize teaching, although it is worth agreeing that the possibilities of its use are still new to both researchers and practitioners [Chiu et al., 2023]. Currently, students can use generative AI tools to support the development of their personal academic skills. Especially popular are solutions in the field of educational software, which contains an AI component that continuously tracks students' work, adjusts feedback and provides guidance. Because this software collects information about the student's performance, it is able to draw conclusions about the student's strengths and weaknesses and suggest remedies, e.g. assigning additional work [ScienceDirect].

Researchers agree that in this situation, addressing the challenges related to learners support, digital exclusion and ethical issues is essential to ensure fair and responsible implementation of AI in open and distance learning [Amin et al., 2025]. Therefore, university centres are already introducing solutions and regulations that, on the one hand, teach how to properly perceive and use AI tools, but do not ignore the ethical aspects of working with artificial intelligence. For example, it is emphasized that in order to prevent plagiarism, students using AI during their studies need to learn and practice traditional academic skills, such as note-taking and clear attribution, which distinguish their own work from any text or material originating from AI tools [Oxford Students].

It seems that, among others, universities from European Union are actively participating in such initiatives. Not without reason, because the leading bodies of this political and economic international organization, which brings together 27 member states, feel responsible for shaping new, digital future. This is confirmed by the fact that in 2024 EU introduced the first in history, most detailed legal regulations regarding artificial intelligence.

The legal regulations concerning artificial intelligence undertaken and introduced by the European Union are closely linked to the concept of sustainable development which is very popular in Europe. Its idea is to improve social well-being in addition to overcome ecological and economic challenges. The EU aims to ensure that the development of technologies, including artificial intelligence, is in line with the principles of social, ecological and economic responsibility. EU regulations, including these regarding AI, focus on safety, ethics and transparency, ensuring that these technologies improve the quality of life and do not harm people and the environment. With an appropriate legal framework, the EU aims to prevent risks related to the use of AI while promoting innovations that will support long-term development, job creation and sustainable economic growth. This approach is in line with the sustainable development goals, which combine technological progress with concern for social good and environmental protection.

With regard to scientific matters, the regulation of the European Commission and the Council of Europe emphasises, among other things, that any event, any research and development activity should be carried out in accordånce with recognised ethical and professional standards for scientific research and should be conducted in accordance with applicable Union law [Regulation (EU) 2024/1689]. It should also be agreed that in this area, interdisciplinary cooperation is needed between computer scientists, educators, psychologists and ethicists, which will make AI tools effectively support educational goals, while maintaining appropriate cultural sensitivity and emotional boundaries [Levkovich et al., 2025]. The challenges are so significant that the economic problem for all of Europe, not just Poland, becomes the lack of employees with appropriate qualifications [Brdulak & Piekarski, 2022]. Following the trend set by the Union, many European universities have already introduced regulatory solutions in the field of AI to prevent unfair benefits and maintain academic integrity. In Poland, the scale of investment in new solutions is high. Well-computerized universities can have from several, to a dozen or so, IT systems in which they process student data, implement business processes and conduct financial and statistical reporting [PCG Academia Homepange]. All universities already use a uniform anti-plagiarism system to detect co-authoring of work by AI. For example, tools such as Turnitin are used to detect plagiarism. However, according to an expert from the Information Processing Center, technologies that allow detecting whether a given text was created using, for example, ChatGPT are not very effective, yet. Adding an appropriate function to the Uniform Anti-plagiarism System (UAS) requires time and additional financial outlays [Bankier.pl].

However, despite institutionalized preventive and training efforts implemented in higher education institutions, students do not always follow these recommendations. For example, A. Ghimire and Y. Qiu's research shows that international students found AI particularly valuable for cultural adaptation and language support, which of course is not a problem itself. At the same time, however, they actively used these tools despite the opposition of the institution seeking to limit their use, which emphasized the gap between student needs and institutional practices [Ghimire and Qiu, 2025].

3 Methods

This article addresses the management challenges universities face in the context of using generative artificial intelligence in knowledge transfer and learning processes. The aim of the article is to diagnose the scope and ways in which students use AI in the learning process, to check whether students accept university guidelines that limit their voluntariness in using AI and whether such acceptability is ethically conditioned or has another basis.

The main research problem was expressed in the form of the question: what are the effects of putting into service and using artificial intelligence systems in the process of transferring knowledge and learning? The main problem was specified through issues that took the form of the following sub-questions:

Do Polish students commonly use AI tools in the process of learning and writing their theses?
What AI tools are currently used by Polish students?
What are the benefits and disadvantages of using AI tools according to students?
Are Polish students aware of the ethical threats related to the use of AI?
Do Polish students accept the current legal regulations related to AI and adopted by their university?

Research hypotheses were not formulated due to the qualitative nature of the study. The authors adopted the principle of conducting a posteriori research in close correlation with theoretical thinking, while recognizing the studied socio-cultural reality as a system (structure) of interconnected elements. Since the purpose of the research did not require determining numerical parameters, it was decided to choose the qualitative method. In its framework, direct, survey methods of collecting data from primary sources were used.

At the beginning, a critical analysis of the literature was carried out based on the methodology of a systematic literature review, i.e. selection of material according to the criteria of basic database, key words, inclusion and elimination [Czakon, 2011]. The research used international literature on the subject. Information was also obtained from the websites of domestic and foreign universities that implemented legal regulations specifying the principles of using AI tools in the study processes.

Next, four focus group interviews were conducted, three with 5 students and one with 4 students, because focus group interviews are used to clarify sensitive and personal issues [Kaczmarczyk, 2014]. The ethical use of AI in the learning process undoubtedly has such character. The idea was also to give respondents the opportunity to speak more broadly. In both cases, the research tool was a questionnaire containing open-ended questions. Personal interviews were conducted in mid- February 2025. The participants were students who took part in a seminar of one of the co-authors of the article. Among them were 4 women and 15 men. The entire seminar group consisted of over 20 people, but three students refused to participate in the study, explaining that they were unable to attend the study during the week due to personal reasons or work obligations (part-time students).

The choice of these people seems justified. Students completing their bachelor's theses seem to be the most suitable group to examine issues related to the use of artificial intelligence in the process of acquiring knowledge and creating written works at university. These are the people who already have experience in research and creating academic texts, not only encounter new technologies in science, but also in everyday life and work. It was considered that their opinions and experiences regarding AI in the context of preparing theses can provide valuable information on the practical application of artificial intelligence in higher education.

The focus group study conducted according to the form was supplemented in each case by free-form interviews (a type of which is a conversation), which were chosen due to their more spontaneous nature. It was assumed that, if the researcher, hearing the answers to previously prepared questions, suddenly discovers a previously unplanned but important from the point of view of research thread, he can follow this new path in order to obtain answers to the questions of interest which are enclosed in the formula of qualitative research: what?, where?, how?, why? These interviews provided valuable insights into experiences, expectations, and opinions regarding the application of AI technology in the higher education.

The final stages of the interviews included transcription, analysis, verification, and reporting. The transcription was post factum, i.e. the text was prepared from previously obtained material. It did not reflect aspects of the statements, such as intonation or sarcastic tone of the statements. However, the focus was on maintaining a balance between the authenticity of the statements and their readability. The analysis of the qualitative research results was conducted in accordance with the orientation of interpretivism, derived from Dilthey, phenomenology and symbolic interactionism.

The supplementary method and autopsy were also used, i.e. participant observation and experience gained from the work of academic lecturers and supervisors of numerous bachelor's and master's theses as well as doctoral dissertations. This observation was public.

4 Empirical Results

The authors of the article intend to share the full conclusions and recommendations resulting from the conducted research with the scientific community in a series of publications. They plan to present in detail the results of their analyses, which can serve as valuable guidelines for future research in the area of the application of artificial intelligence in education, as well as for practitioners interested in implementing new technologies in academic processes. These works will aim to deepen the understanding of the topic and consider the potential implications for the future of science and technology in higher education.

At this point, being aware of the limitations related to the volume of the article, the authors decided to present in its research part, the most important and interesting, in their opinion, conclusions from the research which has been conducted. They can be a proper starting point for further, in-depth studies and research.

The results of the empirical research conducted by the authors of this article lead primarily to the conclusion that, contrary to appearances, Polish students do not yet use AI tools very widely in the processes of learning and preparing papers, or at least they do not admit to do it. As, if contrary to the statements from the literature emphasizing that the use of artificial intelligence tools as assistants in everyday life is becoming more and more common - "just as the use of mobile phones or laptops has become an integral part of everyday life" [Raszyd et al., 2024, p. 374].

This is related to three fundamental issues which reflect a real gap between the theoretical forecasts and actual student practices, suggesting that the penetration of AI into academic environments is still at an early stage. First of all, as the interviews showed, students have developed a way of working over the years that is effective for them and meets their individual needs (using the Internet, as indicated by respondents from each focus group, but not necessarily requiring AI). Many people have adapted their methods of learning and preparing work to their own preferences which allows them to manage their time effectively and achieve their goals. Therefore, although AI can be a useful tool in many cases, students do not feel the need to use it extensively because their existing strategies have proven sufficient and comfortable. From the perspective of many of them, AI support is not necessary because their methods of work are already well-established and adapted to their learning style.

It is also probably not without significance that the interviewed students do not have full knowledge of the currently available commercial AI solutions created for use in various areas. Therefore, they do not use them widely. Only 5 of all the interviewees were able to indicate and name any solutions other than ChatGPT. Among the examples given were Google Gemini, Murf (generating voice), BeautifulAI, PresentationsAI (generating presentations), Galaxy AI (as a solutions dedicated to Samsung smartphones). This observation suggests that the limited adoption of AI tools in the academic environment is not only a matter of students' preferences or attitudes but also of awareness and access to reliable information. Without a broader understanding of the functionalities, potential benefits, and limitations of such technologies, students are less inclined to experiment with them or incorporate them into their study practices.

Third, some students have concerns about their own skills in using AI tools (Fig. 1). These results, somehow contradict the view that AI-based applications are simple and intuitive. Expressed concerns turned out to be at least partially justified, as it turned out that previous attempts to use artificial intelligence ended with unsatisfactory results for some students. An example can be a situation in which one of the students tried to use a tool to generate content for one of the classes (Ecologistics) but the results obtained were imprecise and inadequate to the academic requirements. He considered that this effect resulted not only from the fact that AI is not always able to fully reflect the academic context or the specific requirements of a given job, but also from the personal lack of full knowledge of the tool's functions and capabilities. Another interlocutor summed it up briefly, but accurately, quoting: "with chat, as with any other tool. You also have to know how to use it […] Formulate questions appropriately and edit results - correct, rephrase and adapt to the context".

In this way, unsatisfactory results in using AI (also due to one's own inability) should be considered an emphasized AI disadvantage. An equally important aspect of the study was also to find out about the ethical aspects related to the usage of AI by students. First, they were asked, if they consider using AI assistance to be unethical - for example, as a form of cheating the lecturer/tutor who assigns them to prepare a text. The vast majority of people (15) did not see a major problem in this. As one of the interviewees asked rhetorically: "why not make your life easier by using available and

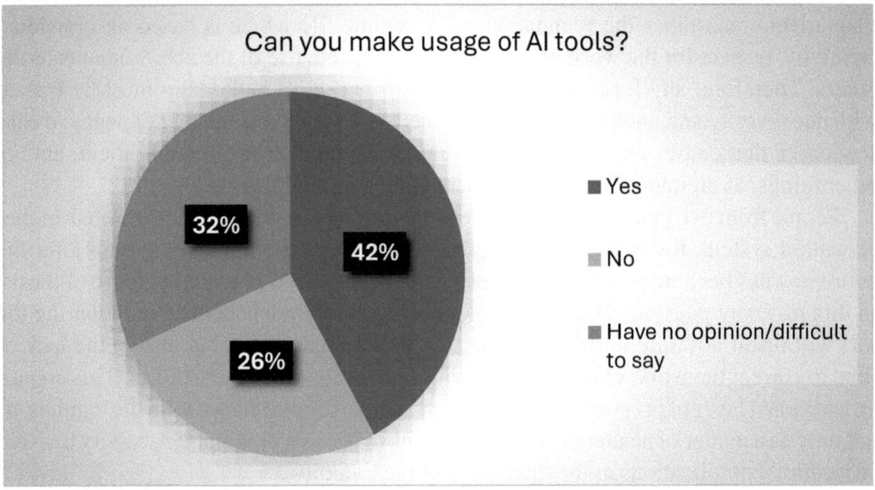

Fig. 1. Interviewees' opinions on their own skills in using AI tools. Source: own elaboration based on research results.

still legal methods?". At the same time, however, an even greater number of interlocutors expressed confidence that they understand steps undertaken by university to limit the use of artificial intelligence. They confirmed that they accept current, operative regulations adopted by the university and were not surprised by the tendency to introduce further.

The conjunction of these results initially caused a kind of cognitive dissonance in conversation's moderator: on the one hand, the interlocutors have no ethical barriers to make usage of artificial intelligence tools, on the other hand, they are not hindered by steps aimed at limiting or prohibiting them from doing so. This apparent contradiction raised further questions about how students actually perceive the coexistence of personal freedom in using AI and institutional restrictions. It suggested that their attitudes may not be based solely on individual preferences, but also on an awareness of the broader academic context in which rules and practices interact. Do they think that such measures are a natural consequence of the rapid technological changes taking place in education and reflect the need to maintain clarity and fairness in the learning processes? Do the respondents treat the regulations as an expected element of academic life, demonstrating a pragmatic approach to the evolving role of AI in higher education, or there is another reason? He obtained an explanation of this matter by spontaneously developing topic.

Perhaps it is wrong, but it turns out that students associate the use of artificial intelligence in writing papers with the possibility of committing plagiarism in them (appropriation of the authorship of someone else's work, i.e. presenting someone else's work as one's own). Plagiarism constitutes a serious violation of academic ethics and copyright law. It is an unacceptable act, contrary to the principles of scientific integrity and intellectual honesty. According to Article 115 of the Copyright and Related Rights Act, this is an offence punishable by a fine, restriction of liberty or imprisonment for up to 3 years [Journal of Laws 1994, No. 24, item 83]. A person who commits plagiarism, not only undermines his credibility but also exposes himself to serious legal and disciplinary consequences, including criminal liability and sanctions stipulated in university regulations.

Plagiarism undermines the foundations of academic life which is based on individual creativity, respect for the work of others and responsible use of the achievements of the others. Therefore, any form of unauthorized attribution of authorship must be treated with due severity and unequivocally condemned. Above all, it is the fear of negative consequences that causes students to accept university regulations, treating them, among other things, as an intention to protect them from potential threats.

People from two panels, also pointed out another aspect – namely, in the Polish higher education system, the work is usually checked in the anti-plagiarism system after the entire text has been prepared, shortly before the deadlines of potential defense of thesis. In this way, any possible disclosure of doubts regarding independence in preparing the text will result not only in the consequences already mentioned but also in the lack of time to correct the work, eliminate fragments borrowed from other studies. This creates an additional layer of pressure on students, who perceive compliance with the regulations not only as a matter of academic honesty but, above all, as a pragmatic necessity to avoid procedural complications at the final stage of their studies.

5 Conclusions

Our study, which consists of an attempt to diagnose the scope and ways in which students use AI in the learning process, is important for better recognition of the role of AI in the learning process of students. The findings enrich the ongoing debate on the role of artificial intelligence in education, pointing out that the process of incorporating AI into students' everyday learning practices is more complex and nuanced than commonly presented in the literature. Study emphasizes that understanding the real patterns of AI usage among students requires going beyond general assumptions and examining actual experiences, attitudes and declared behaviours. Only, such an approach allows to identify both the opportunities and the limitations of integrating artificial intelligence into academic contexts. Therefore, the conclusions presented here should be seen not only as an academic contribution to the theory of digital transformation in education but also as practical guidance for institutions of higher learning. By taking into account students' perceptions and practices, universities will be better prepared to design policies, curricula, and support systems that responsibly and effectively incorporate AI technologies, ensuring that their potential is harnessed in ways that genuinely enhance the quality of teaching and learning.

Taking into account the conclusions drawn from the research and presented in this article will contribute to better use of the potential of AI in education. Considering the issue of AI technology from the perspective of social sciences, including behavioural sciences, will also contribute to prevent potential abuse. This is an important issue because, so far, there are no proven, effective and universal ways to deal with using AI in an unethical manner by students. One should agree with the opinion of the expert and scientist that in the case of using AI, the only thing that can be done at present is to use mind, adhere to the principle of limited trust, remember about self-reflection, and in case of uncertainty: "check, check and check again" [Gierak, 2024]. By relying on ethics and reliability, we avoid many problems. Immoral actions drive out ethical ones, can intensify and deepen over time, and consequently lead to individuals' desensitization and immunity to ethical problems [Pabian & Pabian, 2024]. Moreover, immoral actions of one person can be an excuse for another. For this reason, each must enforce proper, dignified behaviour in the use of AI. This will benefit students and lecturers and will serve for the good image of university.

Disclosure of Interests. The authors have no competing interests to declare that are relevant to the content of this article.

References

Amin, M.R.M., Ismail, I., Sivakumaran, V.M.: Revolutionizing education with artificial intelligence (AI)? Challenges and implications for open and distance learning (ODL). Social Sci. Human. Open **11**, 1–8 (2025)

Bankier.pl: Universities are preparing for a clash with artificial intelligence (2023). https://www.bankier.pl/wiadomosc/Uczelnie-wyzsze-szykuja-sie-na-zderzenie-ze-sztuczna-inteligencja-8516585.html. Accessed 15 Mar 2025

Brdulak, J., Piekarski C.: The role of education – the need to develop digital competence: the cybersecurity perspective. In: Niewielska, M. (ed.) Artificial Intelligence (AI) as a Megatrend Shaping Education. How to Prepare for the Socio-Economic Opportunities and Challenges Presented by Artificial Intelligence?, pp. 181–184. Institute of Educational Research, Warsaw (2022)

Chiu, T.K.F., Xia, Q., Zhou, X., Chai, C.S., Cheng, M.: Systematic literature review on opportunities, challenges, and future research recommendations of artificial intelligence in education. Comput. Educ. Artif. Intell. **4**, 1–15 (2023)

Czakon, W.: Methodology for a systematic literature review. Organ. Rev. **3**(854), 57–61 (2011)

Gierak, A.: Is artificial intelligence ethical?, Foundation for the development of the education system https://www.frse.org.pl/aktualnosci/czy-sztuczna-inteligencja-jest-etyczna, last accessed 05/01/2025

Grand view research: market analysis report. Artificial intelligence market size, share & trends analysis report by solution, by technology (deep learning, machine learning, NLP, machine vision, generative AI), by function, by end-use, by region and segment forecasts (2025–2030). https://www.grandviewresearch.com/industry-analysis/artificial-intelligence-ai-market. Accessed 14 Mar 2025

Ghimire, A., Qiu, Y.: Redefining pedagogy with artificial intelligence: how nursing students are shaping the future of learning. Nurse Educ. Pract. **84**, 1–11 (2025)

Guzik, A., Flisowski, K.: Artificial intelligence in education. Does fear have big eyes?. https://www.alinaguzik.com/single-post/sztuczna-inteligencjaw-edukacji-czy-strach-ma-wielkie-oczy?. Accessed 28 Mar 2025

Journal of Laws 1994 No. 24 item 83. https://isap.sejm.gov.pl/isap.nsf/DocDetails.xsp?id=wdu19940240083. Accessed 15 Feb 2025

Kaczmarczyk, S.: Marketing research. Methodological foundations. PWE Press, Warsaw (2014)

Kotler, P., Kartajaya, H., Setiawan, I.: Marketing 6.0. The Future is immersive, Mtbiznes Press, Warsaw (2024)

Lattie, E.G., Adkins, E.C., Winquist, Shields, C.S., Wafford, Q.E., Graham, A.K.: Digital mental health interventions for depression, anxiety, and enhancement of psychological well-being among college students: systematic review. J. Med. Internet Res. **21**(7), e12869 (2019)

Levkovich, I., Rabin, E., Farraj, R.H., Elyoseph, Z.: Attributional patterns toward students with and without learning disabilities: artificial intelligence models vs. trainee teachers. Res. Dev. Disabil. **160**, 1–14 (2025)

Maurer, C.: Digital marketing in tourism, In: Xiang, Z., Fuchs, M., Gretzel, U., Höpken, W. (eds.) Handbook of e-Tourism. pp. 1–24. Springer, Cham (2021)

Oxford Students: Use of generative AI tools to support learning. https://www.ox.ac.uk/students/academic/guidance/skills/ai-study. Accessed 05 Dec 2024

Pabian, B., Pabian, A.: A sketch for a discussion on the issue of ethics in business. J. Mod. Sci. **1**(55), 236–249 (2024)

Pabian, B., Pabian, A.: Online activity time management – dilemmas and challenges. J. Mod. Sci. **1**(61), 236–249 (2025)

PARP, PFR Group, Report: Labor. market, education, competencies. The use of Artificial Intelligence in education. https://www.parp.gov.pl/component/publications/publication/wykorzystanie-sztucznej-inteligencji-w-edukacji. Accessed 12 Mar 2025

PCG Academia Homepange: The 5 most common mistakes made when purchasing and implementing software at universities, https://pcgacademia.pl/5-najczesciej-popelnianych-bledow-zwiazanych-z-zakupem-i-wdrazaniem-oprogramowania-na-uczelniach/. Accessed 05 Feb 2025

Przegalińska, A.: Collaborative artificial intelligence: the example of virtual assistants and conversational AI. In: Niewielska, M. (ed.) Artificial Intelligence (AI) as a Megatrend Shaping Education. How to Prepare for the Socio-Economic Opportunities and Challenges Presented by Artificial Intelligence?, pp. 181–184. Institute of Educational Research, Warsaw (2022)

Raszyd, K.I., Wesołowska, A., Tomaszewska, K.: Artificial Intelligence in science – how students use AI in higher education. Manag. Acad. **8**(3), 373–400 (2024)

Regulation (EU) 2024/1689 of the European Parliament and of the council of 13 June 2024. https://eur-lex.europa.eu/legal-ontent/EN/TXT/?uri=CELEX%3A32024R1689. Accessed 05 Feb 2025

ScienceDirect, Intelligent tutoring system, https://www.sciencedirect.com/topics/psychology/intelligent-tutoring-system. Accessed 09 Mar 2025

Śluzek, N., Suppan, K.: Artificial intelligence in higher education – opportunity or threat? School Prof. Work **27**, 13–24 (2024)

Integrating Artificial Intelligence into Higher Education: Motivations, Challenges, and the Effectiveness of Personalized Learning Paths

Małgorzata Sobińska[1(✉)], Konrad Liszczyk[1(✉)], Wojciech Sobiński[2], Patrycja Zwiech[3(✉)], and Klaudia Smoląg[4(✉)]

[1] Wroclaw University of Economics and Business, Ul. Komandorska 118/120, 53-345 Wroclaw, Poland
{malgorzata.sobinska,konrad.liszczyk, konrad.liszczyk}@ue.wroc.pl, w.sobinski@student.uw.edu.pl, patrycja.zwiech@usz.edu.pl, klaudia.smolag@pcz.pl

[2] Faculty of Mathematics, Informatics and Mechanics, University of Warsaw, Ul. Stefana Banacha 2, 02-097 Warszawa, Poland

[3] University of Szczecin, Al. Papieża Jana Pawła II 22a, 70-453 Szczecin, Poland
patrycja.zwiech@usz.edu.pl

[4] Czestochowa University of Technology, Ul. Dabrowskiego 69, 42-201 Czestochowa, Poland
klaudia.smolag@pcz.pl

Abstract. This article explores current trends in the integration of artificial intelligence (AI) in higher education, with a focus on personalized learning pathways and factors affecting the motivation and engagement of students and academic staff. A review of recent literature highlights AI-driven tools such as intelligent tutoring systems, adaptive learning models, educational chatbots, and predictive analytics, which can improve flexibility, accessibility, and educational quality, while enabling early identification of learning challenges and providing automated feedback. The article also presents findings from a survey conducted at the Wroclaw University of Economics and Business, examining experiences, expectations, and attitudes toward AI use in teaching. Special attention is given to the demand for more flexible, student-centred learning formats and to the challenges of implementing AI in academic settings. The results show strong support for hybrid models, alongside both optimism and ethical concerns regarding AI's role in education. The paper concludes with practical recommendations for universities aiming to enhance personalization, reduce dropout rates, and improve program quality through responsible use of AI technologies.

Keywords: Artificial Intelligence in Higher Education · Personalized Learning Pathways · Student Engagement and Retention · Flexible Learning Formats

Published by Springer Nature Switzerland AG 2026
M. Pondel et al. (Eds.): AI4KAM 2025, IFIP AICT 776, pp. 15–21, 2026.
https://doi.org/10.1007/978-3-032-18920-2_2

1 Introduction

Modern universities face increasing competition, declining student numbers, and evolving student expectations, which create a strong need to transform traditional teaching models. Flexible and online learning formats are gaining importance in response to the needs of diverse academic and professional audiences [3, 16].

Artificial intelligence supports innovation in teaching, knowledge management, and the design of personalized learning paths. Its implementation can improve educational quality, increase accessibility, and reduce dropout rates [1, 11]. Both students and academic teachers have high hopes for artificial intelligence, as well as concerns about its impact on human interaction, the role of teachers, and data privacy. These topics were explored through a survey conducted for this article.

The literature shows that research on artificial intelligence in higher education is rapidly expanding. Intelligent tutoring systems are among the most prominent applications, functioning like virtual mentors that adapt content to individual needs. They enable real-time progress tracking and personalized feedback [5, 6]. Adaptive learning models allow students to follow individualized learning paths at their own pace [7]. Artificial intelligence also supports predictive analytics, helping to detect risks such as burnout or dropout [1, 11]. Generative AI and educational chatbots are increasingly integrated into online and hybrid learning, supporting tasks such as content explanation and personalized tutoring [4, 18]. Affective artificial intelligence, which considers emotional responses, is also emerging as a way to foster student engagement and prevent emotional or educational crises [8]. Despite these opportunities, challenges remain, including data protection, algorithmic bias, and the need for staff training [1, 6]. Based on the literature, it is important to examine how students and academic staff at Polish universities perceive the integration of artificial intelligence in teaching, what benefits and risks they see, and how teaching format preferences may influence their motivation to continue studying or teaching. Despite the growing body of international research on AI in education, there remains a noticeable gap in empirical studies focused on user perspectives within the context of Polish higher education institutions. The research problem addressed in this article focuses on identifying the following issues:

- Students' and academic staff's attitudes toward the use of AI in teaching,
- The level of acceptance of personalized learning paths,
- The key benefits and concerns related to the use of AI in education,
- Factors influencing engagement and motivation in the context of digital transformation in higher education.

This article examines emerging trends in the application of artificial intelligence in higher education, focusing on its potential to enhance learning personalization, student motivation, and institutional competitiveness. It also presents findings from a survey conducted at the Wroclaw University of Economics and Business, offering an initial assessment of academic stakeholders' readiness to adopt AI-driven educational tools.

2 Research Methodology

The study employed a diagnostic survey methodology using an online questionnaire distributed among students and academic staff at the Wroclaw University of Economics and Business. A total of 310 individuals participated, including 278 students and 32 faculty members. This division enabled the identification of both converging and diverging perspectives on the integration of artificial intelligence and preferences regarding instructional formats. The survey instrument was developed based on a thorough review of contemporary academic literature, with items thematically aligned to areas such as AI-driven personalization, ethical considerations (e.g., data protection and the human role in teaching), and attitudes toward hybrid education. The questionnaire consisted of closed-ended questions (Likert scales and multiple-choice) and one open-ended item. Prior to deployment, a pilot study with 10 respondents ensured clarity and technical reliability. The final version was disseminated via the university's internal communication channels, and participation was voluntary and anonymous. Although the sample exhibited an imbalance between students and faculty, this reflects the structural composition of the university population. Given the exploratory and descriptive nature of the study, the adopted sample structure remains sufficient for capturing general perceptual trends. The analysis was limited to descriptive statistics (frequencies and cross-tabulations) rather than inferential comparisons, in line with the study's diagnostic aims.

3 Results

The analysis of the research results confirms that one of the key challenges of contemporary higher education is the growing demand for flexible forms of teaching, with particular emphasis on hybrid models. A clear preference for hybrid learning was observed among respondents, with 64% of students and nearly 47% of academic staff selecting this format as their preferred mode of instruction. However, notable differences emerged in the preference for fully in-person classes: only 9.35% of students favored this option, compared to as many as 31% of academic teachers (Fig. 1).

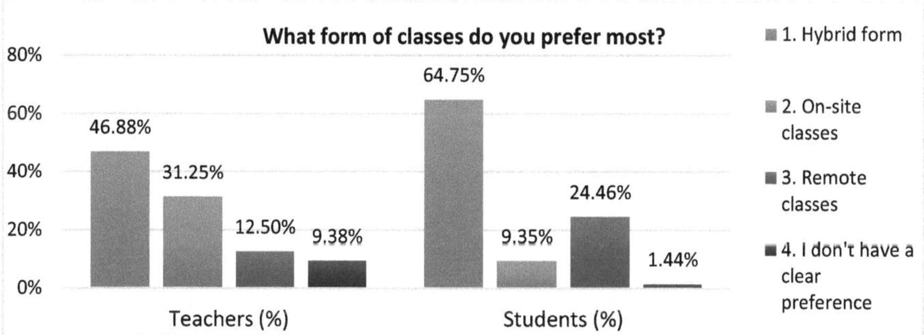

Fig. 1. Distribution of opinions on preferred form of classes

Nearly 70% of academic teachers and more than 70% of students believe that flexibility in choosing the form of classes could increase their motivation to remain at the university (Fig. 2).

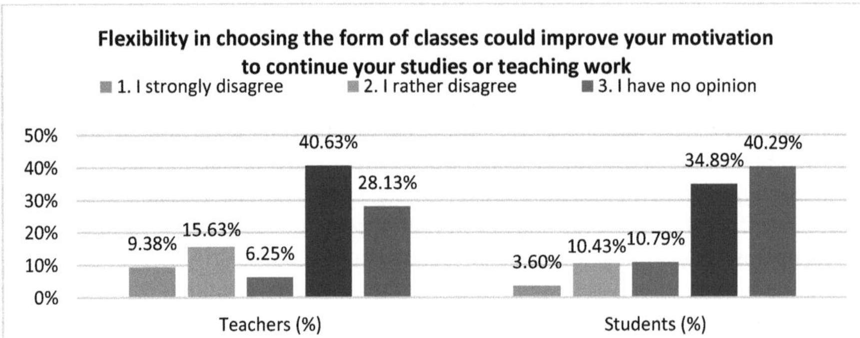

Fig. 2. Distribution of responses from students and academic staff regarding the motivational impact of flexible learning formats

Respondents generally show a positive attitude toward the use of artificial intelligence in the teaching process, recognizing its potential to improve the organization of education and increase its overall effectiveness (Fig. 3).

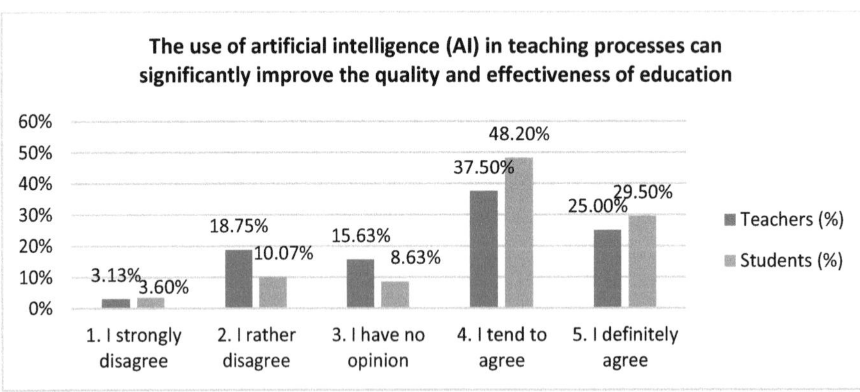

Fig. 3. Distribution of student and academic staff responses to the statement on the perceived impact of AI on teaching quality and effectiveness

Among the main benefits of using AI, respondents indicated: support in creating teaching materials (e.g. generating quizzes, simulations), automatic checking of assignments and tests (time efficiency, fast personalized feedback, objectivity, scalability), data analysis to support teaching decisions (early detection of student difficulties), and personalization of the learning path (individual recommendations, pace adjustment). The distribution of opinions within the surveyed groups is presented in Fig. 4.

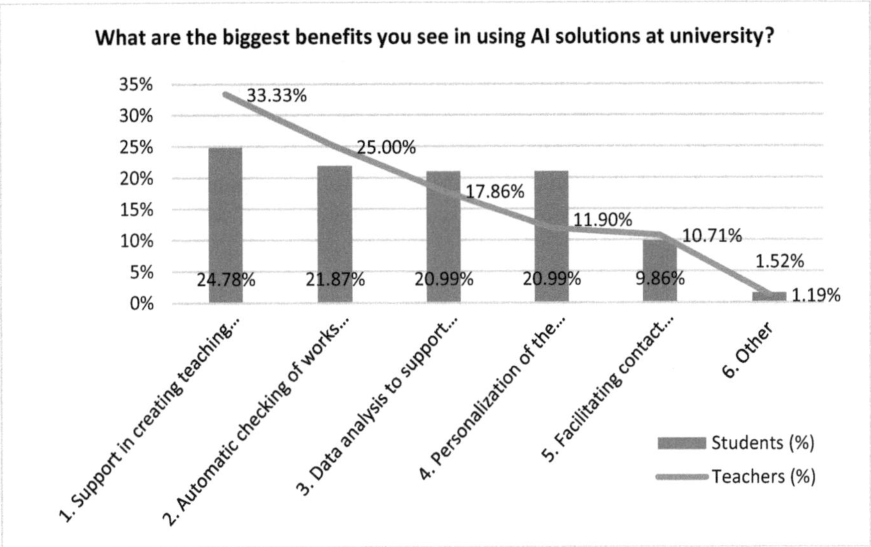

Fig. 4. Top perceived benefits of using AI solutions at university – comparison between students and academic staff

Respondents highlight the importance of acquiring AI-related skills in both professional and academic settings. They emphasize the need to educate people about its benefits and dangers. The importance of interpersonal interaction in education continues despite growing automation. The findings also highlight the importance of enhancing the quality of online instruction and acknowledge the role of AI in assisting student retention via personalized learning. Respondents urge increased spending on technological infrastructure and on the digital skills of instructional faculty.

4 Discussion

The findings of this study confirm that higher education is undergoing a significant shift from traditional face-to-face models toward hybrid and AI-supported approaches. Students strongly favor flexible, hybrid formats that align with their academic, professional, and personal commitments. In contrast, academic staff demonstrated a stronger preference for in-person teaching, often citing concerns about reduced interpersonal interaction and technical challenges. These discrepancies highlight the need for universities to act as learning organizations and equip educators with digital competencies to use AI tools effectively and ethically [7]. While some apprehension remains, both groups recognize the value of AI in enhancing educational efficiency, particularly through early identification of learning difficulties and real-time feedback to support student retention [3, 8]. The perceived weaknesses in current online teaching models point to the necessity of improving e-learning design, as emphasized in instructional technology research [10]. The integration of AI should be approached not merely as a technological upgrade but as part of a broader knowledge management strategy that supports curriculum optimization, personalized learning paths, and institutional knowledge development [5, 14]. This strategy must also emphasize equity, data protection, and human interaction to ensure AI complements, rather than replaces, pedagogical relationships [13, 20]. Although the study was limited to a single university and featured an imbalanced sample across academic roles, it offers a valuable point of departure for further research into AI adoption in higher education and highlights critical areas for institutional development. An extension of the study to additional higher education institutions is planned, accompanied by a multivariate correlation analysis supported by statistical methods.

5 Conclusion

The study confirms that higher education is undergoing a significant shift, with growing demand for flexible, AI-supported teaching models that reflect evolving academic needs. Students favour hybrid formats that support personalization and motivation, while academic staff remain more inclined toward traditional approaches—though both groups acknowledge the value of AI in enhancing educational quality. This article offers a user-informed perspective on AI integration within the underexplored context of Polish universities, highlighting the importance of personalization, digital infrastructure, staff training, and transparent communication. Future research should examine the long-term educational impact of AI, its effectiveness across disciplines, and institutional challenges related to its implementation.

6 Disclosure of Interests.

The authors have no competing interests to declare that are relevant to the content of this article.

References

1. Akavova, A., Temirkhanova, Z., Lorsanova, Z.: Adaptive learning and artificial intelligence in the educational space. In: E3S Web of Conferences (2023). https://doi.org/10.1051/e3sconf/202345106011
2. Alam, A., Mohanty, A.: Educational technology: Exploring the convergence of technology and pedagogy through mobility, interactivity, AI, and learning tools. Cogent Eng. **10**, 2283282 (2023). https://doi.org/10.1080/23311916.2023.2283282
3. Baig, A., Cressler, J.D., Minsky, M.: The future of AI in education: Personalized learning and intelligent tutoring systems. AlgoVista: J. AI Comput. Sci. **1**(1), 592648 (2024). https://doi.org/10.4018/979-8-3693-7220-3.ch013
4. Baskara, F.X.R., Vasudevan, A., Sain, Z.H., Tee, M.: Redefining educational paradigms: integrating generative AI into Society 5.0 for sustainable learning outcomes. J. Infrastruct. Policy Dev. **8**(12), 6385 (2024). https://doi.org/10.24294/jipd.v8i12.6385
5. Batsaikhan, N., Correia, A.-P.: The effects of generative artificial intelligence on intelligent tutoring systems: a systematic review. Smart Learn. Environ. (2024). https://doi.org/10.21428/8c225f6e.33570bb1
6. Crompton, H., Burke, D.: Artificial intelligence in higher education: The state of the field. Int. J. Educ. Technol. High. Educ. **20**, 22 (2023). https://doi.org/10.1186/s41239-023-00392-8

7. Demartini, C.G., Sciascia, L., Bosso, A., Manuri, F.: Artificial intelligence bringing improvements to adaptive learning in education: a case study. Sustainability **16**, 1347 (2024). https://doi.org/10.3390/su16031347
8. Fernández-Herrero, J.: Evaluating recent advances in affective intelligent tutoring systems: a scoping review of educational impacts and future prospects. Educ. Sci. (2024). https://doi.org/10.3390/educsci14080839
9. Gouia-Zarrad, R., Gunn, C.: Enhancing students' learning experience in mathematics class through ChatGPT. Int. Electron. J. Math. Educ. (2024). https://doi.org/10.29333/iejme/14614
10. Halkiopoulos, C., Gkintoni, E.: Leveraging AI in E-learning: personalized learning and adaptive assessment through cognitive neuropsychology—a systematic analysis. Electronics **13**, 3762 (2024). https://doi.org/10.3390/electronics13183762
11. Kuleto, V., et al.: Exploring opportunities and challenges of artificial intelligence and machine learning in higher education institutions. Sustainability **13**, 10424 (2021). https://doi.org/10.3390/su131810424
12. Lim, W.M., Gunasekara, A., Pallant, J.L., Pallant, J.I., Pechenkina, E.: Generative AI and the future of education: Ragnarök or reformation? A paradoxical perspective from management educators. Int. J. Manag. Educ. **21**, 100790 (2023). https://doi.org/10.1016/j.ijme.2023.100790
13. Sajja, R., Sermet, Y., Cwiertny, D., Demir, I.: Platform-independent and curriculum-oriented intelligent assistant for higher education. Int. J. Educ. Technol. High. Educ. **20**, 42 (2023). https://doi.org/10.1186/s41239-023-00412-7
14. Saleem, N., Mufti, T., Sohail, S.S., Madsen, D.Ø.: ChatGPT as an innovative heutagogical tool in medical education. Cogent Educ. (2024). https://doi.org/10.1080/2331186X.2024.2332850
15. Singh, A.P., Saxena, R., Saxena, S.: The future of learning: AI-driven personalized education. Asian J. Curr. Res. (2024). https://doi.org/10.56557/ajocr/2024/v9i49018
16. Slimi, Z.: The impact of AI implementation in higher education on educational process future: A systematic review. University of Deusto (2021). https://doi.org/10.21203/rs.3.rs-1081043/v1
17. Su, J., Yang, W.: Unlocking the power of ChatGPT: a framework for applying generative AI in education. ECNU Rev. Educ. **6**, 355–366 (2023). https://doi.org/10.1177/20965311231168423
18. Van Schoors, R., & Fastré, A.: Learning, teaching and training in the era of Artificial intelligence: challenges and opportunities for evidence-based educational research (2024). https://doi.org/10.13140/RG.2.2.13765.05603
19. Wang, T., et al.: Exploring the potential impact of artificial intelligence (AI) on international students in higher education: generative AI, chatbots, analytics, and international student success. Appl. Sci. (2023). https://doi.org/10.20944/preprints202305.0808.v1
20. Zekaj, R.: AI language models as educational allies: enhancing instructional support in higher education. Int. J. Learn. Teach. Educ. Res. (2023). https://doi.org/10.26803/ijlter.22.8.7

A Study on Awareness of Artificial Intelligence Tools in Primary Education - Case of Poland

Małgorzata Nermend[1](✉) and Małgorzata Łatuszyńska[2]

[1] University of Szczecin, Ogińskiego 16/17, Szczecin 71-431, Poland
lncs@springer.com

[2] University of Szczecin, Cukrowa 8, Szczecin 71-004, Poland

Abstract. The purpose of this article is to study the level of awareness of artificial intelligence tools among elementary school students. The results of the survey indicate that the awareness of artificial intelligence among elementary school students is not yet at a satisfactory level, but the children show great potential and willingness to learn. This means that education about AI in elementary school is not only necessary, but also feasible and promising.

Keywords: Artificial Intelligence · Primary Education · Awareness

1 Introduction

In recent years, artificial intelligence (AI) has begun to play an increasingly important role in education. From intelligent learning support systems to personalized learning platforms to tools that support content creation and progress analysis, AI has the potential to significantly change the way students learn and teachers teach. But the effectiveness of AI-based tools depends not only on their availability or technological sophistication, but also on the level of understanding of their functions and capabilities by users. For elementary school students who are growing up in a digital world, awareness of the existence and operation of AI tools can be the key to a more effective and engaging learning process.

Learning effectiveness is closely linked to the ability to interact with educational material, self-regulate the learning process and select appropriate cognitive strategies. Artificial intelligence (AI)-based tools can support these processes by offering personalized feedback, customization of content to meet individual student needs, or interactive learning environments to enhance motivation, among other things. However, if students are not aware of the existence or capabilities of these technologies, their potential remains untapped. In this context, creating awareness of AI as early as each stage of education can lay the foundation for building digital competence and improving learning effectiveness.

Until recently, the predominant theme in the literature has been research on the introduction of artificial intelligence into secondary and higher education [1–7] or in the context of teachers' experiences and opinions [8–13], however, for several years now, studies have been appearing, often of a review nature, focusing on identifying the level

Published by Springer Nature Switzerland AG 2026
M. Pondel et al. (Eds.): AI4KAM 2025, IFIP AICT 776, pp. 22–33, 2026.
https://doi.org/10.1007/978-3-032-18920-2_3

of understanding of the role and impact of AI on education in elementary schools, for example [14, 15]. They stress that with the increasingly important role of technology in everyday life in the 21st century, AI education should be promoted as early as elementary schools, and not limited to high schools or colleges. However, the awareness of elementary school students and their understanding of artificial intelligence skills are still under-researched [16]. This raises the question - what is the level of awareness and knowledge of artificial intelligence tools among elementary school students? This question is addressed by the purpose of this article, which is to examine the level of familiarity with artificial intelligence tools among elementary school students. By analyzing which tools they are familiar with, how they perceive them, and to what extent they use them in an educational context, the study is intended to lay the groundwork for further analysis on the role of AI awareness in learning processes. And it seems that developing an understanding of AI is not just a part of digital competence today, but can be an important step toward shaping more independent, critically thinking and effective students in a world full of technology.

The remainder of the article is structured as follows: Sect. 2 discusses the theoretical background of the issues raised, Sect. 3 focuses on the methodology used in the research. The next Section presents the results of the research, and the last Section presents the research conclusions, limitations and directions for further work.

2 Theoretical Background

Artificial intelligence (AI), defined after McCarthy as the science and engineering of creating intelligent machines [17], is an interdisciplinary field of science [18]. According to a UNESCO report, knowledge of AI is critical to education in the 21st century [19, 20]. The literature points out that AI teaching significantly supports ethical understanding of AI-generated data[21, 22], as well as prepares students for future work and functioning in society [23, 24]. As Yim [16, 25] points out, the main goals of teaching AI tools skills are to prepare for solving disinformation problems [24] and to promote students' holistic understanding of AI [26].

While past research has focused on AI education in high schools and colleges [22, 27], there has recently been increased interest in teaching AI in elementary schools [22, 25, 28–30], facilitated by the emergence of more age-appropriate learning tools [16, 31–33]. AI literacy instruction is mainly introduced in elementary schools as part of computer science and other science curricula [16, 28, 34, 35], but there are calls for the development of new pedagogical approaches [36–38], which would encompass the entire educational experience of students and support their holistic understanding of AI - not just in selected subjects [18, 26].

One of the foundations for the effective use of AI as an educational tool in elementary schools is the development of awareness of the concept of artificial intelligence among students [16, 29]. However, the literature highlights the widespread lack of such awareness [22, 39–41] and at the same time points out that the level of cognitive development at the elementary school stage places students at an optimal point for effective understanding of AI concepts [16, 29, 30, 35, 42]. In this context, it seems that research is needed to identify the level of awareness and knowledge of tools among elementary school students.

An *ad hoc* search of Scopus, Web of Science, Science Direct, Springer Link and EBSCO databases conducted on April 23, 2025 for the phrase: ((“artificial intelligence” OR AI) AND (education) AND (“primary school” OR “primary students” OR “primary education”) AND (awareness)) in the title, keywords or abstract (for years 2015 to 2025; articles, conference publications and chapters in books; in English only) revealed that few studies to date have examined awareness and knowledge of AI tools among elementary school students. These are summarized in Table 1.

Table 1. Research on awareness and knowledge of artificial intelligence tools among elementary school students.

Author (s)	Title	Short description
Wala [39]	Primary school students’ perceptions of artificial intelligence - for good or bad	A survey of elementary school students in Sweden aged 11–12 (60 students) to determine their cognitive and affective perceptions of AI and its current use
Oruc et al. [43]	Primary School Students’ Views on Artificial Intelligence	A qualitative study of the views of students (25 fourth-graders) in elementary schools in Turkey on artificial intelligence
Kang et al. [44]	Elementary school students’ awareness of the use of artificial intelligence chatbots in violence prevention education in South Korea: a descriptive study	A descriptive study of South Korean elementary school students’ (215 students in the fourth to sixth grades) awareness the use of AI-based chatbots in violence prevention education
Mertala et al. [22]	Finnish 5th and 6th grade students’ pre-instructional conceptions of artificial intelligence (AI) and their implications for AI literacy education	Qualitative survey of 195 Finnish students in grades 5 and 6 to analyze understanding of AI concepts

Source: own elaboration

The table shows that in recent years, there have been surveys among elementary school students in several countries, which in some way related to determining the level of awareness and knowledge of artificial intelligence tools. Of course, there have also been other studies conducted among elementary school students (e.g. [16, 45], but in a different context than the one considered in this study. The surveys tabulated in the table show, among other things, that most students have heard of AI or have had contact with it (e.g., chatbots, voice assistants, educational apps). However, their knowledge is superficial, intuitive and often based on perceptions from the media rather than an understanding of the technology.

Given these observations, it was decided to conduct similar research in elementary schools in Poland. The methodology of this research is presented in the next section.

3 Methodology

A study of the level of awareness, knowledge, experiences and attitudes of elementary school students toward artificial intelligence (AI) tools was conducted using a proprietary survey questionnaire. The research group consisted of students (boys and girls) aged 10–15, attending six public elementary schools in Szczecin. The selected establishments were characterized by a similar number of students, ranging from 500 to 750. The selection of schools was based on the criterion of differentiated educational results, with the basis for the classification being the *Ranking of Elementary Schools Szczecin 2025* published by the WaszaEdukacja.pl educational portal (Ranking of Elementary Schools 2025, n.d.). The ranking took into account the results of eighth-grade examinations and other relevant indicators of educational quality, which made it possible to select schools representing different levels of educational achievement. In addition, as part of the preliminary research, face-to-face interviews were conducted with principals, which provided information on the social structure of the study population. The data obtained indicated that the parents of the students represent diverse social backgrounds, have different levels of education and material status, as well as different expectations from the educational process of their children. Social and economic conditions, as well as the child's close environment, have a significant impact on the child's level of educational awareness, the way he acquires knowledge and his aspirations for education.

The survey questionnaire consisted of 15 closed and semi-open questions divided into four sections. The first concerned students' awareness and knowledge of artificial intelligence (3 questions). The next concerned the use of AI tools (4 questions), while the third referred to opinions and concerns about the use of AI (3 questions). The last part contained 5 questions on views and attitudes about various issues related to AI technology. A link to the survey (https://forms.gle/JwSAfW2tf9JJV68G8) and a QR code for completing it were made available to students through the electronic journal.

The questionnaire was addressed to about 3,000 students of selected elementary schools, covering the population of students in grades IV-VIII, i.e., between the ages of 10 and 15. Prior to the survey, respondents were informed of its purpose and how the results would be used. The survey was anonymous; there was no way to identify participants or assign the number of responses to individual schools. The return rate of the questionnaires was about 6%, while a survey sample of 143 respondents was obtained after the data cleaning process.

The survey was conducted in accordance with the principles set out in the Declaration of Helsinki (2013 version) (World Medical Association, 2013). The scientific experiment was reported to and approved by the Bioethics Committees at Regional Chamber of Physicians in Szczecin. The subjects gave written consent to use the results of the study in the analysis of the scientific paper.

4 Results

Analysis of the results was carried out using structure indicators, which allowed to synthesize the level of awareness and knowledge of students about artificial intelligence. Due to the required volume of the article, the description was limited to the most important issues. In the first part of the survey on students' awareness and knowledge of artificial intelligence (Table2) seventy-two boys and 71 girls took part in the study, bringing the total to 143 participants. Most of them were 12 years old and 13 years old. The vast majority of students surveyed (77.6%) report having knowledge and awareness of artificial intelligence (AI) and confirm that they have encountered the concept. Only 7.7% of respondents admitted that they had not heard of AI before. It is noteworthy that 14.7% of respondents were unable to clearly identify whether they had used AI-based technologies, which may indicate incomplete understanding or uncertainty in identifying AI-based tools.

Interpretation of the term "artificial intelligence" among students varies. The largest group of respondents (59.4%) equates AI with computer programs capable of analyzing data and learning on their own. Another 46, 2% of respondents understand artificial intelligence as robots that can think in a human-like manner. Meanwhile, 27.3% of students identify artificial intelligence with computer games and voice assistants such as Siri or Alexa. It is also worth noting that 11.9% of respondents could not explain the concept of artificial intelligence on their own, which may indicate the need for in-depth education in this area. Analysis of the results indicated that respondents most often came into contact with artificial intelligence through search engines such as Google (39.4%), and a slightly smaller group pointed to social media such as TikTok or YouTube (38%). Much less frequent contact with artificial intelligence-based technologies was in the context of computer games (4.2%) and educational applications (7.7%), while 10.6% of respondents admitted that they had not encountered artificial intelligence or could not clearly state this.

Table 2. Students' awareness and knowledge of artificial intelligence

1. Awareness of AI	Percentage (%)
Had heard of AI	77.6%
Were unsure	14.7%
Hhad not heard of AI	7.7%
2. Interpretation of the term "artificial intelligence" among students	
Self-learning programs:	59.4%
Human-like robots	46.2%
Games/voice assistants	27.3%
No clear idea	11.9%
3. Where Students Encounter AI Content	
Search engines (Google)	39.4%
Social media (TikTok, YouTube)	38%
Computer games	4.2%
Educational applications	7.7%
Has not encountered artificial intelligence or could not tell	0.6%

Source: own elaboration

In the second part (Table3), which concerned the use of artificial intelligence tools as many as 79.7% of respondents said they had used AI-based applications or programs, while 8.4% declared that they had no such experience, and 11.9% were unsure whether the mazedia they used actually used artificial intelligence. Among the most frequently cited tools, students mentioned Chatbots (67.8%), voice assistants (39.9%), online translators (39.2%), and photo and video editing programs (21%). According to students, the use of artificial intelligence in school education could be particularly useful in translating texts into other languages (57.3%), solving mathematical tasks (42.7%), assisting in writing essays (35.7%), and creating presentations (30.8%). However, 18.9% of respondents could not indicate how they could use the capabilities of AI.

Table 3. Use of AI tools

4. Use of AI Technologies	Percentage (%)
Used AI tools	79.7%
Did not use	8.4%
Unsure	11.9%
5. Most Common AI Tools Used	
Chatbots	67.8%
Voice assistants	39.9%
Online translators	39.2%
Photo/video editors	21%
6. Use of AI-based applications	
Used AI based applications or programs	79.7%
Did not use them	8.4%
Are not sure whether the media they used employed AI	11.9%
7. Potential Uses of AI in School	
Translating texts	57.3%
Solving math problems	42.7%
Essay writing	35.7%
Presentation support	30.8%
I don't how could use the capabilities of AI	18.9%

Source: own elaboration

In the third part concerning opinions and concerns regarding the use of AI (Table 4), despite their generally positive attitude, students also expressed some concerns about artificial intelligence. 54.2% of those surveyed thought artificial intelligence was helpful, 14.8% saw it mainly as a threat, while 31% of students could not clearly state their position. The most frequently cited concerns of students were the possibility of misleading the user (43.4%), the risk of replacing people at work (37.8%), and the risks of invasion of privacy (25.9%), with 32.9% of students expressing no concerns about AI. Interest in furthering knowledge about artificial intelligence was shown by 44.1% of the students surveyed, while 29.4% said they were not interested, and 26.6% remained undecided on the issue.

Table 4. Use of AI tools

8. Concerns About AI	Percentage (%)
Artificial intelligence was helpful	54.2%
Mainly as a threat	4.8%
Could not clearly state their position	31%
9. The most frequently cited concerns of students were:	
Misleading the user:	43.4%
Replacing people at work	37.8%
Risks of invasion of privacy	25.9%
No concerns about AI	32.9%
10. Interest in furthering knowledge about artificial intelligence	
Interested	44.1%
Not interested	29.4%
Undecided	26.6%

Source: own elaboration

In the last part of the survey, respondents answered questions about the use of AI in the school context (Table5). When asked about the organization of classes or workshops on AI in schools, 39.2% of students said that such classes were not held and they are not interested in them, 35.7% expressed a desire to participate in similar initiatives, despite the lack of them so far, 14% of students noted that such classes were held, but they did not participate in them, and 11.2% participated in such classes. Regarding the potential use of AI by teachers, 31.7% of respondents believed that it should be used, as it could facilitate learning, while 29.6% indicated the need for limited use of these technologies. 22.5% of respondents had no opinion on the matter, and 16.2% were against the use of AI in the teaching process, favoring the preservation of traditional educational methods. Students also expressed their concerns about the impact of artificial intelligence on the self-learning process. 44.1%, indicated that the development of artificial intelligence could lead to a decrease in the independence of students, who will rely too much on technology. On the other hand, 23.8% of respondents felt that AI will remain only a tool to support learning, not to limit student autonomy, while 32.2% could not express a clear opinion.

Regarding practical applications of AI in the school environment, students most often pointed to the ability to create personalized tests and quizzes tailored to the individual student's level of knowledge (58%), automatic homework prompts (38.5%), creation of interactive presentations (36.4%) and individual online tutoring (21.7%). There were also isolated responses indicating the use of AI in assisting with daily learning, generating codes for programs or explaining topics of interest. Finally, in response to a question about the future of education under the influence of artificial intelligence, 33.6% expressed the belief that changes would occur gradually, 31.5% thought that schools would function in a completely different way, 26.6% could not determine their position, while 8.4% thought that traditional education would remain unchanged.

Table 5. Views and attitudes on various issues related to AI technology.

11. Availability of AI-Related Classes	Percentage (%)
Not available, not interested	39.2%
Not available, but interested	35.7%
Available but did not attend	14%
Attended	11.2%
12. Should Teachers Use AI?	
Yes, it could make learning easier	31.7%
Limited use	29.6%
No opinion	22.5%
No, teaching should be based on traditional methods	16.2%
13. AI and Student Independence Impact on Learning Autonomy	
Yes, because they may rely too much on technology	44.1%
No, because it is a tool that can help with learning	23.8%
No clear opinion	32.2%
14. Practical AI Applications in School	
Ability to create personalized tests and quizzes tailored to the individual student's level of knowledge	58%
Homework prompts	38.5%
Interactive presentations	36.4%
Online tutoring	21.7%
15. Future of education under the influence of artificial intelligence	
Gradual change	33.6%
Radical transformation	31.5%
Don't know	26.6%
No change	8.4%

Source: own elaboration

5 Discussion and Conclusion

The survey allows us to draw some important conclusions about students' knowledge, perceptions and experiences with artificial intelligence (AI). The vast majority of respondents declared familiarity with the concept of AI and contact with technologies using its solutions, which confirms the growing presence of this topic in the daily life of the younger generation. Interpretations of the concept of artificial intelligence varied, with the dominant understanding of AI as advanced computer programs capable of analyzing data and self-learning. However, a significant percentage of students also identified AI with robotics, and some identified it with everyday applications, indicating a rather broad, though not always precise, understanding of the concept. It is noteworthy that about 12% of students could not explain the definition of AI on their own, suggesting the need for more structured education in this area. The results of the survey also showed that while students often use AI-based technologies (e.g., chatbots, online translators or voice assistants), they are not always aware that artificial intelligence is behind their actions. This underscores the importance of developing digital competencies, including the informed and critical use of new technologies. The use of AI in education is perceived rather positively by students - they see benefits in the form of support for translating texts, solving math tasks or creating presentations. However, at the same time, they express concerns about the possible reduction of independence in learning and the risk of dependence on ready-made solutions. They also point to potential risks of privacy violations and misinformation generated by AI. A significant number of respondents were interested in deepening their knowledge of AI, which may provide important guidance for organizers of the education system. The interest in classes devoted to this topic, despite the declared lack of them so far, confirms the need to systematically introduce content related to artificial intelligence into the curricula. Significantly, students also recognized the potential benefits of using AI by teachers, with the prevailing view being that its use should be limited and controlled, rather than completely replacing traditional teaching methods. Students also showed a fairly realistic view of the future of education, with most predicting gradual change rather than a revolutionary shift away from traditional forms of teaching.

It can be concluded that the conclusions of the surveys conducted in Poland largely coincide with those of the studies summarized in Table 1, which show that awareness of artificial intelligence among elementary school students is not yet at a satisfactory level, but children show great potential and willingness to learn. This means that education about AI in elementary school is not only necessary, but also feasible and promising.

Acknowledgments. This study was co-financed by the Minister of Science under the "Regional Excellence Initiative".

Regionalna Inicjatywa Doskonałości

Ministry of Science and Higher Education Republic of Poland

Disclosure of Interests. The authors have no competing interests to declare that are relevant to the content of this article.

References

1. Alimi, A.E., Buraimoh, O.F., Aladesusi, G.A., Babalola, E.O.: University students' awareness of, access to, and use of artificial intelligence for learning in Kwara State. Indonesian J. Teach. Sci. **1**, 91–104 (2021). https://doi.org/10.17509/ijotis.v1i2.38014
2. Tripathi, C.R.: Awareness of artificial intelligence (AI) among undergraduate students. NPRC J. Multidis. Res. **1**, 126–142 (2024). https://doi.org/10.3126/nprcjmr.v1i7.72478
3. Vinothkumar, Dr.K., Saratha, Mrs.M.: Awareness on artificial intelligence tools and applications among post graduate students in madurai district. gengpi. **5**, 2915–2919 (2024). https://doi.org/10.55248/gengpi.5.0224.0557
4. Albayati, H.: Investigating undergraduate students' perceptions and awareness of using ChatGPT as a regular assistance tool: a user acceptance perspective study. Comput. Educ.: Artif. Intell. **6**, 100203 (2024). https://doi.org/10.1016/j.caeai.2024.100203
5. Lee, D., et al.: The impact of generative AI on higher education learning and teaching: a study of educators' perspectives. Comput. Educ.: Artif. Intell. **6**, 100221 (2024). https://doi.org/10.1016/j.caeai.2024.100221
6. Johnson, N., Seaman, J., Seaman, J.: The anticipated impact of artificial intelligence on higher education. OLJ. **28** (2024). https://doi.org/10.24059/olj.v28i3.4646
7. Sluzek, N., Suppan, K.: Artificial intelligence in higher education - opportunity or threat? School - Profession - Work. 13–24 (2024). https://doi.org/10.34767/SZP.2024.01.01
8. Parthiban, V., Ganesh, B.J.: A study on awareness of artificial intelligence (AI) tools among prospective teachers of Namakkal District. Int. J. Innovative Sci. Res. Technol. (IJISRT). 2052–2055 (2024). https://doi.org/10.38124/ijisrt/IJISRT24JUL1812
9. Liu, L.: Survey and analysis of primary school teachers' use of generative artificial intelligence. Lecture Notes Educ. Psychol. Public Media. **74**, 43–52 (2024)
10. Wang, X., Li, L., Tan, S.C., Yang, L., Lei, J.: Preparing for AI-enhanced education: conceptualizing and empirically examining teachers' AI readiness. Comput. Hum. Behav. **146**, 107798 (2023). https://doi.org/10.1016/j.chb.2023.107798
11. Johnson, D.: Exploring academic perspectives: a survey-based study on the impact of artificial intelligence in education (2024). https://osf.io/c6bfv. https://doi.org/10.35542/osf.io/c6bfv
12. Qin, Y., Liu, G., Wu, M.: Good or Bad? Explore the application of ChatGPT in education – based on interviews and user experience analysis. In: 2023 Twelfth International Conference of Educational Innovation through Technology (EITT). pp. 158–163. IEEE, Fuzhou, China (2023). https://doi.org/10.1109/EITT61659.2023.00037
13. Güneyli, A., et al.: Exploring teacher awareness of artificial intelligence in education: a case study from northern Cyprus. EJIHPE. **14**, 2358–2376 (2024). https://doi.org/10.3390/ejihpe14080156
14. Lee, S.J., Kwon, K.: A systematic review of AI education in K-12 classrooms from 2018 to 2023: Topics, strategies, and learning outcomes. Comput. Educ.: Artif. Intell. **6**, 100211 (2024). https://doi.org/10.1016/j.caeai.2024.100211
15. Martin, F., Zhuang, M., Schaefer, D.: Systematic review of research on artificial intelligence in K-12 education (2017–2022). Comput. Educ.: Artif. Intell. **6**, 100195 (2024). https://doi.org/10.1016/j.caeai.2023.100195
16. Yim, I.H.Y.: Artificial intelligence literacy in primary education: an arts-based approach to overcoming age and gender barriers. Comput. Educ.: Artif. Intell. **7**, 100321 (2024). https://doi.org/10.1016/j.caeai.2024.100321
17. McCarthy, J., Minsky, M.L., Rochester, N., Shannon, C.E.: A proposal for the dartmouth summer research project on artificial intelligence, august 31, 1955. AI Mag. **27**, 12 (2006)
18. Southworth, J., et al.: Developing a model for AI Across the curriculum: transforming the higher education landscape via innovation in AI literacy. Comput. Educ.: Artif. Intell. **4**, 100127 (2023)

19. Miao, F., Shiohira, K.: K-12 AI curricula. A mapping of government-endorsed AI curricula. UNESCO Publishing **3**, 60 (2022)
20. UNESCO: Guidance for generative AI in education and research | UNESCO, https://www.unesco.org/en/articles/guidance-generative-ai-education-and-research. Last accessed 22 Apr 2025
21. Benjamin, R.: Race after technology. In: Social Theory Re-Wired. pp. 405–415. Routledge (2023)
22. Mertala, P., Fagerlund, J., Calderon, O.: Finnish 5th and 6th grade students' pre-instructional conceptions of artificial intelligence (AI) and their implications for AI literacy education. Comput. Educ.: Artif. Intell. **3**, 100095 (2022)
23. Zhang, H., Lee, I., Ali, S., DiPaola, D., Cheng, Y., Breazeal, C.: Integrating ethics and career futures with technical learning to promote AI literacy for middle school students: an exploratory study. Int. J. Artif. Intell. Educ. **33**, 290–324 (2023)
24. Ali, S., et al.: Children as creators, thinkers and citizens in an AI-driven future. Comput. Educ.: Artif. Intell. **2**, 100040 (2021)
25. Yim, I.H.Y.: Design of Artificial Intelligence (AI) education for primary schools: arts-based approach. Istes Books. 65–90 (2023)
26. Xia, Q., Chiu, T.K., Chai, C.S.: The moderating effects of gender and need satisfaction on self-regulated learning through Artificial Intelligence (AI). Educ. Inf. Technol. **28**, 8691–8713 (2023)
27. Ng, D.T.K., Leung, J.K.L., Su, M.J., Yim, I.H.Y., Qiao, M.S., Chu, S.K.W.: AI literacy education in primary schools. In: AI literacy in K-16 classrooms. pp. 75–86. Springer (2022)
28. Dai, Y., Chai, C.-S., Lin, P.-Y., Jong, M.S.-Y., Guo, Y., Qin, J.: Promoting students' well-being by developing their readiness for the artificial intelligence age. Sustainability **12**, 6597 (2020)
29. Dai, Y.: Dual-contrast pedagogy for AI literacy in upper elementary schools. Learn. Instr. **91**, 101899 (2024)
30. Dai, Y., Lin, Z., Liu, A., Dai, D., Wang, W.: Effect of an analogy-based approach of artificial intelligence pedagogy in upper primary schools. J. Educ. Comput. Res. **61**, 159–186 (2024)
31. Tlili, A., et al.: What if the devil is my guardian angel: ChatGPT as a case study of using chatbots in education. Smart Learn. Environ. **10**, 15 (2023). https://doi.org/10.1186/s40561-023-00237-x
32. Toivonen, T., Jormanainen, I., Kahila, J., Tedre, M., Valtonen, T., Vartiainen, H.: Co-designing machine learning apps in K-12 with primary school children. Presented at the 2020 IEEE 20th international conference on advanced learning technologies (ICALT) (2020)
33. Williams, R., Ali, S., Alcantara, R., Burghleh, T., Alghowinem, S., Breazeal, C.: Doodlebot: an educational robot for creativity and AI literacy. Presented at the Proceedings of the 2024 ACM/IEEE International Conference on Human-Robot Interaction (2024)
34. Touretzky, D., Gardner-McCune, C., Martin, F., Seehorn, D.: Envisioning AI for K-12: what should every child know about AI? Presented at the Proceedings of the AAAI conference on artificial intelligence (2019)
35. Jia, F., Sun, D., Looi, C.: Artificial Intelligence in science education (2013–2023): research trends in ten years. J. Sci. Educ. Technol. **33**, 94–117 (2024). https://doi.org/10.1007/s10956-023-10077-6
36. Holmes, W., Luckin, R.: Intelligence unleashed: an argument for AI in education. Pearson : UCL Knowledge Lab, London (2016)
37. Rizvi, S., Waite, J., Sentance, S.: Artificial Intelligence teaching and learning in K-12 from 2019 to 2022: a systematic literature review. Comput. Educ.: Artif. Intell. **4**, 100145 (2023)
38. Holmes, W., Bialik, M., Fadel, C.: Artificial intelligence in education promises and implications for teaching and learning. Center for Curriculum Redesign (2019)
39. Walan, S.: Primary school students' perceptions of artificial intelligence - for good or bad. Int. J. Technol. Des. Educ. **35**, 25–40 (2025). https://doi.org/10.1007/s10798-024-09898-2

40. Tedre, M., et al.: Teaching machine learning in K-12 classroom: pedagogical and technological trajectories for artificial intelligence education. IEEE Access **9**, 110558–110572 (2021)
41. Chiu, T.K., Meng, H., Chai, C.-S., King, I., Wong, S., Yam, Y.: Creation and evaluation of a pretertiary artificial intelligence (AI) curriculum. IEEE Trans. Educ. **65**, 30–39 (2021)
42. Yang, W.: Artificial Intelligence education for young children: why, what, and how in curriculum design and implementation. Comput. Educ.: Artif. Intell. **3**, 100061 (2022). https://doi.org/10.1016/j.caeai.2022.100061
43. Oruc, T., Korkmaz, O., Kurt, M.: Primary school students' views on artificial intelligence. IJTES. **8**, 583–601 (2024). https://doi.org/10.46328/ijtes.577
44. Kang, K.-A., Kim, S.-J., Kang, S.R.: Elementary school students' awareness of the use of artificial intelligence chatbots in violence prevention education in South Korea: a descriptive study. Child Health Nurs Res. **28**, 291–298 (2022). https://doi.org/10.4094/chnr.2022.28.4.291
45. Cai, L.: Realities vs expectations: children's perception and imagination of AI. Int. J. Technol. Des. Educ. **34**, 1687–1704 (2024). https://doi.org/10.1007/s10798-024-09879-5

Educational Value of Board Games Using 3D Printing Technology in the Process of Education (Formal and Informal)

Małgorzata Nermend[1], Patryk Wlekły[2](✉), Mariusz Borawski[3], Julia Leżała[2], and Piotr Przetacznik[2]

[1] University of Szczecin, Ogińskiego 16/17, 71-431 Szczecin, Poland
[2] University of Szczecin, Cukrowa 8, 71-004 Szczecin, Poland
patryk.wlekly@usz.edu.pl
[3] Maritime University of Szczecin, Wały Chrobrego 1-2, 70-500 Szczecin, Poland

Abstract. The article presents a study on the educational potential of board games enhanced through 3D printing technology in the context of primary school education. The authors explore how such games can support cognitive, emotional, and social development in children aged 10–13, aligning with constructivist pedagogical frameworks and theories of active and social learning. A prototype board game titled Adventure Trackers was developed using 3D-printed elements and evaluated by 81 students participating in structured group activities inspired by the Polish Scouting Association methodologies. The study applied UX research tools—primarily surveys with Likert scales and open-ended questions—to assess user satisfaction, usability, and engagement. Results show high acceptance levels, with over 90% of participants expressing willingness to replay the game. Statistical analysis, including Cronbach's alpha, Mann–Whitney U test, and logistic regression, confirmed both the reliability of the survey instrument and the significant role of visual design, gameplay curiosity, and component ergonomics in shaping user satisfaction. The research underscores the interdisciplinary potential of combining 3D printing, game-based learning, and user-centered design to create inclusive, motivating, and developmentally beneficial educational tools. Implications for further refinement, including iterative testing and integration of qualitative UX data, are also discussed.

Keywords: 3D printing in education · Primary Education · Educational board game · User Experience · Child-centered design

1 Introduction

Board games are currently experiencing great popularity, both as a form of interesting time spent and as an educational tool. Contemporary research in the field of pedagogy and psychology indicates a wide range of applications of board games in education

Published by Springer Nature Switzerland AG 2026
M. Pondel et al. (Eds.): AI4KAM 2025, IFIP AICT 776, pp. 34–44, 2026.
https://doi.org/10.1007/978-3-032-18920-2_4

because games can support the development of key competencies, such as logical thinking, teamwork, mathematical competence, reading with understanding, as well as emotional development. They can be classified by purpose (educational, strategic, social), mechanics (random, decision-making, cooperative), and age of players. Board games develop analytical, mathematical, linguistic, and memory skills. Examples include the game "Scrabble," which supports the expansion of vocabulary, or "Catan," where players learn resource management and strategic planning.

This raised the question of whether the board game could be enhanced to give it educational value. The authors presented the results of a study aimed at assessing the reception and satisfaction of users in the context of the educational game "Adventure Trackers", created using 3D printing technology. The conducted study made it possible to answer not only the main question of whether the form of the board game enriched with 3D printed elements is suitable for learning through play, but also indicated which visual, functional, and usability aspects require possible improvements. In the context of this study, the first section presents the educational value of board games in terms of cognitive, social, and emotional development of children and adolescents. Based on a review of the literature and empirical examples, it is shown how board games support learning processes, motivation, and social integration. The next section focuses on the methodology used in the study. Further on, the results of the study are presented, and the last section presents the research conclusions, limitations, and directions for further work.

2 Theoretical Background

Modern education is increasingly turning to activation methods, in which students become active participants in the didactic process. Board games are part of the trend of pedagogical constructivism [1, 2], according to which knowledge is constructed as a result of active action, social interaction, and personal involvement. In this context, board games function not only as a teaching tool but also as a means of supporting the psychosocial development of the individual [3]. The use of board games in education finds its justification in several psychological theories. John Dewey's theory of learning by doing assumes that students learn most effectively through active participation and experience [4]. Games, as a form of activity in a safe environment, allow for the exploration of knowledge in an engaging and non-invasive way. Another theory of social learning by Albert Bandura emphasizes the importance of modeling, observation, and interaction [5]. In this view, board games promote learning by observing the behavior of other players, imitating strategies, and solving problems together. Another theory is the constructivism of Jean Piaget and Lev Vygotsky, which points to the student's active construction of knowledge [6]. Vygotsky also emphasized the importance of the zone of immediate development - games can be designed in such a way as to support children in the transition from current to potential levels of cognitive development.

The importance of games in education was recognized in 2012 by the Academic Games Consortium, highlighting their role as a tool for education, research, and social innovation [7]. It brought together 19 universities and representatives from industry and government institutions who expressed the belief that games have the potential to foster

21st century competencies such as critical thinking, problem solving, and creativity [8]. The above is part of the trend of so-called "game-based learning," which assumes that well-designed games can motivate, teach and develop skills in a way that traditional schooling often cannot.

The process of designing educational games is a complex teaching endeavor, the structure of which can be captured within three logically related phases: telic, construction and evaluation [9]. Each has a different but complementary function in shaping the educational and functional value of the final teaching product.

From a cognitive development perspective, board games support the development of strategic and logical thinking, exercise working memory, attention and concentration, develop mathematical and linguistic competence, and teach problem-solving and decision-making in situations of uncertainty. Research [10] shows that children who regularly participate in board games show higher levels of so-called executive functions, such as cognitive control, cognitive flexibility and planning.

In the context of social-emotional development, board games also play a very important role because they teach the principles of fair play, cooperation and empathy, shape the ability to assume different social roles and interpersonal communication, and strengthen emotional resilience by learning to cope with failure and stress [11]. Cooperative games, in which players must work together, teach mutual trust, responsibility for the group and joint conflict resolution. As a result, they can be used as a tool to support the integration of class teams and children with adaptation difficulties.

From a didactic point of view, board games play a multifaceted role in the educational process, performing several functions that support both cognitive and social development of students. First of all, they perform a motivational function, increasing students' interest in the subject of study, as well as stimulating their natural curiosity and inner commitment. This makes the process of acquiring knowledge more dynamic and engaging, which translates into educational efficiency [12].

At the same time, games enable the realization of the diagnostic function, since they create conditions for the teacher to observe individual students' ways of thinking, their problem-solving strategies and interpersonal competence. In this context, the game becomes a tool for informal didactic and social diagnosis, allowing one to identify strengths and areas in need of support. The consolidation function is also not insignificant, which involves the possibility of repeated repetition and processing of educational material in an attractive and less stressful form [13].

In addition, board games perform an integrative function, fostering cooperation between students and promoting pro-social attitudes. They enable the participation of people with diverse learning needs, building a sense of belonging and strengthening communication competencies. As a result, they can provide a space conducive to building relationships, developing soft skills and fostering inclusiveness in the educational environment [14].

Board games are a valuable teaching tool in the context of inclusive education, offering real opportunities to support children with disabilities in the educational process. With the flexibility to modify the rules, teaching materials and level of difficulty, it is possible to create a space that allows students with diverse learning needs to actively participate in team activities. Such adjustments not only facilitate participation but also foster a cooperative environment, strengthen peer bonds and counteract marginalization and social exclusion. The motivational potential of board games also deserves special attention. Thanks to the element of play, which naturally engages and activates students, their interest in the educational process and willingness to make intellectual efforts increases. Educational thematic games can effectively support the teaching of school subjects, providing an alternative working tool [15].

Nowadays, board games are successfully used both in formal and informal education. In school settings, teachers are increasingly willing to use them as part of lessons and extracurricular activities, appreciating their value in implementing the core curriculum and in developing key competencies in students. In out-of-school environments, such as homeschooling, school day care centers or community centers, games play a supporting role in the educational and upbringing process, allowing children to learn through play in an atmosphere of safety and acceptance [16].

Using modern 3D printing technology as an innovative solution in the field of science can be an effective tool to help the teacher. As Thyssen and Meier point out, 3D printing is not only used in medicine or biology, but also significantly in education. They indicated the use of 3D printing as a tool for teaching chemistry, economics, mathematics, as well as science subjects of technical and engineering subjects [17]. 3D printing technology is also being used in educational board games. One example is a game for children to learn a programming language for the visually impaired. The use of 3D printing makes it easy and inexpensive to create new blocks that are attractive to children, which are then placed on the board in a specific way. This allows them to learn the programming language using specific blocks they know [18].

Other important aspects of using this technology to develop soft skills in children such communication skills as well as teamwork. An example is the study by Pantazis et al., who used 3D printing technology to create items with superimposed text in Braille. They indicated that this is a low-cost and effective method of learning this language to communicate with visually impaired people. It also allows for quick and easy production of items with embedded text [19].

The educational process is not only about learning a specific element it is also about the student developing skills. Chien et al. point out that modeling and 3D printing allow for increasing the creativity aspect of students and pupils. The process of designing printable items and related activities translates into an increase in creativity, but also in academic performance in technical as well as science subjects that are related to 3D printing [20]. Bicer and Kwon indicate, in addition to an increase in creativity levels, an increase in problem-solving skills through the creation of new solutions, but also motivation to learn new skills [21, 22].

3D printing, therefore, has potential from both a methodological and didactic point of view, and it is important to gain a fairly accurate insight into teachers' beliefs in these areas. All these aspects indicate that board games can be an integral part of modern education, combining cognitive, emotional and social qualities into one coherent form of teaching activity.

3 Methodology of the Study

3.1 Purpose of the Study

The purpose of the study was to assess user reception and satisfaction in the context of the educational game "Adventure Trackers", created using 3D printing technology. The study aimed to determine whether the form of the board game enriched with 3D printed elements is suitable for learning through play, and what visual, functional and usability aspects require possible improvements. The tool used was Meshy.ai, a web-based platform based on artificial intelligence, which enables the generation of 3D models based on a text description or photo. Models of pawns in various specific shapes, such as a scout, were generated this way. These objects were then prepared and printed using a 3D printer. The source photos used to generate the models were generated using ChatGPT, which was also used to prepare the final visualization of the game.

3.2 Characteristics of the Research Group

The study included 81 participants between the ages of 10 and 13. The choice of this age group was purposeful and based on the assumptions of the educational methodology of the Polish Scouting Association (in polish - ZHP) [23]. In the methodological system of the Polish Scouting Association, scouts aged 10–13 form a separate educational division, in which the basic form of work is a game [24]. For this age group, the game is not seen merely as a form of entertainment, but as a conscious educational tool, supporting the development of such competencies as group cooperation, decision-making, action planning, problem solving or adherence to the rules of fair play. Therefore, the selection of participants from this age range was aimed at adapting the tested game to the expectations and cognitive abilities of the group, for which the game is a natural and methodically sound educational tool. In terms of the gender of the participants, the group consisted of 60 girls (74%), 13 boys (16%), and 8 people (10%) who did not specify their gender or marked a different one.

The survey was anonymous. Personal information was not collected, and participation in the study was completely voluntary. All participants took part based on written consents given by their parents or legal guardians.

3.3 Description of Research Tools

The study used a questionnaire as the main tool, consisting of closed-ended questions (single-choice and Likert scale ratings) and open-ended questions that allowed people to express their own opinions, ratings and suggestions about the game. Microsoft Excel software was used to analyze the data and built-in statistical methods, which were adapted to the data obtained.

3.4 Course of the Study

The research was conducted at school facilities where children attend scout meetings daily. This allowed the participants to take the tests in a natural, familiar environment.

The participants were divided into teams of six, corresponding to the actual structure of scout troops. The selection of the number of participants in a group was not random, but resulted directly from the principles of the small group system, which is the foundation of educational work in the Polish Scouting Association [25]. In scouting, he basic organizational unit is a troop, usually consisting of 5 to 12 people, but it is recommended that the size does not exceed 6-8 members for the sake of maintaining the nature of individual educational work and effective organization of activities.

A troop in scouting reflects the natural social mechanisms of peer group functioning - it includes a fixed composition, internal roles and a leader emerging from the group. Therefore, the adoption of six-person teams in the study was intended both to preserve the authenticity of the conditions of the testing environment and to allow analysis of the game's effectiveness in a setting that corresponds to actual educational work in a scout troop.

During the game, children played in their regular groups at common tables. The conditions of the study were favorable: the room was relatively quiet, although in the background, other children playing independently were present. This allowed a natural atmosphere of teamwork to be maintained, without artificially isolating the participants (Fig. 1).

Fig. 1. Photo of the board game used in the study during gameplay. Source: Own photo.

3.5 Technologies and Materials Used in the Game

The designed game "Adventure Trackers" was made using 3D printing technology. PLA filament, which is a material safe for children and environmentally friendly, was used to produce the game elements. The models were printed using the Prusa MK4 printer, which enables precise reproduction of details and ensures adequate durability of the prints while maintaining high aesthetic quality (Fig. 2).

Fig. 2. Photo of the game board model used in the game. Screenshot from Blender. Source: own elaboration.

All 3D models were designed taking into account the principles of ergonomics and adapting the size and shape of the elements to the motor capabilities of children aged 10–13. The design process included not only visual aspects, such as the attractiveness of the forms and colors, but also functional issues related to the ease of grasping, sliding and mechanical strength of the elements.

4 Description of Research Results

In order to comprehensively assess the reception of the educational game "Adventure Trackers", data were collected by means of a questionnaire filled out after the game. The declarations of the respondents made it possible to find out both the subjective feelings of the participants and to collect objective information about the functionality and attractiveness of the game elements. The second part of the survey was to collect basic statistical data about the respondents like age, gender and seniority. First, the results of the survey were analyzed, including visual, functional and emotional evaluation of the game, as well as opinions on possible improvements. The survey data was collected from all survey participants. Below are the results that were obtained after analyzing the survey responses.

Analysis of the results of a survey of 81 participants showed that the board game was rated very well both visually and functionally. The vast majority of respondents (93.83%, or 76 people) thought the game looked interesting and encouraged gameplay, and almost 90% (73 people) rated its appearance positively on a scale of 1 to 5. In addition, 85.19% (69 people) said the game was well made, and as many as 95.06% (77 people) confirmed that the components were durable and did not break during use. The rules of the game were considered easy to understand in 87.65% of cases (71 people), and 90.12% of respondents (73 people) said it took them little time to learn them.

The gameplay itself was also highly rated - 88.89% of respondents (72 people) found it interesting, although some respondents (8.64%, or 7 people) indicated that it becomes monotonous over time. The game's elements - their look, feel and ergonomics - also received positive evaluations. 77.78% (63 people) indicated that the components were comfortable to use, and 74.07% (60 people) found the colors and shapes attractive. At the same time, some 20–25% of respondents (17 to 20 people depending on the question) noted that some figures or components could be more convenient or refined, especially in terms of size and comfort of use.

In open-ended questions, participants most often praised the tasks in the game (33 people), the Figures (22 people) and the overall impression of the whole (13 people). Difficulties in the game were not encountered by the majority of respondents (56 people), but there were voices regarding the questions being too difficult (9 people) and the board and components being too small (10 people). Among the suggestions for improvement, the most common were the need to increase the size of the board and components (34 people) and to add additional Figures (16 people). Importantly, as many as 90.12% of respondents (73 people) said they would like to play the game again, confirming its positive reception and well-thought-out design.

Several statistical tests were used to analyze the statistical data. The first was the Alpha-Krombach test to check the reliability of the set of closed questions. After recalculating the results obtained for all respondents, a score of 1.17 was obtained, indicating a very high reliability index for the set of questions in the survey. In the next step, using the Mann-Whitney U test, the results from the closed questions were compared between the groups divided by gender (they were unequal). In all cases, the result of the p parameter came out above the alpha threshold of 0.05 (the lowest value for question two - 0.0537), so none of the comparisons are statistically significant. As a third test, an additional logistic regression was conducted against the closed questions. The highest coefficient was achieved by the question about the appearance of the game (1.408 and an odds ratio of more than 4). Such a high coefficient suggests that the appearance of the game as a whole is the most important factor characterizing the willingness to play the game again. Other high coefficients were achieved by questions about gameplay curiosity (0.5 and 1.649) and convenience of use (0.334 and 1.397).

5 Discussion and Conclusion

Analysis of the results of the survey conducted among the participants of the educational game "Adventure Trackers" allows a positive assessment of the game in terms of its visual, functional and emotional aspects. Survey data confirms the high visual attractiveness of the game and the durability and aesthetics of its components, which are important for maintaining user involvement. The high rating for the accessibility of the game's rules and the short time it takes to learn them attest to a properly designed playability interface and a well-chosen level of difficulty. It is worth noting the results of the open-ended questions, which most often emphasized the attractiveness of the tasks and the aesthetics of the figures, confirming the importance of both the content and form of the game.

Despite the generally positive reception, some respondents pointed out the monotony of gameplay with prolonged use and the need for improvements related to board and component size. These observations are important for future iterations of the product, as they show areas of possible optimization. The use of statistical methods, such as the Alpha-Kronbach test, confirmed the high reliability of the research tool, and the lack of significant differences between groups by gender indicates the universal nature of the game's reception. What is particularly noteworthy is that logistic regression revealed that it is primarily the aesthetics of the game, as well as curiosity about gameplay and convenience of use, that are the main predictors of willingness to repeat participation in the game. This may suggest that when designing educational games, it is crucial to provide an enjoyable visual experience and ease of interaction with the components, not just the educational value itself.

The survey indicates that the educational game "Adventure Trackers" was very well received by the participants in terms of visual, functional and emotional aspects. The results of the analysis suggest that visual appeal, convenience of use and interesting tasks have a significant impact on the positive evaluation of the game and willingness to participate again. Despite minor suggestions for improvement, the game was rated as thoughtful, durable and engaging. These results make a strong case for the use of board games as an effective educational tool, provided they are carefully designed with users' needs and preferences in mind.

Further development of research in this direction can be expanded to include additional collection of declarative data using interviews, but also by conducting observations. Based on the declarative data obtained, modifications can be made to the models or rules of the game to enhance the positive user experience of playing this board game. Thus, there are several avenues for further research in this direction.

Acknowledgments. This study was co-financed by the Minister of Science under the "Regional Excellence Initiative".

Regionalna Inicjatywa Doskonałości
Ministry of Science and Higher Education Republic of Poland

Disclosure of Interests. The authors have no competing interests to declare that are relevant to the content of this article.

References

1. Klus-Stańska, D.: Konstruktywizm edukacyjny – niejednoznaczność, kontrowersje, dylematy. PWE **51**, 7–20 (2020). https://doi.org/10.26881/pwe.2020.51.01
2. Rzeźnicka-Krupa, J.: Konstruktywizm i pedagogika różnorodności w kontekście edukacji włączającej. PWE **51**, 153–166 (2020). https://doi.org/10.26881/pwe.2020.51.12
3. Tkaczyk, P.: Grywalizacja: jak zastosować mechanizmy gier w działaniach marketingowych. Wydawnictwo Helion, Gliwice (2012)
4. Ültanır, E.: An epistemological glance at the constructivist approach: constructivist learning in Dewey, Piaget, and Montessori. Int. J. Instruct. (2012)

5. Płaczkiewicz, B.: Teoria społecznego uczenia się a wychowanie dziecka w wieku przedszkolnym. Społeczeństwo Edukacja Język **4**, 129–137 (2016)
6. Chmielewska-Banaszek, D.: Teorie i idee konstruktywistyczne w psychologii. Principia 56: (2012)
7. Professor Mary Flanagan Participates in White House Consortium|Dartmouth(2012). https://home.dartmouth.edu/news/2012/08/professor-mary-flanagan-participates-white-house-consortium. Accessed 7 May 2025
8. Salen Tekinbaş, K.: Quest to Learn: Developing the School for Digital Kids. MIT Press, Cambridge (2011)
9. Słomczyński, M.: Dydaktyczne aspekty projektowania gier. Homo Ludens, pp. 141–151 (2014)
10. Sarı, M.H., Szczygieł, M.: The role of math anxiety in the relationship between approximate number system and math performance in young children. Psychol. Sch. **60**, 912–930 (2023). https://doi.org/10.1002/pits.22794
11. Przetacznik-Gierowska, M., Tyszkowa, M.: Psychologia rozwoju człowieka. T. 1: Zagadnienia ogólne, Wyd. 5. Wydawnictwo Naukowe PWN, Warszawa (2009)
12. Alotaibi, M.S.: Game-based learning in early childhood education: a systematic review and meta-analysis. Front. Psychol. **15**, 1307881 (2024). https://doi.org/10.3389/fpsyg.2024.1307881
13. Sun, L., Kangas, M., Ruokamo, H., Siklander, S.: A systematic literature review of teacher scaffolding in game-based learning in primary education. Educ. Res. Rev. **40**, 100546 (2023). https://doi.org/10.1016/j.edurev.2023.100546
14. Dell'Angela, L., Zaharia, A., Lobel, A., et al.: Board games on emotional competences for school-age children. Games Health J **9**, 187–196 (2020). https://doi.org/10.1089/g4h.2019.0050
15. Trimlett, G., Barton, E., Baum, C., et al.: Teaching board game play to young children with disabilities. J. Posit. Behav. Interv. **24**, 109830072098528 (2021). https://doi.org/10.1177/1098300720985287
16. Cos, P., Doyen, A.L., Duflos, M., Giraudeau, C.: Board games in the elementary classroom: teachers' perspectives. Educ. Res. **67**(2), 190–211 (2025). https://doi.org/10.1080/00131881.2025.2467067
17. Thyssen, C., Meier, M.: 3D Printing as an element of teaching—perceptions and perspectives of teachers at German schools. Front Educ **8**, 1233337 (2023). https://doi.org/10.3389/feduc.2023.1233337
18. Jorge, C.S.P., Albuquerque, P.L., Carvalho, F.: The use of 3D printing for the development of a learning tool for the visually impaired. Int. J. Innov. Educ. Res. **9**(9), 387–398 (2021). https://doi.org/10.31686/ijier.vol9.iss9.3353
19. Pantazis, A., Priavolou, C.: 3D printing as a means of learning and communication: the 3Ducation project revisited. Telematics Inform. **34**, 1465–1476 (2017). https://doi.org/10.1016/j.tele.2017.06.010
20. Chien, Y.-H., Chu, P.-Y.: The different learning outcomes of high school and college students on a 3D-printing STEAM engineering design curriculum. Int. J. Sci. Math Educ. **16**, 1047–1064 (2018). https://doi.org/10.1007/s10763-017-9832-4
21. Bicer, A., Nite, S.B., Capraro, R.M., et al.: Moving from STEM to STEAM: The effects of informal STEM learning on students' creativity and problem solving skills with 3D printing. In: 2017 IEEE Frontiers in Education Conference (FIE), pp 1–6 (2017)
22. Kwon, S.-H., Lee, Y.-J., Kwon, Y.-J.: An active learning approach to investigate the ecosystem of tide flats using 3D modeling and printing. J. Biol. Educ. **54**, 88–97 (2020). https://doi.org/10.1080/00219266.2018.1546760

23. Podstawy wychowawcze ZHP(2009). https://gkzhp.sharepoint.com/:w:/r/sites/dokumentyzhp/_layouts/15/Doc.aspx?sourcedoc=%7B678BBA23-C3C3-441E-ADB1-2897783900E7%7D&file=Podstawy%20wychowawcze%20TEKST%20JEDNOLITY%20po%20zmianach%20z%20XXXIX%20i%20XLII%20Zjazdu%20ZHP%20oraz%20po%20zmianach%20wynikacj%25u0105cych%20z%20Uchwa%25u0142y%20nr%2085XL%20Rady%20Naczelnej%20ZHP.docx&action=default&mobileredirect=true. Accessed 6 May 2025
24. Regucka-Kwaśnik A (2003) Harcerki i harcerze: zarys metodyki. Horyzonty : na zlec. Głównej Kwatery ZHP, Warszawa
25. Czechowska, L., Gackowski, S., Kluszczyńska, M., et al.: NIC NIE ZASTĄPI ZASTĘPÓW. Warszawa (2014)

AI Chatbot as Frontline Advisor in University Recruitment: Enhancing Candidate Experience

Bartłomiej Balsamski(✉), Jakub Kanclerz, Dariusz Put, and Janusz Stal

Krakow University of Economics, 31-510 Krakow, Poland
Balsamb@uek.krakow.pl

Abstract. This study explores the potential of generative artificial intelligence (GenAI) in enhancing the university recruitment process by integrating AI-driven chatbots to assist applicants. Focused on Krakow University of Economics, the research evaluates five language models for their effectiveness in providing accurate responses to admission queries. Utilizing a blend of lexical, semantic, and model-as-judge evaluation methods, the study demonstrates varied performance across models, with significant implications for enhancing candidate experience and streamlining communication. The findings highlight the need for iterative improvements and domain expert interventions in refining AI responses, aiming to optimize the integration of GenAI in higher education settings. The research contributes to the understanding of AI applications in academia and suggests practical approaches to improve information dissemination during critical recruitment phases.

Keywords: Large language models · generative artificial intelligence · higher education · student recruitment

1 Introduction

Annually, thousands of candidates engage in the university admission process, seeking detailed information on study programs, curricula, tuition fees, and required documents. The dispersal of this data across numerous platforms complicates access, prompting students to frequently contact university staff, which increases the need for additional personnel during peak periods. Literature suggests significant potential for the application of generative artificial intelligence (GenAI) in higher education. Nonetheless, Pit, Linden, and Mendoza [11] observe that its adoption is limited by policymakers' reservations, resulting in scant practical implementations despite a generally positive research outlook. This study aims to address this gap by utilizing AI-driven solutions to enhance service efficiency for university applicants and streamline information dissemination. The primary research questions are:

1. Which large language model (LLM) provides the best real-time responses to university applicants?

Published by Springer Nature Switzerland AG 2026
M. Pondel et al. (Eds.): AI4KAM 2025, IFIP AICT 776, pp. 45–51, 2026.
https://doi.org/10.1007/978-3-032-18920-2_5

2. To what extent does developing a proprietary knowledge base for context-aware answering influence the relevance of generated responses?

To address the research questions, the recruitment process at the Krakow University of Economics (KUE), Poland, which involves the use of GenAI, was examined. The subsequent section provides the research background. This is followed by a description of the research methodology, and then a presentation of the results from the initial phase of the study. Finally, the research plan for evaluating the accuracy of LLMs responses and the impact of the study on both theory and practice is outlined.

2 Research Background

The integration of LLMs into task-oriented dialogue (TOD) or user response (UR) systems enhances the naturalness and coherence of interactions, thereby increasing the efficacy of user assistance [7]. Incorporating domain-specific documents into the fine-tuning processes of LLMs results in more accurate response generation, leveraging training datasets filled with relevant documents. Empirical research underscores the enhanced performance of such fine-tuned LLMs within UR systems. For instance, employing document retrieval mechanisms during response generation allows models to incorporate external information in real-time, thus improving response accuracy [9]. Additionally, fine-tuning LLMs with specialized content such as medical literature has yielded models proficient in delivering reliable medical advice [8]. A notable study by Alghisi et al. [1] assessed various adaptation techniques for LLMs in dialogue settings, including fine-tuning and retrieval-augmented generation (RAG), revealing that integrating external knowledge sources significantly augments the models' response relevancy. An et al. [2] demonstrated that models trained exclusively on responses, termed Response Tuning (RT) models, effectively address a broad spectrum of user queries. Moreover, Chow et al. [3] enhanced the Best-of-N (BoN) sampling method, where a verifier selects the most suitable among multiple generated responses, improving the overall quality of responses from LLMs [13].

Responses generated by LLMs necessitate validation through diverse methods: (1) automatic benchmarking that evaluates LLM performance on specific tasks using predefined inputs and metrics (lexical like BLEU and ROUGE assess word overlap between generated and reference texts and/or semantic like BERTScore and Cosine Distance measure semantic similarity by comparing text vector representations [5]), (2) LLM-based evaluation that uses models to judge text fluency, coherence and relevance, either independently or against reference texts [14] and (3) human evaluation that involves human raters who assess response quality and rank outputs according to set criteria [5]. Fangkai et al. [4] apply these methods, combining lexical, semantic metrics, and Keyword/Span-Hit-Rate (KHR) to verify keyword presence in responses. Selecting suitable models and fine-tuning methods is crucial for system reliability. This paper investigates the application of LLMs in user-response systems, focusing on the Retrieval-Augmented Generation (RAG) approach with predefined documents and the Internet searches to enhance response quality. It reviews various evaluation methods and the outcomes of fine-tuned LLMs.

3 Research Method

The research was conducted based on five LLMs: (1) OpenAI (gpt-4o), (2) Llama (llama3.1:8b), (3) DeepSeek (deepseek-r1:14b), (4) Claude (claude-3-5-sonnet-20241022) and (5) BIELIK (Bielik-11B-v2.2-Instruct), a Polish language model trained on Polish texts. It demonstrates a strong performance across a range of NLP tasks for utilizing to answer questions in Polish [10]. All LLMs to respond received the same prepared prompt. For surveys without additional sources, its content read: "You are a helper in the student service center, Answer questions precisely and concisely" and for questions with extended context: "You are a helper in the student service center, Answer questions precisely and concisely. When answering, use only the following context. If it doesn't answer the question, answer 'I don't know the answer'." To evaluate the accuracy of LLMs responses we applied two methods of automatic benchmarking (lexical and semantic) and model-as-a-judge. All methods rely on a pre-established set of reference answers for specific questions. As the lexical method we used ROUGE, which calculates the number of matching n-grams between the reference text and the generated text [4]. As semantic metric we chose BERTScore (BERT), which utilizes models to evaluate the semantic similarity between the generated text and the reference text [15]. In model-as-a-judge method (LLMAJ), LLMs can serve as judges to assess text quality, taking into consideration aspects such as fluency, coherence, consistency, and relevance [14]. Using multiple evaluation metrics rather than relying on a single one provides a more comprehensive view of a model's performance, especially for complex tasks such as question answering. Additionally, metrics come with their own limitations, but when used together, they help counterbalance one another's weaknesses and provide a more reliable and comprehensive assessment. Selected metrics follow recommendations from Microsoft [14], Huggingface [5] and domain LLM related metrics mentioned by Fangkai et al. [4].

4 Research Results

Krakow University of Economics (KUE), a prominent business-oriented state university in Poland, addressed communication challenges with prospective students by deploying an AI-driven chatbot. Initially, a rule-based chatbot was used, operating on predefined responses [12]. However, this system required frequent manual updates and expansions of its knowledge base to maintain response accuracy and relevance. This study evaluates the effectiveness of AI-based chatbots in providing accurate, context-aware responses to the admission process. The research utilized 126 questions and answers written in Polish. With this number of questions tested, the quantity represents a sufficient sample size for many research contexts [6]. Questions were aimed at implementing a rule-based chatbot, collected by the recruitment department from historical inquiries on a web-based platform and refined for accuracy and conciseness. To ensure uniformity across tested models, a common interface, the UML Abstract Domain LLM class was developed.

This interface standardized interactions across different models to maintain consistency throughout the experiment, as detailed in the data structure for model-generated responses (Table 1).

Table 1. The expected data structure for evaluated model-generated answer.

Column	Type	Range	Comment
id	int	1.. n	A unique identifier for each entry
q	text		The input question for which an answer is generated
a	text		The reference (ground truth) answer
model	text		Model-{variant}, variant of model used (e.g., Model-Base, Model-Context-Aware, Model-ISCA)
answer	text		The response generated by the model
ROUGE	float	[0,1]	Score being lexical overlap between the generated and reference answers with ROUGE metric
BERT	float	[0,1]	Score assessing semantic similarity using BERTScore
LLMAJ	float	[0,1]	Score provided by an LLM-based evaluator, indicating the overall quality of generated response

The evaluation process was structured to facilitate the substitution of different model implementations without altering the experimental framework or its parameters (Fig. 1). To evaluate the accuracy of the responses, five language models were employed, each utilized in three distinct answering contexts:

- Baseline Answering (BA): Models responded to questions without any prior provision of additional information or domain-specific knowledge [4].
- Context-Aware Answering (CAA): Models were initially fine-tuned with external domain-specific knowledge, including content from university websites, regulations, and relevant documents, before answering the questions.
- the Internet Search-based Context Answering (ISCA): This phase used Retrieval-Augmented Generation (RAG), a method to enhance the performance of Large Language Models (LLMs) (OpenAI 2024). For ISCA, a Google search was conducted using the precise question. The top five results were processed from HTML to plain text, segmented into 800-character chunks with a 200-character overlap. Cosine distance was calculated to determine the relevance between the question embeddings and each text chunk. The five most relevant chunks were concatenated to serve as the contextual input for the model's response.

The results for each question in relation to each reference answer for each model input variant (BA, CAA, and ISCA) across selected metrics (ROUGE, BERT, and LLMAJ) have been presented in Table 2. The data represent the obtained results of similarity of answers returned by LLMs with the reference answers using statistical measures: arithmetic means and standard deviations. On the top of that, ANOVA tests for statistical significance for variance differences were computed pairwise between all evaluated models.

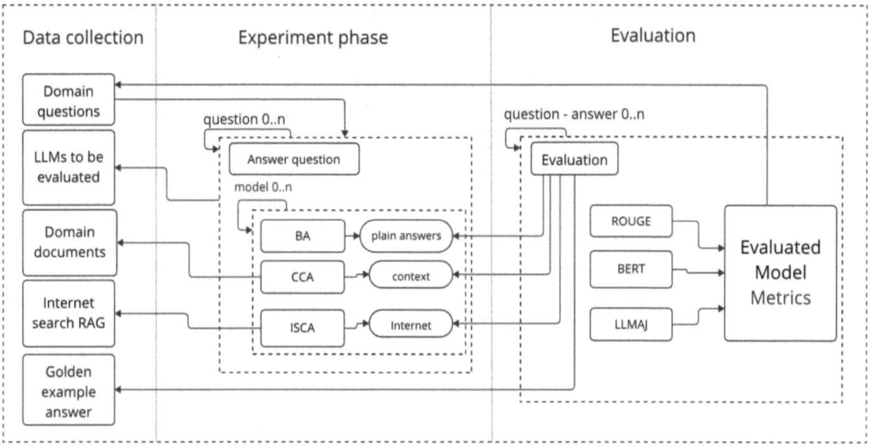

Fig. 1. A structured pipeline for evaluating LLMs.

Table 2. Similarity values of LLMs responses with reference answers.

Model	ROUGE mean	BERT mean	LLMAJ mean	ROUGE std	BERT std	LLMAJ std
bielik (CAA)	0.09	0.65	0.48	0.03	0.02	0.10
bielik (BA)	0.19	0.70	0.60	0.07	0.03	0.09
bielik (ISCA)	0.31	0.75	0.66	0.22	0.08	0.16
claude (CAA)	**0.41**	**0.78**	0.66	0.25	0.09	0.23
claude (BA)	0.28	0.74	**0.76**	0.25	0.07	0.13
claude (ISCA)	0.30	0.73	0.58	0.26	0.09	0.24
deepseek (CAA)	0.06	0.60	0.30	0.08	0.03	0.11
deepseek (BA)	0.03	0.57	0.42	0.02	0.02	0.15
deepseek (ISCA)	0.07	0.60	0.48	0.07	0.04	0.17
llama (CAA)	0.11	0.66	0.42	0.04	0.03	0.11
llama (BA)	0.13	0.67	0.46	0.12	0.05	0.10
llama (ISCA)	0.21	0.68	0.48	0.20	0.09	0.33
openai (CAA)	**0.47**	**0.81**	**0.70**	0.33	0.12	0.27
openai (BA)	0.13	0.68	0.58	0.07	0.03	0.15
openai (ISCA)	0.31	0.74	0.56	0.35	0.13	0.32

5 Discussion and Conclusion

The initial research findings demonstrate that OpenAI's GPT-4o model is the most effective LLM for communicating with university applicants, as evidenced by the highest semantic similarity scores according to the BERT metric: OpenAI GPT-4o (0.81), Cloude

(0.78) and BielikAI (0.75). Nevertheless, ANOVA tests for variance among the models results in high p-values accordingly – GPT-4o vs. Cloude (104.53), GPT-4o vs. BielikAI (16.56) and Cloude vs. BielikAI (9.76) – indicating that these differences are not statistically significant (p-value > 0.05). The ROUGE metric reveals that LLMs from OpenAI (0.47) and Cloude (0.41) adhere more strictly to quoting exact sentences from provided contexts compared to the self-hosted Bielik AI (0.31). However, ANOVA test does not suggest statistically significant difference between them. Despite these similarities, the calculated average standard deviation for the LLMAJ metric (0.17) compared to ROUGE (0.15) and BERT (0.06) indicates a higher variability, suggesting that while the responses are semantically similar, they exhibit lexical differences and maintain accuracy in most instances. For LLMAJ metric, statistical significance of differences between groups are confirmed between GPT vs Cloude (0.02), GPT vs BielikAI (0.01) and not confirmed Cloude vs BielikAI (0.06).

As for the impact on theory, the study confirms LLMs' ability to effectively perform summarization and paraphrasing tasks. Additionally, the analysis shows that models using predefined document sets as contextual inputs (CAA) outperform those relying on the Internet search results for context (ISCA), highlighting the importance of thorough data preparation and the significance of the data curation stage in planning work with automated question-answering systems. Moreover, ANOVA test confirms that models with predefined context provide answers that are more accurate (higher mean) and have statistically significant difference. GPT-CCA vs GPT-ISCA reach 0.002 p-value that confirm significant difference, None of the evaluated models score more than 0.8 on average for the BERT, LLM, or ROUGE metrics, suggesting that while the LLM-generated answers provide valuable insights, there is room for improvement, especially in the implementation of critical, decision-supportive solutions. To address these gaps, further model tuning, prompt engineering, and an expanded range of evaluation metrics should be considered.

Future phases of our research will include: (1) evaluating the accuracy of LLM responses using the expert method with university admissions staff assessing responses across three model configurations: BA, CAA, and ISCA and (2) assessing LLMs' accuracy in foreign languages to improve interactions with international candidates. This article also does not address the issue of ethical AI. However, we acknowledge the importance of this topic and, given its significance, it will be incorporated as one of the steps in our future research.

The practical implementation of the concepts presented in this publication demonstrates that access to a comprehensive dataset and the Internet enables models to generate effective responses for prospective university applicants. Although the research is based on university recruitment data, its findings may also be relevant for the deployment of corporate chatbots – particularly in cases where the underlying knowledge base is well-documented.

The iterative model development process is essential, involving regular updates with more accurate data and expert review to refine responses. High-performing responses should be used to further improvement of the model. These findings can guide the creation of chatbots tailored to specific needs, suitable for administrative or governmental settings in higher education.

Acknowledgments. The article presents the result of the Project no 020/ZIS/2024/POT financed from the subsidy granted to the Krakow University of Economics.

References

1. Alghisi, S., Rizzoli, M., Roccabruna, G., Mousavi, S.M., Riccardi, G.: Should we fine-tune or RAG? Evaluating different techniques to adapt LLMs for dialogue. Proceedings of the 13th International Conference on Natural Language Generation, pp. 180–197 (2024)
2. An, S., Kim, M., Kim, H.: Revealing the Inherent Instructability of Pre-Trained Language Models (2025). arXiv:2410.02465
3. Chow, Y., et al.: Inference-aware fine-tuning for Best-of-N sampling in large language models. ArXiv. **2412**, 15287 (2024)
4. Fangkai, Y., et al.: Empower Large Language Model to Perform Better on Industrial Domain-Specific Question Answering (2023). arXiv:2305.11541v3
5. Fourrier, C.: Let's Talk about LLM Evaluation (2025). 16 Feb 2025. https://huggingface.co/blog/clefourrier/llm-evaluation
6. Hair, J.F., Black, W.C., Babin, B.J., Anderson, R.E.: Multivariate Data Analysis, 8th edn. Cengage Learning, United Kingdom (2018)
7. Hu, Z., Feng, Y., Luu, A.T., Hooi, B., Lipani, A.: Unlocking the Potential of User Feedback: Leveraging Large Language Model as User Simulator to Enhance Dialogue System (2023). arXiv:2306.09821
8. Lee, D.-H., Pujara, J., Sewak, M., White, R., Jauhar, S.: Making large language models better data creators. Proceedings of the 2023 Conference on Empirical Methods in Natural Language Processing, pp. 15349–15360 (2023)
9. Lewis, P., et al.: Retrieval-Augmented generation for knowledge-intensive NLP tasks. Adv. Neural. Inf. Process. Syst. **33**, 9459–9474 (2020)
10. Ociepa, K., Flis, Ł., Wróbel, K., Gwoździej, A., Kinas, R.: Bielik 7B v0.1: A Polish Language Model - Development, Insights, and Evaluation (2024). arXiv. https://doi.org/10.48550/arXiv.2410.18565
11. Pit, P., Linden, T., Mendoza, A.: Generative artificial intelligence in higher education: one year later. AMCIS 2024 Proceedings, 11 (2024). https://aisel.aisnet.org/amcis2024/is_education/is_education/11
12. Singh, J., Joesph, M.H., Jabbar, K.A.: Rule-based chatbot for student enquiries. J. Phys.: Conf. Ser. **1228**(1), 012060. IOP Publishing (2019)
13. Stiennon, N., et al.: Learning to summarize with human feedback. Adv. Neural. Inf. Process. Syst. **33**, 3008–3021 (2020)
14. van Schaik, T.A.: A list of metrics for evaluating LLM-generated content (2024). https://learn.microsoft.com/en-us/ai/playbook/technology-guidance/generative-ai/working-with-llms/evaluation/list-of-eval-metrics
15. Zhang, T., Kishore, V., Wu, F., Weinberger, K.Q., Artzi, Y.: BERTScore: Evaluating Text Generation with BERT (2020). arXiv, 24 Feb 2020. https://doi.org/10.48550/arXiv.1904.09675

Retrieval-Augmented Generation for Organisational Communication: Designing and Deploying an AI Assistant for University-Student Interaction

Maciej Pondel(✉), Iwona Chomiak-Orsa, Małgorzata Sobińska, Wojciech Grzelak, Artur Kotwica, Andrzej Małowiecki, Kamila Łuczak, and Andrzej Greńczuk

Wroclaw University of Economics and Business, Komandorska 118/120, Wroclaw, Poland
maciej.pondel@ue.wroc.pl

Abstract. This paper presents the design and implementation of an AI assistant based on Retrieval-Augmented Generation (RAG) and Large Language Models (LLMs) to enhance organisational communication, focusing on university-student interaction. The system automates responses to frequently asked questions by leveraging semantic search and a dynamic knowledge base built from real user queries. Topic modelling using Latent Dirichlet Allocation (LDA) was applied to cluster questions into thematic areas, improving relevance and response quality. Deployed in a university setting, the assistant significantly reduced the administrative workload and improved user experience through faster and more accurate information delivery. The proposed architecture is modular, scalable, and adaptable to various organisational contexts, including public institutions and enterprises. This research demonstrates the potential of AI-driven assistants in supporting knowledge management, real-time communication, and digital transformation in complex environments.

Keywords: Artificial Intelligence (AI) · AI Assistant · Retrieval Augmented Generation (RAG) · Large Language Models (LLMs) · Knowledge Management · Topic Modelling · Organisational Communication

1 Introduction

In the era of rapid technological advancement, integrating artificial intelligence (AI) into enterprise-client communication is becoming a cornerstone of modern business strategies. This paper presents a concept for utilising a large language model (LLM) to develop an AI assistant to enhance communication processes between organisations and their stakeholders. This initiative is aligned with the broader vision of the "smart company" concept, which emphasises intelligent business operations, data-driven decision-making, and personalised stakeholder engagement. The goal is to lay new foundations for business intelligence in customer relationship management through innovative natural language processing technologies.

Published by Springer Nature Switzerland AG 2026
M. Pondel et al. (Eds.): AI4KAM 2025, IFIP AICT 776, pp. 52–61, 2026.
https://doi.org/10.1007/978-3-032-18920-2_6

The proposed solution is particularly relevant for large enterprises across various business sectors, nonprofit organisations, and governmental institutions. Its beneficiaries include end-users, such as customers and citizens, and internal personnel responsible for customer service and administrative support. By deploying AI-driven systems, these entities can streamline information flows, reduce operational burdens, and improve the overall quality of user interactions. This study explores the feasibility and practical implementation of such a system using a real-world example from a university environment.

The paper aims to answer the following research questions:

RQ1. How can AI-based agents support organisations with knowledge management and improve organizational communication?
RQ2. What technologies can be used to build an AI-based assistant, and what might be the obstacles to utilizing them?

To provide a comprehensive overview, the paper is structured as follows: the next section presents the theoretical foundations and related work concerning the theme of the paper. The subsequent section describes the methodology used to collect user data and construct the knowledge base, including the architecture of the developed AI assistant. This case study illustrates the deployment of the assistant in a university setting. A key contribution of this research is proposing a novel method for building a structured knowledge base from unstructured user queries and associated responses using topic modelling techniques, specifically Latent Dirichlet Allocation (LDA). This method enables the transformation of scattered question-and-answer datasets into coherent thematic clusters that support personalised and accurate AI-generated responses. The discussion and conclusion section reflects on the findings, challenges, and implications for broader organisational contexts, summarises the results, and outlines future research and development directions.

2 Related Works

Modern knowledge management approaches must address technological and organisational factors, continuously improving knowledge management systems as organisations shift to business models incorporating information and communication technologies (ICT) and AI tools. AI can optimise processes, automate tasks, and enhance information quality and human interaction, as shown by various studies (Bachiller et al., 2022; Custodio-Bachiller et al., 2022).

Organisations face significant challenges in safely integrating new technologies, particularly AI, into their business models while minimising negative impacts on information security and society. AI mechanisms can optimise processes that were previously handled less efficiently. Scholars emphasise the advantages of collaboration between humans and robots and specific AI applications, stating that this helps individuals better understand their environment and respond appropriately (Tarafdar et al., 2020). Studies by Wamba-Taguimdje et al. show that AI improves outcomes at organisational and process levels (Wamba-Taguimdje et al., 2020). AI enhances business process management by providing greater flexibility and automation (Zebec, 2019). AI, particularly involving LLMs, can enhance business process management in education and other domains

(Chakraborti et al., 2020). Loureiro et al. provide a review of AI in business, identifying trends and challenges, suggesting future research directions, while Ribeiro et al. show that machine learning and AI have been successfully applied in various sectors for different tasks (Loureiro et al., 2021; Ribeiro et al., 2021).

There is a need for a holistic understanding of how AI technologies create business value. Researchers emphasise understanding the factors influencing AI adoption, propose further research on AI utilisation,. Researchers propose further research on AI utilisation and highlight the role of developing non-technical resources, including AI-employee integration strategies (Dwivedi et al., 2021; Enholm et al., 2022; Chowdhury et al., 2023).

Building chatbots is a crucial aspect of using AI to improve organizational communication. As a computer programme, the chatbot allows users to have a conversation with a system that simulates a human conversation (Pondel et al., 2024). Chatbots empower users to get fundamental information from their chosen area and handle simple matters such as making an application or handling an order. The key difference between an AI assistant and a chatbot is the ability to use reasoning to provide users with action or decision support recommendations. Researchers have also questioned the validity of implementing AI assistants in various areas, including healthcare (Thirunavukarasu, 2023). In the study, Thirunavukarasu indicates that GPT-4 prepared answers to patients' social media questions with greater accuracy than doctors performing these tasks in their spare time (Thirunavukarasu, 2023).

AI assistants can significantly improve user experience (UX) from both the enterprise and customer perspectives. Primarily, AI assistants allow users to receive the immediate support necessary to complete selected activities (Virvou, 2023). They can improve accessibility by providing alternative communication methods through various image, sound, voice, and video analysis technologies (Virvou, 2023). In addition, AI assistants can use sentiment analysis to identify the tone of a user's speech or text and, in turn, generate customised responses, ultimately improving UX and user satisfaction (Aslam, 2023). The aspect concerns personalisation - researchers confirm that AI can solve personalised and complex problems through algorithms (Affandi et al., 2025).

A distinct research gap remains in understanding how LLMs and Retrieval-Augmented Generation (RAG) frameworks can effectively operationalize enterprise-client communication. While numerous studies have focused on AI's capabilities in business process management, enterprise resource planning (ERP) systems, and user interaction through chatbots, thereThere is a noticeable scarcity of empirical research specifically addressing how AI assistants employing LLMs and RAG architectures contribute to dynamic knowledge structuring and personalised communication within enterprise stakeholder relations. Current research tends to emphasise the technical capacities of AI and general process automation. It lacks in-depth investigation into how these tools can support knowledge creation, contextual adaptation, and human-AI collaboration in real-time communication settings, particularly in large, diversified organisations and public institutions. Another key underexplored area involves the practical application of AI-driven assistants in academic environments, such as universities, where the flow of complex and frequently changing information between institutions and stakeholders (students, candidates, administration) presents unique challenges and opportunities for

AI-supported knowledge mediation. The article's case study on AI-assisted student communication highlights this gap, yet the topic remains relatively novel and insufficiently theorised or evaluated in broader organisational contexts.

3 Methodology

The research was performed by applying several research methods and techniques, such as a literature review, a scientific experiment, which was conducted based on the questionnaire provided to the university students, which provided the input, and a computer simulation for preparing the agent based on the student input. Building the AI assistant consisted of architecture design, agent construction, and initial testing. Based on the research procedure, the results are being prepared for the agent-based solution, which will assist students (and study candidates) with answering the commonly asked questions in the Office for Student Affairs.

3.1 Research Procedure

The research was performed by applying several research methods, such as literature analysis based on a systematic literature review (Lenart-Gansiniec, 2021; Mazur & Orłowska, 2018; Orłowska et al., 2017) performed on the Scopus database, a scientific experiment, which was conducted based on the questionnaire provided to the university students which provided the input and computer simulation for preparing the agent based on the student input. This research method combination was required to receive results for such a complicated subject (**Fig. 1**).

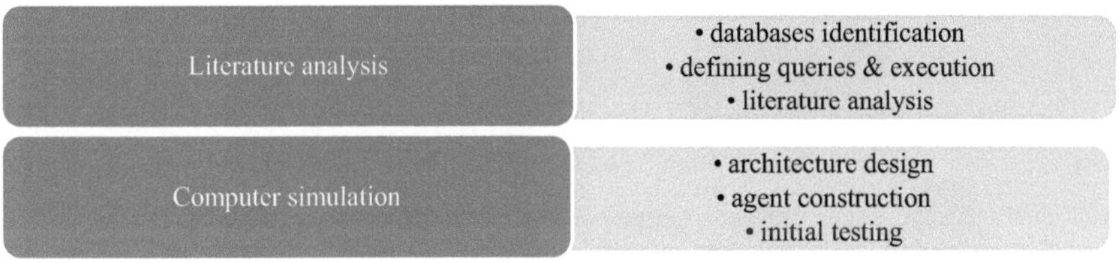

Fig. 1. Research procedure. Source: own elaboration.

3.2 Case Study: AI Assistant in University-Student Communication

As part of the conducted study, a **chatbot** was created to support communication between students/candidates and the university (Pondel et al., 2024). This study aimed to identify ways modern technologies - and AI systems in particular - can streamline information processes and ensure the efficient and rapid exchange of information. The chatbot was designed to allow users to get answers to questions related to administrative, organisational, and technical issues within the university. The specification of this study was mainly focused on implementing a dynamic, adaptive technology that responds to inquiries by providing precise answers based on a knowledge base prepared based on questions collected from students and candidates in a survey.

3.3 AI Assistant Architecture

The AI Assistant architecture uses a modular and scalable system to facilitate efficient, AI-driven communication between users, primarily students and candidates, and an educational institution. Users interact through a front-end User Widget, which captures their queries and passes them to an Assistant Agent responsible for orchestrating communication with underlying AI services. The Question Embedding Agent transforms incoming queries into semantic vectors using OpenAI's text-embedding-ada-002 API, enabling a Vector Embedding Database (e.g., Qdrant) to perform a semantic search for relevant knowledge (Pondel et al., 2024). These results are supplemented or refined using a GPT-4 model to generate accurate, context-aware answers. A Knowledge Embedding Agent and Content Management System (CMS) work together to update the knowledge base, ensuring content managers can continuously improve response quality. This architecture leverages LangChain, Python, and Strapi technologies to build a robust RAG pipeline that delivers fast, relevant, and personalised support (Fig. 2).

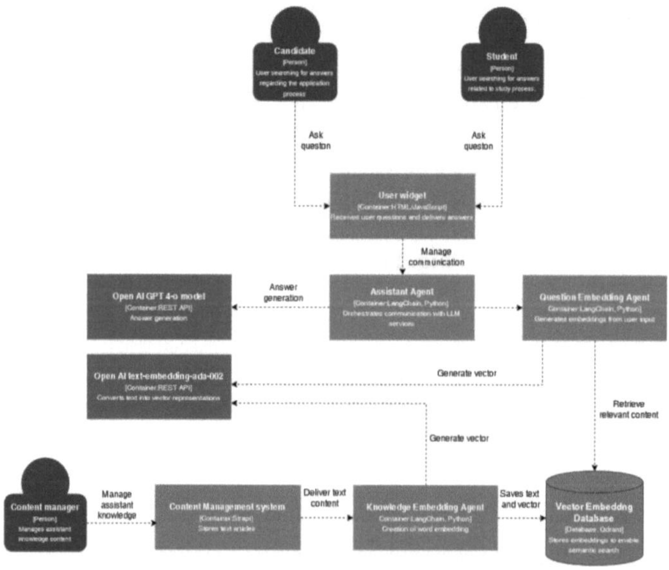

Fig. 2. AI Assistant architecture. Source: own elaboration.

3.4 Topic Modelling

In the next phase, **topic modelling techniques** were applied to analyse a dataset consisting of questions submitted by students. The objective was to identify prevalent thematic structures and group similar queries into coherent clusters. The methodology incorporated Natural Language Processing (NLP) tools, primarily leveraging the spaCy and Gensim Python libraries. SpaCy facilitated text preprocessing and transformation into numerical vectors, with native support for the Polish language. Gensim was employed to perform topic modelling and evaluate model quality. The chosen algorithm for topic modelling was Latent Dirichlet Allocation (LDA) - a probabilistic generative model that represents documents as mixtures of latent topics, where each topic is defined by a specific word distribution. The research followed a structured sequence of preprocessing and modelling steps:

1. **Tokenisation** – Conversion of all questions into tokens (individual words).
2. **Cleaning** – Removal of punctuation (e.g., ".", ",", "?", "-") and stopwords (e.g., "also", "such", "rather").
3. **Lemmatisation** – Reduction of words to their base form (e.g., "goes" → "go").
4. **Vectorisation** – Transformation of questions into a bag-of-words and dictionary representation suitable for modelling.
5. **Model selection** – Construction of 200 LDA models, each with a different number of topic clusters. Coherence scores (ranging from 0 to 1) were calculated for each model to assess the interpretability and semantic consistency of the topics. The coherence evaluation showed that models with two and ten clusters had the highest scores (0.64 and 0.63, respectively), with ten clusters selected as the optimal balance between interpretability and thematic granularity.
6. **Final modelling** – A final LDA model was trained to classify all questions into ten thematic clusters. Each question was assigned to a cluster based on the highest posterior probability of topic association.

4 Results

The application of the final LDA model resulted in the creation of ten distinct thematic clusters, each representing a major area of concern or interest among student queries.

The highest coherence value is observed for the LDA model with two topic clusters, equaling 0.64, suggesting that the number of groups is too small (Fig. 3). The second highest coherence value, 0.63, is found for the LDA model with ten clusters. Given the relatively small difference between these coherence values, we ultimately decided to divide the dataset into ten clusters.

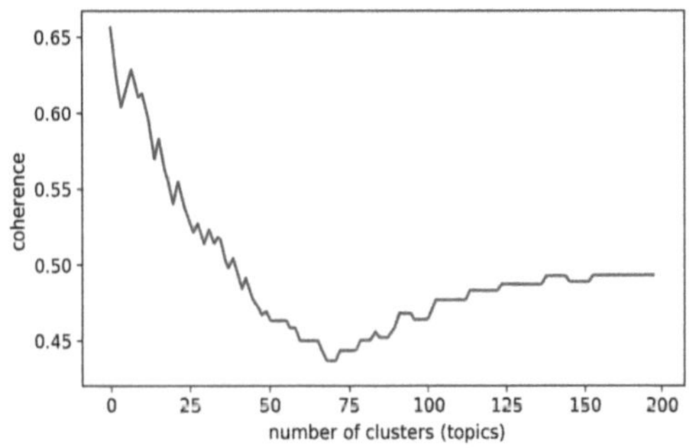

Fig. 3. Thematic coherence for different numbers of clusters. Source: own elaboration.

In addition, the size of each topic cluster is presented, highlighting the relative popularity of the topics (Fig. 4). Every bar represents the size of a single thematic cluster. The most numerous group was the second group, which means that the second topic discovered by the algorithm is the most interesting for students. The calculated probability of a single question belonging to the thematic cluster is different for each question, and within the topic group, there are questions that have different values of belonging probability to the group.

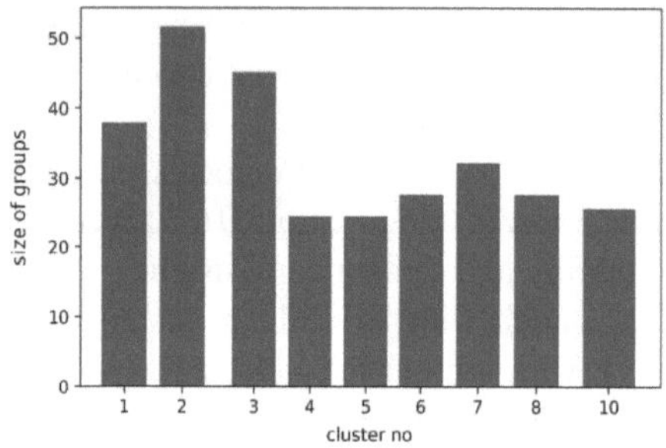

Fig. 4. Number of students' questions in every cluster. Source: own elaboration.

These findings demonstrate the effectiveness of topic modelling using LDA for structuring large volumes of qualitative data in a meaningful and interpretable way, making it a valuable approach for understanding user needs and designing responsive AI systems. The calculated probability of a single question belonging to the thematic cluster is different for each question, and within the topic group, there are questions that have different values of belonging probability to the group. The range of probabilities for every thematic group is shown (Fig. 5).

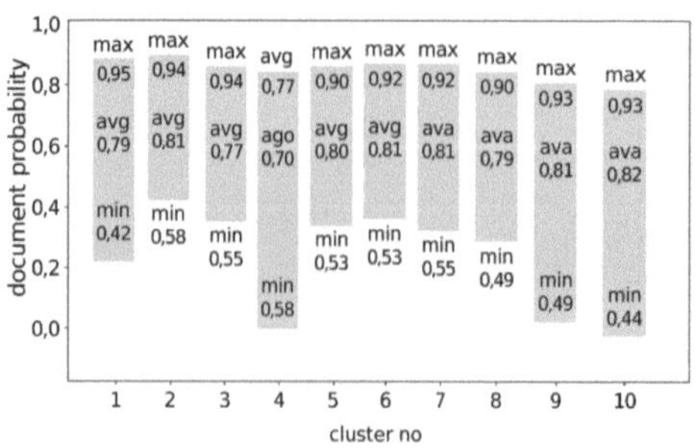

Fig. 5. Range of probabilities. Source: own elaboration.

The max value on the graph represents the maximum probability of a single question belonging to the thematic cluster. In each cluster, the maximum probability value is very high and close to 1, which means the LDA model is very confident in its classification. The minimum value on the bars represents the minimum probability of a single question belonging to the thematic cluster. The average value is the mean probability for the entire thematic cluster. The model's classifications were both coherent and reliable. The range of probabilities suggests that the LDA model is a promising method.

In summary, cluster distribution revealed that the second cluster contained the largest number of questions, indicating a particularly high level of student interest in that topic. For each question, the LDA model calculated a probability score reflecting its degree of association with the assigned cluster. The range of probabilities within clusters varied, with maximum values close to 1.0, signifying high confidence in classification accuracy. Minimum and average probabilities were also analysed, showing consistent classification performance across all ten groups. The study demonstrates that the proposed method of generating a knowledge base using example user questions and domain-specific articles is an effective foundation for building AI-powered assistants. This approach not only enables the structure of unstructured information but also supports real-time, personalised communication with stakeholders.

5 Discussion and Conclusion

The research aimed to design and implement an AI assistant based on RAG and LLMs to enhance organisational communication, especially university-student interaction. One key observation is the critical role of high-quality, domain-specific training data—in this case, real user questions—which ensures relevance and accuracy in generated responses (Pondel et al., 2024). Gathering the knowledge base from the unit's documentation, questions from students, formulating appropriate answers by the AI Assistant, and analyzing them allows organizations to better build and structure knowledge management (RQ1). The use of the LDA made it possible to assign all questions to clusters based on the highest posterior probability of topic association (RQ2). Additionally, topic modelling (LDA) integration effectively identified thematic clusters, allowing for more structured and context-aware knowledge management (RQ1, RQ2). Challenges remain, particularly in maintaining the actuality of the knowledge base and handling ambiguous or highly specific user queries (RQ2). The reliance on LLMs also raises computational cost, scalability, and data privacy issues, especially in institutional environments such as universities (RQ2), which may be indicated as a limitation.

As further research, it should be indicated to test and build the assistant in terms of support not only in organizational issues at universities, but also, for example, as support for supervisors of Bachelor's and Master's theses. Despite these challenges, the project demonstrates that AI assistants can significantly improve communication efficiency and user satisfaction with appropriate design and continuous feedback loops, supporting broader digital transformation goals in educational and business contexts.

The study demonstrates that the proposed method of generating a knowledge base using example user questions and domain-specific articles is an effective foundation for building AI-powered assistants in organisational contexts. This approach not only enables the structure of unstructured information but also supports real-time, personalised communication with stakeholders. An example of the graphical user interface (GUI) of the assistant is presented (Fig. 6).

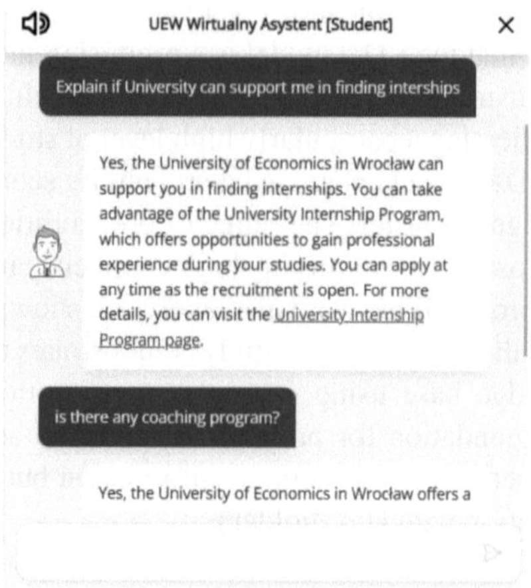

Fig. 6. Example of an AI Assistant to assist university students

Source: own elaboration.

The presented interface demonstrates its practical usability and user-friendly design. This implementation confirms the method's practicality and adaptability, offering a replicable blueprint for similar AI-driven initiatives in public institutions and enterprise environments.

(Aslam, 2023). By leveraging LLMs and an RAG framework, the developed assistant was successfully implemented in a university setting to enhance communication between students and administrative services. The assistant's architecture shows an integrated and scalable solution that combines semantic search, natural language understanding, and dynamic content management. In addition, the simplicity and usability of the graphical interface of the presented solution make it easier and potentially more likely to be used by students.

Acknowledgments. This study was funded by the UEW Internal Grant.

Disclosure of Interests. The authors have no competing interests to declare that are relevant to the content of this article.

References

Affandi, S., Ishaq, M.I., Raza, A., Talpur, Q., Ahmad, R.: AI assistant is my new best friend! Role of emotional disclosure, performance expectations and intention to reuse. J. Retail. Consum. Serv. **82**, 104087 (2025). https://doi.org/10.1016/j.jretconser.2024.104087

Aslam, F.: The impact of artificial intelligence on Chatbot technology: a study on the current advancements and leading innovations. Eur. J. Technol. **7**(3), 62–72 (2023). https://doi.org/10.47672/ejt.1561

Bachiller, Y., Busch, P., Smith, S.: Innovation through process automation: Factors for success. ACIS 2022 Proceedings, pp. 1–11 (2022). http://acis.aaisnet.org/acis2022/

Chakraborti, T., et al.: From robotic process automation to intelligent process automation: – emerging trends. Asatiani, W.A. et al. (eds.), Business Process Management: Blockchain and Robotic Process Automation Forum (T. 393, s. 215–228). Springer International Publishing (2020). https://doi.org/10.1007/978-3-030-58779-6_15

Chowdhury, S., et al.: Unlocking the value of artificial intelligence in human resource management through AI capability framework. Hum. Resour. Manag. Rev. **33**(1), 100899 (2023). https://doi.org/10.1016/j.hrmr.2022.100899

Custodio-Bachiller, Y., Busch, P., Guzman, G.: Process managing the study abroad experience: an innovative framework. J. Organ. Knowl. Manage. **2022**, 3–42 (2022). https://doi.org/10.5171/2022.489606

Dwivedi, Y.K., et al.: Artificial Intelligence (AI): multidisciplinary perspectives on emerging challenges, opportunities, and agenda for research, practice and policy. Int. J. Inf. Manage. **57**, 101994 (2021). https://doi.org/10.1016/j.ijinfomgt.2019.08.002

Enholm, I.M., Papagiannidis, E., Mikalef, P., Krogstie, J.: Artificial Intelligence and business value: a literature review. Inf. Syst. Front. **24**(5), 1709–1734 (2022). https://doi.org/10.1007/s10796-021-10186-w

Lenart-Gansiniec, R.: Systematyczny przegląd literatury w naukach społecznych. Wydawnictwo Naukowe Scholar Spółka z Ograniczoną o (2021). https://doi.org/10.7366/9788366849006

Loureiro, S.M.C., Guerreiro, J., Tussyadiah, I.: Artificial intelligence in business: state of the art and future research agenda. J. Bus. Res. **129**, 911–926 (2021). https://doi.org/10.1016/j.jbusres.2020.11.001

Mazur, Z., Orłowska, A.: Jak zaplanować i przeprowadzić systematyczny przegląd literatury. Polskie Forum Psychologiczne **23**(2), 2 (2018). https://doi.org/10.14656/PFP20180202

Orłowska, A., Mazur, Z., Łaguna, M.: Systematyczny przegląd literatury: Na czym polega i czym różni się od innych przeglądów? Ogrody Nauk i Sztuk **7**(7), 7 (2017). https://doi.org/10.15503/onis2017.350.363

Pondel, M., et al.: AI Tools for Knowledge Management – Knowledge Base Creation via LLM and RAG for AI Assistant. In: Hernes, M., Wątróbski, J., Rot, A. (eds.), Emerging Challenges in Intelligent Management Information Systems (Vol. 1218, pp. 3–15). Springer Nature Switzerland (2024). https://doi.org/10.1007/978-3-031-78468-2_1

Ribeiro, J., Lima, R., Eckhardt, T., Paiva, S.: Robotic process automation and artificial intelligence in Industry 4.0–A literature review. Procedia Comput. Sci. **181**, 51–58 (2021). https://doi.org/10.1016/j.procs.2021.01.104

Tarafdar, M., Beath, C.M., Ross, J.W.: Using AI to enhance business operations. W Mit Sloan Management Review, How AI Is Transforming the Organization (s. 67–86). The MIT Press (2020). https://doi.org/10.7551/mitpress/12588.003.0015

Thirunavukarasu, A.J.: How can the clinical aptitude of AI assistants be assayed? J. Med. Internet Res. **25**, e51603 (2023). https://doi.org/10.2196/51603

Virvou, M.: Artificial intelligence and user experience in reciprocity: contributions and state of the art. Intell. Decis. Technol. **17**(1), 73–125 (2023). https://doi.org/10.3233/IDT-230092

Wamba-Taguimdje, S.-L., Fosso Wamba, S., Kala Kamdjoug, J.R., Tchatchouang Wanko, C.E.: Influence of artificial intelligence (AI) on firm performance: the business value of AI-based transformation projects. Bus. Process. Manag. J. **26**(7), 1893–1924 (2020). https://doi.org/10.1108/BPMJ-10-2019-0411

Zebec, A.: Cognitive BPM: Business Process Automation and Innovation with Artificial Intelligence. CEUR Workshop Proceedings, pp. 51–59 (2019)

Managing Academic E-Learning Using Artificial Intelligence: Integrating Intelligent Systems with the Moodle Platform

Tomasz Turek(✉) and Tomasz Walasek

Czestochowa University of Technology, Dabrowskiego 69, Czestochowa 42-201, Poland
{tomasz.turek,tomasz.walasek}@pcz.pl

Abstract. This article explores the integration of artificial intelligence (AI) within the Moodle platform to enhance academic e-learning. It outlines the potential of AI to personalize learning, automate assessment, and provide predictive insights that support both students and educators. By examining current AI plugin capabilities and proposing a structured, three-level model of AI use in Moodle-analytics, content generation, and prediction-the paper highlights practical benefits and technical considerations. The proposed approach aims to guide institutions toward effective, ethical, and scalable AI adoption in digital education environments.

Keywords: Artificial Intelligence (AI) · Moodle · E-learning

1 Introduction

Over the years, Moodle has established itself as one of the most important learning management systems (LMS) in the world. Its openness, flexibility, and active community have made it the foundation of digital education at many universities. However, the rapid development of technology, particularly artificial intelligence (AI), presents new challenges for e-learning systems while opening up completely new possibilities.

The evolution of Moodle-from a simple repository of educational content to a comprehensive platform supporting personalized learning - takes on new meaning in the context of integration with AI tools. With new features such as dedicated AI plugins, Moodle can not only support teachers in content creation and assessment automation, but also analyze student data to predict educational problems and personalize the teaching process [1].

This article attempts to systematically present the benefits of integrating AI with the Moodle platform, with a particular focus on the three-level model of intelligent technology use: analytics, content generation, and prediction. The aim is not only to demonstrate the technological potential, but also to highlight the practical and ethical aspects of implementing AI in an academic environment.

Published by Springer Nature Switzerland AG 2026
M. Pondel et al. (Eds.): AI4KAM 2025, IFIP AICT 776, pp. 62–69, 2026.
https://doi.org/10.1007/978-3-032-18920-2_7

Effective management of LMS platforms in the age of AI requires a strategic approach that encompasses not only technological integration but also pedagogical alignment and organizational readiness. Administrators must ensure that AI tools are implemented in ways that enhance teaching and learning while maintaining system reliability, data security, and equitable access for all users.

2 Context of the Topic

The contemporary landscape of higher education is inextricably linked with e-learning, a domain where Learning Management Systems (LMS) have become the foundational infrastructure for digital pedagogy. Within this ecosystem, the Moodle platform has firmly established its preeminent position as a leading global LMS, recognized for its widespread adoption and open-source flexibility [2, 3]. It has become the de facto digital campus for countless institutions. However, this established paradigm now confronts a profound disruption precipitated by the rapid mainstreaming of generative artificial intelligence (AI), which poses novel challenges to traditional notions of academic integrity and instructional design.

This technological shift has created a significant challenge, one powerfully articulated in the recent Leading Through Disruption report [4]. The study reveals a stark dichotomy: while a majority of students actively use generative AI, faculty engagement remains minimal, and institutional preparedness is alarmingly low. The report underscores that while institutional leaders recognize the immense potential of AI to personalize learning (91%), most remain in a state of diagnostic paralysis, unprepared to train students or support faculty in this new era [4]. This gap highlights an urgent need for structured frameworks that can guide the effective and ethical integration of AI into established educational practices.

The deployment of artificial intelligence in educational environments raises concerns related to transparency, accountability, bias, and the autonomy of both students and teachers. Algorithms used to analyze student behavior or generate learning paths must be transparent and explainable to avoid reinforcing existing inequalities or making opaque decisions that affect educational outcomes. Additionally, the potential for surveillance or excessive monitoring through AI tools must be carefully balanced with students' right to privacy and agency. Ensuring that AI systems respect diversity, avoid discriminatory outputs, and are governed by clear ethical guidelines is essential to foster trust and fairness in digital education ecosystems. To address these concerns, educational institutions should establish clear regulations governing the responsible use of AI tools by both teachers and students. Such guidelines should define acceptable practices, clarify boundaries regarding content generation and assessment, and reinforce adherence to academic integrity and data protection standards.

As noted in the introduction the primary objective of this paper is therefore to delineate the key opportunities for integrating artificial intelligence with the Moodle platform and to systematically demonstrate the resultant pedagogical benefits. Empirical evidence increasingly shows that the strategic use of AI tools within e-learning environments can lead to significant gains in student performance and engagement [student]. To achieve this objective, this article will demonstrate how integrating AI into Moodle yields significant advantages across three critical domains of the educational experience.

First, it enables the deep personalization of the learning journey. By leveraging AI algorithms, Moodle can transition from a static content repository to an adaptive environment. This facilitates the diagnosis of individual learning styles and allows for the personalization of the learning process, an approach that has been empirically validated to significantly improve both educational success and student satisfaction [5]. Second, AI automates and enriches the processes of assessment and engagement. Students in online settings require timely feedback to maintain motivation, a common challenge in distance education. AI integration addresses this by enabling predictive models to identify at-risk students early and by supporting dynamic online evaluation methods, which are a core function of modern LMS platforms [6]. Finally, it provides scalable, intelligent support. By creating predictive models based on student digital footprints, institutions can automate the identification of students who may need additional support, thereby reducing educator workload and creating more effective learning environments.

3 The Place of AI-Moodle in (e-)learning Models

Information and communication technology solutions have long supported learning processes. They make teaching more effective, and the benefits are mutual: both for teachers and for pupils and students. Technology is changing the teaching process and, in some cases, becoming a critical element. This was the case during the COVID-19 pandemic, when, within a few weeks, most countries around the world were paralyzed to a greater or lesser extent. Education was moved online, and teaching processes were only possible thanks to e-learning. Moodle-based solutions played a huge role in this, especially in higher education.

The role of technology in teaching is also presented through the context of models such as TPACK, ADDIE, and SAM. These approaches assume IT support for teaching processes, which opens up a wide range of possibilities for LMS and Moodle.

The traditional TPACK (Technological Pedagogical Content Knowledge) model assumes that the effective use of technology in education requires a combination of technological (TK), pedagogical (PK), and content knowledge (CK) [7]. Equipping Moodle with additional AI capabilities will cause the components of the TPACK model to evolve towards new challenges:

• Technological Knowledge (TK) - Integrating AI with Moodle requires the teacher to be familiar with both the LMS platform and artificial intelligence tools (e.g., AI plugins, content generation functions).
Pedagogical Knowledge (PK) - AI in Moodle allows for personalization of teaching, automation of assessment, and individualization of support. Teachers should therefore be able to design activities that use AI to support the learning process, rather than just automating tasks.
Content Knowledge (CK) - AI can generate content, quizzes, and summaries, but teachers must be able to assess their accuracy and relevance to the learning objectives. This means that the verification and selection of AI-generated materials is a key competence.

As a result of applying AI, the TPACK model changes its form and transforms into a model that can be referred to as AI-TPACK – Fig. 1.

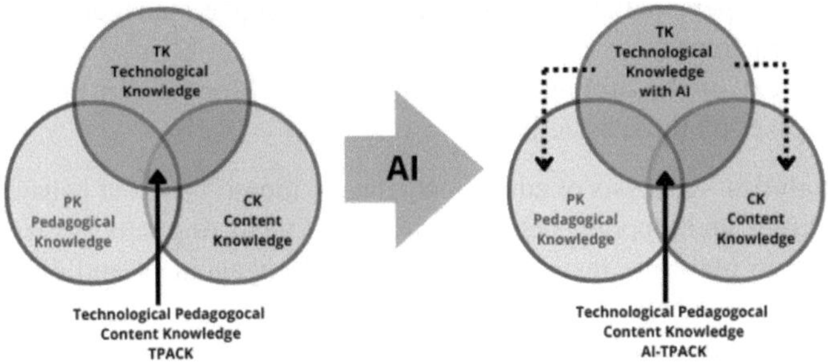

Fig. 1. The evolution of the TPACK model to AI-TPACK

In this model, AI solutions, which would traditionally be classified as TK (Technological Knowledge), have a strong and direct impact on PK (Pedagogical Knowledge) and CK (Content Knowledge). In the traditional model, TK played a supporting role for teachers in the distribution of knowledge and interaction with students. Thanks to AI in LMS and Moodle, competencies, forms of cooperation, and course implementation rules are significantly improved [8].

Similar changes are taking place in ADDIE cascade models. This is a structure for designing instructional systems, whose name comes from the classic phases: Analysis, Design, Development, Implementation, Evaluation. Virtually each of these phases can use AI in the preparation of courses on Moodle [9].

In the **Analyze** phase, AI can automatically analyze data on training needs, participant profiles, and the results of previous training. Thanks to advanced algorithms, large data sets can be processed quickly, allowing for precise definition of objectives and target groups.
In the **Design** phase, AI supports the creation of personalized and adaptive training materials. It can generate content, scenarios, and interactions tailored to different levels of participant proficiency.
In the **Develop** phase, AI automates the creation of materials, e.g., by generating texts, quizzes, simulations, or multimedia. AI tools can also support prototyping and rapid testing and refinement of content, reducing training production time.
In the **Implement** phase, AI can support the personalization of the participant experience by providing personalized recommendations and real-time support, e.g., through chatbots. AI systems monitor progress, adjusting the pace and difficulty of the materials to individual needs.
In the **Evaluate** phase, AI automates the collection and analysis of evaluation data, generating accurate reports and insights. It can identify patterns in participant results, suggest areas for improvement, and recommend course modifications.

In the context of TPACK and ADDIE, reference should also be made to the SAM (Successive Approximation Model) [10]. The SAM is an iterative, agile instructional design model that emphasizes rapid development and continuous improvement of training materials through multiple iterations. Moodle, equipped with AI solutions, supports the iterative and flexible process of course creation according to the SAM model, enabling rapid prototyping, personalization of learning, and effective management and analysis of educational outcomes.

In the context of the aforementioned models, AI-Moodle can therefore support teaching processes on three basic levels:

Level 1 - **Analytics** – analysis of educational data to identify student behavior patterns, predict difficulties, and support teaching decisions. By analyzing activity logs and results, teachers can gain real-time insight into the level of engagement and effectiveness of individual course elements.
Level 2 - **Content Generation** - automatic creation and adaptation of teaching content (e.g., quizzes, materials, summaries) using generative language models. Such solutions allow for faster preparation of personalized educational resources that match the student's level of knowledge and learning style.
Level 3 - **Prediction** - building predictive models that allow for early detection of students at risk of failure and recommending personalized learning paths. These models can integrate various data sources (attendance, grades, activity) and generate alerts to support educational intervention before difficulties arise.

AI-Moodle significantly evolves traditional teaching models such as TPACK, ADDIE, and SAM by integrating artificial intelligence, which not only supports the educational process technologically, but also influences pedagogy and teaching content, enabling personalization, automation, and prediction in e-learning.

4 Utilization of AI Plugins in the Latest Moodle Platform

Theoretical models could not be implemented without the practical integration of Moodle with AI tools and plugins. This section of the paper will present the basic outlines of such an implementation.

In recent years, the integration of artificial intelligence (AI) tools within learning management systems (LMS) has become increasingly prevalent. Moodle, one of the most widely used open-source LMS platforms, has incorporated a dedicated AI subsystem in its latest versions, enabling access to diverse generative language models and AI-driven functionalities. This development allows educational institutions to enhance the teaching and learning process by supporting educators in content creation, assessment automation, and providing personalized student assistance [11].

Moodle's AI integration is architected around two primary plugin types: provider plugins and placement plugins. Provider plugins facilitate the connection to external AI services via APIs, while placement plugins define the contexts and locations within the Moodle interface where AI functionalities are made available to users, such as content editors, discussion forums, or chat blocks. This modular design allows for independent development and flexible extension of AI capabilities within the platform [12, 13].

Administrators can configure multiple AI providers within Moodle, supporting services like OpenAI, Microsoft Azure AI, or other large language models (LLMs) compatible with standard API protocols. Each provider instance can be customized with specific parameters such as model type, response creativity (temperature), and API credentials. This flexibility enables tailored use of AI models for different pedagogical tasks-e.g., using one model for generating course content and another for text analysis or summarization.

Practical applications of AI plugins in education include automatic generation of instructional materials such as lecture outlines, assignment descriptions, and quiz questions, thereby reducing teacher workload while maintaining consistent style and quality. Additionally, AI-driven image generation tools (e.g., DALL-E or Stable Diffusion) integrated via plugins can produce visual aids that enhance comprehension of complex subjects. Students benefit from AI-supported personalized summaries of lengthy course texts or can generate sample answers to practice critical thinking and writing skills.

Several AI integration plugins are available within the Moodle community ecosystem. For example, the AI Connector plugin establishes secure API communication with AI service providers without altering core Moodle interfaces, serving as a foundational layer for higher-level modules – Fig. 2 [14]. Meanwhile, the AI Manager plugin supports multi-tenant management, allowing distinct configurations for different faculties or institutional units within a single Moodle installation, thus enabling scalable and context-specific AI deployment.

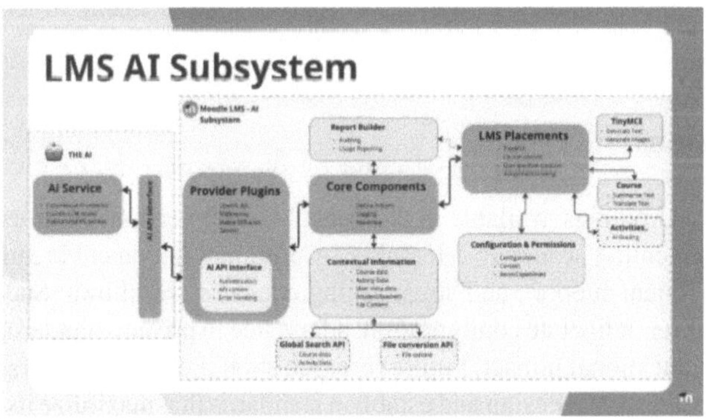

Fig. 2. Integration of AI services with Moodle LMS

From a pedagogical perspective, plugins such as OpenAI Chat Block and Corolair AI Teaching Assistant introduce interactive chatbot assistants directly into courses [15]. These virtual tutors offer 24/7 support by answering student questions based on course content or providing hints for assignments. This feature not only alleviates repetitive queries for instructors but also ensures learners receive timely, contextualized guidance. Furthermore, AI can assist in preliminary grading of written assignments by evaluating language style and identifying factual inconsistencies, though expert oversight remains essential.

The deployment of AI plugins in Moodle also necessitates careful attention to data privacy and security. Educational institutions must ensure that personal student data transmitted to external AI services comply with legal frameworks such as the General Data Protection Regulation (GDPR). Moreover, administrators should monitor API usage limits, cost implications, and system performance impacts associated with large-scale AI query volumes.

Despite the growing availability of AI plugins for Moodle, their implementation is not without significant barriers and limitations. One of the primary challenges is the technical complexity of configuring AI services, which often requires advanced knowledge of APIs, data formats, and server infrastructure. Many institutions also face legal and organizational constraints related to data protection regulations, particularly when sensitive student data is processed by external AI providers. Furthermore, not all Moodle installations are updated to the latest versions that support AI subsystems, which limits compatibility and functionality. Financial constraints may also arise, as some AI services involve usage-based fees that can escalate with large-scale adoption. Lastly, there is often a lack of institutional policy or support for training educators, which hampers effective and responsible use of these tools in everyday teaching practice.

AI integration in education brings considerable benefits, but it also introduces several critical challenges. Verifying the accuracy and relevance of AI-generated content remains a concern, as does the potential for misuse, including academic dishonesty or overreliance on automated tools. Ensuring equitable access to AI resources across different student populations is equally important to avoid widening existing educational gaps. Moreover, there is a pressing need for empirical research to evaluate the actual impact of AI-supported learning compared to traditional methods, particularly in terms of student engagement, motivation, and academic outcomes.

5 Conclusions

In conclusion, AI plugins available in the latest Moodle versions offer extensive possibilities for augmenting academic e-learning by automating content creation, delivering personalized student support, and streamlining educator workflows. Successful implementation requires deliberate configuration, adherence to privacy standards, and ongoing evaluation of educational impact. Future research should explore best practices for integrating AI in pedagogical design and establish standards that maximize its benefits while mitigating risks in academic environments.

The integration of artificial intelligence with the Moodle platform offers significant benefits for academic e-learning, including content automation, personalized support, and improved course management. The proposed three-level model-analytics, content generation, and prediction - provides a clear framework for the effective application of artificial intelligence [16].

However, successful implementation requires not only technical configuration but also attention to ethics, data privacy, and teacher training. As AI tools become more embedded in education, institutions must ensure responsible use while fostering innovation. With a thoughtful approach, AI can enhance the quality, accessibility, and efficiency of digital learning environments.

This study is a contribution to further discussion and research on LMS management using AI. At present, it has not been possible to conduct empirical research, as Moodle 4.5, which allows for the addition of AI subsystems, was released in October 2024. Many universities had already started the academic year at that time. In most cases, it was therefore only possible to update university LMSs after the end of the teaching process. The current version is Moodle 5.0 [17].

References

1. Tarik, T., Zakaria, T., Souhaib, A.: Enhancing socio-pedagogical engagement in Moodle through AI-driven personalized learning. Cadernos De Educação Tecnologia E Sociedade **17**(se4), 63–73 (2024)
2. Magro, M.J., Sharp, J.H., Ryan, K., Ryan, S.D.: Investigating ways to use facebook at the university level: a delphi study. Issues Inform. Sci. Inf. Technol. **10**, 295–311 (2013)
3. Afolabi, O.S., Ajani, O.A.: South African rural students' adoption and use of learning management systems. Int. J. Res. Bus. Soc. Sci. (2147-4478), **12**(4), 410–421. (2023)
4. Leading Through Disruption: Higher Education Executives Assess AI's Impacts on Teaching and Learning. https://www.aacu.org/research/leading-through-disruption. Last accessed 09 May 2025
5. Amiri, M., Montazer, G., Mousavi, E.: Educational Customization by Homogenous Grouping of e-Learners based on their Learning Styles (2024). https://doi.org/10.48550/arXiv.2408.12619
6. Mohammed, M.A.E., et al.: Leveraging artificial intelligence for enhanced electronic course design and student achievement: Unlocking the potential of AI in education. Res. J. Adv. Hum. **6**(2) (2025)
7. Filippidi, A., Karachristos, C., Lavidas, K., Komis, V., Karacapilidis, N.: Leveraging cognitive and social engagement in blended learning through an AI-augmented pedagogical framework. In: Kubincová, Z., Caruso, F., Kim, Te., Ivanova, M., Lancia, L., Pellegrino, M.A. (eds.) Methodologies and Intelligent Systems for Technology Enhanced Learning, Workshops - 13th International Conference. MIS4TEL 2023. LNNS, vol. 769, pp. 28–38. Springer, Cham (2023). https://doi.org/10.1007/978-3-031-42134-1_3
8. Ye, L., Ismail, H.H., Aziz, A.A.: Innovative strategies for TPACK development in pre-service English teacher education in the 21st century: a systematic review. Forum Linguistic Stud. **6**(6), 274–294 (2024)
9. Lee, G.-G., Mun, S., Shin, M.-K., Zhai, X.: Collaborative learning with artificial intelligence speakers: pre-service elementary science teachers' responses to the prototype. Sci. Educ. **34**(2), 847–875 (2025)
10. Hey, A.: Ultimate Guide to the Successive Approximation Model (SAM). Coursebox, https://www.coursebox.ai/blog/successive-approximation-model. Last accessed 09 May 2025
11. Seidel, N.: Architecture for gradually AI-Teamed adaptation rules in learning management systems. In: International Conference on Computer Supported Education, CSEDU-Proceedings, 1, pp. 243–254, CSEDU, Porto, Portugal (2025)
12. Strategies to Integrate AI in Moodle. https://mindfieldconsulting.com/strategies-to-integrating-openai-llama-microsoft-copilot-and-gemini-in-moodle. Last accessed 09 May 2025
13. AI in Moodle LMS: what can it do and how to enable it for your Moodle site. https://www.catalyst-au.net/blog/ai-in-moodle-lms-what-can-it-do-and-how-to-enable-it-for-your-moodle-site. Last accessed 09 May 2025
14. AI providers. https://docs.moodle.org/500/en/AI_providers. Last accessed 09 May 2025
15. Bachiri, Y.-A., Mouncif, H., Bouikhalene, B.: Artificial Intelligence empowers gamification: optimizing student engagement and learning Outco1[15mes in E-learning and MOOCs. Int. J. Eng. Pedagogy **13**(8), 4–19 (2023)
16. Daws, R.: AI in education: balancing promises and pitfalls. https://www.artificialintelligence-news.com/news/ai-in-education-balancing-promises-and-pitfalls. Last accessed 09 May 2025
17. https://download.moodle.org/releases/latest/. Last accessed 09 May 2025

AI Significance in Higher Education: A Comparative Study Across the Polish Universities

Anna Sołtysik-Piorunkiewicz[1](✉), Witold Chmielarz[2], and Magdalena Kossowska-Lai[3]

[1] University of Economics in Katowice, 1 Maja Str. 50, 40-287 Katowice, Poland
anna.soltysik-piorunkiewicz@uekat.pl

[2] University of Warsaw, Krakowskie Przedmieście 26/28, 00-927 Warsaw, Poland

[3] Jagiellonian University, Prof. S. Łojasiewicza Str. 4, 30-348 Kraków, Poland

Abstract. The paper aims to show artificial intelligence (AI) in higher education based on a study in a Polish environment. In this study, the awareness of AI, particularly ChatGPT, was shown across Polish universities. This research was carried out in three selected universities: the University of Economics in Katowice (UE Kat), the University of Warsaw (UW), and the Jagiellonian University in Krakow (JU). Data were collected using a mixed method of PAPI (Paper and Pencil Interviewing) and CAPI (Computer Assisted Personal Interview). The surveys were distributed in a partially targeted manner in the academic environment to randomly selected groups of students. After that the collected data were processed: statistical indicators, distance measures, and tests were calculated. Thanks to the analysis of the results, it was possible to identify the level of artificial intelligence awareness in the studied environment and understand how respondents qualify for its usage. The originality of the research lies in the comprehensive treatment of the topic and its presentation in terms of gender and university location. The results may be useful to university authorities and researchers dealing with the applications of artificial intelligence in economic, and social life.

Keywords: AI · ChatGPT · students · higher education institution · university

1 Introduction

Higher education faces multiple challenges today. One of them is applying the basics of the Education 4.0 paradigm, which tries to address the opportunities and needs of Industry 4.0. Education 4.0 itself relates to the concept of learning by doing – this means that education should encourage students to discover and learn experimentally. However, the implementation levels are in the very early stages of development; in most locations, this is still in the testing phase. Artificial intelligence (AI) can play a vital role in the education 4.0 paradigm application process by supporting the identification of student performance factors, answering their routine questions, and helping to implement

Published by Springer Nature Switzerland AG 2026
M. Pondel et al. (Eds.): AI4KAM 2025, IFIP AICT 776, pp. 70–83, 2026.
https://doi.org/10.1007/978-3-032-18920-2_8

personalised learning, predictive modelling, and learning analytics. It brings multiple challenges, such as the difficulty in defining student outcomes and how to redefine the Education 4.0 paradigm to recognise creative and innovative students in the right way [36]. Three significant areas for higher education need urgent attention: the language of AI; issues of labour, and accountability; and the implications for learning and teaching contexts beyond single innovations. This means that we can understand more deeply how AI works by researching various actors included in the process so that teachers, students, machines, administrators, and corporations impact each other from a learning and teaching perspective. The authors outline that this leads to a research agenda for higher education studies which is to develop a more nuanced language for AI; and to test how AI-driven technologies change accountability and how this can influence teaching and learning relationships [8, 21].

Therefore, by conducting a study in three large academic institutions in Poland, based on current knowledge, we have developed research questions that are tested throughout the study.

RQ1: Is there awareness and understanding between the basic categories of artificial intelligence, and if so, to what extent?
RQ2: What are the theoretical basic advantages and usefulness of using artificial intelligence in university life?
RQ3: Are there differences between opinions on the possibilities of artificial intelligence applications according to respondents' gender?
RQ4: Are there differences between respondents' opinions on the possibilities of using artificial intelligence regionally, that is, between respondents from Katowice, Kraków, and Warsaw?
RQ5: Are there differences between respondents in their opinion on the use of ChatGPT based on gender and/or location?

2 Literature Review

Currently, there is a large movement that highlights the need to incorporate AI tools into the education sector. Among the vast number of tools, ChatGPT is considered a powerful and versatile resource that can bring multiple benefits [2]. It is an AI-driven conversational model, created by OpenAI, that can deliver human-like responses [14]. Universities must rethink how to use AI tools so that they can develop student capacity for their use and develop unique skills within this group that, in the graduate employment market, cannot be delivered by ChatGPT or other AI tools [26, 34].

The literature available on ChatGPT is scarce, highly speculative, and hypothetical. ChatGPT is bringing about a revolution within higher education through deep learning models that generate human-like content. This integration raises many concerns within the academic community about academic integrity, plagiarism detection, and how critical thinking skills are affected. There is a call for urgent empirical research that can provide more insight into how to maximize the opportunities this tool brings and how we should build strategies and practices that could be widely acceptable by the academic community [1, 10, 11]. Furthermore, there is an urgent need to dive deeper into understanding its core users within the given system, that is, the perceptions, interactions, consequences, and concerns of students or academics connected with ChatGPT [5, 15, 23, 31, 35]. The use of AI tools in education requires careful study of its risks, ethical implications, and benefits for all stakeholders due to its high complexity. It calls for collaboration between all stakeholders within academia to develop policies, frameworks, and guidelines. This can help promote inclusivity, transparency, and a sense of responsibility. We must unlock the full potential of AI in education by ensuring that we have created a fair and equitable learning environment and mitigate all risks for included stakeholders at the same time [31].

The current findings show that students have limited exposure and experience with ChatGPT. However, quick learners and early adopters within this group find new technologies easy to apply and become skilled at using them at a very fast pace [32, 33]. The results also indicate that university students tend to adopt AI-based tools in their learning process [3, 17]. It is vital to understand what the current state across university locations is, as in the future, those who use AI tools will be more efficient than those who do not. To create equal opportunities in education, the use of AI must be included in the curricula of higher education institutions [24]. This can facilitate students and attract them for more experience, as well as revolutionising the conventional teaching environment, bringing enhanced effectiveness in student education [3]. It is recommended to explore students as a stakeholder group within academia with a distinction of academic background, discipline, age groups, and cultural contexts for future studies [9].

Generative Artificial Intelligence (GenAI), distinguished by its capacity to autonomously generate content resembling human creation, has profoundly transformed numerous sectors, notably education, media, healthcare, governance, and research. Although the transformative potential of generative AI, exemplified by OpenAI's ChatGPT and similar large language models (LLMs), is indisputable, their swift implementation elicits substantial ethical considerations that necessitate thorough examination and regulation [7].

A principal ethical concern is the potential dissemination of biases and misinformation. Generative AI systems rely extensively on vast datasets, frequently embedding intrinsic societal biases, which can perpetuate discriminatory practices, misrepresentations, or deliberate dissemination of misinformation through authentic-looking synthetic media [30]. For instance, deepfakes—highly realistic, AI-generated synthetic media—constitute a significant threat to truth and democratic principles, necessitating the development of ethical frameworks centered on transparency and accountability [4].

The incorporation of generative artificial intelligence into higher education further exemplifies the ethical complexities involved. While it augments learning through personalized content and enhances efficiency [25], enduring concerns pertain to academic integrity, the over-dependence on AI-generated materials, and the potential degradation of critical thinking abilities. The peril of plagiarism and the covert use of AI in student evaluations pose challenges to conventional educational methodologies, underscoring the necessity for explicit institutional policies and effective plagiarism detection mechanisms [23, 38].

The domain of data privacy and protection represents an essential ethical facet. The management of sensitive information by generative AI tools presents privacy challenges necessitating rigorous adherence to data protection regulations, thereby underscoring the importance of human oversight in AI implementations [38]. Moreover, the algorithmic biases inherent in AI systems have the potential to intensify societal disparities, rendering the ethical deployment of generative AI imperative for ensuring equity and inclusivity [4].

Within the domain of academic research, generative AI provides substantial support to scholars, yet simultaneously introduces ethical challenges related to authorship, accuracy, and cognitive considerations. The manifestation of "hallucinations" in AI-generated content—information that appears semantically credible yet is factually incorrect—threatens the integrity of scholarly research. Addressing these concerns requires the advancement of AI literacy among researchers to ensure a comprehensive evaluation of AI-generated outputs before their inclusion in academic endeavors [12].

To effectively address these ethical challenges, it is imperative to engage in multidisciplinary collaboration among policymakers, educators, technologists, and researchers to formulate comprehensive guidelines and ethical frameworks. Suggested methodologies encompass the incorporation of AI ethics into education curricula, the integration of human oversight within AI processes, and the promotion of AI literacy to improve responsible AI application across various sectors [30].

In conclusion, the ethical considerations related to generative AI necessitate the formulation of an overarching ethical framework that emphasizes transparency, accountability, human rights, and fairness. This approach is essential for harmonizing AI advancements with societal values, safeguarding democratic institutions, and ensuring a sustainable and ethically robust incorporation of AI into daily life.

3 Research Methodology

3.1 Research Materials

This research was carried out in three selected universities: the University of Economics in Katowice (UE Kat), the University of Warsaw (UW), and the Jagiellonian University in Krakow (JU). Data were collected using a mixed method of PAPI (Paper and Pencil Interviewing) and CAPI (Computer Assisted Personal Interview). The surveys were distributed in a partially targeted manner in the academic environment to randomly selected groups of students. After being completed, they were transferred to an Excel spreadsheet, where preliminary data processing was performed: statistical indicators, distance measures, and tests were calculated. To assess the diversity of data, basic statistics were calculated.

3.2 Research Procedure

The collected data were made comparable by calculating the shares of respondents from each center relative to the total number, and their variance and standard deviation of data were examined.

- Variance, i.e. a measure of the variability of the results obtained. It informs about the variation in the results during the observation. We examine whether the results are more concentrated around the mean, i.e. whether there are small differences between the mean and individual results, or whether the results are widely dispersed, i.e. there is a large difference between individual results and the average.

$$s^2 = \frac{\sum_{i=1}^{n}(x_i - \tilde{x})^2}{n-1} \tag{1}$$

where: s^2 – variance, x˜ – mean, xi next observation in the sample, n – size of the set

Standard deviation is a statistical measure of the dispersion of a data set relative to its mean and is calculated as the square root of the variance.

$$s = \sqrt[2]{\frac{\sum_{i=1}^{n}(x_i - \tilde{x})^2}{n-1}} \tag{2}$$

where: s - standard deviation, x˜ - mean, x_i another observation in the sample, n - size of the set.

The identified research gap concerned the differences between views on the essence and possibilities of using artificial intelligence in the operating conditions of a university.

To determine the number of differences between the three regional centres, the following distance measures were calculated:

– Urban distance (Manhattan, City block) - i.e. the sum of absolute differences between the same attributes in the next sample. The impact of single large differences is attenuated, so they are more difficult to detect in the population. It is calculated according to the formula:

$$Cd = \sum_{i=1}^{n} |x_i - y_i| \tag{3}$$

where: Cd – urban distance, x_i, y_i – compared attributes,

– Euclidean distance - the square root of the sum of attribute differences squared, geometric distance in multidimensional space, The Euclidean distance formula is calculated as follows.

$$Ce = \sqrt[2]{\sum_{i=1}^{n} (x_i - y_i)^2} \tag{4}$$

where: Ce – Euclidean distance, x_i, y_i – compared attributes

Additionally, a research hypothesis was put forward.

3.3 Research Methods

Therefore, the research hypothesis was formulated as H0 about the existence of discrepancies between the opinions of the respondents in three academic centres on the assessment of the place and role of artificial intelligence in university life, against the hypothesis H1 that such differences do not exist, with an assumed probability of 0.05.

To prove the truth of this hypothesis, the α significance level was calculated for the right-sided F-Snedecor distribution. The F-Snedecor test was used here to compare the degree of significance of two data series and compare it with the p-value determined based on the test values (i.e. compare the test results of the F-calculated test value with the tabulated F_{kr} value). If $p \leq \alpha$, we reject the hypothesis, and if $p \geq \alpha$, we accept the hypothesis.

A study on the potential use of ChatGPT was conducted at three selected universities: the University of Warsaw (UW), the Jagiellonian University (JU), and the University of Economics in Katowice (UE Kat). A total of 616 people took part in the survey conducted at the turn of 2023 and 2024: 237 questionnaires were obtained from the University of Warsaw, 220 from the Jagiellonian University, and 103 questionnaires from the University of Economics in Katowice. A mixed method of PAPI (Paper and Pencil Interviewing) and CAPI (Computer Assisted Personal Interview) was used, which allowed for fast, reliable, safe, and high response. Data were collected purposefully and randomly in the academic environment of specific universities, where lecture or tutorial groups were randomly selected in which students completed surveys. After completing them traditionally or electronically, they were initially merged into an Excel spreadsheet and made comparable. Then, basic statistical indicators (means, variances, standard deviations) and urban and Euclidean distance measures and test values were calculated.

4 Research Analysis

4.1 Research Sample Analysis

A total of 616 people participated in the survey at the turn of 2023 and 2024 (63% of women, more than 36% of men, and almost 1% of people who declared a different gender - the mentioned 1% is not included in further analyses). In Warsaw, questionnaires were collected among Polish (237) and English-speaking students. Considering the 616 responses, 103 questionnaires were from UE Kat, 220 from JU, and 293 from UW.

Due to the nature of the research conducted, the vast majority (92%) were people aged 18–25 (5% aged 26–35 and less than 3% in other age groups). The surveys were completed by 3% more women than men aged 18–25 and more than 2% of women aged 26–35. Therefore, despite the high variance values (at a level of 14%) and the standard deviation (on average 37%) in both gender groups, caused by the predominance in one age group, both the urban distance (7%) and the Euclidean distance (0.18%) were small, and the right F-Snedecor test did not exceed the tabulated critical value. At the UW, 57% of women and 43% of men completed the survey, for the UE Kat there were almost 50% women and just over 50% men, at the JU - 75% women and 25% men. The largest difference in terms of urban distance, equal to 52%, occurred between JU and UE Kat.

Most of the respondents (31% on average) come from cities with more than 500,000 inhabitants. And the least (5% on average) from cities with 10–30 thousand inhabitants. The greatest differences (13% urban distance, 1.7% Euclidean distance) between women and men occur in the category of cities with more than 500,000 inhabitants (men - 41%; women – 28%). Therefore, we can find a relatively high value of variance and standard deviation in both groups. At UW and JU, most of the respondents came from cities with more than 500,000 inhabitants; at UE Kat most came from cities with a population of 100,000 to 500,000 inhabitants. The biggest difference between academic centres was between JU and UE Kat and was up to 85% in terms of urban distance (Euclidean distance 0.11).

Due to the level of studies at which the students were surveyed, most of them had secondary education (66%) or a bachelor's degree (24%). More than 9% of the respondents have higher education. In the research sample analysed, there were 13% more men with secondary education than women and 9% more women with bachelor's or engineering education. The majority (58%) of the people who participated in the survey declared themselves as students, 41% as working students, and the survey was also completed by 3 employees. Among students, there were 12% more people with the student status and a similar percentage (13%) among female students - in the working student category. At the UW and UE Kat most people have secondary education, and those at the Jagiellonian University have bachelor's/engineering degrees. The biggest difference is between UE Kat and JU: 65% urban distance (0.10 - Euclidean distance). At the UW and the UE Kat there are more non-working students, while at the Jagiellonian University, 71% of respondents are working students.

This structure is probably the reason why 57% of all respondents believe their financial situation is good and 26% that it is very good. Only 18% declare a lower standard of living. Men assess their financial situation as worse. The urban situation of men compared to women is 3% lower for the financial status of 'very good', 3% lower for the financial status of 'good', and almost 7% higher for the financial status of 'sufficient'.

4.2 Results Analysis

The survey consisted of 18 theoretical questions and six demographic questions. These were divided into two groups. The first was about theoretical problems, and the second was about practical ones that were related to the possibility of using ChatGPT in university practice. In this article, we will focus on the theoretical problems section and their significance in Higher Education.

The first group of questions - theoretical - concerns the initial assessment of the level of students' knowledge about artificial intelligence and chatbots, with particular emphasis on ChatGPT. It contained 10 questions with 42 criteria to choose from Likert scale 1–5 [20].

It starts with a general question about the knowledge of the artificial intelligence concept before starting the overall research. On average, 98% of the respondents responded positively, obtaining the same result regardless of gender or region. Most, over 99.5% of positive responses, were sent from the UW, and the least (97%) from the JU. Therefore, the largest absolute (urban) distance occurred between the UW and the JU and reached nearly 5%. Due to the respondents' focus on one answer, there were large differences in the calculated variance (30%) and standard deviation (50%). However, slight differences appeared between the results obtained in individual academic centres (the calculated right-sided F-Snedecor (FS) test was much lower than the tabulated FS test.

The next question was aimed at understanding the respondent's definition of artificial intelligence (AI), that is, what they consider AI. Four different definitions are provided, obtained from the most common ones on the Internet. Respondents rated which of these definitions corresponded to their ideas and to what extent. Those were obtained by imitating the behaviour of students - if they don't know something, they enter an unfamiliar term into the search engine and assume one of the first solutions displayed in the response to be true. On average, the distribution of the answers obtained was similar (22–29%), but the most recognition (29%) was gained by the first definition, which states that it is… *The ability of an IT system to correctly interpret data from external sources, learn based on them and use this knowledge to perform specific tasks and achieve goals through flexible adaptation...* [16]. It should be noted that artificial intelligence was assigned to an IT system, showing a comprehensive approach, not only to computers or the software itself, as defined in other definitions. Both women and men did not differ in their views here. No significant differences were observed in the regional system. This time, the distribution of opinions between the definitions of AI was relatively even, hence the small values of variance (0.08–012%) and standard deviation in each case. Small differences in variance in individual locations also result in low calculated values of the FS test. Second (25%) was the definition of AI as a… *Mathematical and logical model of analysed problems and their implementation in the form of information systems intended to implement specific functions commonly considered to be components of intelligence...* [6]. However, in this approach, the focus is on the mathematical and logical models that make up the basis of the software. The largest differences between universities from different cities occurred in the case of academic centres in Katowice and Warsaw (urban distance 7%), and the smallest between universities in Warsaw and Krakow (over 3 times less).

A similar situation occurred in the first case when asked about chatbots. On average, 96% of the respondents admitted that they had heard of this concept before taking the survey (96% of women and 97% of men). The diversity indicators between the academic centres behaved similarly to the first case. The respondents' concentration on one answer out of three resulted in high values of variance (30%) and standard deviation, including low indicators of urban distance (the highest of 3%) and Euclidean distance.

In the next question, respondents were asked to identify the basic source of information about chatbots. The most common (63% of the responses) was undoubtedly the Internet. The second place (21%) were the friends of the respondents, which shows that the discussions are taking place there, and the third place (8%) was the university where they study. The greatest difference in the views of women and men occurred concerning the Internet as a source of information about chatbots (men's opinions predominated by 9%). Friends were indicated as a source by 5% more women than men. Since the remaining categories were at a similar level, the variance and standard deviation were not high, but the values of urban distance - 19% and Euclidean distance - 0.72% were higher than in the previous cases. In all locations, the option 'I learned from the Internet' was in the first position. Most of the respondents who believed so, 77% came from Katowice, and the least, 53% from Kraków. Due to the variation in responses at each location, the variance is on the order of 8% and the standard deviation is approximately 29%. The largest variation expressed in urban distance (45%) occurred between universities in Katowice and Kraków. The most similar answers occurred in the UW and UE Kat. The calculated values of the FS test did not exceed the tabulated values.

The logical continuation of these theoretical considerations was the question of knowledge and awareness of the definition of the chatbot phenomenon [22]. Respondents were given the choice to choose five definitions of chatbots (downloaded from the Internet). In this case, individual responses to chatbot definitions were more or less evenly distributed (between 19 and 20%). Similarly, small differences occurred between the opinions of women and men. The regional differences also appeared to be small, the urban distance did not exceed 4%. However, there was a large variation in the variances of the individual categories in the academic centres. This meant that the differences between the Warsaw and Katowice data turned out to be statistically significant (FScalculated = 7.8061 > FStabled = 6.3882). The differences between Warsaw and Kraków universities were similarly significant (FScalculated = 5.3751 > FStabled = 5.0503).

In this situation, there were no statistically significant differences between the views of both genders. Among the definitions presented, the most popular was the one on the Oracle website [27–29], which states that it is… *A data processing model* based *on the use of services, a computer programme that simulates and processes human conversation (written or spoken), enabling people to interact with digital devices, as if they were communicating with a real person...* It is interesting that the remaining definitions do not differ much from the one chosen by the majority of respondents and are the least accurate.

Since the vast majority of people had heard of it before participating in the research and were able to correctly define the chatbot category, they can also theoretically determine what their most important advantages are. Detailed results are presented in Fig. 1. Of the ten positive features mentioned, the most appreciated were: increasing the speed of response to customer notifications (16–20%), reducing the average time of handling notifications (13–14%), and increasing the number of received and processed notifications (11–14%). The least importance is given to the number of reports submitted to the consultant (2–5%) and the improvement of internal and external communication (3–4%). The first place among the advantages was determined by the attitude of the men, who had 4.31%, and the result of the UW - 19.56%. Since there was no difference between the responses of women/men and the locations of individual universities, the urban and Euclidean distances did not exceed a significant value in any of the cases analysed. In turn, low and similar variance values did not result in higher FS test values than the tabulated values.

A similar formal procedure was also used, as in the case of the previous two definitions, in the third case of the currently most popular chatbot, ChatGPT. First, respondents were asked if they had ever heard of such a concept. And, as before, 95% responded positively to these questions. This resulted in a high internal diversity of individual options, both in general, by gender and location. The urban distance is the largest between the answers from the UE Kat and the JU (7.9%), and slightly smaller between the answers from the UE Kat and the UW - 6.5%. The value of the FS test is high but shows similar levels of variance in all cases analysed.

When it comes to the place where respondents heard/saw information about ChatGPT for the first time, online media was again in the first place (on average 61%; women - 55%, men - 65%; UW - 70% and JU - 46%). In terms of gender, this category also has the greatest diversity, reaching 10%, with a structural predominance of men's opinions. A slightly lower preponderance of women's opinions (8%) occurs in the second most popular category, citing friends as the source of knowledge about ChatGPT (23% on average). In the third place (9%), universities were the primary place to acquire knowledge about ChatGPT. Due to the small values for the workplace, the media such as TV, radio, the press, and other sources, the variance (6%) and the standard deviation (24%) are relatively high. The highest level of regional differentiation between universities expressed by urban distance occurs between the UE Kat and the JU at 48% (Euclidean distance 3.25%).

To determine what respondents believed to be ChatGPT, five definitions were presented to choose from and compare, retrieved after entering the phrase "ChatGPT" on the Internet. The distribution of responses for these five cases was even by gender and region and amounted to 16%-24%, respectively. The highest share of responses was obtained by the definition [37], which defines ChatGPT as a special... *Chatbot used to generate responses to data entered by the user. This model was developed based on large data sets so that it can conduct conversation and engage in a variety of topics, from general conversations to specific areas of knowledge...* The worst result was the definition that ChatGPT is... *a virtual person who can provide a sensible answer to very diverse questions and solve simple tasks...*[18, 19]. The differences between the statements of women and men were again small in this case, so the urban distance and the Euclidean distance were also low (below 3% and below 0.05%, respectively). The variation in urban distance between individual academic centres did not exceed 4.5%, and the largest difference occurs between the UE Kat and the JU - 4.25% (Euclidean distance 0.03%). Due to the even share of responses, variances, and standard deviations were low in each of the analyses.

In response to the question about potential fields of application, the largest number of people assigned ChatGPT to help them in tasks related to answering questions (on average 23%, the same percentage of women and men) and summarising the collected data and drawing conclusions (on average 20%, women 20%, men 21%). The urban distance did not exceed 10% (the Euclidean distance was also low and had a value of – 0.14%).

Slightly larger differences occurred in the regional system. The first two places were dominated by the same answers as in the gender distribution, with the highest share in the JU in the first case (24%) and the lowest in the second (19%). In both sections, the fewest people (2%) mentioned generating an X tweet as a potential application field. The largest differences in terms of urban distance again appeared between the UE Kat and the JU - 16% (Euclidean distance 0.002). The calculated values of the FS test are not higher than those tabulated; therefore, the existing differences are not statistically significant in the cases analysed.

In summary, there was a similarity of views on the concept of AI, chatbots, and the ChatGPT chatbot by gender and region. Common definitions of these concepts have indeed been used, but among Internet users, these common formulations, most frequently found on the Internet, are the most popular. The material presented also shows that people frequently browse the Internet in search of information about new technologies and pass it on to friends.

There was also agreement in the approach to the theoretical advantages of using chatbots, both to the gender divisions of the survey participants and between individual academic centres.

However, statistically significant differences occurred in defining the concept of chatbots. Due to the large differences in the variance of individual categories in academic centres, the differences between the data from UW and UE Kat and UW and UJ turned out to be statistically significant (the calculated value of the FS test was greater than that tabulated for a specific number of degrees of freedom, with a probability of 0.05).

5 Conclusion

The presented research results lead to the following conclusions in terms of semantic awareness and theoretical knowledge of issues related to AI and its applications.

The study of awareness of the existence and functioning of AI was carried out at three levels of categories related to this concept:

(1) The most general, i.e. AI systems,
(2) As part of AI systems, tools supporting thematic interaction with the user, i.e. chatbots,
(3) A special case of these tools, i.e. ChatGPT.

Due to the broad and widespread current discussion about the role and place of AI in social and economic life, almost all respondents had heard about these categories before participating in the study.

Because the Internet was the main source of information on AI, chatbots, and ChatGPT (60–70% of the respondents), the definitions of the related concept taken for further research were found among the most frequent ones on the Internet.

At the first level in the definitions of AI, it was assigned to the IT systems and the algorithms that operate in them.

In the second case, chatbot definition, attention was paid to communication software that allows interaction in a way similar to the way of contact with a human,

Thirdly, the majority were willing to define ChatGPT as a chatbot that generates answers to questions asked on a given topic, based on large databases.

The differences in all these cases, both regional and by gender, were minimal except for statistically significant differences in the definition of chatbots.

Among the advantages of chatbots from a theoretical point of view, operational advantages were highlighted: increasing the speed and volume of service and reducing the time of service execution.

Among the advantages of ChatGPT is assistance with tasks related to answering questions, summarising the collected data, and drawing conclusions.

Respondents, regardless of gender and location, have a high and very similar awareness of the existence, importance, and theoretical advantages and conditions of the use of AI systems and their specific group: chatbots, and in particular ChatGPT. They obtain their knowledge on this subject mainly from the Internet and from friends, and hence they are satisfied with common, although superficial, definitions existing on the Internet, and if they have many definitions of this phenomenon to choose from, they choose those whose characteristics are similar or complementary.

The results allowed us to positively answer the first research question on the existence and knowledge of the existence and use of AI among students. However, there was no confirmation of the significance of the differences in students' opinions about AI, either by gender or by location. In the latter case, only 10% of the responses obtained met the significance conditions determined by the Fisher-Snedecor test.

Generally speaking, the students demonstrate a rational approach to the phenomenon of AI, which has reemerged with the spread of ChatGPT. In the 1990s, similar software was called intelligent intermediaries (bots) [13, 39], then the phenomenon of AI began to be attributed, on the one hand, to Business Intelligence systems (especially considering the capabilities of Business Analytics), and then to industrial automation and robotization systems. Therefore, ChatGPT fits into the development trends of IT systems, and as another "fashion" attracting attention at the moment, it will probably quickly turn into interest in the next generation of systems that respond to an increasingly greater extent like a human. However, even if it is a temporary phenomenon that supports (like other systems) various areas of professional and social life, as shown by the above research, it is at least accepted and used most reasonably despite the observed errors in the algorithm of its operation. According to the students, openness to new technologies is the basic condition for using artificial intelligence in practice. And not - as some people thought, immediately - but over at least ten years. It seems that respect for this period results from both an accurate assessment of the economic situation of the organisation and the awareness that new technologies in the field of artificial intelligence may appear quickly that are much superior to today's solutions.

The study focuses only on one of the users that are located at three selected universities in Poland, i.e. students. The group of research and administrative workers was not included in this study. Undoubtedly, it would be useful to extend these analyses to foreign universities and include more stakeholders to generalise the conclusions.

Our findings indicate possible further directions of research development toward the inclusion of all university employees and foreign universities, increasing the number of analysed scientists. We also believe that this research can constitute the basis for the construction and verification of a model of universities' adaptation to modern information technologies and their acceptance. This is a promising research direction, but it would require reformatting the survey questionnaire and adapting it to the requirements of the technology acceptance models, i.e. TAM, and acceptance theories with the unified theory of acceptance and use of technology (UTAUT) or D&M models.

References

1. Adiguzel, T., Kaya, M.H., Cansu, F.K.: Revolutionizing Education with AI: Exploring the Transformative Potential of ChatGPT. Contemp. Educ. Technol. **15**(3), 1–13 (2023). https://doi.org/10.30935/cedtech/13152
2. Ahadi, N., Zanjanab, A.G., Sorooshian, S., Monametsi, G., Virutamasen, P., Wongpreedee, K., Deebhijarn, S., Taghipour, A.: The outlook of ChatGPT, an AI-based tool adoption in academia: applications, challenges, and opportunities. In: TENCON 2023 - 2023 IEEE Region 10 Conference (TENCON), pp. 15–20 (2023). https://doi.org/10.1109/TENCON58879.2023.10322527
3. Algerafi, M.A.M., Zhou, Y., Alfadda, H., Wijaya, T.T.: Understanding the factors influencing higher education students' intention to adopt artificial intelligence-based robots. IEEE Access **11**, 99752–99764 (2023). https://doi.org/10.1109/ACCESS.2023.3314499

4. Al-kfairy, M., Mustafa, D., Kshetri, N., Insiew, M., Alfandi, O.: Ethical challenges and solutions of generative AI: an interdisciplinary perspective. Informatics **11**(3), 58 (2024)
5. Amani, S., White, L., Balart, T., Arora, L., Shryock, K.J., Brumbelow, K., Watson, K.L.: Generative AI Perceptions: A Survey to Measure the Perceptions of Faculty, Staff, and Students on Generative AI Tools in Academia (2023). http://arxiv.org/abs/2304.14415
6. Angulska, W.: Sztuczna inteligencja (2020). https://prezi.com/p/lxgafi9blcoj/sztuczna-inteligencja/. Accessed 2 Jan 2024
7. Bautista, Y.J.P., Theran, C., Aló, R.: Ethical Considerations of Generative AI: A Survey Exploring the Role of Decision Makers in the Loop. AAAI Spring Symposium Series (2024)
8. Bearman, M., Ryan, J., Ajjawi, R.: Discourses of artificial intelligence in higher education: a critical literature review. High. Educ. **86**(2), 369–385 (2023). https://doi.org/10.1007/s10734-022-00937-2
9. Chan, C., Yuk, K., Hu, W.: Students' Voices on Generative AI: Perceptions, Benefits, and Challenges in Higher Education (2023). http://arxiv.org/abs/2305.00290
10. Coderslab: ChatGPT- co to jest i jak z tego korzystać? (n.d.). https://coderslab.pl/pl/blog/chatgpt-co-to-jest-i-jak-z-tego-korzystac. dostęp 20 Nov 2023
11. Europarl: Sztuczna inteligencja: co to jest i jakie ma zastosowania? (2020). https://www.europarl.europa.eu/news/pl/headlines/society/20200827STO85804/sztuczna-inteligencja-co-to-jest-i-jakie-ma-zastosowania. Accessed 2 Jan 2024
12. Farangi, M.R., Nejadghanbar, H., Hu, G.: Use of generative AI in research: ethical considerations and emotional experiences. Ethics & Behavior (2024). Advance online publication
13. Gilbert, D.: Intelligent agents: the right information at the right time. IBM Intellligent Agent White Paper, IBM Corporation, Research Triangle Park (1997). http://www.networking.ibm.com/iag/iaghome.html
14. Hasanein, A.M., Sobaih, A.E.E.: Drivers and consequences of ChatGPT use in higher education: key stakeholder perspectives. Eur. J. Invest. Health Psychol. Educ. (EJIHPE) **13**(11), 2599–2614 (2023). https://doi.org/10.3390/ejihpe13110181
15. Hosseini, M., et al.: An exploratory survey about using ChatGPT in education, healthcare, and research. PLoS ONE **18**(10), 1–14 (2023). https://doi.org/10.1371/journal.pone.0292216
16. Kostrzewa, A.: Platforma inteligencji wizualnej – nowe narzędzie dla marketingu (2021). https://mitsmr.pl/b/platforma-inteligencji-wizualnej---nowe-narzedzie-dla-marketingu/PKWEbuCwu. Accessed 2 Jan 2024
17. Kordowski, M.: Algorytm NLP (2022). https://marcinkordowski.com/algorytm-nlp/. Accessed 20 Nov 2023
18. Kruczkowska, E.: Chat GPT – przełom czy odgrzewane kotlety? (2023). https://pfr.pl/blog/chat-gpt-przelom-czy-odgrzewane-kotlety.html. Accessed 1 Jan 2024
19. Kruczkowska, E.: Co to jest Chat GPT, jak i do czego go używać? (2023). https://startup.pfr.pl/pl/aktualnosci/co-jest-chat-gpt-jak-i-do-czego-go-uzywac/. Accessed 20 Nov 2023
20. Likert, R.: A technique for the measurement of attitudes. Archiv. Psychol. **140**, 55 (1932)
21. Livberber, T., Ayvaz, S.: The Impact of artificial intelligence in academia: views of Turkish academics on ChatGPT. Heliyon **9**(9), e19688 (2023). https://doi.org/10.1016/j.heliyon.2023.e19688
22. Maryl-Wójcik, M., Puczko, M.: Czy chatboty zastąpią człowieka (n.d.). https://www.kopernik.org.pl/sites/default/files/2022-04/Scenariusze_KEI_czy_czatboty_zastapia_czlowieka_20220413.pdf
23. Michel-Villarreal, R., Vilalta-Perdomo, E., Salinas-Navarro, D.E., Thierry-Aguilera, R., Gerardou, F.S.: Challenges and opportunities of generative AI for higher education as explained by ChatGPT. Educ. Sci. **13**(9), 856 (2023). https://doi.org/10.3390/educsci13090856

24. Neumann, M., Rauschenberger, M., Schön, E.-M.: "We Need To Talk About ChatGPT": The Future of AI and Higher Education. In: 2023 IEEE/ACM 5th International Workshop on Software Engineering Education for the Next Generation (SEENG), pp. 29–32. IEEE, Melbourne, Australia (2023). https://doi.org/10.1109/SEENG59157.2023.00010
25. Nguyen, K.V.: The use of generative AI tools in higher education: ethical and pedagogical principles. J. Acad. Ethics (2025). Advance online publication
26. Opteam: Chatboty - co to jest i jak są wykorzystywane? (2023). https://opteam.pl/firma/baza-wiedzy/artykuly-eksperckie/chatboty---co-to-jest-i-jak-sa-wykorzystywane-30.html. Accessed 2 Jan 2024
27. Oracle: Czym jest sztuczna inteligencja? Dowiedz się więcej o sztucznej inteligencji (n.d.a). https://www.oracle.com/pl/artificial-intelligence/what-is-ai/. Accessed 2 Jan 2024
28. Oracle: Jaka jest różnica między asystentem cyfrowym a chatbotem? (n.d. b). https://www.oracle.com/pl/chatbots/what-is-a-digital-assistant/difference/. Accessed 2 Jan 2024
29. Oracle Developer: Building Digital Assistants and Chatbots (n.d.). https://developer.oracle.com/technologies/chatbots.html. Accessed 2 Jan 2024
30. Paul, R.K., Sarkar, B.: Generative AI and ethical considerations for trustworthy AI implementation. Int. J. Artif. Intell. Mach. Learn. **2**(1), 95–102 (2023)
31. Perera, P., Lankathilake, M.: AI in Higher Education: A Literature Review of ChatGPT and Guidelines for Responsible Implementation **VII**, 307–314 (2023). https://doi.org/10.47772/IJRISS.2023.7623
32. Strzelecki, A.: To use or not to use ChatGPT in higher education? A study of students' acceptance and use of technology. Interact. Learn. Environ. (2023a). https://doi.org/10.1080/10494820.2023.2209881
33. Strzelecki, A.: Students' acceptance of ChatGPT in higher education: an extended unified theory of acceptance and use of technology. Innov. High. Educ. (2023b). https://doi.org/10.1007/s10755-023-09686-1
34. Sullivan, M., Kelly, A., Mclaughlan, P.: ChatGPT in higher education: considerations for academic integrity and student learning. J. Appl. Learn. Teach. **6**(1), 17 (2023). https://doi.org/10.37074/jalt.2023.6.1.17
35. Tominc, P., Rožman, M.: Artificial intelligence and business studies: study cycle differences regarding the perceptions of the key future competences. Educ. Sci. **13**(6), 580 (2023). https://doi.org/10.3390/educsci13060580
36. Udvaros, J., Forman, N.: Using artificial intelligence in education 4.0: POUŽITIE UMELEJ INTELIGENCIE VO VZDELÁVANÍ 4.0. J. Technol. Inf. Educ. **15**(1), 1–19 (2023). https://doi.org/10.5507/jtie.2023.004
37. Vincent, J.: AI-generated answers temporarily banned on coding Q&A site Stack Overflow (2022). https://www.theverge.com/2022/12/5/23493932/chatgpt-ai-generated-answers-temporarily-banned-stack-overflow-llms-dangers. Accessed 31 Dec 2022
38. Williams, R.T.: The ethical implications of using generative chatbots in higher education. Front. Educ. **8**, 1331607 (2024)
39. Wooldridge, M., Jennings, N.R.: Intelligent agents: theory and practice. Knowl. Eng. Rev. **10**(2), 115–152 (1995)

AI for Smart Cities, Public Sector, and Government Innovation

Acquiring Smart People for Smart City 3.0 Using Rough Set Theory

Karol Kuczera[1(✉)] and Damian Dziembek[2]

[1] University of Szczecin, Cukrowa Str. 8, 71-004 Szczecin, Poland
karol.kuczera@usz.edu.pl
[2] Czestochowa University of Technology, Armii Krajowej Str. 19 B, 42-200 Czestochowa, Poland
damian.dziembek@pcz.pl

Abstract. In the paper an approach has been proposed that can support city management in line with Smart City 3.0 in the context of attracting and retaining human resources capable of identifying and undertaking ICT initiatives. Smart City 3.0, unlike previous versions of this concept, put a lot of emphasis on human and social capital. Here, the creative involvement of residents is to dominate. These are to be smart civic cities, open to the active participation of residents and encouraging such attitudes. The administration should create conditions and opportunities to harness people's potential. The technology is rather a tool. And people need power to act. This strategic, long-term vision focuses on attracting and retaining human resources, critical in building the Smart City concept. In this context, the question is asked: How to acquire Smart People to build Smart Cities?

An approach has been proposed that allows to discover knowledge about what combination of people's characteristics and the image of the city makes people want to link their careers with the city or want to leave it. This knowledge may be important for decision-makers and indicate what image the city should build to attract the desired people. The research focused on well-educated young people who were making decisions about their career path after graduation. The flexibility and applicability of the proposed approach allows it to be used also for other target groups. The survey was conducted among management students from the University of Szczecin and Częstochowa University of Technology. A total of 286 students participated in the study.

The discovery of knowledge hidden in data was carried out in accordance with the Rough Set Theory. The Rough Set Theory enables the induction of decision rules that take a form that is easy to understand and interpret: if premise, then conclusion. Additionally, the rules have qualitative characteristics. Example rules and their interpretations are presented in the article. In this part of the research arguments and evidence have been provided for the usefulness of the rough set theory in this field.

Keywords: Smart City 3.0 · Rough Set Theory · (Smart) Human resources

Published by Springer Nature Switzerland AG 2026
M. Pondel et al. (Eds.): AI4KAM 2025, IFIP AICT 776, pp. 87–98, 2026.
https://doi.org/10.1007/978-3-032-18920-2_9

1 Introduction

The Smart City phenomenon has been developing since the turn of the 20th and 21st centuries, as a continuation of movements known as Smart Growth since the 1990s. Recently, Smart City has found an ally in Artificial Intelligence. This even resulted in the appearance of the concept of an artificially intelligent city in the literature as the potential successor of the widespread Smart City phenomenon. This is intended to indicate a strong connection between the smartness of the city and the use of viable technological solutions, including artificial intelligence [1]. AI stands alongside technologies like the Internet of Things and Big Data. Over the years, it has covered various domains of life to address complex challenges and improve the quality of life for individuals and communities. Healthcare, governance, environment, transportation, energy, safety and security, infrastructure, or education can be pointed out [2].

However, as the concept has matured, there has been a clear shift in emphasis. Smart City 1.0, the earliest phase, was inspired by IT companies seeking to generate demand for their technologies. There was not always much consideration given to whether cities actually needed these technologies or whether they effectively solved important problems. Smart City 2.0 reflects the key role of public administration, which identifies urban problems and coordinates actions to solve them using available technologies. Greater awareness and purposefulness are highlighted here. The IoT is glorified, including sensors, indicators, controllers and the accompanying data flows and optimisation algorithms. As a result, people are pushed into the background and lose out to technocracy. Criticism of previous, technologically dominated versions led around 2015 to Smart City 3.0, also known as Human Smart Cities [3]. Here, the creative involvement of residents is to dominate. These are to be smart civic cities, open to the active participation of residents and encouraging such attitudes. The administration's role is evolving from independently seeking solutions to creating conditions and opportunities to harness people's potential. This can be understood as involving residents in using technology, including education in this area, and encouraging and creating conditions for them to develop their own solutions, for example, through open data. The idea here is that people themselves know best what problems they face and often know how to solve them. They need to be given the power to act. The following characteristics are emphasised: Human-centricity [4, 5] and Co-creation [6]. This means that activities focus on the needs of residents, businesses, employees, students and non-profit organisations. Solutions are developed in a participatory manner, with the active involvement of citizens in identifying problems and implementing solutions. Technology is intended to involve people in urban affairs, activate them in co-governance, and strengthen social capital and bonds between residents. Social actors are partners in implementing modern smart city solutions [7].

There are many definitions of Smart City [8, 9, 10, 11]. An interesting analysis of the concept and its components can be found in [2]. These authors point out that the absence of a universally accepted definition of smart cities poses challenges for stakeholders seeking to develop and implement smart city initiatives. They also show that the words most frequently appearing in the definitions are: quality, technology, information and life.

An example definition could be: "A smart city is a city engaged in multiple initiatives to create better environmental, social and economic conditions and increase its attractiveness and competitiveness. It is an integrated system in which human and social capital interact, using technology to effectively achieve sustainable development and a high quality of life." [12].

According to The Brussels-Capital Region's definition of the smart city, "it is a city which uses smart solutions, based on data and certain technologies, which can lead to improved quality of life in a region." Quality of life is pointed out as a primary objective of Smart City, and data is the means to achieve it. "Technologies are not an end in themselves, but rather a tool in the design of the smart city of the Brussels-Capital Region." [13] Data can be considered as directly collected by sensors and sent to computing centres, but also as data needed to know what affects quality, what problems and needs people have, how they perceive the city and how it should be designed to improve quality of life. How quality is understood, what its components are and how the satisfaction of needs is perceived. Of course, data should be used to induce knowledge necessary for decision-makers to make proper decisions.

This strategic, long-term vision focuses on attracting and retaining human resources, critical in building the Smart City concept. Young, well-educated people open to modern technologies and involved in their community's life are particularly important here. People with these characteristics are likelier to "work well" with smart solutions, especially in the current version of Smart City 3.0. This view corresponds, among other things, to the concept of "The Smart City System", where Smart People come first [14]. Good decisions lead to a satisfying quality of life in the city. A good quality of life attracts and retains people who want to connect their future with the city. The good quality of a city's human capital influences good decisions regarding improving the quality of life and satisfaction with that quality. In this way, the cycle of interdependence closes. Hence, it is important to ask how to determine the directions of urban development while considering residents' quality of life.

Potential actions could be identifying needs and preferences and satisfaction with their fulfilment in a complex urban area, as well as linking perceptions of the city with the desire to stay or the intention to leave. Due to, among other things, the complexity and multifaceted nature of such a perspective about such a large entity as a city, this is not a trivial task.

2 Research Methodology

Concerning the issues signalled, the authors assumed two objectives for the study. The first is a proposed approach that can support city management in line with Smart City 3.0 by attracting and retaining the right human resources—qualified human resources who are open and ready to engage in co-creation and participation in finding problems and proposing solutions. It was assumed that no universal answers would be appropriate for all interested cities. Towns, cities and agglomerations operate in different conditions and face various challenges. Therefore, the proposed approach should be applicable in different places and situations. Accordingly, an attempt was made to identify the image of two medium-sized Polish cities, Szczecin and Częstochowa, in the eyes of students and the influence of the image on decisions concerning the place of pursuing a professional career after graduation. The research question was asked what image of the city might contribute to a desire to pursue a career in that city and what image induces a desire to leave. It should lead to discovering how to build a city's image, considering many different aspects and factors, to attract and retain qualified human resources ready to develop a Smart City. We also want to see if different cities should meet different needs to attract the desired human capital.

Naturally, the target group is management students, who can be seen as critical future decision-makers who will influence the shaping and development of the space in which they operate. The survey was conducted among students of all years of full-time undergraduate and postgraduate studies in management at the University of Szczecin and the Częstochowa University of Technology. The study used a survey questionnaire tool, the design of which was primarily inspired by a publication by M. Gębarowski and E. Glińska [15]. The technique used was CAWI, which PAPI replaced in individual cases (technical difficulties). In the case of Szczecin, 147 completed questionnaires were obtained out of 340 students, and in the case of Częstochowa, 139 out of 235.

The second aim of our study is to provide arguments and evidence for the usefulness of the rough set theory in this field. The theory of rough sets, published by Z. Pawlak in 1982 [16, 17], is a response to the problem of "imperfect" knowledge. From a logical point of view, rough set theory is a mathematical approach to fuzzy concepts. At the same time, from a practical perspective, it is a new method of data analysis that helps to search for relationships among data, reduce data redundancy, determine the weights of individual data components, or derive decision rules from data [18]. Simultaneously, Pawlak [19] pointed out that data analysis should be treated as a special instance of inductive inference, which, starting from certain partially known facts about reality and by way of generalisations, strives to create knowledge about a world broader than the one being the starting point for the considerations. Such efforts aim to draw probable (possible) conclusions, while the validation of hypotheses is performed through experiments. The set theory terminology is used to describe "rough operations on sets, as well as exact operations on rough sets. (…) The proposed method is based on an imperceptible relation that glues together all the objects that one could not distinguish by available means of observation (measurement) or expression (language)." [17].

The growing complexity of the world and the need to operate with rapidly growing sets of empirical data, which are by nature imprecise and incomplete, as well as the development of computational technologies, favour the continuous growth of the popularity of this theory and the tools in which it is implemented. Moreover, this development has resulted in it being widely considered to be one of the fastest-developing artificial intelligence methods, and R. Tadeusiewicz [20] calls it a Polish island in the artificial intelligence archipelago. We also refer readers to Wang et al. who mean that "the achievements of rough sets in application fields are low-hanging fruit in many domains (…) They show that the demands of practical use in many real-life fields are one of the driving forces promoting the development of rough set theory." [21] They give many examples. The rough set theory has such features as the ability to analyse imprecise, incomplete, or uncertain data. It enables data reduction and attribute selection, classification, and pattern discovery. It also does not require a specific data distribution. These features make it an essential part of AI development. For years there has been interest in its applications in artificial intelligence and cognitive sciences. The rough set theory is known for its usefulness in various machine learning techniques like clustering, feature selection and rule induction [22].

The proposed procedure is based on searching for knowledge in the data and finding its repeatability. Potential rules revealed are representations of knowledge about behaviour and ongoing phenomena. As a result of the rough set theory application, we obtain decision rules that fit well with the human mind (data fit versus mental fit [23]). The decision rules provide an explicit representation of knowledge, which adds to a better understanding and explanation of the decision problems under consideration.

The collected data formed a decision table containing 45 conditional variables (C1–C45) and a decision variable (D). The conditional variables defined the characteristics of the respondents and elements of the city's image, including the satisfaction of needs in various areas. Table 1 shows all considered conditional variables.

The decision variable (D) referred to the willingness to stay or leave the city after completing education.

The collected data operated on various scales. For example, variables C9–C21 were obtained from questions in the form of a semantic differential on a 7-point scale. Opposing statements were juxtaposed, allowing respondents to choose one and the strength of their identification with it. Questions regarding satisfaction with need fulfilment (C22 C33) and statements describing the city (C34–C45) used a 5-point scale. The primary decision table was discretized. Most variables took values in the range 1–3. This procedure, which reduces the level of detail in the source data, is expected to increase the likelihood of discovering repeatability in the data, i.e. the induction of decision rules.

The rough set theory makes it possible to identify data redundancy in information tables and eliminate it, i.e. create attribute reducts. Hence, the next step was an attempt to reduce the number of conditional attributes. Potential reducts still allow to identify combinations of condition attributes leading to the selected target (dropout, enrolled, graduate). Unfortunately, no reducts could be created, which means that all the assumed conditional attributes are relevant and none of them can be omitted. Further inference will be made using the full set of 45 conditional attributes.

The search for rules was based on the Explore algorithm, which allows the minimum requirements for rules to be determined. The adopted tuning parameters were: Maximum length = 3, Minimum relative strength = 20% and Minimum discrimination level = 80%.

The research was performed using ROSE2 [24, 25] software developed by the Laboratory of Intelligent Decision Support Systems of the Poznan University of Technology.

3 Results and Discussion

First, an attempt was made to induce rules concerning Szczecin. 36% of respondents indicated that after graduation, they plan to pursue their career in Szczecin, 34% plan to leave Szczecin, and 29% have not yet decided. Among the rules obtained, the vast majority concerned the decision-making class regarding staying in the city. Thus, many recurring combinations of perceptions of the city encourage people to stay. Single recurring

Table 1. Table captions should be placed above the tables.

Symbols for conditional variables	Names of conditional variables
	Characteristics of students
C1	Sex
C2	Bachelor's or Master's student
C3	Year of study in the current program
C4	Total number of years of study in Szczecin/Częstochowa
C5	Origin, Szczecin/Częstochowa, voivodship, different voivodship/country
C6	Size of the place of origin
C7	Total length of work (in years)
C8	Are you currently working?
	Image of the city:
C9	Friendly vs Unfriendly
C10	Romantic vs Mundane
C11	Safe vs Unsafe
C12	Clean vs Dirty
C13	Interesting vs Boring
C14	Vibrant vs Sleepy
C15	Pretty vs Ugly
C16	Sophisticated vs Kitschy
C17	Natural vs Artificial
C18	Open vs Closed
C19	Full of harmony vs Full of disorder
C20	Modern vs Backward
C21	Metropolitan vs Parochial
C22	Overall satisfaction with life in Szczecin/Częstochowa
	Satisfaction with fulfilling the needs by Szczecin/Częstochowa in the field of:
C23	Sport and recreation
C24	The quality of higher education
C25	Culture
C26	Entertainment
C27	Develop your passions
C28	Professional self-fulfilment

(*continued*)

Table 1. (*continued*)

Symbols for conditional variables	Names of conditional variables
C29	Meeting interesting people
C30	Safety
C31	Health care
C32	Public transport
C33	Housing conditions
	To what extent do you agree or disagree with the statement about Szczecin/Częstochowa:
C34	Main academic centre
C35	A city to which I will always gladly return in my thoughts
C36	A city attractive to students
C37	A city inhabited by open and generous people
C38	A well-managed city
C39	A European city
C40	A city with prospects for young people
C41	A city where you can find an attractive job
C42	A city typical of western/southern Poland
C43	A city where provincial people live
C44	A city where nothing interesting happens
C45	A city that everyone would like to escape from

combinations of city characteristics that encourage people to leave or remain undecided were discovered.

Choosing rules for interpretation is worth referring to the qualitative characteristics of the rules, which can be seen in the first square bracket. They denote respectively:

Cover – the number of examples from the table consistent with the rule premise, the number of examples covered by the rule.

Support – the number of examples consistent with the premise and conclusion of the rule, the number of positive examples covered.

Coverage factor – the ratio of the number of positive examples covered (support) to the number of examples consistent with the conclusion in the dataset.

Certainty factor – the ratio of the number of positive covered examples to the number of covered examples. It expresses the conditional probability that an object belongs to the decision class specified by the decision rule, given that it satisfies the rule's conditions. If a decision rule uniquely determines a decision in terms of conditions – the rule is certain. Otherwise, the decision rule is uncertain (possible).

It should be noted that the coverage factor is related to inverse decision rules (if conclusion, then premise), which can be used to give explanations (reasons) for the decision. To say to what extent a given decision results from the conditional part of the rule. It shows the conditional probability of reasons for a given decision [26].

Sometimes, the strength of the rule is also used to show how many examples from the entire dataset are described by the rule; in other words, how many objects support the rule.

Due to the high certainty factor, attention was drawn to four similar rules with three conditional variables, including two identical ones, which took the following form:

$$\text{if C4} = 3\ \&\ \text{C5} = 1\ \&\ \text{C15} = 1,\ \text{then D} = 1;\ [12,\ 11,\ 20.75\%,\ 91.67\%]\ [11,\ 1,\ 0] \tag{1}$$

$$\text{if C4} = 3\ \&\ \text{C5} = 1\ \&\ \text{C18} = 1,\ \text{then D} = 1;\ [13,\ 12,\ 22.64\%,\ 92.31\%]\ [12,\ 1,\ 0] \tag{2}$$

$$\text{if C4} = 3\ \&\ \text{C5} = 1\ \&\ \text{C21} = 1,\ \text{then D} = 1;\ [13,\ 12,\ 22.64\%,\ 92.31\%]\ [12,\ 0,\ 1] \tag{3}$$

$$\text{if C4} = 3\ \&\ \text{C5} = 1\ \&\ \text{C22} = 3,\ \text{then D} = 1;\ [12,\ 11,\ 20.75\%,\ 91.67\%]\ [11,\ 1,\ 0] \tag{4}$$

Considering the names of the encoded variables and the values they take, they can be decoded to the following form:

If the total number of years of study in Szczecin = 3 (C4 = 3) and I come from Szczecin (C5 = 1) and I perceive Szczecin as nice (C15 = 1), then I plan to pursue my career in Szczecin (D = 1).

If the total number of years of study in Szczecin = 3 (C4 = 3) and I come from Szczecin (C5 = 1) and I perceive Szczecin as open (C18 = 1), then I plan to pursue my career in Szczecin (D = 1).

If the total number of years of study in Szczecin = 3 (C4 = 3) and I come from Szczecin (C5 = 1) and I perceive Szczecin as metropolitan (C21 = 1), then I plan to pursue my career in Szczecin (D = 1).

If the total number of years of study in Szczecin = 3 (C4 = 3) and I come from Szczecin (C5 = 1) and I rate my overall satisfaction with life in Szczecin as high (C22 = 3), then I plan to pursue my career in Szczecin (D = 1).

The rules indicate that Szczecin residents who, in most cases, complete their first degree after three years of studying in Szczecin and positively evaluate the city in terms of the characteristics specified in the rules, plan to remain in Szczecin. These rules are possible with a certain factor above 90%. This means that over 90% of respondents described by the rule's premise agree with its conclusion. In case of (1) and (4), the conditional part of the rule describes the views of 12 people, 11 of whom agree with the rule's conclusion, intend to stay, one has not made a decision. No one plans to leave (the distribution of examples to decision classes is shown in the second square bracket). At the same time, the number of positive covered examples translates into a 20.75% coverage factor (11 out of 53 people planning to stay). So, the reversed rule describes the view of 20.75% of people who plan to stay.

One more rule with a certainty factor above 90% should be noted:

$$\text{if C5} = 1\ \&\ \text{C13} = 1\ \&\ \text{C40} = 1,\ \text{then D} = 1;\ [12,\ 11,\ 20.75\%,\ 91.67\%]\ [11,\ 1,\ 0] \tag{5}$$

It can be interpreted as:

If I come from Szczecin (C5 = 1) and perceive Szczecin as interesting (C13 = 1) and consider Szczecin a city with prospects (C40 = 1), then I plan to pursue my career in Szczecin (D = 1).

The rule has quality parameters similar to those of the previous ones.

The following rule is also noteworthy:

$$\text{if } C5 = 1 \ \& \ C41 = 1 \text{ then } D = 1; \ [19, \ 16, \ 30.19\%, \ 84.21\%] \ [16, \ 3, \ 0] \quad (6)$$

If I come from Szczecin (C5 = 1) and I believe that attractive jobs can be found in Szczecin (C41 = 1), then I plan to pursue my career in Szczecin (D = 1).

The rule is characterised by lower certainty – 84.21% (19 people are described by the premise, 16 of them are consistent with the conclusion, three people have not decided). However, it has a higher coverage factor of 30.19%. This means that 30% of people who plan to pursue their career in Szczecin think in line with the rule's premise. Another advantage is its short length, containing only two conditional variables, making it more readable and interpretable.

It is also worth referring to the rules with a conclusion concerning leaving after graduation. The rule with the highest certainty (91.67%) took the following form:

$$\text{if } C35 = 3 \ \& \ C40 = 3 \text{ then } D = 3; \ [12, \ 11, \ 21.57\%, \ 91.67\%] \ [0, \ 1, \ 11] \quad (7)$$

And means:

If it is not true that Szczecin is a city to which I will always gladly return in my thoughts (C35 = 3) and I believe that it is not a city with prospects for young people (C40 = 3), then I do not plan to tie my future to Szczecin (D = 3).

21.57% of the surveyed population who declare their willingness to leave perceive Szczecin this way. Among the 11 people described by the rule and wishing to leave (1 person has not made a decision), there is only one person from Szczecin. In conjunction with rules (1)–(6), this may indicate that Szczecin cannot attract young, educated people, but is better at retaining Szczecin residents. This may be confirmed by the rule:

$$\text{if } C5 = 2 \ \& \ C16 = 3 \text{ then } D = 3; \ [15, \ 13, \ 25.49\%, \ 86.67\%] \ [1, \ 1, \ 13] \quad (8)$$

If I am from the province but not from Szczecin (C5 = 2) and I perceive Szczecin as kitschy (not sophisticated) (C16 = 3), then I do not plan to tie my future to Szczecin (D = 3).

A certainty factor of 86.67% shows that almost 90% (13 out of 15) of respondents who perceive Szczecin in this way want to leave after completing their studies. At the same time, as many as 25.46% of those declaring their intention to leave see it this way.

In order to verify the assumption that different combinations of the degree to which residents' needs are met have different effects on their attitudes in different cities, an attempt was made to discover knowledge in the form of rules among respondents from Częstochowa. Also, particular attention was paid to rules with high quality parameters, looking for dominant, strong behaviour patterns. Unlike in Szczecin, many certain rules were discovered (all cases described by a premise are also consistent with the conclusion), of which about 95% concerned the desire to remain in the city after completing education. It should be noted that only 18% of respondents want to stay in Częstochowa, while 48% want to leave. It can be concluded that the small group that wants to stay in the city has similar views, while the large group that wants to leave the city is motivated by various reasons. Also, unlike in Szczecin, none of the rules contained a conditional variable relating to origin, so being a resident or coming to study does not determine the conclusion about behaviour.

Among the rules with the highest coverage factor, there are two similar rules:

$$\text{if C37} = 1 \ \& \ \text{C41} = 1 \& \ \text{C42} = 1, \text{ then D} = 1; \ [8, 8, 32.00\%, 100.00\%] \ [8, 0, 0] \tag{9}$$

$$\text{if C38} = 1 \ \& \ \text{C41} = 1 \& \ \text{C42} = 1, \text{ then D} = 1; \ [8, 8, 32.00\%, 100.00\%] \ [8, 0, 0] \tag{10}$$

They can be interpreted as follows:

If I perceive the city as inhabited by open-minded and generous people (C37 = 1) and I believe that attractive jobs can be found in the city (C41 = 1) and I consider the city to be typical of southern Poland (C42 = 1), then I plan to pursue my career in Częstochowa (D = 1).

If I perceive the city as well-managed (C38 = 1) (…) the rest as above.

The coverage factor of 32% shows that one-third of those who want to stay see the city as described by the rules. There were only eight people, but we must remember that only 25 out of 142 interviewed declared they would stay.

It is also worth referring to the rules concerning the reasons for leaving. Two such rules were discovered (based on the satisfaction parameters indicated above). As before, the rules are similar to each other. They largely describe the behaviour of the same respondents.

$$\text{if C20} = 3 \ \& \ \text{C26} = 1 \ \& \ \text{C31} = 2, \text{ then D} = 3; \ [14, 14, 20.90\%, 100.00\%] \ [0, 0, 14] \tag{11}$$

$$\text{if C26} = 1 \ \& \ \text{C27} = 1 \ \& \ \text{C31} = 2, \text{ then D} = 3; \ [14, 14, 20.90\%, 100.00\%] \ [0, 0, 14] \tag{12}$$

They can be read respectively as:

If I perceive the city as backward (C20 = 3) and rate my satisfaction with the fulfilment of my entertainment needs as low (C26 = 1) and rate my satisfaction with the fulfilment of my healthcare needs as average (C31 = 2), then I do not plan to tie my future to Częstochowa (D = 3).

If I rate my satisfaction with the fulfilment of my entertainment needs as low (C26 = 1) and I rate my satisfaction with the fulfilment of my hobbies as low (C27 = 1) and I rate my satisfaction with the healthcare needs being met as average (C31 = 2), then I do not plan to tie my future to Częstochowa (D = 3).

Looking at individual rules, each of them indicates how one-fifth of those who want to leave the city described the image of the city and the reason for leaving – coverage factor equals 20.90% (14 out of 67). Looking at the rules as a whole, they apply to almost a quarter of those planning to leave (16 out of 67). Hence, they can provide valuable knowledge for decision-makers about the image of the city and the reasons for the loss of human resources, which are valuable in the context of the discussed phenomenon of Smart City 3.0.

4 Conclusion

Following the adopted objective, an approach has been proposed that can support city management in line with Smart City 3.0 in the context of attracting and retaining human resources capable of identifying and undertaking ICT initiatives. As shown, different cities need different actions, different needs to be met, and different images to be built. The proposed approach is flexible, allowing for the development of customised solutions that consider a city's own set of metrics, including both conditional and decision variables. A framework for action has been outlined, and its usefulness has been demonstrated through examples, including the use of rough set theory for knowledge discovery in the area under consideration.

Regular repetition of such activities, including in a smart automated manner, will enable dynamic and intelligent monitoring of the situation and adjustment of city management priorities and directions to changing conditions. In this approach, the proposed solution can be the foundation of a dynamically reacting system that identifies and suggests the changing preferences of Smart People. Of course, as our research has shown, such a system must use as input local data describing local conditions.

Acknowledgments. This study was co-financed by the Minister of Science under the "Regional Excellence Initiative" Program.

Disclosure of Interests. The authors have no competing interests to declare that are relevant to the content of this article.

References

1. Yigitcanlar, T., et al.: Can Building "Artificially Intelligent Cities" safeguard humanity from natural disasters, pandemics, and other catastrophes? An urban scholar's perspective. Sensors **20**, 2988 (2020)
2. Gracias, J.S., Parnell, G.S., Specking, E., Pohl, E.A., Buchanan, R.: Smart cities—a structured literature review. Smart Cities **6**, 1719–1743 (2023)
3. Inteligentne i Zrównoważone Miasta w Teorii i Praktyce Zarządzania. (IGSMiE PAN & WSEI Kraków, 2022). https://doi.org/10.33223/smartcity/22/1
4. Giela, M.: The human element in the context of smart cities. Manag. Syst. Prod. Eng. **31**, 300–305 (2023)
5. Podgórniak-Krzykacz, A. Przywojska, J. Skoncentrowana na ludziach transformacja miast w kierunku inteligentnego zrównoważonego rozwoju. Władza Sądzenia, 70–87 (2023). https://doi.org/10.18778/2300-1690.25.05
6. Winkowska, J. Analiza wdrożeń smart city w Polsce i na świecie. Akad. Zarządzania **3** (2021)
7. Rudewicz, J.: Przemysł i technologie wobec wdrożenia wizji miasta inteligentnego (smart city). Stud. Ind. Geogr. Comm. Pol. Geogr. Soc. **33** (2019)
8. Sharif, R.A., Pokharel, S.: Smart city dimensions and associated risks: review of literature. Sustain. Cities Soc. **77**, 103542 (2022)
9. SMACC_Definitions_of_a_Smart_City
10. Tahmasseby, S.: The implementation of smart mobility for smart cities: a case study in Qatar. Civ. Eng. J. **8**, 2154–2171 (2022)
11. Winkowska, J., Szpilko, D., Pejić, S.: Smart city concept in the light of the literature review. Eng. Manag. Prod. Serv. **11**, 70–86 (2019)

12. Monzon, A.: Smart cities concept and challenges: bases for the assessment of smart city projects. In: Helfert, M., Krempels, K.-H., Klein, C., Donnellan, B., Gusikhin, O. (eds.) Smart Cities, Green Technologies, and Intelligent Transport Systems. CCIS, vol. 579, pp. 17–31. Springer, Cham (2015). https://doi.org/10.1007/978-3-319-27753-0_2
13. The Brussels Smart City strategy (2025)
14. Vinod Kumar, T.M. Dahiya, B. Smart economy in smart cities. In: Vinod Kumar, T.M. (ed.) Smart Economy in Smart Cities, pp. 3–76. Springer Singapore, Singapore (2017). https://doi.org/10.1007/978-981-10-1610-3_1
15. Gębarowski, M., Glińska, E.: Wizerunek Białegostoku i Rzeszowa w opiniach studentów. Przegląd Organ. **2**, 38–43 (2011)
16. Pawlak, Z. Rough Sets. Int. J. Comput. Inf. Sci. **11** (1982)
17. Pawlak, Z. Some remarks about rough sets. Inst. Comput. Sci. Pol. Acad. Sci. (1982)
18. Greco, S., Matarazzo, B., Słowinński, R.: Decision rule approach. In: Multiple Criteria Decision Analysis: State of the Art Surveys, vol. 78, pp. 507–555. Springer-Verlag, New York (2005)
19. Pawlak, Z. Rough sets. basic notions. Inst. Comput. Sci. Pol. Acad. Sci. (1981)
20. Tadeusiewicz, R.: Archipelag sztucznej inteligencji. Część II. Napędy Sterow. (2021)
21. Wang, G., Mao, H., Liu, C., Zhang, Z., Yang, L.: Rough set approximations based on a matroidal structure over three sets. Appl. Intell. **53**, 13082–13109 (2023)
22. Mahajan, P., Kandwal, R., Vijay, R.: Rough set approach in machine learning: a review. Int. J. Comput. Appl. **56**(10) (2012)
23. Weihs, C. Sondhauss, U.M.: Combining mental fit and data fit for classification rule selection. In: Schwaiger, M., Opitz, O. (eds.) Exploratory Data Analysis in Empirical Research, pp. 188–203. Springer Berlin Heidelberg, Berlin, Heidelberg (2003). https://doi.org/10.1007/978-3-642-55721-7_21
24. Predki, B., Slowinski, R., Stefanowski, J. Susmaga, R. Wilk, Sz.: ROSE – software implementation of the rough set theory. In: Polkowski, L., Skowron, A. (eds.) Rough Sets and Current Trends in Computing, Lecture Notes in Artificial Intelligence, vol. 1424, pp. 605–608, Berlin (1998)
25. Predki, B., Wilk, Sz.: Rough set based data exploration using ROSE system. In: Ras, Z. W.,Skowron, A. (eds.) Foundations of Intelligent Systems, Lecture Notes in Artificial Intelligence, vol. 1609, pp. 172–180, Berlin (1999)
26. Pawlak, Z.: Rough set theory and its applications. J. Telecommun. Inf. Technol., 7–10 (2002). https://doi.org/10.26636/jtit.2002.140

Multi-criteria Assessment of DESI Levels for Managing AI Implementation in Smart Cities of European Union and Neighboring Countries

Kesra Nermend[1], Laura Brancu[2], Jarosław Woźniak[3](✉), and Waldemar Tarczyński[4]

[1] University of Szczecin, Cukrowa 8, 71-004 Szczecin, Poland
kesra.nermend@usz.edu.pl
[2] West University of Timisoara, Blv. V. Pârvan 4, 300223 Timişoara, Romania
laura.brancu@e-uvt.ro
[3] West Pomeranian University of Technology, Ul. Żołnierska 49, 71-210 Szczecin, Poland
jaroslaw.wozniak@zut.edu.pl
[4] University of Szczecin, Mickiewicza 64/66, 71-101 Szczecin, Poland
waldemar.tarczynski@usz.edu.pl

Abstract. This study presents a dynamic multi-criteria assessment of the digital readiness of European Union and selected neighboring countries for implementing artificial intelligence (AI) in smart cities. Using Vector Measure Construction Method (VMCM), a synthetic indicator was developed based on DESI data for years 2020–2024. The analysis revealed significant differences in digital maturity and allowed to distinguish both leaders and idlers in the digital transformation process. Countries such as Ireland, Hungary and Greece demonstrated the most dynamic improvements, driven by growth in e-government, e-commerce, and online financial services. On the other hand, countries like Poland and Slovakia showed measurable progress but remained at an intermediate level of readiness. The findings emphasize the importance of multidimensional development, policy coherence, and infrastructural investment for successful AI adoption in urban governance. Our VMCM approach proved effective in identifying not only the state but also the trajectory of digital transformation, offering valuable insights for decision-makers and planners in the context of smart urban development.

Keywords: Artificial intelligence · Smart cities · Digital transformation · DESI · Multi-criteria analysis VMCM

1 Introduction

The contemporary development of artificial intelligence (AI) constitutes one of the key challenges and priorities of the European Union's digital policy. AI is gradually permeating various aspects of everyday life—from smart home devices, through healthcare, education, and transport, to public administration and security [1–6]. However, the effective

Published by Springer Nature Switzerland AG 2026
M. Pondel et al. (Eds.): AI4KAM 2025, IFIP AICT 776, pp. 99–109, 2026.
https://doi.org/10.1007/978-3-032-18920-2_10

implementation of AI-based solutions depends not only on technological progress itself but also on the societal, institutional, and infrastructural readiness for their adoption. A crucial instrument for measuring level of this readiness is Digital Economy and Society Index (DESI)—an indicator developed by the European Commission that synthetically reflects the level of digitalization in EU member states in four key areas: human capital, connectivity, integration of digital technology in business, and digital public services [7, 8]. Research shows that countries with higher DESI scores also exhibit greater activity in the implementation of AI solutions [4, 9–12]. This is due, among others, to better access to digital infrastructure, higher levels of citizens' digital skills, more advanced e-services, and greater adoption of cutting-edge technologies such as big data, cloud computing, and the Internet of Things (IoT) [13]. These components provide the necessary foundation for developing and implementing AI algorithms across both the public and private sectors. A high DESI score also translates into greater societal readiness to engage with AI in everyday life. For example, in countries like **Denmark, Estonia**, and **Finland**—leaders in the DESI ranking—there is visible development of AI-powered smart solutions in homes (IoT-based energy management), healthcare (diagnostic support algorithms), education (personalized e-learning platforms), and transport (real-time traffic management systems) [14–16]. In contrast, countries with lower digital maturity, despite increasing interest in AI, face greater implementation barriers, stemming from limited digital skills, weak infrastructure, and low trust in new technologies.

This paper seeks to assess the extent to which DESI reflects the actual readiness of EU countries to implement AI solutions, and what relationships exist between the index's components and the real-world adoption of AI in social and economic life. The analysis may serve as a basis for further development of readiness measurement tools and for the formulation of public policy recommendations that support digital transformation aligned with the principles of responsible and sustainable development.

2 Theoretical Background

Artificial intelligence (AI) constitutes a fundamental component of modern smart cities, which are increasingly conceptualized as complex cyber-physical systems integrating digital infrastructure with the physical and social dimensions of urban environments [17, 18]. In this context, AI functions as an advanced analytical and decision-making mechanism that enables the processing and interpretation of large-scale data streams originating from distributed sources such as IoT networks, environmental sensors, video surveillance systems, smart transport infrastructures, and energy grids.

The integration of these data sources via AI algorithms allows the dynamic management of urban resources, event forecasting, and the optimization of administrative and operational processes—forming the backbone of contemporary urban structures oriented toward sustainability and quality of life [19–23]. A prerequisite of AI usage as a tool for smart city governance is the accurate assessment of EU member states' readiness for digital transformation. The **Digital Economy and Society Index (DESI)**, published annually by the European Commission, evaluates digital progress in five main dimensions: human capital, connectivity, digital technology integration in enterprises, online public services, and—since 2021—ICT research and development [24–27]. These indicators directly affect a country's or region's capacity to implement Smart City concepts, understood as integrated urban management systems based on data, AI, and IoT. DESI is currently one of the most comprehensive tools for measuring digitalization in the EU. However, prior research using this index has been largely static and comparative, limiting its utility for strategic change management. To address this, numerous studies have employed methods from multi-criteria decision analysis (MCDA/MCA) and statistical techniques [28–42]. In this study, a dynamic approach is proposed, using the Vector Measure Construction Method (VMCM), which enables the analysis of DESI evolution over time [43, 44]. This method allows for the assessment not only of the current digitalization status but also of the direction, intensity, and consistency of changes across years. It thus enables the identification of countries demonstrating high-quality digital transformation governance as well as those requiring strategic correction.

The authors aim to fill a research gap by applying a dynamic analysis of the DESI index using the VMCM method, which has been largely overlooked in existing literature. It is worth to point out that most prior studies have focused on static analyses, ignoring temporal changes that are crucial for managing digital transformation. There is also a lack of studies linking specific DESI components to countries' actual capabilities in implementing AI within smart cities. This article includes a detailed analysis of indicators such as internet access and use, e-commerce and e-government activity, R&D investments, ICT sector employment, and actions related to data protection and training. This approach provides a comprehensive assessment of digital readiness quality and the effectiveness of policy implementation in EU countries.

3 Methodology

The present study aims to assess the digital readiness of EU member states for the implementation of AI-based applications within the smart city framework. The assessment is based on DESI component indicators obtained from Eurostat and the European Commission for the multi-year period 2020–2024. The year 2020 was adopted as the reference point—the baseline year—for constructing the pattern and anti-pattern, which are essential for subsequent evaluation and comparison.

The Vector Measure Construction Method (VMCM) was applied as the analytical tool to capture the dynamics of change over time and to determine the relative level of development of each country in comparison to reference benchmarks. This approach enables the identification of both digitalization leaders and countries that require systemic support in preparing for AI implementation in urban environments. Decision variables used are as follows:

x_1 – Percentage of people using the internet, broken down by frequency and NACE[1] Rev. 2 classification
x_2 – Individuals: frequency of internet use
x_3 – Percentage of enterprises with internet access, broken down by NACE Rev. 2 activity
x_4 – Percentage of households with internet access
x_5 – Percentage of people shopping online.

[1] Statistical Classification of Economic Activities in the European Community (according to Eurostat database).

x_6– Percentage of people from a financial services background
x_7 – Percentage of people who made online purchases for private use
x_8 – Percentage of households with internet access at home
x_9 – Percentage of enterprises (by size class) that provided ICT training for employees
x_{10} – Percentage of people who have never used the internet
x_{11} – Percentage of people using the internet to buy goods or services
x_{12} – Percentage of people using the internet for online banking
x_{13} – Percentage of people using the internet to seek health information
x_{14} – Percentage of people using the internet to take online courses
x_{15} – Total value of e-commerce sales by enterprises, divided by NACE Rev. 2 activity
x_{16} – Total research and development (R&D) expenditures by implementation sector
x_{17} – Employment rate in technology and knowledge-intensive sectors, disaggregated by gender and NACE Rev. 2 classification
x_{18} – Number of highly skilled workers (HRST[2]) by category (e.g. professionals, technicians), gender and age group
x_{19} – Percentage of people engaging in e-government activities (e.g. filing tax returns, voting) via government websites
x_{20} – Percentage of people who are concerned about privacy and personal data protection

Analyzed selected countries (decision variants) were: Austria (AT), Belgium (BE), Bulgaria (BG), Croatia (HR), Cyprus (CY), Czechia (CZ), Denmark (DK), France (FR), Germany (DE), Hungary (HU), Ireland (IE), Italy (IT), Latvia (LV), Lithuania (LT), Luxembourg (LU), Malta (MT), Netherlands (NL), Poland (PL), Portugal (PT), Romania (RO), Slovakia (SK), Slovenia (SI), Spain (ES), Sweden (SE), Estonia (EE), Finland (FI), Greece (EL), and Norway (NO), Bosnia and Herzegovina (BA), Serbia (RS). Data referred to year 2020–2024 and were obtained from the Eurostat database.

The Vector Measure Construction Method (VMCM) is a relatively recent multicriteria decision-making technique introduced in 2009 [43], it is explained in details in [44]. According to this method we observed that almost all criteria acted as stimulants in the investigation, namely criteria $x_1 - x_9$ and $x_{11} - x_{20}$ had stimulating character (benefits) and x_{10} was destimulating (no benefit). Here we extended VMCM method by assigning weights to diagnostic variables using correlation between the variables. Namely, to determine the weights of the criteria, the Spearman rank correlation coefficient was applied separately for each year included in the analysis. Subsequently, the arithmetic mean of the correlation values across all years was calculated to obtain a unified measure. These averaged correlation values were then subtracted from one, in order to reduce the influence of variables that showed strong mutual correlations. This approach aimed to minimize the dominance of highly interrelated indicators in the subsequent stages of the multicriteria evaluation. Values of weights of the criteria are: x_1 – [0.61], x_2 –[0.32], x_3 –[0.82], x_4 – [0.34], x_5 –[0.32], x_6– [0.57], x_7 – [0.30], x_8 – [0.34], x_9 – [0.45], x_{10} –[0.32], x_{11} –[0.30], x_{12} – [0.33], x_{13} –[0.40], x_{14}– [0.42], x_{15} – [0.48], x_{16} – [0.47], x_{17} – [0.84], x_{18} –[0.32], x_{19} –[0.65], x_{20} – [0.41].

Table 1. Ranking of EU and neighboring countries in 2020 and 2024 concerning the level of digital readiness in the dynamic approach of the VMCM method

Country	Measure value	Class	Country	Measure value	Class
Pattern form 2020 and Data (2020)			Pattern form 2020 and Data (2024)		
Finland	0.058	1	Finland	0.068	1
Denmark	0.055	1	Netherlands	0.064	1
Sweden	0.050	1	Denmark	0.063	1
Norway	0.048	1	Norway	0.058	1
Netherlands	0.048	1	Sweden	0.058	1
Germany	0.041	2	Ireland	0.056	1
Ireland	0.038	2	Czechia	0.047	2
Estonia	0.034	2	Belgium	0.043	2
Belgium	0.032	2	Estonia	0.042	2
France	0.032	2	Germany	0.041	2
Spain	0.029	2	Austria	0.039	2
Czechia	0.028	2	Spain	0.039	2
Austria	0.028	2	Luxembourg	0.037	2
Slovenia	0.027	2	Slovenia	0.033	2
Cyprus	0.023	2	Hungary	0.031	2
Luxembourg	0.022	2	Malta	0.031	3
Hungary	0.016	3	France	0.030	3
Slovakia	0.012	3	Cyprus	0.027	3
Malta	0.011	3	Poland	0.026	3
Latvia	0.011	3	Slovakia	0.026	3
Poland	0.011	3	Lithuania	0.023	3
Lithuania	0.007	3	Latvia	0.021	3
Italy	0.005	3	Portugal	0.021	3
Croatia	0.005	3	Italy	0.016	3
Portugal	−0.001	3	Croatia	0.014	3
Greece	−0.008	4	Greece	0.010	3
Serbia	−0.013	4	Serbia	0.003	4
Bulgaria	−0.023	4	Romania	−0.008	4
Romania	−0.031	4	Bosnia and Herzegovina	−0.011	4
Bosnia and Herzegovina	−0.033	4	Bulgaria	−0.014	4

4 Results

Table 1 shows a comparative analysis of the readiness levels of European Union member states and EFTA countries to implement artificial intelligence (AI) within the framework of smart city development, using a synthetic measure based on the Digital Economy and Society Index (DESI). The examination draws on data from years 2020–2024 and applies the Vector Measure Construction Method (VMCM), with year 2020 adopted as the reference point for all comparative evaluations. First, countries were classified into four groups based on their level of digital maturity and readiness to implement AI. In 2024, six countries were assigned to Class 1 (best prepared), nine to Class 2, eleven to Class 3, and four to Class 4 (least prepared). When compared to the baseline year 2020, most countries maintained stable positions, although several showed notable changes in both classification and synthetic index values. Countries such as Ireland, Hungary, and Greece significantly improved their positions. Ireland advanced from Class 2 to Class 1, increasing its DESI value from 0.0384 in 2020 to 0.0558 in 2024, representing a growth of 45.4%. The key contributing indicators included: Financial activities over the internet (2020 onwards) (+73.9%), Individuals - internet activities.2 (+56.9%), and Privacy and protection of personal data (+36.1%). Hungary moved from Class 3 to Class 2 with a DESI increase from 0.0163 to 0.0314 (+93.1%). The primary factors were: Financial activities over the internet (+27.0%), Individuals - internet activities (+22.5%), and Privacy and protection of personal data (+20.9%). Greece, although still below the EU average, advanced from Class 4 to Class 3, increasing its DESI from –0.0083 to 0.0104 (+225%). This improvement was driven by: Value of e-commerce sales by NACE Rev. 2 activity (+58.6%), Financial activities over the internet (+50.8%), and Individuals - internet activities (+37.1%). In contrast, Cyprus and France experienced a decline in their classification. Cyprus dropped from Class 2 to Class 3, despite a DESI increase of 17.9% (from 0.0226 to 0.0266), indicating a relatively slower pace of digital progress compared to other nations. France fell from Class 2 to Class 3 and showed a decrease in DESI of 7.4% (from 0.0321 to 0.0297), suggesting stagnation in digital policy. Key indicators included: Financial activities over the internet (+114.0%), Individuals - internet use (+21.8%), and E-government activities of individuals via websites (+10.1%). However, indicator analysis for France also revealed limited progress or even regression in e-commerce and persistently low uptake of public e-services and civic digital engagement. The analysis also considered countries that maintained stable classification and recorded only modest DESI changes. Austria and Belgium, for instance, remained in Class 2 throughout the period. Their DESI growth reached 40.4% and 33.2%, respectively, indicating stable but not transformative digital progress. Special attention was given to four countries at a moderate level of readiness, which remained in Class 3 between 2020 and 2024: Italy, Croatia, Slovakia, and Poland. Italy increased its DESI from 0.0049 to 0.0157 (+219.7%). Key growth areas included: Financial activities over the internet (+45.4%), E-commerce sales by NACE Rev. 2 activity (+37.2%), and Individuals - internet activities.2 (+26.8%). Similarly, Croatia improved its DESI from 0.0048 to 0.0137 (+187%) but did not advance to a higher class. The main contributing indicators were: Financial activities over the internet (+34.3%), Individuals - internet activities

[2] Human Resources in Science and Technology (according to Eurostat database).

(+18.1%), and GERD by sector of performance (+13.4%). Slovakia saw an increase from 0.0124 to 0.0256 (+105.8%), with strongest contributions from: Financial activities over the internet (+118.1%), Individuals - internet activities.2 (+44.4%), and Internet purchases by individuals (+29.5%). Poland's DESI increased from 0.0109 to 0.0259 (+137.0%). Key drivers included: Enterprises providing ICT training (+48.4%), Individuals - internet activities.1 (+20.7%), and E-government engagement (+20.5%). Both Slovakia and Poland made significant progress, but not enough to reach a higher classification. Their improvements were primarily associated with expanding e-commerce, growing digital skills, and increased public engagement with e-government and data protection. A shared characteristic of these countries is consistent improvement in digital infrastructure and services, combined with a need to intensify strategic efforts in areas such as interoperability, digital inclusion, and AI policy integration. In 2024, considering both DESI values and top classification (Class 1), the best-prepared countries for AI implementation in smart cities were Ireland, Denmark, Sweden, Finland, the Netherlands, and Estonia. Compared to 2020, these countries maintained a high pace of digital development, with DESI increases exceeding 30%, alongside consolidation of digital transformation policies, system interoperability, and support for digital education and innovation. Meanwhile, the least-prepared countries (Class 4) Bosnia and Herzegovina, Bulgaria, Serbia, and Montenegro despite some improvement in DESI scores (e.g. Bosnia and Herzegovina from –0.0333 to –0.0112, or + 66%), remained at a very low level of digital maturity. Common barriers include: low-quality digital infrastructure, underdeveloped e-government, limited digital skill development, and the absence of coherent AI strategies. The findings suggest that the actual level of readiness for AI deployment in smart cities should be assessed not only based on absolute DESI values but also in relation to the dynamics of change from the baseline year. Countries that adopted accelerated and strategic approaches achieved higher classifications, whereas those progressing more slowly lost competitive position despite absolute DESI growth.

This analysis underscores the necessity of continuous adaptation and investment in the digital sector, especially in the context of urban policy planning and digital transformation foundations essential for the effective integration of AI into future smart cities.

5 Discussion and Conclusion

The dynamic analysis using the VMCM-based synthetic measure revealed significant differences in the readiness of EU and neighboring countries to implement artificial intelligence (AI) in urban environments. By referencing 2020 as the baseline, the study enabled not only classification of countries by digital maturity but also identified the main sources of both positive and negative changes over the 2020–2024 period.

Countries such as Ireland, Denmark, Sweden, Estonia, and Finland were characterized by high levels of digitalization combined with stability and balance. These countries have transitioned from a phase of rapid growth to a phase of consolidation and integration of advanced digital solutions-including AI-into public services and urban infrastructure. Their high VMCM values, accompanied by moderate dynamics, suggest digital maturity conducive to long-term optimization of smart city management processes. Next group, that is countries such as Ireland, Hungary, and Greece demonstrated substantial progress in their classifications. Between 2020 and 2024, their digital indicators increased on average by 90–225%, reflecting successful development efforts in areas such as online financial services, citizen internet activity, e-commerce, and data protection. Specifically: Ireland's strongest improvements occurred in Financial activities over the internet (+73.9%), Individuals - internet activities.2 (+56.9%), and Privacy and protection of personal data (+36.1%). Hungary's growth was driven by Financial activities over the internet (+27.0%), Individuals - internet activities (+22.5%), and Privacy and protection of personal data (+20.9%). In Greece: Value of e-commerce sales (+58.6%), Financial activities over the internet (+50.8%), and Individuals - internet activities (+37.1%) were the top contributors. Poland, Italy, Slovakia, and Croatia, despite achieving VMCM increases exceeding 100% relative to 2020, remained in Class 3. Their digital transformation, while noticeable, requires further enhancement. In Poland, the main growth factors were ICT training by enterprises (+48.4%), Individuals - internet activities.1 (+20.7%), and E-government usage (+20.5%). Slovakia showed gains in Financial activities over the internet (+118.1%), Individuals - internet activities.2 (+44.4%), and Online purchasing (+29.5%). Italy advanced thanks to Financial activities over the internet (+45.4%), E-commerce sales (+37.2%), and Individuals - internet activities.2 (+26.8%). In Croatia, key factors included Financial activities over the internet (+34.3%), Internet activities (+18.1%), and GERD by sector of performance (+13.4%).

At the opposite end, countries such as Bulgaria, Bosnia and Herzegovina, Serbia, and Montenegro remain characterized by low absolute levels of digital readiness, despite some relative improvements. The primary reasons for this include:

- low digital engagement of enterprises and limited ICT adoption in the private sector,
- underinvestment in R&D and a small share of HRST,
- poor quality of digital transformation and limited access to e-government services,
- low public trust in technology and weak data protection practices,
- fragmented digital skills policy and limited broadband infrastructure.

To enhance AI readiness in urban environments, these countries should adopt comprehensive digital strategies focusing on:

- developing core digital infrastructure (internet access, broadband),
- strengthening digital education and skills across all levels of society,
- increasing R&D investment and building national data ecosystems,
- strengthening institutions responsible for data protection and public trust,
- fostering public-private partnerships in digital urban transformation.

The findings confirm that the success of digital transformation and AI implementation in smart cities depends not only on technology's availability but also on coherent digital policy, effective governance, and institutional capacity. In this context, the VMCM approach provides a valuable tool for diagnosing development potential, monitoring progress, and supporting strategic decision-making in urban digitalization.

Acknowledgments. This study was co-financed by the Minister of Science under the "Regional Excellence Initiative".

Regionalna Inicjatywa Doskonałości
Ministry of Science and Higher Education Republic of Poland

Disclosure of Interests. The authors have no competing interests to declare that are relevant to the content of this article.

References

1. Wirtz, B.W., Weyerer, J.C., Geyer, C.: Artificial intelligence and the public sector—applications and challenges. Int. J. Public Adm. **42**, 596–615 (2019)
2. Chander, B., Pal, S., De, D., Buyya, R.: Artificial intelligence-based internet of things for industry 5.0. In: Artificial Intelligence-Based Internet of Things Systems, pp. 3–45 (2022)
3. Nikitas, A., Michalakopoulou, K., Njoya, E.T., Karampatzakis, D.: Artificial intelligence, transport and the smart city: definitions and dimensions of a new mobility era. Sustainability **12**, 2789 (2020)
4. Topol, E.J.: High-performance medicine: the convergence of human and artificial intelligence. Nat. Med. **25**, 44–56 (2019). https://doi.org/10.1038/s41591-018-0300-7
5. Demir, F.: Innovation in the Public Sector. Smarter States, Services and Citizens. Springer, Cham (2022)
6. Zawacki-Richter, O., Marín, V.I., Bond, M., Gouverneur, F.: Systematic review of research on artificial intelligence applications in higher education – where are the educators? Int. J. Educ. Technol. High. Educ. **16**, 39 (2019). https://doi.org/10.1186/s41239-019-0171-0
7. Digital Economy and Society Index (DESI) 2022. https://akademiacyfryzacji.gs1.pl/wp-content/uploads/2022/08/desi_2022__poland__eng.pdf
8. European Commission. Digital Economy and Society Index (DESI) 2023, Publications Office of the European Union. Dostępne na (2023). https://digital-strategy.ec.europa.eu/en/policies/desi. European Commission
9. Brock, J.K.-U., Von Wangenheim, F.: Demystifying AI: what digital transformation leaders can teach you about realistic artificial intelligence. Calif. Manag. Rev. **61**, 110–134 (2019)
10. Mihai, F., Aleca, O.E., Gheorghe, M.: Digital transformation based on AI technologies in European Union organizations. Electronics **12**, 2386 (2023)
11. Aly, H.: Digital transformation, development and productivity in developing countries: is artificial intelligence a curse or a blessing? Rev. Econ. Polit. Sci. **7**, 238–256 (2022)
12. Szajt, M., Smoląg, K., Chomiak-Orsa, I.: Digitization in European Union countries. Convergence analysis based on DESI. Procedia Computer Science. 246, 5448–5457 (2024)
13. Thillaiarasu, N., Lata Tripathi, S., Dhinakaran, V.: Artificial Intelligence for Internet of Things: Design Principle, Modernization, and Techniques. CRC Press, Boca Raton (2022). https://doi.org/10.1201/9781003335801
14. Gulati, S., Sousa, S., Lamas, D.: Design, development and evaluation of a human-computer trust scale. Behav. Inf. Technol. **38**, 1004–1015 (2019)
15. Borowiecki, R., Siuta Tokarska, B., Maroń, J., Suder, M., Thier, A., Żmija, K.: Developing digital economy and society in the light of the issue of digital convergence of the markets in the European Union countries. Energies **14**, 2717 (2021)
16. Kubicek, H., Noack, T.: Different countries-different paths extended comparison of the introduction of eIDs in eight European countries. Ident. Inf. Soc. **3**, 235–245 (2010)
17. Batty, M., et al.: Smart cities of the future. Eur. Phys. J. Spec. Top. **214**, 481–518 (2012). https://doi.org/10.1140/epjst/e2012-01703-3

18. Kitchin, R.: The ethics of smart cities and urban science. Phil. Trans. R. Soc. A **374**, 20160115 (2016). https://doi.org/10.1098/rsta.2016.0115
19. Alam, M., Khan, I.R.: Application of AI in smart cities. In: Industrial Transformation, pp. 61–86. CRC Press (2022)
20. Kamruzzaman, M.: New opportunities, challenges, and applications of edge-AI for connected healthcare in smart cities. In: Presented at the 2021 IEEE Globecom Workshops (GC Wkshps) (2021)
21. Townsend, A.M.: Smart cities: Big data, civic hackers, and the quest for a new utopia. WW Norton & Company (2013)
22. Mora, L., Deakin, M., Reid, A.: Strategic principles for smart city development: a multiple case study analysis of European best practices. Technol. Forecast. Soc. Chang. **142**, 70–97 (2019). https://doi.org/10.1016/j.techfore.2018.07.035
23. Luusua, A., Ylipulli, J., Foth, M., Aurigi, A.: Urban AI: understanding the emerging role of artificial intelligence in smart cities. AI Soc. **38**, 1039–1044 (2023)
24. Cioacă, S.-I., Cristache, S.-E., Vuță, M., Marin, E., Vuță, M.: Assessing the impact of ICT sector on sustainable development in the European Union: an empirical analysis using panel data. Sustainability **12**, 592 (2020). https://doi.org/10.3390/su12020592
25. Zhao, K., O'Mahony, M., Qamar, A.: Bridging the gap in creative economy and ICT research: a regional analysis in Europe. Appl. Econ. **52**, 3153–3166 (2020). https://doi.org/10.1080/00036846.2019.1706718
26. Savulescu, C.: Dynamics of ICT development in the EU. Procedia Econ. Finan. **23**, 513–520 (2015). https://doi.org/10.1016/S2212-5671(15)00552-3
27. Billon, M., Marco, R., Lera-Lopez, F.: Innovation and ICT use by firms and households in the EU: a multivariate analysis of regional disparities. ITP. **30**, 424–448 (2017). https://doi.org/10.1108/ITP-05-2015-0098
28. Nermend, M., Singh, S., Singh, U.S.: An evaluation of decision on paradigm shift in higher education by digital transformation. Procedia Comput. Sci. **207**, 1959–1969 (2022). https://doi.org/10.1016/j.procs.2022.09.255
29. Ogonowski, P.: Application of VMCM, to investigate, the dynamics of changes information society development. Procedia Comput. Sci. **176**, 3182–3190 (2020). https://doi.org/10.1016/j.procs.2020.09.131
30. Marchewka, A.: Applying Vector Measure Construction Method (VMCM) to analyze change dynamics of investment attractiveness of the region. Procedia Comput. Sci. **192**, 3252–3261 (2021). https://doi.org/10.1016/j.procs.2021.09.098
31. Karczmarczyk, A., Wątróbski, J., Jankowski, J., Ziemba, E.: Comparative study of ICT and SIS measurement in Polish households using a MCDA-based approach. Procedia Comput. Sci. **159**, 2616–2628 (2019). https://doi.org/10.1016/j.procs.2019.09.254
32. Singh, S., Singh, U.S., Nermend, M.: Decision analysis of e-learning in bridging digital divide for education dissemination. Procedia Comput. Sci. **207**, 1970–1980 (2022). https://doi.org/10.1016/j.procs.2022.09.256
33. Yılmaz, B.: Determining the digitalization levels of leading countries in logistics performance index: an application with CRITIC-TOPSIS approach. Verimlilik Dergisi **59**, 431–450 (2025). https://doi.org/10.51551/verimlilik.1541480
34. Nermend, M., Singh, U.S., Singh, S.: Educational background as decision criteria in selection of private sector healthcare service providers. Procedia Comput. Sci. **207**, 4651–4659 (2022). https://doi.org/10.1016/j.procs.2022.09.529
35. Łatuszyńska, M., Nermend, K.: Energy Decision Making: Problems, Methods, and Tools—An Overview. Energies **15**, 5545 (2022). https://doi.org/10.3390/en15155545
36. Ogonowski, P.: Integrated AHP and TOPSIS method in the comparative analysis of the internet activities. Procedia Comput. Sci. **207**, 4409–4418 (2022). https://doi.org/10.1016/j.procs.2022.09.504

37. Piwowarski, M., Nermend, K.: Issues of multi-criteria methods applicability supporting complex business process decision-making in management. Procedia Comput. Sci. **207**, 4161–4170 (2022). https://doi.org/10.1016/j.procs.2022.09.479
38. Hajduk, S.: Multi-criteria analysis of smart cities on the example of the Polish cities. Resources **10**, 44 (2021). https://doi.org/10.3390/resources10050044
39. Piwowarski, M., Nermend, M., Nermend, K.: Psychophysiological research in the design of useful interfaces for e-learning materials. In: Hernes, M., Wątróbski, J. (eds.) Emerging Challenges in Intelligent Management Information Systems, pp. 215–227. Springer, Cham (2024). https://doi.org/10.1007/978-3-031-66761-9_18
40. Nermend, K., Grima, S.: Quality of life in the European Union. In: Grima, S., Romānova, I., Noja, G.G., Dorożyński, T. (eds.) Contemporary Studies in Economic and Financial Analysis, pp. 177–200. Emerald Publishing Limited (2024). https://doi.org/10.1108/S1569-375920240000115011
41. Singh, S., Singh, U.S., Nermend, M.: Sustainability in a digitized era analyzing the moderation effect of social strata and digital capital dependence on digital divide. Sustainability **14**, 14508 (2022). https://doi.org/10.3390/su142114508
42. Piwowarski, M., Borawski, M., Nermend, K.: The problem of non-typical objects in the multidimensional comparative analysis of the level of renewable energy development. Energies **14**, 5803 (2021). https://doi.org/10.3390/en14185803
43. Nermend, K.: Vector Calculus in Regional Development Analysis. Physica-Verlag HD, Heidelberg (2009). https://doi.org/10.1007/978-3-7908-2179-6
44. Nermend, K.: Multi-Criteria and Multi-Dimensional Analysis in Decisions: Decision Making with Preference Vector Methods (PVM) and Vector Measure Construction Methods (VMCM). Springer Nature Switzerland, Cham (2023). https://doi.org/10.1007/978-3-031-40538-9
45. Nermend, K.: Metody analizy wielokryterialnej i wielowymiarowej we wspomaganiu decyzji. Wydawnictwo Naukowe PWN, Warszawa (2017)
46. Grabiński, T.: Metody statystyki międzynarodowej. PWE, Warszawa (1988)
47. Kukuła, K.: Metoda unitaryzacji zerowanej. Wydaw. Naukowe PWN, Warszawa (2000)
48. Nermend, K.: Taxonomic vector measure of region development (TWMRR). Polish J. Environ. Stud. (2007)
49. Nermend, K.: Vector Calculus in Regional Development Analysis: Comparative Regional Analysis Using the Example of Poland. Physica-Verlag, Berlin (2009)

Social Trust in Artificial Intelligence as a Condition for the Acceptance of Smart City Applications

Magdalena Rzemieniak[1], Dorota Jelonek[2(✉)], Lilla Knop[3], and Agnieszka Bitkowska[4]

[1] Lublin University of Technology, Lublin, Poland
[2] Czestochowa University of Technology, Częstochowa, Poland
dorota.jelonek@pcz.pl
[3] Silesian University of Technology, Gliwice, Poland
[4] Warsaw University of Technology, Warsaw, Poland

Abstract. The purpose of this article is to analyze the impact of social trust on the acceptance of artificial intelligence (AI) technologies in the context of smart city applications. The article adopts a research model based on three key components of trust: security, anonymity and ethical use of data. The survey was conducted using a quantitative method (CAWI) among 1,500 residents of provincial cities in Poland. PLS-SEM (Partial Least Squares - Structural Equation Modeling) structural equation modeling was used in the analysis. The results indicate that all factors have a significant impact on the overall level of social trust, which in turn determines the declared willingness to use AI applications in smart city.

Keywords: social trust · AI · smart city · urban decision-making

1 Introduction

For over a decade, the development of the concept of smart cities has been the focus of attention for researchers and practitioners involved in urban planning, public policy, and modern technologies. Dynamic urbanization, growing demand for public services, and the need for rational management of limited resources mean that local governments are increasingly turning to innovative digital solutions. In this context, artificial intelligence (AI) is seen as a key element enabling the automation of administrative processes, the personalization of municipal services, and the optimization of resource consumption (Allam & Dhunny, 2019; Duan et al., 2019; Son et al., 2023). Its potential is evident in the areas of intelligent transport and energy management systems, as well as in the prediction of environmental hazards, public safety monitoring, and spatial planning.

At the same time, the use of AI in urban spaces raises a number of social, legal, and ethical challenges. The literature indicates that the success of AI-based technology implementations does not depend solely on their technical or economic effectiveness,

Published by Springer Nature Switzerland AG 2026
M. Pondel et al. (Eds.): AI4KAM 2025, IFIP AICT 776, pp. 110–123, 2026.
https://doi.org/10.1007/978-3-032-18920-2_11

but to a large extent on the degree of social acceptance (Ahmad et al., 2022; Allam & Dhunny, 2019). The trust of residents in the institutions responsible for data collection and processing is becoming one of the key conditions for the success of smart city initiatives. A lack of trust can result in resistance to new solutions, limited civic participation, and, consequently, the failure of entire modernization programs.W literaturze zaufanie społeczne wobec systemów sztucznej inteligencji definiowane jest jako przekonanie obywateli, że instytucje publiczne działają w sposób przejrzysty, odpowiedzialny i zgodny z zasadami etyki, a ich działania nakierowane są na dobro wspólne (Ziller & Andreß, 2021; Deng & Yu, 2021). Badacze podkreślają, że zaufanie to opiera się na trzech filarach: bezpieczeństwie danych, ich anonimowości oraz etycznym sposobie ich wykorzystania.

An analysis of data security, anonymity, and ethical data use leads to the conclusion that building public trust should be an integral part of the design and implementation of AI-based solutions in smart cities. This challenge is interdisciplinary in nature, combining issues from the fields of computer science, law, ethics, sociology, and management sciences.

The aim of this article is to identify and empirically verify a model of social trust in the use of artificial intelligence in smart cities. The proposed model assumes that the level of trust among residents is a function of perceived data security, anonymity, and the belief that local authorities will use the technology ethically. The analyses carried out aim to deepen knowledge about the conditions for social acceptance of AI technology in urban spaces and to indicate practical implications for decision-makers responsible for implementing innovative urban solutions.

2 Research Background and Hypothesis

2.1 Social Trust in Artificial Intelligence

In recent years, the importance of artificial intelligence (AI) as a key tool supporting the development of so-called smart cities has been increasingly emphasized. These cities are defined as complex urban ecosystems in which the use of modern information and communication technologies aims to optimize management processes, improve the quality of life of residents, and increase the efficiency of resource use (Allam & Dhunny, 2019; Duan et al., 2019; 2020; Son et al., 2023). The application of AI in this context concerns, among other things, areas such as intelligent transport systems, energy management, urban traffic forecasting, environmental monitoring, and the optimization of public services. Increasing attention is also being paid to the role of artificial intelligence in supporting the decision-making processes of local authorities, which, thanks to analytical tools, can respond more efficiently to the needs of residents, predict potential crises, and allocate available resources more effectively.

However, technological development alone does not automatically guarantee successful implementation in urban areas. As research shows, a key condition for the effective integration of AI in smart cities is a high level of public trust (Allam & Dhunny, 2019; Ahmad et al., 2022). This trust can be defined as the belief of citizens that public institutions use AI tools in a transparent, responsible, and ethical manner, always acting in the interests of residents (Ziller & Andreß, 2021; Deng & Yu, 2021). A lack of such trust can lead to resistance to new technological solutions, limiting their social acceptance and, consequently, to the failure of entire smart city projects.

Social trust in artificial intelligence (AI) is crucial for its acceptance and effective integration into society. Research shows that trust in AI depends on many factors, including user characteristics, AI system properties, and the context of interaction. An adequate level of trust can accelerate the implementation of AI, while a lack of trust leads to its rejection or misuse.

The main factors influencing trust in AI are technological factors, i.e., reliability, transparency, explainability, security, and fairness of AI systems (Kaur et al., 2022; Thiebes et al., 2020; Ferrario et al., 2019) and social and psychological factors (Gillath et al., 2021, Riedl, 2022), as well as context and experience (Omrani et al., 2022; Riedl, 2022). The challenge and threat is excessive or insufficient trust—too much trust leads to AI abuse, while too little trust leads to its rejection (Zerilli, 2022), ethics and justice—lack of transparency, algorithmic bias, and threats to human autonomy can undermine trust (Thiebes et al., 2020; Kaur et al., 2022) and context and experience – the possibility of testing AI, previous experiences, and the way AI is presented (Omrani et al., 2022). Public trust in AI is a complex phenomenon that depends on technical, social, and contextual characteristics. Building trust requires not only improving technology, but also taking psychological and ethical factors into account. Proper trust management is essential for the safe and acceptable integration of AI into society.

Based on a review of the literature, public trust in the context of AI application in smart cities can be considered as a construct based on three key components:

- *Data security* – refers to citizens' belief that data collected and processed within urban systems is adequately protected against unauthorized access, theft, or manipulation. In the face of a growing number of cyberattacks and privacy breaches, ensuring data security is becoming one of the most important factors in building public trust.
- *Anonymity* – refers to residents' belief that data used by AI-based systems cannot be directly linked to a specific person. Guaranteeing anonymity is particularly important when collecting sensitive data, such as health information or data on residents' mobility. The lack of this guarantee can lead to a sense of threat and reduce the willingness to share data.
- *Ethical use of data* – this is related to the belief that data collected by municipal systems is not abused or used in a manner contrary to the interests of citizens. This includes, among other things, preventing the use of data for commercial purposes without residents' consent, avoiding discriminatory practices, and ensuring that algorithms used in city management operate in a fair and transparent manner.

All these elements remain closely interlinked and form the foundation on which long-term public acceptance for the use of artificial intelligence in urban spaces can be built. In practice, this means that local governments, when implementing AI-based solutions, should not only focus on technical aspects, but also on building relationships with residents, ensuring transparent communication, and creating mechanisms for social control. Only the combination of high-quality technology with responsible management and strong social trust can guarantee that smart cities will become a real response to contemporary urbanization challenges.

2.2 Data Security and Anonymity in Smart City Applications

Smart city applications are employed across various sectors, including utilities, healthcare, transportation, entertainment, and government services. However, users frequently express concerns regarding security and loss of privacy, driven by the awareness of risks such as data breaches or unauthorized access (Van Zoonen, 2016; AlDairi & Tawalbeh, 2017).

Research on security in smart city applications has developed along two primary strands. The first examines security in the context of studies of social acceptance of smart city applications using dedicated technology acceptance models (TAM) (Dirsehan & van Zoonen, 2022; Babdullah et al., 2017; Van Heek, et al., 2016; Julsrud & Krogstad, 2020). Scholars have investigated attitudes toward the use of smart cards for access to local facilities and for micropayments for basic services (Belanche-Gracia et al., 2015), as well as the acceptance of surveillance in public space and in private smart city environments (Van Heek, et al., 2016). Findings confirm that perceived security and anonymity have a significant impact on the use of smart city applications (Fabrègue & Bogoni, 2023).

The second strand focuses on ensuring the security of smart city users' data at every stage of the data management process (Zhang et al., 2017), from data acquisition and collection, through processing, to data sharing (Barnaghi, et al., 2015; Fang et al., 2020). Guaranteeing authenticity, integrity, and the secure transmission of data requires the implementation of appropriate frameworks, cryptographic algorithms, and communication protocols (Huerta & Salazar, 2018; Zang et al., 2017).

For users, anonymity i.e., ensuring that personal data cannot be traced to specific individuals, is crucial. Privacy breaches can erode social trust and discourage citizens from participating in smart city initiatives. Accordingly, the latest security systems are employed to protect data and ensure anonymity (Braun et al., 2018; Kramer et al., 2019), including blockchain (Mora et al., 2018) and artificial intelligence (Huerta & Salazar, 2018).

In the context of the above considerations, the authors proposed the following hypotheses:

H1: Residents trust that data collected by AI systems in smart cities is secure.

H2: Residents trust that data collected by AI systems in smart cities is anonymous.

2.3 Ethical Use of Data in Smart Cities

The use of data in smart city technologies raises important ethical issues (Chang, 2021) related to privacy, data collection, dataveillance and geo-surveillance, as well as data uses such as social sorting and anticipatory governance (Kitchin, 2016). Central in this regard are privacy and surveillance (Clever et al., 2018; Phillips & Jiao, 2023; Kitchin, 2016; Bibri & Allam, 2022), together with consent, transparency, and data ownership (Gilman et al., 2024; Mark & Anya, 2019). Smart-city applications require the collection of large volumes of data, which raises serious concerns about privacy and the observance of civil rights and the principle of justice. Urban solutions based on intelligent technologies should serve all residents, regardless of their social status or level of technological proficiency. Potential abuses in monitoring citizens are a recurring ethical challenge. The same holds for obtaining consent, ensuring transparency of data practices, and clarifying data ownership. There is strong consensus that data should remain under the control of citizens or municipalities, rather than private firms. Transparent communication about how data are used and who has access to them is essential for building social trust. Three main areas of ethical mechanisms can be distinguished:

1. Ethical frameworks and principles
   - Contemporary approaches are shifting away from technology and automation toward value-sensitive design that incorporates law, culture, and public participation (Helbing et al., 2021; Singh et al., 2023).
   - Ethical frameworks grounded in moral theories are being proposed, encompassing the relationships among residents, businesses, and public administration (Ziosi et al., 2022).
2. Data governance and accountability
   - Key mechanisms include transparency, consent to data processing, data ownership, and accountability (König, 2021; Ahmad et al., 2022).
   - •The need for democratic control over data and for limiting the dominance of private entities is emphasized (König, 2021).
3. Social inclusion and participation
   - Ethical mechanisms include the active engagement of citizens in decision-making processes, equitable access to benefits, and counteracting digital exclusion (Helbing et al., 2021).
   - The importance of justice, transparency, and the adaptability of urban policies is underscored (Ziosi et al., 2022).

In light of the above considerations, the authors proposed the following hypothesis: H3: Residents trust that data collected by AI systems in smart cities is used ethically.

3 Methodology

3.1 Research Sample and Data Collection

The study was conducted in the first quarter of 2025 using the CAWI (Computer-Assisted Web Interviewing) method, among 1,500 residents of provincial capital cities in Poland, with consideration given to the population size of each city Table 1).

The selection criteria included age (18+), familiarity with at least one AI-based application (e.g., transport, security, energy), and consent to participate in the study. Prior to the main survey, a pilot study was conducted to assess the clarity of the questionnaire items and the duration of the interview. After data verification and the removal of incomplete questionnaires, the final dataset consisted of 1,490 valid responses.

Subsequently, Partial Least Squares Structural Equation Modeling (PLS-SEM) was employed for data analysis. PLS-SEM is particularly well-suited for theory development and complex models, especially when the sample size is relatively small and the data distribution deviates from normality (Sarstedt et al., 2022). The analysis was conducted

Table 1. Information about structure of respondents

Voivodeship	City	Population	Population share	Number of respondents
Lower Silesian Voivodeship	Wroclaw	673 743	0.092098	138
Kuyavian-Pomeranian Voivodeship	Torun	194 771	0.026624	40
Lublin Voivodeship	Lublin	329 565	0.04505	68
Lubusz Voivodeship	Zielona Gora	138 932	0.018991	28
Lodz Voivodeship	Lodz	652 015	0.089128	134
Lesser Poland Voivodeship	Krakow	806 201	0.110204	165
Masovian Voivodeship	Warsaw	1 861 599	0.254472	382
Opole Voivodeship	Opole	126 077	0.017234	26
Subcarpathian Voivodeship	Rzeszow	197 268	0.026966	40
Podlaskie Voivodeship	Bialystok	291 688	0.039872	60
Pomeranian Voivodeship	Gdansk	487 371	0.066621	100
Silesian Voivodeship	Katowice	279 190	0.038164	57
Holy Cross Voivodeship	Kielce	182 295	0.024919	37
Warmian-Masurian Voivodeship	Olsztyn	167 311	0.022871	34
Greater Poland Voivodeship	Poznan	538 439	0.073602	110
West Pomeranian Voivodeship	Szczecin	389 066	0.053184	80
	Total	7 315 531	1	1.500

using SmartPLS 4 software, while descriptive statistics were generated using IBM SPSS Statistics 21.

3.2 Measures

The questionnaire included multi-item measures for each construct, developed based on previous research. Social trust was measured using items related to key dimensions of social trust, including acting in the interest of citizens, acting transparently, responsibly, and in accordance with ethical standards (Justwan et al., 2018; di Napoli, 2019). Data security referred to the belief that personal data are adequately protected (Fabrègue & Bogoni, 2023). Anonymity encompassed the belief that data cannot be linked to an identifiable individual (Braun et al., 2018; Krämer et al., 2019). Ethical data use referred to the belief that data are not misused or exploited against the interests of residents (Bibri & Allam, 2022; Gilman et al., 2024).

A seven-point Likert scale, ranging from 'strongly disagree' (1) to 'strongly agree' (7), was used to assess each questionnaire item related to the constructs described above. All variables were measured using Likert-type scales. The constructs were developed based on prior literature (e.g., Justwan et al., 2018; di Napoli, 2019), and included: data security (3 items), data anonymity (3 items), and ethical data use (3 items). Social trust was modeled as a latent variable composed of these three constructs. AI acceptance was measured as the declarative willingness to use smart city applications (3 items). The constructs and items are presented in Table 2.

Table 2. Constructs and items

Constructs & Items	Statement
Data Security	
BD1	Data collected by smart city applications are adequately protected against unauthorized access
BD2	AI-based urban systems are an easy target for data breaches
BD3	The city employs effective procedures for responding to data security incidents
Data Anonymity	
A1	Information collected by smart city applications does not allow the identification of a specific individual
A2	Prior to being shared, data are anonymized in a way that prevents re-identification
A3	Based on data from smart city applications, authorities or companies can easily determine a person's identity
Ethical Use of Data	
EUD1	Residents' data are used solely for purposes aligned with the public interest
EUD2	Rules governing data use (who has access and for what purpose) are communicated clearly and transparently
EUD3	Data are not used in ways that may harm residents or their groups
AI Acceptance (declarative willingness to use)	
AIA1	Residents are willing to use AI-based municipal applications (e.g., transport, energy, public services)
AIA2	Residents would recommend the use of AI-based smart city applications to other residents
AIA3	Residents do not avoid smart city solutions if they involve artificial intelligence

4 Results

4.1 Measurement Model

All constructs in the measurement model were defined as reflective. Accordingly, the constructs were evaluated based on outer loadings, internal consistency reliability, convergent validity, and discriminant validity (Well et al., 2005; Liu et al., 2018). All outer loadings are significant and exceeded the recommended threshold of 0.715. Subsequently, for all latent constructs, the values of Cronbach's alpha ex-ceeded the threshold of 0.70, confirming good internal consistency reliability. Next, the average variance extracted (AVE) values were examined to ensure convergent validity. For all constructs, the AVE values sur-passed the recommended threshold of 0.50.

Finally, discriminant validity was confirmed using the Heterotrait-Monotrait ratio of correlations (HTMT), applying the recommended cut-off value of 0.90. The HTMT criterion assesses whether a latent variable represents a conceptually distinct construct and evaluates the extent to which the items accurately reflect the intended underlying construct (Russo & Stol, 2021).

The quality assessment of the measurement model was conducted in several stages. First, the outer loadings were analyzed, which exceeded the threshold of 0.715 for all indicators and were statistically significant. This indicates that each of the observed indicators significantly and strongly reflects its latent variable, and the constructs are correctly operationalized.

In the next step, internal reliability was verified using Cronbach's alpha coefficient and the CR (Composite Reliability) coefficient. Both Cronbach's alpha and CR values exceeded the threshold of 0.70, which indicates high consistency and stability of measurement within individual constructs. This is particularly important in research on technology perception, where conceptual and operational consistency is a prerequisite for accurate conclusions.

Convergent validity was assessed based on average variance explained (AVE). All AVE values exceeded the threshold of 0.50, indicating that the constructs explain a significant portion of the variance of their indicators. In practice, this means that the adopted theoretical operationalizations are adequate and the latent variables have an appropriate empirical representation.

The final stage of verification was the assessment of differential validity, i.e., the degree to which the constructs are distinct from each other. The HTMT (Heterotrait-Monotrait ratio of correlations) indicator was used, whose values for all pairs of constructs were below the recommended limit of 0.90. This result confirms that the analyzed constructs—data security, anonymity, and ethical use of data—reflect distinct, yet related, aspects of AI acceptance.

In summary, the results of the measurement model evaluation confirm its high psychometric quality and provide a solid basis for further analysis of the structural model.

4.2 Structural Model

The structural model was evaluated using a bootstrap procedure with 5,000 replications, which enabled precise estimation of confidence intervals and significance tests for path coefficients. Before analyzing the relationships between constructs, an assessment of the collinearity of explanatory variables was performed. The VIF values ranged from 1.178 to 2.219, which is well below the threshold of 5. This result excludes the risk of multicollinearity and thus confirms the stability of the coefficient estimates.

Analysis of the structural model showed that:

- data security has a significant positive impact on the acceptance of artificial intelligence ($\beta = 0.285$, $p < 0.001$),
- anonymity also has a significant positive impact on the acceptance of AI ($\beta = 0.263$, $p < 0.001$),
- the ethical use of data proves to be the strongest predictor of the acceptance of artificial intelligence ($\beta = 0.385$, $p < 0.001$).

All research hypotheses (H1, H2, and H3) were confirmed, indicating consistency between theoretical assumptions and empirical results. It is worth noting that the impact of ethical data use was not only significant but also stronger than the other effects, which emphasizes the importance of normative and regulatory aspects in the process of building trust in artificial intelligence in smart cities.

Furthermore, the model analysis showed that the explanatory variables contribute significantly to explaining the variance in AI acceptance. The high level of fit and significant causal paths indicate that the proposed model has high explanatory and application value (Fig. 1 and Table 3).

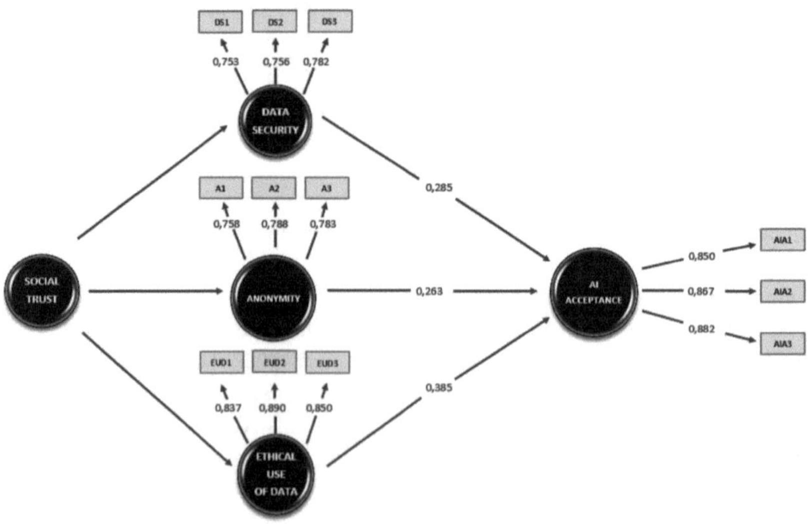

Fig. 1. General structural model

Table 3. Results of the structural model analysis

Hypotheses		Path coefficient	*t*-value	*p*	Decision
H1	Data security - > AI Acceptance	0.285	3.736	0.000	Supported
H2	Anonymity- > AI Acceptance	0.263	4.620	0.000	Supported
H3	Ethical Use of Data - > AI Acceptance	0.385	5.228	0.000	Supported

5 Discussion

Cities are increasingly relying on artificial intelligence (AI)-based systems for urban management, public services, and decision-making. As a result, establishing and maintaining social trust—between citizens, institutions, and technology—has become essential for broad acceptance and effective implementation. Social trust is thus a key factor in the successful integration of AI in smart cities.

Research confirms that residents' trust in the security of collected data is a prerequisite for the acceptance of AI applications in smart cities (H1). This finding aligns with previous studies emphasizing the importance of safeguarding user data throughout all stages of data management in smart city applications—from acquisition and storage to processing and sharing (Kuan Zhang et al., 2017; Barnaghi, Bermudez-Edo, & Tönjes, 2015; Fang et al., 2020). The confirmation that ensuring the anonymity of data collected by smart city applications positively influences the acceptance of AI solutions (H2) is consistent with findings from Braun et al. (2018), and Krämer et al. (2019). Hypothesis H3, which posited a positive impact of the ethical use of data collected by AI applications in smart cities, was also supported. This result is in line with studies conducted by Gilman et al. (2024), Mark & Anya (2019), and Deng & Yu (2021), which broadens the discussion on broader applicability and potential cross-cultural comparisons.

The study results confirm that public trust is a fundamental factor determining the acceptance of AI in the context of smart cities. Citizens are more accepting of new technologies when they are confident that data is secure (H1), anonymous (H2), and used ethically (H3). A positive verification of hypothesis H1 demonstrates that data protection is a necessary condition for building acceptance of AI. This aligns with previous findings (Kuan Zhang et al., 2017; Barnaghi et al., 2015; Fang et al., 2020), according to which data security breaches can lead to a decline in public trust and reduced participation in smart city solutions. The results related to hypothesis H2 indicate that residents expect their data to be treated in a manner that prevents their identification. This confirms previous findings (Braun et al., 2018; Krämer et al., 2019), according to which anonymity plays a key role in overcoming fears of surveillance and citizen profiling. The strongest effect was observed for hypothesis H3 – ethical use of data. This result emphasizes that citizens pay attention not only to technical issues (security, privacy) but also to ethical norms and the transparency of institutional activities.

This confirms the findings of Gilman et al. (2024), Mark & Anya (2019), and Deng & Yu (2021), indicating that ethics in data management is a key factor in the long-term acceptance of technology.

The study results also allow for practical implications:

- For city decision-makers – the implementation of AI-based solutions should be linked to the creation of clear data security standards, technology certification, and mandatory security audits.
- For public policymakers, it is essential to establish legal regulations that guarantee citizens full anonymity in data collection and processing processes, as well as provide sanctions for violations.
- For technology developers, companies developing AI systems should implement the principles of privacy by design and ethics by design to ensure that solutions comply with both technical requirements and the ethical expectations of society.
- For civil society, the study results highlight the need for educational campaigns that will increase citizens' awareness of their rights related to personal data and the ability to control its use.

6 Conclusions

The primary objective of this article was to develop and empirically validate a model that highlights the influence of social trust on the acceptance of artificial intelligence (AI) technologies in the context of smart city applications. The findings clearly demonstrate that three key components—data security, anonymity, and the ethical use of application data—exert a significant and positive effect on public trust toward AI-based urban solutions. These dimensions together form a coherent framework, reinforcing the argument that technological efficiency alone does not guarantee societal acceptance; rather, it must be complemented by transparent, responsible, and ethically sound governance practices.

The analysis conducted contributes to the existing body of literature in several ways. First, it provides an integrated perspective on social trust by considering simultaneously the impact of data security, anonymity, and ethical use of data. Previous studies have tended to examine these factors in isolation, whereas the proposed model demonstrates the synergistic nature of their combined influence. Second, the study extends the theoretical discourse on smart city governance by linking the concept of social trust to concrete technological determinants, thereby bridging the gap between abstract ethical debates and measurable, empirically validated constructs.

From a practical standpoint, the results of this study offer important implications for policymakers, urban planners, and technology developers. The successful implementation of AI in smart cities depends not only on the deployment of advanced algorithms or digital infrastructure, but equally on fostering citizen confidence through transparent data management and ethical accountability. Local authorities should prioritize the creation of institutional safeguards that guarantee data protection, ensure anonymity, and establish clear principles for the fair and just use of information. Equally important is the role of communication strategies that explain the rationale behind data collection and the intended benefits for the community, thereby reducing resistance and enhancing civic engagement.

Nevertheless, the study is not without limitations. The sample was restricted to respondents representing provincial capital cities in Poland, which may limit the generalizability of the results to other national or cultural contexts. Moreover, the reliance on self-reported, declarative data entails a risk of bias associated with subjective perceptions rather than observed behavior. Future research should therefore focus on a broader and more diverse range of cities, including smaller urban centers as well as international comparisons, to capture potential cultural or socio-economic differences in attitudes toward AI. Additionally, longitudinal studies could provide insights into the dynamic evolution of social trust over time, particularly in response to technological progress, policy changes, or emerging ethical challenges.

In conclusion, this study underscores the pivotal role of social trust in shaping the acceptance of AI technologies in smart city environments. By demonstrating that trust is anchored in perceptions of data security, anonymity, and ethical use, it provides a robust framework for both academic inquiry and practical policymaking. Ultimately, the integration of AI into urban governance will succeed only if technological innovation is coupled with ethical responsibility, institutional transparency, and genuine respect for citizens' rights and expectations.

References

Ahmad, K., Maabreh, M., Ghaly, M., Khan, K., Qadir, J., Al-Fuqaha, A.: Developing future human-centered smart cities: critical analysis of smart city security, data management, and ethical challenges. Comput. Sci. Rev. **43**, 100452 (2022)

AlDairi, A., Tawalbeh, L.A.: Cyber security attacks on smart cities and associated mobile technologies. Procedia Comput. Sci. **109**, 1086–1091 (2017)

Allam, Z., Dhunny, Z.A.: On big data, artificial intelligence and smart cities. Cities **89**, 80–91 (2019)

Babdullah, A., Rana, N.P., Ali, A.A., Dwivedi, Y.K., Lal, B.: Assessing consumer's intention to adopt mobile internet services in the Kingdom of Saudi Arabia. In: AMCIS 2017, Boston, USA, 10–12 August 2017 (2017)

Barnaghi, P.M., Bermudez-Edo, M., Tönjes, R.: Challenges for quality of data in smart cities. J. Data Inf. Qual. **6**(2–3), 1–4 (2015)

Belanche-Gracia, D., Casaló-Ariño, L.V., Pérez-Rueda, A.: Determinants of multi-service smart-card success for smart cities development: a study based on citizens' privacy and security perceptions. Gov. Inf. Q. **32**(2), 154–163 (2015). https://doi.org/10.1016/j.giq.2014.12.004

Bibri, S.E., Allam, Z.: The Metaverse as a virtual form of data-driven smart cities: the ethics of the hyper-connectivity, datafication, algorithmization, and platformization of urban society. Computat. Urban Sci. **2**(22), 715–727 (2022)

Braun, T., Fung, B.C., Iqbal, F., Shah, B.: Security and privacy challenges in smart cities. Sustain. Cities Soc. **39**, 499–507 (2018)

Chang, V.: An ethical framework for big data and smart cities. Technol. Forecast. Soc. Chang. **165**, 120559 (2021)

Deng, X., Yu, M.: Scale of cities and social trust: evidence from China. Int. Rev. Econ. Financ. **76**, 215–228 (2021)

Di Napoli, I., Dolce, P., Arcidiacono, C.: Community trust: a social indicator related to community engagement. Soc. Indic. Res. **145**(2), 551–579 (2019). https://doi.org/10.1007/s11205-019-02114-y

Dirsehan, T., van Zoonen, L.: Smart city technologies from the perspective of technology acceptance. IET Smart Cities **4**(3), 197–210 (2022)

Duan, Y., Edwards, J.S., Dwivedi, Y.K.: Artificial intelligence for decision making in the era of Big Data–evolution, challenges and research agenda. Int. J. Inf. Manag. **48**, 63–71 (2019)

Fabrègue, B.F., Bogoni, A.: Privacy and security concerns in the smart city. Smart Cities **6**(1), 586–613 (2023)

Fang, W., Cui, N., Chen, W., Zhang, W., Chen, Y.: A trust-based security system for data collection in smart city. IEEE Trans. Ind. Inf. **17**(6), 4131–4140 (2020)

Ferrario, A., Loi, M., Viganò, E.: In AI we trust incrementally: a multi-layer model of trust to analyze human-artificial intelligence interactions. Phil. Technol. **33**, 523–539 (2019)

Gillath, O., Ai, T., Branicky, M., Keshmiri, S., Davison, R., Spaulding, R.: Attachment and trust in artificial intelligence. Comput. Hum. Behav. **115**, 106607 (2021)

Gilman, E., et al.: Addressing data challenges to drive the transformation of smart cities. ACM Trans. Intell. Syst. Technol. **15**(5), 88 (2024)

Helbing, D., et al.: Ethics of smart cities: towards value-sensitive design and co-evolving city life. Sustainability **13**(20), 11162 (2012)

Huerta, J., Salazar, P.: Audit process framework for data protection and privacy compliance using artificial intelligence and cognitive services in smart cities. In: Paper Presented at the 2018 IEEE International Smart Cities Conference, ISC2 2018 (2019)

Julsrud, T.E., Krogstad, J.R.: Is there enough trust for the smart city? Exploring acceptance for use of mobile phone data in Oslo and Tallinn. Technol. Forecast. Soc. Chang. **161**, 120314 (2020)

Justwan, F., Bakker, R., Berejikian, J.: Measuring social trust and trusting the measure. Soc. Sci. J. **55**, 149–159 (2018)

Kaur, D., Uslu, S., Rittichier, K., Durresi, A.: Trustworthy artificial intelligence: a review. ACM Comput. Surv. (CSUR) **55**, 1–38 (2022)

Kitchin, R.: The ethics of smart cities and urban science. Phil. Trans. R. Soc. A **374**, 20160115 (2016)

König, P.: Citizen-centered data governance in the smart city: from ethics to accountability. Sustain. Cities Soc. **75**, 103308 (2021)

Krämer, M., Frese, S., Kuijper, A.: Implementing secure applications in smart city clouds using microservices. Futur. Gener. Comput. Syst. **99**, 308–320 (2019)

Liu, J., Milojev, P., De Zúñiga, H., Zhang, R.: The global trust inventory as a "Proxy Measure" for social capital: measurement and impact in 11 democratic societies. J. Cross Cult. Psychol. **49**, 789–810 (2018)

Mark, R., Anya, G.: Ethics of using smart city AI and big data: the case of four large European cities. ORBIT J. **2**(2), 1–36 (2019)

Mora, O.B., Rivera, R., Larios, V.M., Beltran-Ramirez, J.R., Maciel, R., Ochoa, A.: A use case in cybersecurity based in blockchain to deal with the security and privacy of citizens and smart cities cyberinfrastructures. In: Paper Presented at the 2018 IEEE International Smart Cities Conference, ISC2 2018 (2019)

Omrani, N., Rivieccio, G., Fiore, U., Schiavone, F., Agreda, S.: To trust or not to trust? An assessment of trust in AI-based systems: concerns, ethics and contexts. Technol. Forecast. Soc. Chang. **181**, 121763 (2022)

Phillips, C., Jiao, J.: artificial intelligence & smart city ethics: a systematic review. In: IEEE International Symposium on Ethics in Engineering, Science, and Technology (ETHICS), West Lafayette, IN, USA, pp. 1–5 (2023)

Riedl, R.: Is trust in artificial intelligence systems related to user personality? Review of empirical evidence and future research directions. Electron. Mark, **32**, 2021–2051 (2022)

Singh, M., Murukannaiah, P., Singh, M., Murukannaiah, P.: Toward an ethical framework for smart cities and the internet of things. IEEE Internet Comput. **27**, 51–56 (2023)

Son, T.H., Weedon, Z., Yigitcanlar, T., Sanchez, T., Corchado, J.M., Mehmood, R.: Algorithmic urban planning for smart and sustainable development: systematic review of the literature. Sustain. Cities Soc. **94**, 104562 (2023)

Thiebes, S., Lins, S., Sunyaev, A.: Trustworthy artificial intelligence. Electron. Mark. **31**, 447–464 (2020)

Ullah, Z., Al-Turjman, F., Mostarda, L., Gagliardi, R.: Applications of Artificial Intelligence and Machine learning in smart cities. Comput. Commun. **154**, 313–323 (2020)

Van Heek, J., Arning, K., & Ziefle, M.: How fear of crime affects needs for privacy & safety: acceptance of surveillance technologies in smart cities. In: Paper Presented at the SMARTGREENS 2016 -Proceedings of the 5th International Conference on Smart Cities and Green ICT Systems, pp. 32–43 (2016)

Van Zoonen, L.: Privacy concerns in smart cities. Gov. Inf. Q. **33**(3), 472–480 (2016)

Zang, L., Yu, Y., Xue, L., Li, Y., Ding, Y., Tao, X.: Improved dynamic remote data auditing protocol for smart city security. Pers. Ubiquit. Comput. **21**(5), 911–921 (2017)

Zerilli, J., Bhatt, U., Weller, A.: How transparency modulates trust in artificial intelligence. Patterns **3**(4), 100455 (2022)

Zhang, K., Ni, J., Yang, K., Liang, X., Ren, J., Shen, X.: Security and privacy in smart city applications: challenges and solutions. IEEE Commun. Mag. **55**, 122–129 (2017)

Ziosi, M., Hewitt, B., Juneja, P., Taddeo, M., Floridi, L.: Smart cities: mapping their ethical implications. SSRN Electron. J. **1**(5), 10 (2022)

Barriers to the Implementation of Artificial Intelligence Applications in Smart Cities

Magdalena Rzemieniak[1](✉), Lilla Knop[2], and Narendra Kumar[3]

[1] Lublin University of Technology, Lublin, Poland
magda.rzemieniak@gmail.com
[2] Silesian University of Technology, Gliwice, Poland
[3] NIMS University Jaipur, Jaipur, India

Abstract. Artificial intelligence (AI) plays an increasingly significant role in the development of smart cities, offering opportunities to enhance the efficiency of urban infrastructure management, transportation, public safety, and residents' quality of life. Despite its promising potential, the implementation of AI applications faces numerous barriers. The aim of this article is to identify the limitations associated with the deployment of AI technologies within the smart city framework. The article poses a key research question: what are the limitations in implementing AI technologies in smart cities? The study was conducted using a quantitative method (CAWI) among 1,500 residents of voivodeship capital cities in Poland. The analysis incorporated secondary data obtained through desk research (a review of scientific literature) as well as primary data collected via a diagnostic survey. To process the primary data, factor analysis was applied to identify and categorize key factors influencing the barriers to implementing AI applications in smart cities. The results indicate that barriers to AI implementation in Polish smart cities can be classified into four main categories: technological and infrastructural, social and perceptual, legal and regulatory, and organizational and managerial.

Keywords: AI · smart city · barriers to implementing AI applications

1 Introduction

The development of the concept of smart cities is one of the key directions of contemporary socio-economic transformation, based on the integration of digital technologies with urban infrastructure. The aim of this process is not only to improve the quality of life of residents, but also to increase the efficiency of resource use, optimize city management processes, and strengthen its competitiveness on the national and international arena. Among the technologies that play a special role in the implementation of the smart city concept, artificial intelligence (AI) occupies a central position. Thanks to its capabilities in process automation, event prediction, big data analysis, and decision support, AI is becoming the foundation of the next generation of cities – Smart City 4.0.

Despite growing expectations for the application of artificial intelligence, the practical implementation of such solutions faces numerous barriers. The literature on the

Published by Springer Nature Switzerland AG 2026
M. Pondel et al. (Eds.): AI4KAM 2025, IFIP AICT 776, pp. 124–138, 2026.
https://doi.org/10.1007/978-3-032-18920-2_12

subject emphasizes that these obstacles are not limited to technological aspects, but also include social, regulatory, and organizational dimensions. These difficulties affect both the pace of integration of innovative systems into the urban fabric and the level of public acceptance of new technologies. As a result, although cities are increasingly undertaking initiatives in the spirit of digital transformation, their effectiveness remains varied and often fragmented.

Definitions of smart cities vary depending on the geopolitical and institutional context (OECD, 2020). Regardless of the perspective adopted, however, the common denominator is the desire to use digital innovation to improve the quality of public services and ensure sustainable development. In Polish cities, especially in provincial capitals, the idea of a smart city is becoming increasingly widespread, although its practical implementation is still in its early stages. This phenomenon points to the need to identify and systematize the barriers that hinder the implementation of artificial intelligence in the urban environment.

The importance of this issue is also highlighted by research on the evolution of the smart city concept – from the early stages, which focused on the introduction of basic digital technologies (Smart City 1.0), to the current stage of Smart City 4.0, in which AI, big data, and the personalization of urban services play a key role (Kumar et al., 2021; Allam & Dhunny, 2019). In this context, artificial intelligence not only supports the development of infrastructure and a knowledge-based economy, but also becomes a tool for supporting the implementation of global goals such as Agenda 2030 and the UN Sustainable Development Goals.

This article focuses on analyzing barriers to the implementation of artificial intelligence applications in Polish cities. The main research question is to identify the limitations that hinder the effective use of AI technology in urban practice. The issue addressed is both cognitive and applicational in nature—on the one hand, it allows for a better understanding of the mechanisms influencing the pace of digital transformation in cities, and on the other, it provides insights for public decision-makers and management practitioners who are responsible for implementing modern technological solutions.

2 Research Background and Questions

2.1 Barriers to the Implementation of AI Technologies

The dynamic development of artificial intelligence technology and the growing availability of implementation models bring both opportunities and significant limitations for the functioning of cities. The literature emphasizes that contemporary metropolises face not only technological challenges, but also legal, organizational, and social problems that directly affect the effectiveness and security of AI-based systems. These solutions are implemented within six recognized areas that make up the smart city concept: smart management, smart economy, smart mobility, smart environment, smart life, and smart people (Belaïd et al., 2023).

However, the integration of artificial intelligence into urban systems is hampered by numerous barriers. The most commonly cited include: inadequate technical infrastructure, limited financial resources, cybersecurity and privacy risks, lack of public trust, as well as regulatory and organizational issues (Chiordi et al., 2022; Fabregue, 2024;

Dong & Liu, 2023; Jonek-Kowalska & Wolniak, 2023). These barriers significantly slow down the process of digital transformation in cities, limiting the potential benefits that could be achieved by residents through the use of artificial intelligence.

The literature emphasizes that these limitations can be divided into three main categories:

1. *Organizational and financial barriers* - from an institutional perspective, organizational inertia and resistance to implementing innovation are particularly significant. City administrations, burdened by traditional procedures, often lack sufficient flexibility to effectively adapt to dynamic technological changes (Rjab et al., 2023). The high costs of implementing and maintaining AI-based systems also remain a significant problem. Combined with limited local budgets, especially in smaller urban centers, this leads to limited investment capacity (Wang et al., 2021). Furthermore, implementing innovative solutions requires human resources with appropriate competencies, the lack of which is one of the key barriers to development (Das, 2025).
2. *Technological and infrastructural barriers* – on the technical level, a significant limitation is the lack of adequate digital infrastructure, including reliable internet connections, data centers, and the computing resources necessary to process large urban data sets (Alwar et al., 2024; Wang et al., 2021). At the same time, concerns about cybersecurity and privacy protection are growing, necessitating the development of advanced security mechanisms (Fabrègue & Bogoni, 2023). Another significant issue is the integration of heterogeneous systems and technologies – from IoT devices to legacy infrastructure – which creates significant technical difficulties in the process of city management (Alwar et al., 2024).
3. *Social and Regulatory Barriers* – the social aspect encompasses both regulatory and ethical issues. Ambiguous or insufficient legal regulations regarding the use of artificial intelligence, data management, and ethical standards generate uncertainty and constitute an obstacle to the widespread implementation of new solutions (Fabregue, 2024; Deng & Yu, 2021). A lack of legal transparency, in turn, fosters increased social skepticism, leading to limited acceptance and low levels of citizen engagement in smart city initiatives. Excessive focus on the technological dimension, while neglecting the social, cultural, and environmental consequences, may undermine the achievement of strategic goals related to the development of smart cities (Allam & Dhunny, 2019).

An analysis of barriers to the implementation of artificial intelligence in smart cities shows that this problem is multidimensional and concerns technological, financial, organizational, regulatory, and social aspects. On the one hand, AI combined with the Internet of Things paves the way for effective management of urban resources, reduction of energy consumption, and improvement of the quality of public services (Alwar et al., 2024). On the other hand, however, the lack of adequate digital infrastructure, high implementation costs, and limited availability of expertise hinder the full exploitation of technological potential (Wang et al., 2021; Das, 2025).

Social and regulatory issues also remain an important factor. Insufficiently precise regulations on data protection, security, and the ethics of AI applications contribute to legal uncertainty, which limits both public trust and the willingness of administrations to invest in innovation (Fabregue, 2024; Fabrègue & Bogoni, 2023). Research shows that without adequate dialogue with residents and without taking socio-cultural factors into account, implementations can lead to the alienation of citizens rather than strengthening their participation in city management (Becker et al., 2023; Deng & Yu, 2021).

At the same time, it is increasingly pointed out that an excessive focus on technology at the expense of reflection on its social and environmental impact can lead to the phenomenon of "technological determinism," which hinders the achievement of sustainable development goals (Allam & Dhunny, 2019). In this context, a holistic approach is needed, combining digital solutions with social and environmental dimensions, in line with the concept of inclusive and citizen-oriented urban development (OECD, 2020; Deguchi, 2020).

In summary, the limitations of AI implementation in smart cities cannot be analyzed solely from a technical perspective. They are the result of complex interactions between institutional, economic, technological, and social conditions, which must be taken into account simultaneously in the process of designing and implementing solutions. As research indicates, the future of smart cities will increasingly depend on the ability to create inclusive, ethically grounded, and sustainable models for the use of artificial intelligence (Belaïd et al., 2023; Rjab et al., 2023; Jonek-Kowalska & Wolniak, 2023).

This article addresses a key research question: What are the limitations in implementing AI technologies in smart cities?

2.2 Smart City 4.0

The smart city concept has undergone significant evolution – from early, technology-oriented solutions to current models that integrate advanced technologies, social participation, and sustainable development. Each generation of smart cities has introduced new priorities and methods of city management – see Table 1.

Key trends and challenges related to the evolution of smart cities include:

- Moving from solutions imposed by providers to models co-created with residents and other stakeholders (Zwick & Spicer 2023; Trencher 2019).
- Growing emphasis on sustainability, inclusiveness and adaptability of urban policies (Park & Yoo, 2022; Safiullin et al., 2019).
- Implementing advanced technologies (IoT, AI, cloud, cyber-physical systems) as tools rather than ends in themselves (Allam & Dhunny, 2019; Yun & Lee, 2019; Safiullin et al., 2019).
- Coexistence of different models in practice – cities often combine elements of several generations (Zwick & Spicer 2023; Park & Yoo, 2023).

Moreover, fourth-generation smart cities, in addition to incorporating features of previous stages, place strong emphasis on aligning with the United Nations Sustainable Development Goals (SDGs) and actively collaborate with both public and private entities to enhance efficiency and explore innovative financing models. Smart City 5.0 represents a sustainable and proactive urban model (Gamayanto & Nurhindarto, 2020; Deguchi, 2020). It introduces a new citizen-centric design paradigm for future cities, where urban environments function as dynamic marketplaces that connect service providers with citizens as consumers (Becker et al., 2023).

Table 1. Smart city generations

Generation	Key Features	Sample sources
Smart city 1.0	Dominance of technology providers, implementation of solutions by IT companies, emphasis on infrastructure and automation, low involvement of residents	Zwick & Spicer 2023; Trencher 2019; LugoSantiago, 2020
Smart city 2.0	The role of public administration, data management, improvement of city services, greater openness to the needs of residents, but still centralization of decisions	Zwick & Spicer 2023; Trencher, 2019
Smart city 3.0	Active citizen engagement, co-creation of solutions, emphasis on participation and local innovation, transition from top-down to bottom-up	LugoSantiago, 2020
Smart city 4.0	ntegration of Industry 4.0 technologies (IoT, AI, big data), platformization, self-organization, digital participation, sustainable development, co-creation of value by all stakeholders	Kumar et al., 2021; Allam & Dhunny, 2019;Yun & Lee, 2019; Safiullin et al., 2019; Postránecký & Svítek, 2017

3 Methodology

3.1 Research Sample and Data Collection

The survey was conducted in the first quarter of 2025 using the CAWI (Computer-Assisted Web Interviewing) method, which allows data to be collected in the form of online questionnaires (Table 2). The use of this technique made it possible to reach a wide and diverse group of respondents, while ensuring the standardization of the research tool and reducing errors related to the interaction between the interviewer and the respondent (Babby, 2013).

The research sample consisted of 1,500 residents of provincial capitals in Poland, which allowed for the inclusion of diverse regional contexts and took into account both large metropolitan centers and smaller provincial capitals. The criteria for selecting respondents were:

- age – minimum 18 years old,
- declared knowledge of at least one artificial intelligence application used in an urban area (e.g., transportation systems, public safety, energy),
- voluntary consent to participate in the study.

Thanks to these criteria, a research sample with an appropriate level of technological awareness was obtained, which increases the reliability and accuracy of the responses obtained (Bryman, 2016).

Table 2. Information about structure of respondents

Voivodeship	City	Population	Population share	Number of respondents
Lower Silesian Voivodeship	Wroclaw	673,743	0.092098	138
Kuyavian-Pomeranian Voivodeship	Torun	194,771	0.026624	40
Lublin Voivodeship	Lublin	329,565	0.04505	68
Lubusz Voivodeship	Zielona Gora	138,932	0.018991	28
Lodz Voivodeship	Lodz	652,015	0.089128	134
Lesser Poland Voivodeship	Krakow	806,201	0.110204	165
Masovian Voivodeship	Warsaw	1,861,599	0.254472	382
Opole Voivodeship	Opole	126,077	0.017234	26
Subcarpathian Voivodeship	Rzeszow	197,268	0.026966	40
Podlaskie Voivodeship	Bialystok	291,688	0.039872	60
Pomeranian Voivodeship	Gdansk	487,371	0.066621	100
Silesian Voivodeship	Katowice	279,190	0.038164	57
Holy Cross Voivodeship	Kielce	182,295	0.024919	37
warmińsko-mazurskie – Warmian-Masurian Voivodeship	Olsztyn	167,311	0.022871	34
Greater Poland Voivodeship	Poznan	538,439	0.073602	110
West Pomeranian Voivodeship	Szczecin	389,066	0.053184	80
	Total	7,315,531	1	1,500

Before starting the actual study, a pilot study was conducted to check the comprehensibility of the survey questions, verify the logic of the questionnaire, and estimate the time needed to complete it. The pilot study allowed for minor modifications to be made, which increased the clarity of the tool and reduced the risk of misinterpretation of the questions (Creswell, 2018).

After the data collection stage was completed, the data was verified and incomplete or inconsistent questionnaires were removed. Ultimately, 1,490 correctly completed questionnaires were qualified for analysis. This level of completeness (99.3% of the original sample) attests to the high quality of the empirical material collected and minimizes the risk of systematic errors (Fowler, 2014).

The respondents came from 16 provincial capitals, with their distribution reflecting the population proportions of individual regions (Table 1). The largest groups were residents of Warsaw (n = 382), Krakow (n = 165), and Poznań (n = 110), while the smallest samples came from Opole (n = 26) and Zielona Góra (n = 28). Taking this structure into account made it possible to achieve spatial representativeness and identify perceptual differences between residents of metropolitan areas and smaller centers (Jonek-Kowalska et al., 2023).

The use of the CAWI method in research on barriers to the implementation of artificial intelligence in cities was considered particularly appropriate for several reasons. First, it allows for effective outreach to people who actively use digital technologies, which is important when analyzing issues related to AI. Second, the standardization of the tool promotes data comparability, and the anonymous nature of the survey reduces the risk of conformist or socially desirable responses (Dillman et al., 2014).

In subsequent stages of the study, the collected empirical data were subjected to exploratory factor analysis (EFA) in order to identify the main groups of barriers perceived by the respondents. The use of PCA (Principal Component Analysis) with Varimax rotation made it possible to reduce the dimensions and isolate the factors with the highest explanatory value (Hair et al., 2019).

3.2 Measures

This study employs factor analysis, a multivariate statistical technique that reduces a large set of observed variables into a smaller set of underlying latent factors. These factors represent shared variances among the variables, capturing patterns where certain variables cluster together based on their correlations. The two main methodological approaches are:

- Exploratory Factor Analysis (EFA): An inductive approach applied when researchers have no prior assumptions about the factors or their structure. Its goal is to explore the data to identify the number and nature of the latent factors.
- Confirmatory Factor Analysis (CFA): A deductive approach used to validate a theoretical framework. Researchers specify the expected factor structure beforehand, and the analysis tests the hypothesis that the predetermined model fits the observed data.

The purpose of using Exploratory Factor Analysis (EFA) was to identify the main groups of barriers which, in the opinion of respondents, hinder the implementation of artificial intelligence applications in smart cities. This made it possible not only to identify individual problems, but also to identify their broader context and systemic links.

The research questionnaire was developed based on a review of the literature and previous empirical studies on the implementation of technology in cities. It included 18 statements reflecting potential barriers, which were divided into four areas: technological and infrastructural, social and perceptual, legal and regulatory, and organizational and managerial. Respondents made their assessments based on a five-point Likert scale (1 – strongly disagree, 5 – strongly agree), which allowed for the quantitative capture of subjective opinions and attitudes.

Principal Component Analysis (PCA) was used to reduce variables and identify hidden factors. PCA is one of the most commonly used techniques in social and management research (Hair et al., 2019). The choice of PCA was justified by the need to simplify the data structure and detect response patterns that best explain the variation in the perception of barriers.

Varimax rotation with Kaiser normalization was used, which allowed for maximum differentiation of factor loadings and facilitated their interpretation (Bryman, 2016). The criterion adopted for factor extraction was an eigenvalue greater than 1, which corresponds to the classic Kaiser rule.

Before proceeding with the actual analysis, diagnostic tests were performed to verify the adequacy of the data. The KMO (Kaiser-Meyer-Olkin) index reached a value of 0.829, which indicates very good sample quality and high correlation between variables, justifying the use of factor analysis. In addition, Bartlett's sphericity test showed statistical significance ($p < 0.001$), confirming that the correlation matrix differs significantly from the unit matrix and is suitable for further exploration.

Based on data obtained from 1,490 respondents representing 16 provincial capitals in Poland, four main factors were identified, which together explained 68.3% of the total variance. The high level of explained variance and the stability of the obtained factor structure indicate the reliability of the adopted model and its consistency with previous theoretical findings regarding barriers to AI implementation in smart cities (Wang et al., 2021).

In the following sections of the study, these factors were subjected to detailed interpretation, and their content-based assignment to specific areas (technological and infrastructural, social and perceptual, legal and regulatory, and organizational and managerial) allowed for the construction of a coherent model of barriers in the process of AI implementation in Polish cities.

3.3 Results

After rotation, four factors were identified, which together explained 68.3% of the total variance. The high level of explained variance indicates that the obtained factor structure reliably reflects the respondents' perception of barriers. Below is a detailed description of each factor along with its key variables (factor loadings > 0.60). Factor 1: Technological and infrastructural barriers (23.5% of variance) (Fig. 1).

- Lack of interoperability of urban systems (loading: 0.81)
- Insufficient quality of urban data (0.77)
- Gaps in digital and computational infrastructure (0.74)
- Low technological maturity of cities (0.69)

Factor 2: Social and perceptual barriers (17.8% of variance)

- Lack of social trust in AI (0.82)
- Concerns about data privacy (0.79)
- Insufficient understanding of technology by residents (0.72)
- Fears of automation and job loss (0.66)

Factor 3: Legal-Regulatory Barriers (14.2%)

- Lack of clear regulations regarding AI (0.83)
- Uncertainty regarding legal liability (0.76)
- Constraints resulting from GDPR and data protection regulations (0.69)

Factor 4: Organizational and Managerial Barriers (12.8%)

- Insufficient competencies within public administration (0.78)
- Lack of coordination between departments within the office (0.74)
- Short-term planning of technological investments (0.70)

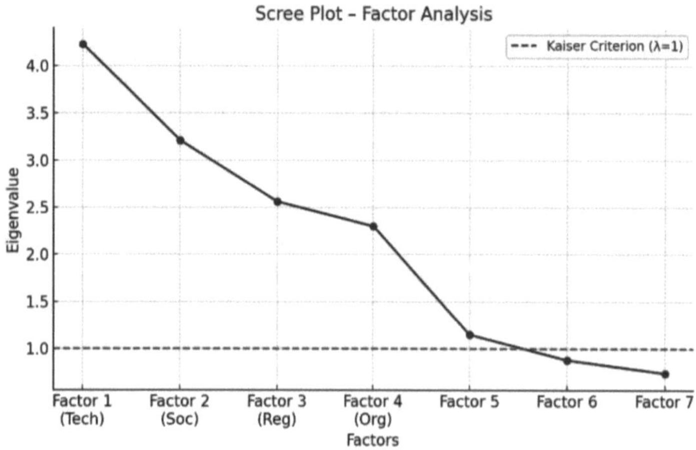

Fig. 1. Scree plot – factor analysis

The factor related to technological and infrastructural barriers had the greatest impact on explaining data variability. The highest loads were obtained by variables such as: lack of interoperability of urban systems (0.81), low data quality (0.77), gaps in digital and computing infrastructure (0.74), and low level of technological maturity of cities (0.69). This result indicates that residents perceive limitations resulting from the incompatibility of digital infrastructure with advanced AI solutions. This problem is consistent with previous studies, which emphasize that the lack of system integration and immature data architecture are among the main barriers to AI implementation in cities (Wang et al., 2021; Alwar et al., 2024).

The second most important factor was social and perceptual barriers. The most strongly correlated variables were: lack of public trust in artificial intelligence (0.82), concerns about data privacy (0.79), insufficient understanding of the technology by residents (0.72), and fear of automation and job losses (0.66). These results indicate that social acceptance is a prerequisite for the effective implementation of AI, and its absence can lead to resistance to new solutions. Similar observations appear in studies by Fabregue (2024) and Deng & Yu (2021), which emphasize that uncertainty and lack of knowledge about technology reinforce public skepticism.

The third factor identified was legal and regulatory barriers. The highest scores were given to variables relating to the lack of clear regulations on the use of AI (0.83), uncertainty regarding legal liability for decisions made by algorithms (0.76), and restrictions resulting from personal data protection regulations, in particular the GDPR (0.69). These results emphasize that legal regulations are not keeping pace with the dynamic development of technology, which creates interpretation problems and discourages local governments from taking investment risks. Similar conclusions are drawn by Deguchi (2020) and Fabrègue & Bogoni (2023), who point to the need to develop a coherent regulatory framework at the national and international levels.

The final factor was organizational and managerial barriers. Key variables included: lack of relevant competencies in public administration (0.78), insufficient coordination between departments (0.74), and short-term planning of technological investments (0.70). Although this factor explained the smallest part of the variance, its practical significance is important. Insufficient organizational preparation and the lack of a stable digital development strategy can effectively prevent the implementation of even the most advanced technologies. Rjab et al. (2023) emphasize that the digital transformation of cities requires a new model of leadership, interdepartmental cooperation, and the development of an organizational culture conducive to innovation (Table 3).

Table 3. Factor loading table

Variable	F1 Tech	F2 Soc	F3 Reg	F4 Org
Lack of interoperability of systems	0.81			
Privacy concerns		0.79		
Lack of clear legislation			0.83	
Insufficient competence of the administration				0.78
Concerns about automation and job loss		0.66		
Fragmentation of urban governance				0.74

Factor analysis revealed that the perceived barriers to the implementation of AI in smart cities can be grouped into four distinct areas. This indicates that an effective AI implementation policy in urban environments must consider diverse aspects – ranging from IT infrastructure modernization to social education and the adaptation of legal frameworks.

The factor analysis carried out allowed for the identification of four main areas of barriers: technological and infrastructural, social and perceptual, legal and regulatory, and organizational and managerial. These results clearly indicate that the policy of implementing AI in cities cannot be limited solely to the modernization of technical infrastructure. Equally important are activities related to education and social communication, clarification of the legal framework, and strengthening of organizational competencies in public administration. Only a systemic approach that takes all four dimensions into account can ensure the effective and sustainable implementation of AI solutions in the urban environment.

4 Discussion

The analyses confirm that the implementation of artificial intelligence in smart cities faces multidimensional constraints, which can be considered in terms of organizational, technological, and social barriers. The results are consistent with earlier findings by researchers working on smart cities, who point out that the process of digital transformation of urban space is not solely a technological issue, but rather the result of complex interactions between economic, legal, institutional, and cultural conditions (Allam & Dhunny, 2019; Fabregue, 2024).

Firstly, in the area of organizational and financial barriers, there is a strong correlation between the level of economic development of a given city and its ability to adapt innovative solutions. Research confirms that local government units with limited budgetary resources are less likely to decide to implement costly technologies, even if the potential long-term benefits outweigh the costs (Wang et al., 2021). In this context, the lack of sufficient digital skills in local government structures remains a significant problem. As Das (2025) notes, the shortage of specialists capable of effectively designing, implementing, and monitoring AI systems leads to local governments' dependence on external providers, which in turn may limit their independence and ability to plan for the long term.

Secondly, research results indicate the importance of technological and infrastructural barriers. In particular, the lack of adequate quality telecommunications infrastructure—especially stable broadband Internet, modern data centers, and compatible integration platforms—is a serious limitation in the implementation of artificial intelligence-based systems (Alwar et al., 2024). At the same time, it should be emphasized that technological developments in the area of smart cities generate new threats to personal data protection and cybersecurity. Fabrègue and Bogoni (2023) point out that the mass collection and processing of data on residents' lives can lead to privacy violations and, in extreme cases, to the erosion of public trust in public institutions.

Importantly, the results of the analyses confirm that the problem of integrating diverse urban systems remains one of the key challenges. The literature emphasizes that the heterogeneity of IoT devices, the dispersion of data platforms, and the presence of legacy systems hinder the creation of coherent digital ecosystems (Alwar et al., 2024). The lack of interoperability standards results in high adaptation and technical costs, which limits the scalability of solutions in urban spaces.

Thirdly, the identified social and regulatory barriers clearly indicate the need for a stronger link between technological processes and the social context. Fabregue (2024) emphasizes that the lack of clear regulations on data management and the use of artificial intelligence algorithms leads to legal uncertainty, which in turn can result in low levels of social acceptance. The results of research by Deng and Yu (2021) show that the level of public trust in municipal institutions is a factor determining residents' openness to data sharing and the use of digital services. In this context, it can be observed that technological implementations that ignore the aspect of social participation encounter particularly strong resistance from citizens.

It is worth noting that the results obtained fit into broader theoretical concepts, such as the idea of Society 5.0, which emphasizes the need to create human-centered technological solutions (Deguchi, 2020). Similar conclusions are presented by Becker et al. (2023), pointing out that the future of smart cities depends on the extent to which residents are involved in the process of designing and evaluating technologies. This means that the successful implementation of AI in an urban environment depends not only on technical aspects, but also on the ability of public institutions to build dialogue and cooperation with citizens.

The discussion of the results also allows us to identify important directions for further research. First, it seems necessary to conduct in-depth comparative analyses covering different regional contexts. As Belaïd et al. (2023) point out, there are significant differences in socio-economic conditions between regions, which translates into different models of smart city technology implementation. Secondly, research should take greater account of the perspective of residents, their expectations, level of trust, and willingness to cooperate with public institutions. Social research methodologies (Babbie, 2013; Bryman, 2016; Creswell, 2018) offer a rich set of tools that can be used to better understand the social dynamics accompanying the implementation of new technologies.

Furthermore, the limitations observed indicate the need to develop interdisciplinary research combining technical, economic, and social perspectives. As noted by the OECD (2020), building smart cities requires not only the implementation of innovative technologies, but also the creation of inclusive development models that allow for a balance between economic efficiency and the well-being of residents.

In summary, the results of the study show that the limitations in the implementation of artificial intelligence in smart cities are systemic and complex. Overcoming them requires comprehensive measures, including infrastructure development, securing stable financing, developing a transparent regulatory framework, and increasing public participation. An integrated approach can contribute to a fuller exploitation of the potential of AI in building more sustainable, efficient, and resident-friendly cities.

5 Conclusions

The factor analyses carried out clearly indicate that the barriers to the implementation of artificial intelligence applications in smart cities are multidimensional and interrelated. This means that they cannot be considered in isolation, but require a systemic and integrated approach that will allow the existing limitations to be effectively overcome. The results confirm that although technological obstacles—such as lack of interoperability or insufficient digital infrastructure—are the most visible, social, regulatory, and organizational factors are equally important (Alwar et al., 2024; Fabregue, 2024). Neglecting these dimensions can significantly reduce the effectiveness of even the most advanced technical solutions, as confirmed by previous studies on the implementation of AI in an urban context (Rjab et al., 2023; Jonek-Kowalska & Wolniak, 2023).

The results of the study emphasize the need to simultaneously develop digital infrastructure, build public trust, create a clear legal framework, and strengthen institutional competencies. It is precisely this combination of technological, social, legal, and organizational aspects that corresponds to contemporary trends in the development of the smart city concept, which emphasizes the need for inclusiveness and a focus on the well-being of residents (Allam & Dhunny, 2019; De-guchi, 2020).

Based on the analyses conducted, several key practical conclusions were formulated. First, the implementation of artificial intelligence in cities should be based on holistic development strategies that integrate technology with social and environmental perspectives. Second, public-private partnerships and cross-sector cooperation mechanisms play an important role in supporting both financing processes and the transfer of knowledge and skills (Be-laïd et al., 2023; OECD, 2020). Thirdly, it is crucial to develop strategies for interoperability and standardization of systems that will enable the efficient integration of diverse platforms and devices within urban digital ecosystems (Alwar et al., 2024).

The conclusions also include recommendations concerning public policy and management practices. It is essential to conduct educational programs for residents that will increase the level of awareness and social acceptance of AI-based solutions (Becker et al., 2023). Another priority is the harmonization of legal regulations, particularly in the areas of personal data protection, the ethics of algorithm use, and the transparency of decision-making processes. Finally, it is necessary to strengthen the digital competencies of local administrations in order to reduce dependence on external providers and enable independent design and evaluation of AI-based systems (Das, 2025).

In summary, the effective implementation of artificial intelligence in smart cities requires a long-term perspective that combines technological development with social inclusiveness, legal stability, and institutional accountability. Only such an approach will allow the full potential of AI to be exploited to build sustainable, innovative, and resident-friendly cities.

A key limitation of the presented research is its geographical restriction to Polish cities. To verify and enrich the results obtained, it would therefore be advisable to expand the research perspective to include large urban centers in other countries. Furthermore, incorporating case studies of smaller cities that are actively developing smart city solutions through various support programs in future analyses could provide valuable supplementary data. Additionally, the article uses the terms "barriers" and "limitations" interchangeably, which warrants clarification. It is important to distinguish between these terms: a limitation most often refers to an imposed limit or restriction that defines the framework of action (e.g., budgetary or time constraints) and does not necessarily represent an obstacle to be overcome, but rather a context. In contrast, a barrier is a specific obstacle or challenge that actively blocks progress and must be overcome (e.g., legal or technological barriers). This distinction is crucial for accurately diagnosing problems and developing appropriate strategies to overcome them.

References

Allam, Z., Dhunny, Z.A.: On big data, artificial intelligence and smart cities. Cities **89**, 80–91 (2019)

Alwar, B., Edgardo, H., Faustino, E.: Integrating IoT and artificial intelligence for sustainable smart city development: a case study approach. Inf. Technol. Eng. J. **9**(2), 80–90 (2024)

Babbie, E.: Podstawy badań społecznych. PWN, Warszawa (2013)

Becker, J., et al.: City 5.0: citizen involvement in the design of future cities. Electron Markets **33**, 10 (2023)

Bryman, A.: Social Research Methods. Oxford University Press, Oxford (2016)

Belaïd, F., Amine, R., Massie, C.: Smart cities initiatives and perspectives in the MENA region and Saudi Arabia. In: Smart Cities: Social and Environmental Challenges and Opportunities for Local Authorities, pp. 295–313. Springer, Cham (2023)

Chiordi, S., Desogus, G., Garau, C., Nesi, P., Zamperlin, P.: A preliminary survey on smart specialization platforms: evaluation of European best practices. In: Proceedings of the International Conference on Computational Science and Its Applications, Malaga, Spain, 4–7 July 2022, pp. 67–84. Springer, Heidelberg (2022)

Creswell, J.W.: Research design: Qualitative, quantitative, and mixed methods approaches. SAGE, Los Angeles (2018)

Das, D.: Digital technology and AI for smart sustainable cities in the global south: a critical review of literature and case studies. Urban Sci. **9**(3), 72 (2025)

Deguchi, A.: From smart city to society 5.0. In: Society 5.0: A People-Centric Super-Smart Society, pp. 43–65. Springer, Singapore (2020)

Deng, X., Yu, M.: Scale of cities and social trust: evidence from China. Int. Rev. Econ. Financ. **76**, 215–228 (2021)

Dillman, D.A., Smyth, J.D., Christian, L.M.: Internet, phone, mail, and mixed-mode surveys: The tailored design method. Wiley, Hoboken (2014)

Dong, L., Liu, Y.: Frontiers of policy and governance research in a smart city and artificial intelligence: an advanced review based on natural language processing. Front. Sustain. Cities **5**, 1199041 (2023)

Fabregue, B.: Artificial intelligence governance in smart cities: a European regulatory perspective. J. Auton. Intell. **7**(2) (2024)

Fabrègue, B., Bogoni, A.: Privacy and security concerns in the smart city. Smart Cities **6**(1), 586–613 (2023)

Fowler, F.J.: Survey research methods. SAGE, Los Angeles (2014)

Gamayanto, I., Nurhindarto, A.: Developing smart city 5.0 framework to produce competency. J. Appl. Intell. Syst. (2020)

Hair, J.F., Black, W.C., Babin, B.J., Anderson, R.E.: Multivariate data analysis. Harlow: Pearson (2019)

Jonek-Kowalska, I., Wolniak, R.: Smart cities in Poland: Towards sustainability and a better quality of life? Abingdon: Taylor & Francis (2023)

Kumar, S.S., Sudhir Bale, A., Matapati, P.M., Vinay, N.: Conceptual study of artificial intelligence in smart cities with industry 4.0. In: 2021 International Conference on Advance Computing and Innovative Technologies in Engineering (ICACITE), pp. 575–577 (2021)

LugoSantiago, J.A.: Is There Such a Thing as the Smart City 1.0, 2.0, or 3.0? Leadership and Strategic Foresight in Smart Cities, pp. 33–42. Palgrave Macmillan, Cham (2020)

OECD. Smart Cities and Inclusive Growth: Building on the outcomes of the 1st OECD Roundtable fon Smart Cities and Inclusive Growth. ECD Regional Development Papers, No. 91, OECD Publishing, Paris (2020)

Park, J., Yoo, S.: Evolution of the smart city: three extensions to governance, sustainability, and decent urbanisation from an ICT-based urban solution. Int. J. Urban Sci. **27**, 10–28 (2022)

Postránecký, M., Svítek, M.: Conceptual Model of Complex Multi-agent System Smart City 4.0, pp. 215–226 (2017)

Rjab, A., Mellouli, S., Corbett, J.: Barriers to artificial intelligence adoption in smart cities: a systematic literature review and research agenda. Gov. Inf. Q. **40**, 101814 (2023)

Safiullin, A., Krasnyuk, L., Kapelyuk, Z.: Integration of Industry 4.0 technologies for "smart cities" development. IOP Conf. Ser. Mater. Sci. Eng. **497**, 012089 (2019)

Trencher, G.: Towards the smart city 2.0: empirical evidence of using smartness as a tool for tackling social challenges. Technol. Forecast. Soc. Chang. **142**, 117–128 (2019)

Wang, K., Zhao, Y., Gangadhari, R., Li, Z.: Analyzing the Adoption Challenges of the Internet of Things (IoT) and Artificial Intelligence (AI) for Smart Cities in China. arXiv, abs/2205.01067 (2021)

Yun, Y., Lee, M.: Smart city 4.0 from the perspective of open innovation. J. Open Innov. Technol. Market Complex. **5**(4), 92 (2019)

Zwick, A., Spicer, Z.: Examining the smart city generational model: conceptualizations, implementations, and infrastructure Canada's smart city challenge. Urban Affairs Rev. **60**, 1229–1253 (2023)

Selected Discrimination Problems in the Application of Artificial Intelligence in Public Services

Kinga Flaga-Gieruszyńska(✉)

Faculty of Law and Administration, University of Szczecin, Narutowicza 17a, 70-240, Szczecin, Poland
kinga.flaga-gieruszynska@usz.edu.pl

Abstract. Artificial intelligence (AI) is increasingly being applied in public services such as social care, education, healthcare, and the justice system. While its implementation can enhance the efficiency and accessibility of these services, it also poses significant risks related to the discrimination of certain social groups. This issue often stems from biases embedded in training data, errors in algorithm design, and a lack of transparency in AI-driven decision-making. Examples from countries using such solutions show that the improper application of algorithms can lead to the systemic exclusion of minorities, individuals of lower economic status, or those from specific ethnic backgrounds. In response to these challenges, the European Union is developing regulations (the AI Act) aimed at ensuring that the use of AI complies with human rights and introduces obligations concerning transparency, human oversight, and risk analysis. To avoid discrimination, it is essential to design and implement technology responsibly, based on ethical principles, public consultation, and effective institutional oversight. Only such an approach can guarantee fair and equal treatment of citizens in the era of public service automation.

Keywords: Artificial Intelligence · Discrimination · Public E-services

1 Introduction

Modern public administration and the justice system are increasingly turning to artificial intelligence (AI) in the management of public services. Algorithms are used to analyze benefit applications, support officials in decision-making, predict public health needs, and even assist in managing education systems and the judiciary. These solutions are intended to enhance the efficiency, speed, and consistency of administrative decisions and other rulings. AI is viewed as a tool that can help reduce the operating costs of public institutions while improving the quality of citizen services.

However, as the use of algorithms in the public sector expands, so do concerns and controversies surrounding their impact on civil rights and social equality. Growing evidence shows that AI systems – even when designed with the intention of neutrality – can result in unintended forms of discrimination. This typically stems from errors in

Published by Springer Nature Switzerland AG 2026
M. Pondel et al. (Eds.): AI4KAM 2025, IFIP AICT 776, pp. 139–150, 2026.
https://doi.org/10.1007/978-3-032-18920-2_13

training data, which may reflect historical biases, the underrepresentation of certain social groups in datasets, and algorithmic designs that lack sufficient transparency and oversight. Particularly troubling are cases in which AI-driven decisions affect citizens' access to essential services such as social assistance, education, healthcare, or justice. In such instances, even seemingly minor algorithmic errors can lead to real, unjust consequences, deepening existing social inequalities and undermining trust in public institutions.

In light of these issues, this paper focuses on analyzing the problem of discrimination arising from the use of artificial intelligence in public services. Its aim is not only to highlight examples of such phenomena and their root causes but also to understand the mechanisms behind them. Additionally, the text seeks to offer recommendations for the responsible, fair, and ethical implementation of AI technologies in public institutions. Only through critical reflection and effective risk management can we harness the potential of these tools without exposing citizens to unfair treatment.

2 Areas of Application of Artificial Intelligence in Public Services

Artificial intelligence is becoming an integral part of public governance. Thanks to its ability to process vast amounts of data and learn from patterns, AI systems support decision-making processes in both central and local administration. Their use spans a wide range of administrative functions – from document analysis and personnel decision support to the automation of citizen interactions. These solutions aim to improve the accessibility, efficiency, and quality of services, yet each implementation carries risks related to potential discrimination, loss of decision-making transparency, and diminished citizen control over personal data.

One of the earliest and most frequently adopted areas of AI implementation is social welfare, particularly systems that support the management of social benefits. Algorithms analyze data on income, place of residence, employment history, or household size to predict the risk of abuse, automatically grant benefits, or identify individuals for inspection. A notable example is the SyRI system (System Risico Indicatie) in the Netherlands, which was designed to identify individuals who might be abusing the social welfare system. The system analyzed data from various public databases (including housing, employment, and education) to detect "suspicious" patterns. However, the algorithm turned out to be non-transparent and disproportionately focused on neighborhoods inhabited by people of lower socioeconomic and migrant backgrounds, which led to systemic discrimination. In 2020, a Dutch court ruled that SyRI violated the fundamental standards of the European human rights protection system. This marked an important step toward safeguarding some of the most marginalized social groups. It was one of the first instances in Europe where the state use of risk assessment software for "predictive policing" was legally challenged, and it may have broader implications for how such technologies are scrutinized in public services [8, 10]. Systems like SyRI demonstrate that even with good intentions—such as combating fraud—there is a risk of mass stigmatization and unfair treatment. These types of systems often rely on low-quality data, the interpretation of which may be flawed or incomplete, especially in the case of individuals on the margins of eligibility (e.g. single mothers, migrants, or persons with disabilities).

Another important area is education and the management of the school system. In education systems, AI is applied in both administrative processes (e.g. school placements, performance analysis of institutions) and didactic functions – for instance, through personalized learning platforms that adapt content to the student's abilities. AI is also increasingly used to assess written assignments or tests. However, automated grading may favor students who follow specific linguistic patterns or styles that align with dominant cultural norms. During the COVID-19 pandemic in the United Kingdom, the government used an algorithm to calculate final grades for students who were unable to sit exams. The system relied on historical data from schools and statistics rather than students' individual achievements. As a result, pupils from less prestigious schools – often located in poorer regions – received significantly lower grades than those predicted by their teachers. This led to massive public protests and the government ultimately reversed its decision [5]. The case of the British student grading algorithm illustrates that such systems are not always capable of properly accounting for individual abilities, leading to unfair assessments of students from smaller towns or lower-ranking schools.

In healthcare, artificial intelligence is used in a variety of applications, including the analysis of medical imaging (e.g. X-rays, CT scans), disease risk prediction based on patient history, treatment planning, hospital resource management, and patient triage in emergency departments. These systems enable faster diagnoses and more efficient use of resources. However, they can be flawed due to unrepresentative data – particularly concerning women, the elderly, or ethnic minorities. An example of this is AI used to assist in medical decision-making by identifying which patients should receive priority care. In the United States, the company called Optum developed an algorithm intended to flag patients with the greatest healthcare needs. However, an analysis showed that the system significantly underestimated the needs of Black patients because it used healthcare costs as a proxy for "health needs," and historically, African Americans have spent less on healthcare due to limited access [9]. This case demonstrates that seemingly neutral indicators – like healthcare costs – may reflect not actual health status, but rather existing inequalities in access to medical services.

Artificial intelligence is also employed by law enforcement and the judiciary. Police departments use predictive systems to identify areas or individuals deemed more "prone" to committing crimes, while courts apply tools that support risk assessments for recidivism. In theory, these tools are meant to improve the objectivity of decisions, but in practice – as revealed by ProPublica's investigation on the COMPAS system – they can reinforce racial biases embedded in historical data [2]. Furthermore, the opacity of such algorithms complicates legal defense and may violate the right to a fair trial.

In the area of financial control and tax policy, tax administrations and oversight institutions use AI to analyze tax declarations, select cases for audits, and detect fraud and abuse schemes. These systems process hundreds of thousands of documents in real time, identifying anomalies that are difficult for humans to detect. However, algorithms may favor larger entities or focus on specific social groups. One example is Belgium, where the tax fraud detection system primarily targeted microenterprises and low-income citizens, raising concerns about the selectivity and fairness of such an approach. The lack of clear selection criteria and the inability to appeal automatic decisions create a risk of potential abuse of power [3]. This lack of transparency may raise concerns about equal treatment of taxpayers and compliance with the principle of non-discrimination.

In cities, AI supports traffic management, environmental monitoring, urban planning, and emergency response systems. For instance, algorithms can control traffic lights, forecast traffic volume, or identify real-time needs for public transport. During the COVID-19 pandemic, AI was used to monitor the spread of the virus and to optimize the locations of testing and vaccination sites. However, a risk in such applications is the misalignment of actions with the actual needs of marginalized communities – for example, by placing support points in districts inaccessible to the elderly, people with disabilities, or individuals with limited mobility [12].

3 Mechanisms and Consequences of Discrimination

One of the main sources of discrimination in AI systems lies in the training data on which algorithms are built. AI algorithms learn from the data we provide, and this data is usually the result of human activity—it may contain existing biases, discrimination, or social inequalities. If the historical data on which AI systems are based reflects incomplete or unequal representation of different social groups, the algorithm may replicate these inequalities, leading to unfair and discriminatory outcomes. A notable example of this is the predictive algorithm used by child welfare authorities in Allegheny County (Pennsylvania, USA), designed to support decisions on initiating investigations into potential child neglect. The algorithm analyzed administrative data (including information on parents' employment, their use of social support services, and legal proceedings) to assess the "risk" associated with a particular family. Research showed that the model was harsher toward poorer families and those from ethnic minorities, as they appeared more frequently in administrative datasets—not necessarily because the children were actually in greater danger. The data only included cases already registered in the system, creating a vicious cycle: the more data about a certain group, the more often it was flagged as "high risk" [7]. A similar issue may arise with the European Travel Information and Authorisation System (ETIAS), an EU digital system for granting travel authorizations to visa-exempt third-country nationals, which uses a profiling algorithm. Scholars have noted that ETIAS is part of a broader digital surveillance infrastructure the EU has been building for years to monitor third-country nationals. At the same time, it is argued that the algorithmic process behind ETIAS acts as a mechanism of differential exclusion, potentially having a negative impact on certain groups of foreign travelers [6]. In the context of public services – where the goal is to ensure social justice – such inequalities may result in the perpetuation of existing social problems. If algorithms are based on data already containing systemic inequalities, they will not be able to go beyond those patterns, potentially leading to further marginalization of groups that are already on the periphery of the system. This not only leads to unjust decisions but also deepens existing social and economic divides.

In AI systems used in public services, a phenomenon known as proxy discrimination often occurs. This refers to situations where the algorithm does not directly use legally protected characteristics such as race, gender, or ethnicity, but relies on variables that are statistically correlated with those traits. These types of variables, while formally neutral, can, in practice, act as "proxies" for those characteristics, leading to discriminatory outcomes (Table 1).

Table 1. The relationships between the main sources of risk associated with the use of artificial intelligence in public services and their potential social consequences.

Sources of Risk	Mechanisms of Impact	Social Consequences
data bias	leads to erroneous algorithmic decisions	the discrimination of citizens, loss of public trust
lack of algorithmic transparency	prevents oversight and appeals	the perception of unfairness, reduced trust in institutions
lack of human oversight and accountability	leads to automated decisions without contextual analysis	the risk of systemic errors, misuse, and lack of ethical control
insufficient data security and protection	enables data leaks or manipulation	privacy violations, reputational damage to institutions
data bias	leads to erroneous algorithmic decisions	the discrimination of citizens, loss of public trust

In practice, this means that the algorithm may, for example, consider variables such as postal code, education level, employment history, or use of social services. These variables, however, may strongly correlate with social class, ethnicity, or gender. As a result, AI systems may make decisions that favor individuals from more privileged areas or backgrounds, even if such criteria are not explicitly included in the model. A good example is the social fraud detection system in Rotterdam, which used various risk indicators. Although variables like ethnicity were not formally included, the system relied on data linked to that characteristic, such as place of residence, marital status, number of children, and employment type. In practice, this meant that low-income individuals, single mothers, and people from migrant backgrounds were more frequently flagged as potential fraudsters [4]. In 2020, the system's operation was halted following public protests and interventions from human rights organizations. This phenomenon is particularly difficult to detect because proxy variables are often regarded as "objective" and seemingly unrelated to protected characteristics. However, without a thorough analysis of the correlations between variables and their influence on the algorithm's decisions, systemic discrimination can be inadvertently introduced into the model.

Another significant mechanism leading to discrimination in the use of artificial intelligence in the public sector is the lack of transparency in algorithms, often referred to as the "black box" problem. This means that even if we know the input data and the algorithm's output, we cannot clearly understand how the algorithm processes the information and makes decisions. This issue is especially apparent in the case of more complex models, such as neural networks, which are not intuitively interpretable.

The lack of transparency in the context of public services leads to a situation where citizens cannot understand why they have been classified in a particular way, such as being labelled as a high-risk individual. This complicates the process of challenging decisions or effectively using appeal mechanisms. At the same time, public authorities may also be unable to explain their decisions, which violates the principle of administrative transparency and weakens public trust in institutions. An example of such a situation was recorded in the United Kingdom, where the Department for Work and Pensions (DWP) used algorithms to assess benefits applications. Social organizations reported that individuals whose applications were denied did not receive any justification for the decision, and the workings of the system were neither publicly available nor understandable to the users. The lack of information on which data were considered and how they influenced the decision made it difficult to seek justice. An internal analysis conducted by the DWP revealed that the AI system in use showed statistically significant differences in how people were treated based on age, disability, marital status, and nationality. Despite this, the DWP did not publish full information about how the algorithm worked, making it impossible to independently assess its fairness and effectiveness [11]. The lack of transparency in how algorithms function in the public sector not only undermines citizens' trust in institutions but also makes it difficult to identify and correct potential errors or biases in decision-making systems. To ensure fairness and accountability, it is essential to introduce mechanisms that allow citizens to understand and challenge decisions made by AI, as well as to ensure independent oversight of these systems.

In the public sector, artificial intelligence systems are increasingly used to support administrative decision-making. Although automation can increase the efficiency and speed of processes, the lack of adequate human oversight over the operation of these systems can lead to serious consequences, including discrimination and violations of citizens' rights. Another example is the so-called "benefit scandal" in the Netherlands, referring to the fact that the tax office (Belastingdienst) used an automated system to detect irregularities in the allocation of childcare benefits (kinderopvangtoeslag). The system classified applicants as potential fraudsters based on internal rules and algorithmic risk profiles, without providing real human oversight. As a result, some families, often of foreign origin, were wrongly accused of fraud and forced to repay significant sums. In many cases, this led to serious financial and personal problems, and even to the removal of children from their parents [1]. A key criticism of the system was the lack of the ability to explain or appeal the decisions, as they were made automatically and were not subject to immediate verification by an official (Table 2).

The analyzed examples indicate that the assessment of algorithmic discrimination in public services largely depends on the technical conditions underpinning the operation of artificial intelligence systems. Discrimination does not stem solely from the quality of input data but also from the structure of models, the methods of training and validation, and the degree of automation in decision-making processes. Understanding these aspects helps reveal how specific design choices can lead to unintended social consequences, such as exclusion or the reinforcement of existing inequalities.

One of the fundamental problems lies in the structure of the model, especially in systems based on deep learning. Due to their complexity, such models operate in a way that is difficult to interpret—a phenomenon commonly referred to as the black box problem. Millions of parameters and hidden decision layers make it impossible even for the system's designers to fully explain why the algorithm arrived at a particular outcome.

Table 2. Comparative overview of selected AI systems used in public administration and their documented discriminatory effects.

System	Context	Type of Model	Observed Bias	Consequences
Allegheny County Predictive Risk Model (USA)	child welfare – assessing risk of neglect and prioritizing interventions	predictive model trained on administrative data (social services, employment, legal records)	bias against low-income and minority families due to overrepresentation in datasets; creation of feedback loop	public criticism; academic review highlighting systemic bias; ongoing reform of data governance practices
European Travel Information and Authorisation System (ETIAS, EU)	automated travel authorization for visa-exempt nationals	profiling algorithm analyzing traveler data against risk indicators	potential for differential exclusion and disproportionate targeting of certain nationalities	scholarly critique of algorithmic surveillance; calls for stronger human oversight and transparency
Rotterdam Social Fraud Detection System (Netherlands)	detection of welfare and social benefit fraud	risk scoring using socio-economic indicators correlated with ethnicity and gender	proxy discrimination – low-income, migrant, and single-parent households disproportionately flagged	suspended in 2020 after protests and NGO intervention; ongoing audits and legal inquiries
Department for Work and Pensions Algorithm (United Kingdom)	automated assessment of benefit applications	AI classification model (details undisclosed)	disparate impact based on age, disability, marital status, and nationality; lack of explainability	civil society pressure for transparency; internal review confirmed disparities
Childcare Benefits System (Belastingdienst, Netherlands)	detection of irregularities in childcare benefit applications	rule-based + risk profiling system without human oversight	systemic discrimination against foreign-origin families; wrongful fraud accusations	major political scandal; public apologies and government resignation; reforms to algorithmic accountability

In practice, this means that it is often impossible to verify which input data had a decisive influence on the result and which may have contained hidden biases. This lack of transparency is particularly problematic in the public sector, where administrative decisions must be justified and open to scrutiny. To address this challenge, researchers have developed methods for increasing model interpretability, such as LIME (Local Interpretable Model-Agnostic Explanations) and SHAP (SHapley Additive exPlanations), which make it possible to partially reconstruct the logic behind algorithmic decisions and identify the most influential variables.

Another crucial factor involves the methods of classification and training used in machine learning models. These systems learn by minimizing error relative to training data, which means that the patterns embedded in the data are reproduced in the model. If historical datasets reflect existing social inequalities, the algorithm will not only replicate but may also amplify them—a process known in the literature as bias amplification. Consequently, decision-making systems based on historical data may systematically assign lower scores to certain social groups—such as residents of economically disadvantaged areas, people with lower educational attainment, or women—even when these variables are not directly relevant to the intended purpose of the model. For this reason, the design and testing of algorithms must include techniques such as data balancing, robustness testing, and sensitivity audits, which help identify cases of disproportionate or unequal treatment across groups.

An equally significant source of bias arises from insufficient validation of input data. In many administrative projects, this stage is treated superficially, and datasets are often aggregated from multiple public registries of varying quality. The absence of checks for sample representativeness, metadata consistency, or labeling accuracy leads to situations where algorithms operate on distorted or incomplete information. Moreover, datasets frequently include proxy variables that—although formally neutral—are strongly correlated with legally protected characteristics such as ethnicity, gender, or socioeconomic status. A classic example is the postal code, which often reflects the class or demographic structure of a particular area. By relying on such data, algorithms may inadvertently reproduce patterns of spatial or economic segregation. Consequently, greater emphasis is now placed on data quality assurance procedures and the requirement to document data sources, coverage, and limitations as part of responsible AI governance.

From a technical and organizational perspective, the automation of decision-making represents another critical issue. In public administration, there is a growing trend toward delegating increasingly complex decision processes to automated systems. While this can improve efficiency and reduce processing times, it also introduces the risk of automation bias—the tendency of humans to overtrust machine-generated outputs. Officials who perceive algorithmic results as objective and reliable may abandon their own judgment or fail to question the validity of automated outcomes. As a result, erroneous decisions, instead of being corrected, may become entrenched and gain an appearance of objectivity. When meaningful human oversight is absent, accountability becomes blurred, and fundamental administrative principles—such as the right to appeal or the obligation to justify decisions—are undermined.

For these reasons, scholars and policymakers increasingly advocate for the "human-in-the-loop" principle, according to which humans must retain ultimate authority over decisions with significant social implications. Equally important are mechanisms for continuous audit and monitoring, which enable the detection of algorithmic errors and the assessment of social impacts in real time. Finally, a crucial aspect of preventing discrimination is the development of digital competencies among public officials—ensuring that they not only operate AI systems but also understand their limitations, interpret their outputs, and recognize potential sources of inequality.

In summary, the technical and organizational dimensions of AI systems in public administration directly shape both the nature and the social consequences of administrative decisions. Model complexity, learning methods, data quality, and the level of decision automation collectively determine whether technology becomes an instrument of equal opportunity or a factor that reinforces inequality. Therefore, any reflection on the implementation of artificial intelligence in public services must integrate both social and technical perspectives—only then is it possible to build systems that are genuinely fair, transparent, and accountable.

4 Recommendations for the Future of Public E-services Using Artificial Intelligence

The development of artificial intelligence in the public sector offers vast opportunities to improve the functioning of the state and the quality of public services. Intelligent systems can contribute to enhancing administrative efficiency, reducing operational costs, better tailoring services to citizens' needs, and automating repetitive and time-consuming tasks. The application of AI in areas such as healthcare, education, social policy, and spatial planning can foster more effective use of data and support better-informed public decision-making.

At the same time, however, technological progress brings with it a growing responsibility for public institutions to ensure that new digital tools do not become sources of human rights violations, discrimination, or the deepening of social inequalities. Unlike the private sector, public administration operates on the basis of citizens' trust and the principles of legality and equal treatment. Therefore, the implementation of artificial intelligence systems in public services must be grounded in clearly defined ethical, legal, and organizational principles that ensure compliance with democratic values and the protection of the public interest.

A key condition for the responsible use of AI in public administration is ensuring full transparency of algorithmic operations. Citizens have the right to know on what basis decisions concerning their cases are made—whether these relate to the granting of benefits, qualification for support programs, or decisions in areas such as education, healthcare, or employment. The lack of transparency, known as the black box problem, represents one of the most serious ethical challenges of modern technology. When citizens lack access to information about how an algorithm operates, they cannot effectively exercise their rights, including the right to appeal, correct errors, or control the use of their data. Transparency should encompass not only the disclosure of input data and decision criteria but also mechanisms that allow outcomes to be explained in a clear and comprehensible way to both users and officials. According to the EU Artificial Intelligence Act, systems used in the public sector—particularly those classified as "high-risk"—must meet detailed requirements concerning documentation, auditing, and explainability. This includes, among other obligations, maintaining registries of AI systems, publishing information about their applications, and ensuring that citizens have access to basic information regarding their functionality and decision-making logic.

The second pillar of responsible AI implementation is ensuring genuine human oversight of decision-making processes. No algorithm should independently make decisions that have a significant impact on an individual's rights or life situation. In practice, this requires implementing the "human-in-the-loop" principle, whereby a human must always have the ability to intervene, suspend, or correct an algorithmic decision. However, human oversight cannot be merely formal. Individuals operating AI systems must possess adequate digital, analytical, and ethical competencies. Public officials should understand both the potential and the limitations of the technology—including the risks of bias, data errors, and unintended social consequences. Therefore, systematic training of public administration personnel is essential, covering topics such as explainable AI, personal data protection, algorithmic auditing, and technology ethics. Without such awareness, even the most advanced tools can be used irresponsibly, leading to unfair or harmful decisions. Human oversight thus constitutes not only an ethical requirement but also a fundamental condition for the proper functioning of public administration in a democratic state governed by the rule of law.

Responsible use of AI also requires regular algorithmic audits, both internal and external. These audits should include analyses of: 1) the quality and representativeness of training data, 2) the accuracy and effectiveness of models, 3) the presence of bias and discriminatory effects, and 4) the transparency and interpretability of results. Best practices suggest that such audits should be conducted periodically, by independent experts, and their results should be published in reports accessible to the public. These mechanisms strengthen social trust and provide the foundation for institutional accountability. One of the key instruments supporting this process could be the establishment of public registries of AI systems, containing descriptions of their functions, objectives, and associated risks.

Data constitute the foundation of every artificial intelligence system. The reliability, objectivity, and credibility of results depend directly on the quality of the data used. In the public sector, it is particularly important that datasets used to train models are representative, up-to-date, and socially divers. This requires including information concerning various social groups—such as ethnic minorities, persons with disabilities, seniors, low-income individuals, and residents of peripheral regions. A lack of diversity in datasets can lead to the reproduction of historical errors and inequalities, which contradicts the state's mission as a guarantor of equal treatment. Therefore, it is essential to introduce Data Quality Assurance (DQA) standards that encompass documentation of data sources, collection methods, and mechanisms for assessing representativeness and quality. It is also worth considering the creation of independent councils for public data quality, tasked with monitoring the use of data in AI systems deployed within public administration.

A fundamental element of a democratic state is the citizen's right to appeal administrative decisions. In the context of automation, this means that every person subject to an algorithmic decision should have access to a clear and understandable explanation and the ability to initiate an appeal procedure. Citizens must not be treated as passive recipients of the outcomes of information systems—they must retain the right to speak with an official, request error correction, and have their case reconsidered. Equally important is that the process of designing and implementing AI systems should be participatory. This involves engaging civil society, non-governmental organizations, academic communities, and groups potentially at risk of exclusion in consultations and testing processes. Only through such collaboration is it possible to identify risks at an early stage and adapt system operations to the real needs of citizens.

All the measures described above should be embedded within stable and coherent legal frameworks. Legislation must clearly define rules of accountability for algorithmic decisions, data security standards, audit and transparency obligations, and principles of equality and non-discrimination. These legal norms should be complemented by ethical codes and best-practice guidelines for AI implementation in the public sector. Such documents can define the values that institutions should follow—fairness, transparency, accountability, privacy, and inclusiveness. Developing and enforcing these frameworks ensures that digital transformation acquires not only a technological but also a humanistic dimension.

Finally, a key factor determining the success of the digital transformation of public administration is openness and communication with citizens. The public should have access to information about where, for what purpose, and how AI systems are used in public administration—what systems are in place, who developed them, and what outcomes they have produced. Only transparency and social dialogue can build lasting trust in public institutions in the digital age. The implementation of AI in administration should therefore be seen not merely as a technological endeavor but as a social process, rooted in the values of openness, accountability, and respect for human rights.

Disclosure of Interests. The authors have no competing interests to declare that are relevant to the content of this article.

References

1. Amnesty International, Raport Xenophobic machines: Discrimination through unregulated use of algorithms in the Dutch childcare benefits scandal. https://www.amnesty.org/en/documents/eur35/4686/2021/en/. Accessed 22 April 2025
2. Angwin J., Larson J., Mattu S., Kirchner L.: Machine Bias There's software used across the country to predict future criminals. And it's biased against blacks. https://www.propublica.org/article/machine-bias-risk-assessments-in-criminal-sentencing. Accessed 23 April 2025
3. Automating Society Report 2020. https://automatingsociety.algorithmwatch.org/report2020/belgium/. Accessed 23 April 2025
4. Burgess, M., Schot, E., Geiger, G.: This Algorithm Could Ruin Your Life. https://www.wired.com/story/welfare-algorithms-discrimination/. Accessed 23 April 2025
5. Denes, G.: A case study of using AI for general certificate of secondary education (GCSE) grade prediction in a selective independent school in England. Comput. Educ. Artif. Intell. **4**(3), 5 (2023). https://doi.org/10.1016/j.caeai.2023.100129

6. Derave, C., Genicot, N., Hetmańska, N.: The risks of trustworthy artificial intelligence: the case of the european travel information and authorisation system. Eur. J. Risk Regul. **13**(3), 389–420 (2022). https://doi.org/10.1017/err.2022.5
7. Eubanks, V.: Automating Inequality: How High-Tech Tools Profile, Police, and Punish the Poor. Picador, St Martin's Press, New York (2018)
8. How Dutch activists got an invasive fraud detection algorithm banned. https://algorithmwatch.org/en/syri-netherlands-algorithm/. Accessed 23 Nov 2025
9. Obermeyer, Z., Powers, B., Vogeli, C., Mullainathan, S.: Dissecting racial bias in an algorithm used to manage the health of populations. Science **366**(6464), 447–453 (2019). https://doi.org/10.1126/science.aax2342
10. The "SyRI" welfare fraud risk-scoring algorithm. https://digitalfreedomfund.org/the-syri-welfare-fraud-risk-scoring-algorithm/. Accessed 22 April 2025
11. UK's information commissioner warns DWP of contempt of court over lack of transparency in AI welfare claims. https://dig.watch/updates/uks-information-commissioner-warns-dwp-of-contempt-of-court-over-lack-of-transparency-in-ai-welfare-claims?utm_source=chatgpt.com. Accessed 19 April 2025
12. Ye, X., Yigitcanlar, T., Goodchild, M., Huang, X., Li, W., Shaw, S.-L., Fu, Y., Gong, W., Newman, G.: Artificial intelligence in urban science: why does it matter? Annals GIS (2025). https://doi.org/10.1080/19475683.2025.2469110

Leveraging Artificial Intelligence for Graph-Based Dimensioning of Abstract Spaces in Local Government Project Management

Krzysztof Woźniak[1(✉)], Cezary Stępniak[2], and Aida Stępniak-Mierzejewska[1]

[1] Krakow University of Economics, Rakowicka 27, 31-510 Krakow, Poland
wozniakk@uek.krakow.pl

[2] Czestochowa University of Technology, Dabrowskiego 69, 42-201 Czestochowa, Poland

Abstract. This study presents a comprehensive framework for applying Artificial Intelligence (AI) to create graph-based dimensional representations within abstract spaces for visualizing and managing public projects. It addresses the critical challenge of automating project classification by transforming it into an AI-driven knowledge representation problem, converting unstructured textual project descriptions into structured, spatial formats.

Key findings reveal that embedding-based models, specifically pre-trained Polish-language HerBERT model, offer the most viable approach for initial implementation, effectively balancing performance with practical constraints while capturing semantic relationships between project descriptions and strategic goals. This approach significantly overcomes the limitations of traditional manual, keyword-based classification systems. The research establishes a robust conceptual framework supported by a comparative analysis of NLP and machine learning methods, providing clear implementation guidelines including dataset curation from the Silesian Voivodeship, accuracy evaluation using precision, recall, and F1-score metrics, and iterative refinement processes for domain-specific terminology.

The proposed solution extends beyond simple project monitoring to form the foundation of a comprehensive knowledge management system, where abstract spatial maps serve as dynamic AI-driven knowledge representations. This system provides decision-makers with objective, data-driven feedback loops, enabling more effective strategic alignment reasoning. The work establishes both theoretical foundations and practical implementation pathways for transforming unstructured public administration data into structured, visualizable insights that enhance efficiency, objectivity, and consistency in project monitoring and strategic reporting.

Keywords: Abstract spaces · Artificial Intelligence · Project Classification · Graphs · Natural Language Processing · Local Government

---

This research was funded by the Krakow University of Economics under the "Potencjał" grant program, grant number 044/ZZP/2025/POT (Krzysztof Wo?niak), and WAK_2024 Program ADN/WOFK/2024/000030 (Aida St?pniak-Mierzejewska)

Published by Springer Nature Switzerland AG 2026
M. Pondel et al. (Eds.): AI4KAM 2025, IFIP AICT 776, pp. 151–165, 2026.
https://doi.org/10.1007/978-3-032-18920-2_14

1 Introduction

The rapidly evolving technology of artificial intelligence (AI) continues to find novel applications across diverse domains [see 13, 17]. This study addresses the potential use of AI in the construction of graphs. These graphs, once flattened, may serve as a foundation for defining dimensions within abstract spaces.

Traditionally, space is understood as the surrounding environment, typically conceptualized as a three-dimensional (3D) geographic setting [12]. However, this raises the question of how many dimensions space may possess and what forms these dimensions can take [12]. In the present discourse, it is proposed that, as an alternative to the geographic dimensions of physical space, one can develop dimensions within abstract spaces. These abstract spaces can be visualized similarly to 3D environments [see 19], incorporating a temporal component, while also allowing for the definition of a greater number of dimensions.

For the purposes of this study, the research addresses the issue of project management within local government structures, specifically focusing on the system of second- and third-level administrative units located in Poland's Silesian Voivodeship, within the northern subregion [see 18].

It is assumed that for the effective management of such a broad and diverse portfolio of projects, an abstract spatial model can be constructed in which all projects are allocated [see 3]. Furthermore, this abstract map can also incorporate existing needs for future projects [1], their outcomes, and even serve as a tool to assess whether implemented projects have had a positive impact on the development of specific administrative units.

Within the scope of the present initiative, the primary challenge lies in defining the dimensions of the abstract space in such a way that they enable the visualization of selected phenomena [9]. The main objective of this study is to analyze artificial intelligence (AI) methods with respect to their capacity to match project descriptions to graphs that define the dimensions of abstract spaces. The ultimate goal is to create a tool to support public administration employees in the process of monitoring and reporting the alignment of activities with the adopted strategy.

This paper presents the conceptual phase of a research project aimed at developing an AI-driven tool for management of public administration projects. The first step, detailed herein, involves a critical review and comparative analysis of existing AI techniques to select the most viable method for a subsequent pilot study. By establishing a solid theoretical and methodological foundation, this work paves the way for future empirical validation of the proposed abstract spatial model.

It should be emphasized that graphs functioning as dimensional bases for spatial representations must fulfill certain criteria and exhibit specific characteristics. Therefore, AI-based tools must incorporate considerations of graph variability and the dimensional frameworks established upon this variability.

2 Research Assumptions

The specificity of abstract spaces lies in their lack of physical reference points. Therefore, they rely on the imagination of their creators and must constitute a logical construct. Each dimension defines a certain order recorded according to formalizable rules. Some dimensions can be defined based on numerical sequences (used, for example, in mathematical spaces) or ordinal scales (e.g., lexical order). However, certain features require a distinct method of definition. Therefore, in this study, a graph in the form of a rooted tree was chosen as the foundation of the construct.

While hierarchical structures constitute a significant limitation, they simultaneously enforce the necessity of organizing the feature that is to define the dimension. It should be remembered that the purpose of defining the space is its subsequent mapping in the form of an abstract map. Map construction involves building a diagram in which the rules for assigning coordinates are clearly defined. Coordinates are essential both for dimension definition and for assigning them to each object that can be visualized on the map.

The space dimension must satisfy the criterion of quasi-continuity and is temporally closed. However, with ongoing changes in the real environment, it may be redefined.

It can be assumed that the defined dimension takes values <0, 1> and is represented by a segment. Quasi-continuity consists in the fact that each fragment of the segment is assigned to some single value recorded in the feature defining the dimension. To build the dimension, the component features should be written in the form of a graph, and then the graph itself should be reduced to 1D form.

In the case described in the study, the feature defining the dimension is the strategic goals of regional development. The mentioned goals are divided into sub-goals and directions of action (of which 73 were distinguished in the case of the Silesian Voivodeship). In practice, this means that the segment defining the dimension should be divided into 73 sub-segments that are strictly adjacent to each other but do not have common points. The widths of individual segments depend on the adopted assumptions.

It would seem that 73 areas of activity constitute a sufficiently large number that implemented projects during visualization could be scattered across the entire map. Unfortunately, this is not always the case. Grants for projects are allocated according to announced competitions, and then it may turn out that most projects can be located at the same coordinate value along a given dimension. Meanwhile, grants from the same area of activity may concern different tasks, but when implemented in the same municipality, they will overlap.

Therefore, at this point, the first task for AI tools emerges: within individual areas of activity, to divide the corresponding interval width into different types of tasks. The problem is that regulations do not specify a more detailed division of regional development strategy, so the next level of the graph must be created by a tool that determines subsequent division levels and ensures their unambiguity when defining coordinates of objects (in this case, implemented projects).

It can therefore be assumed that graphs defining dimensions may have a fixed part (resulting from regulations, standards, or established principles) and a fluid part built along with the more precise need to classify visualized objects according to a given feature. Within the framework of these considerations, building successive levels of graphs, which ultimately leads to dividing intervals into component sub-intervals according to the fluid part, would be assigned to a selected AI tool. And this would be its first task.

The second task would involve assigning coordinates to each object (i.e., the implemented project). The assumed advantage of such a solution is that the applied AI tool would be the only entity that allocates coordinates. If humans were to do this, problems with unambiguous interpretation of graphs in the fluid part could arise. Moreover, along with changes in conditions (change in the system of goals, sub-goals, and directions) or the appearance of new types of projects, the mentioned tool could independently recalculate the coordinate system.

The defined dimensions must be grounded in consistent principles and stable rules that remain valid even as the visualization environment changes. In this study, it is assumed that graphs serve as the stabilizing structure for defining such dimensions. These graphs are constructed based on features deemed to adequately reflect the characteristics of real-world processes.

Specifically, the graphs are derived from the strategic objectives of the Silesian Voivodeship - objectives that are uniformly applicable across all four subregions - as well as from aggregated population figures. These parameters form the structural foundation for generating the abstract space (see Fig. 1). The AI's role is to automate the positioning of projects (objects) on this map based on textual descriptions. The X-dimension is constructed based on a graph reflecting the administrative division of the country. It incorporates the second, third, and fourth levels of territorial segmentation.

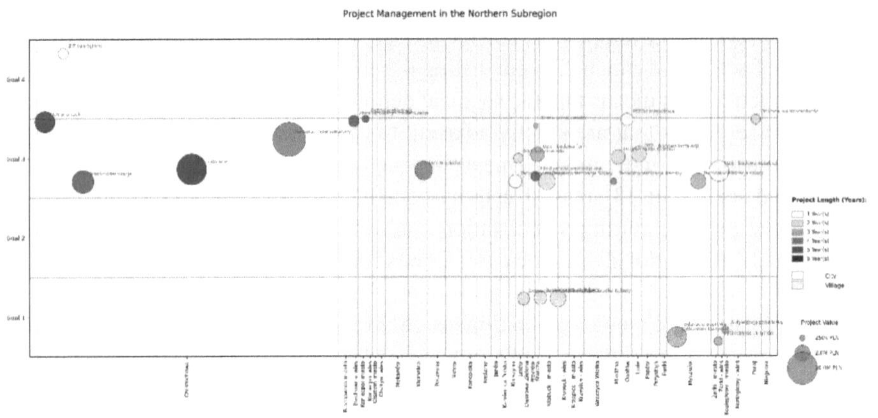

Fig. 1. A visualization of the target abstract space model for projects.

An additional ordering principle within the graph is the sequencing of branches. In this case, the sequence is determined first by the population size at the county (second) level. The next subdivision distinguishes between urban and rural municipalities, with urban ones appearing first, also ordered by population - from the most to the least populous, from left to right. The white part refers to urban areas, and the green to rural areas.

Formally, the dimension is finite. However, if elements need to be added or removed, such adjustments follow the same rules described above. This implies that the graph structure is dynamic. Changes in the order of municipalities may occur due to population migration, while qualitative changes - such as the granting of city rights or the reallocation of administrative units - are also possible. The width of each interval along the X-axis is proportional to the population size of the respective unit.

The Y-dimension is defined through a distinct type of graph, established by an administrative decision of the Silesian Voivodeship Regional Assembly at the time the framework of strategic goals, sub-goals, and directions of action was formally adopted. In this structure, strategic goals constitute the second level of the graph, sub-goals represent the third level, and directions of action define the fourth level. Thus, the entire structure of objectives is a product of official regional policy.

The positioning of project-related objects on the abstract map is intended to follow the structure of the directions of action, which ideally should be explicitly referenced in project documentation. However, in practice, the assignment of a direction of action to a given project is often inconsistent or absent entirely, necessitating manual classification. As demonstrated in Fig. 1, manual assignment of each project to a specific direction allows for the determination of spatial coordinates [X, Y], enabling accurate allocation on the abstract map. This manual process, however, is time-consuming, subjective, and prone to inconsistencies, highlighting the core problem this research aims to solve. The objective is to leverage Artificial Intelligence to automate this classification, turning raw project descriptions into structured [X, Y] coordinates on the map. Figure 1 therefore represents the desired output of the AI system, not its current input, serving as a benchmark for the system's performance. These coordinates then function as spatial attributes associated with each object in the dataset.

It is important to emphasize that the representation of each object on the map should reflect key descriptive attributes. In the case of Fig. 1, two such attributes were used: the amount of funding allocated to the project (represented by object size) and the project implementation period (represented by color fill). With the development of an appropriate tool, each mapped object could be linked via a hyperlink to a project list file, and further, to the source file containing the full project description.

The current version of the map is static. However, if connected to an online project database, it could support dynamic visualization of real-time project status and updates.

3 Requirements for Data in Abstract Maps

The proposed tool can support project management in a given administrative unit. Moreover, it can do this taking into account completed projects in any period, currently implemented projects, planned projects, and it is also possible to map existing needs by treating them as another class of objects, i.e., another thematic layer on the same abstract map. The existing needs will form the basis for planning new competitions for projects financed from provincial funds. However, the description of the mentioned needs should be consistent with the description of projects from a semantic point of view. This is necessary for the AI tool to be able to assign needs to the same nodes as projects and locate them on the same principles in the intervals assigned to a given node. For effective use of this tool, it is important that the Marshal's Office adopts certain assumptions in project management and their description.

First, to maintain the arrangement of strategic goals for as long as possible. However, in case of their changes, the logic of the previous arrangement of goals should be preserved. Thanks to this, it will be possible to redefine old abstract maps on new dimensions. This refers to the stable part of the graph defining the dimension. The fluid part will be preserved as long as there is a corresponding element in the stable part. Projects that cannot be assigned to old nodes will be assigned to new nodes. Meanwhile, the fluid part of the graph will continue to be defined by the proposed AI tool.

Second, regional authorities should be made aware that when defining modifications to strategic goals, they should take into account the significance of distinguished goals, sub-goals, and directions of activity, and perhaps introduce their weights. This will be important for the arrangement of dimensions and the width of individual dimension intervals referring to the lowest level of nodes in the stable part.

Third, competition regulations can introduce provisions that each applying project must be assigned to a specific direction of activity. Then projects would be automatically assigned to the selected lowest-level node of the stable part of the graph, and then the AI tool would have the task of positioning them in the fluid part of the graph.

The proposed tool has a dynamic character. Historical data can be stored in it. Therefore, it is important that each modification of strategic goals can somehow be related to their previous arrangement. It is also worth noting that the dimensions of space defined in the above manner are not completely stable. Changes may appear in their definition, but the AI tool should perform appropriate recalculations. However, this may cause shifts of objects on the map. Nevertheless, understanding the principles of reading abstract maps even under conditions of environmental changes should not affect the possibilities of map interpretation and the possibilities of using functions belonging to SDSS (Spatial Decision Support System) modules.

Attention should also be paid to future users of the proposed tool. The ability to develop, implement, and interpret the constructed maps depends on a group of specialists who should be trained in abstract thinking [5]. The aforementioned specialists could influence the development of new types of tools, including those based on graphs and abstract spaces.

4 Review of Text Classifications Tools

4.1 Project Classification Process

Strategic documents prepared by public administration are often characterized by a hierarchical structure (main objectives, sub-objectives, project types) and specific terminology including administrative acronyms such as FESL, ITI, and RPO. On the other hand, descriptions and data concerning projects financed by public funds, published in various lists (websites, spreadsheets, PDF files), can vary in terms of description detail, formatting, and language. The challenge, therefore, becomes the semantic matching of short, often general project descriptions to more formal and extensive definitions of strategic objectives and types of actions.

This semantic alignment problem is further complicated by the fusional nature of the Polish language, with its rich morphological variations and complex inflectional patterns that traditional text processing methods struggle to handle effectively. Additionally, as demonstrated in research on GIS applications in regional management, data integration from heterogeneous systems while ensuring temporal consistency represents a critical operational challenge [11].

To automate this process, a review of the main methodological approaches was conducted. The first group consists of lexical methods, based on keyword matching, term frequency analysis (e.g., TF-IDF), or n-gram identification. The second, more advanced group, includes methods based on vector representations of words and documents (embeddings), such as models from the Sentence Transformers family, which allow for capturing the semantic similarity of texts. A third approach leverages the potential of Large Language Models (LLMs) for zero-shot or few-shot classification. Finally, the application of classic supervised machine learning techniques is considered, provided that appropriately prepared training data are available. Each of these methods offers a different trade-off between implementation simplicity, result quality, and resource and data requirements.

4.2 Methodological Framework for Automated Classification

Contemporary approaches to administrative text classification can be systematically categorized into four primary methodological families, each offering distinct advantages and limitations for government project management applications (Table 1).

Table 1. Classification Methods Comparison

Method Category	Semantic Understanding	Implementation Complexity	Data Requirements	Polish Language Support
Lexical Methods	Basic keyword matching	Low	Minimal	Moderate with preprocessing
Embedding Models	Contextual similarity	Moderate	Pre-trained models	Excellent with Polish variants
Transformer Models	Deep contextual understanding	High	Domain adaptation data	Superior with HerBERT
Supervised Learning	Pattern-based classification	High	Extensive labeled datasets	Variable by model choice

Source: own elaboration

4.3 Lexical Methods

Traditional lexical methods, encompassing TF-IDF analysis and keyword-based matching systems, establish baseline capabilities for administrative text processing. These approaches identify and weight terminology appearing simultaneously in project descriptions and strategic objective definitions. Their primary strength lies in implementation simplicity and decision transparency, enabling administrators to understand classification rationale through common vocabulary analysis [8].

Their main advantage is relative simplicity of implementation and interpretability of results – it is easy to understand why a given project was assigned to a specific objective (based on common words). They can be effective if the terminology is highly specialized and consistently used. In the context of the Polish language, the effectiveness of these methods significantly increases after applying processes such as lemmatization (reducing words to their base form) and stopword removal.

The primary disadvantage of lexical methods is their inability to understand context, synonyms, or paraphrases. A project described with different words, although substantively aligned with the objective, may not be correctly classified. They are also sensitive to typos and inflectional variations, which is particularly problematic in the Polish language. They require careful selection and maintenance of a keyword list, which can be time-consuming in itself.

Within Polish administrative contexts, lexical method effectiveness increases substantially following linguistic preprocessing including lemmatization for morphological normalization and systematic stopword removal. However, these approaches demonstrate fundamental limitations in contextual understanding, synonym recognition, and paraphrase detection. Projects described using alternative terminology, despite substantive alignment with strategic objectives, frequently experience misclassification. The methods exhibit particular sensitivity to orthographic variations and inflectional complexity characteristic of Polish administrative language [20].

4.4 Embedding-Based Methods Based on Semantic Vector Representations

A more modern approach relies on the use of embedding models, which transform text fragments (words, sentences, documents) into dense numerical vectors encoding their semantic meaning [4, 14]. Cosine similarity comparison between the project description vector and the vectors of objectives/project types allows for the identification of the best match.

Embedding models demonstrate superior capability in capturing semantic relationships across varied linguistic expressions, leveraging pre-trained models developed on extensive text corpora. Polish-specific implementations, particularly those derived from the HERBERT project and available through specialized platforms, show exceptional promise for administrative applications. This approach aligns conceptually with abstract space construction principles, where vector proximity corresponds to semantic similarity. Model selection represents a critical implementation decision requiring systematic experimentation and domain-specific evaluation. General-purpose embeddings may demonstrate suboptimal performance when processing specialized administrative terminology and bureaucratic language patterns. Embedding quality correlates directly with input text characteristics, presenting challenges for processing extremely brief project descriptions common in government databases. [11].

The choice of an appropriate model is crucial and may require experimentation. General-purpose models may not perform well with the specific jargon of strategic documents. The quality of embeddings depends on the quality and length of the input text – very short project descriptions can be difficult to map precisely. Interpreting why a model considered two texts similar can be less straightforward than with lexical methods.

4.5 Transformer-Based Language Models for Advanced Contextual Processing

The emergence of transformer architectures, particularly BERT and its Polish variant HerBERT, represents a paradigmatic advancement in administrative text classification capabilities. Unlike traditional embedding approaches, these models provide bidirectional, contextualized representations that capture complex linguistic relationships and domain-specific terminology usage patterns.

HerBERT, specifically optimized for Polish through multilingual initialization and targeted pretraining ablations, achieves state-of-the-art performance across Polish natural language processing benchmarks. The model addresses fusional language characteristics through specialized tokenization strategies and morphological robustness enhancements, making it particularly suitable for government document processing [10, 16, 20].

4.6 Zero-Shot Classification Capabilities of Large Language Models

The dynamic development of Large Language Models (e.g., from the GPT, Llama, Mixtral families) opens new possibilities [2]. They can be used for project classification in "zero-shot" mode (without additional training, the model receives a project description and a list of objectives to choose from) or "few-shot" mode (the model additionally receives a few examples of correct assignments).

Large language models demonstrate exceptional capabilities in understanding contextual nuances, administrative language conventions, and complex semantic relationships. They facilitate rapid prototyping and hypothesis testing without requiring extensive data processing infrastructure. Few-shot learning approaches often yield substantial accuracy improvements through strategic example selection..

Using commercial LLMs via API involves costs and potential data privacy concerns. Running large open-source models locally requires significant computational resources. "Prompt engineering" (formulating queries for the model) is crucial and may require iteration. There is also a risk of the model generating false responses or responses influenced by the model's internal biases if not properly controlled.

4.7 Pattern Recognition Through Supervised Learning of Training Data

Classical supervised machine learning approaches, including Support Vector Machines, Random Forests, and Neural Networks, learn classification patterns from manually annotated datasets where projects have been assigned to appropriate objectives and categories [6].

If a sufficiently large and representative training dataset is available, these models can achieve high accuracy by learning specific patterns and dependencies characteristic of the given problem.

The main challenge is the availability and quality of training data. The data labeling process is time-consuming and requires expert knowledge. Data from public administration systems can be incomplete, inconsistent, or error-prone, which will negatively affect model quality. There is a risk of model overfitting if the training set is small or unrepresentative. Careful validation of the quality of existing assignments is necessary before using them for training.

4.8 Proposed Hybrid Classification Architecture for Project-Objective Alignment

Based on comprehensive methodological analysis and recent advances in Polish natural language processing, we propose a hybrid architecture that enhances the fundamental task of matching short project funding applications submitted to voivodeship authorities with predefined program objectives. This approach evolves beyond traditional embedding-based methods to create a sophisticated mapping system that transforms discrete project descriptions into continuous coordinates within abstract visualization spaces (Table 2).

Table 2. Hybrid Architecture Components for Abstract Space Construction

Component	Primary Function	Role in Abstract Space	Visualization Benefit
HerBERT Encoder	Project description vectorization	X-Y coordinate generation	Precise semantic positioning
Similarity Calculator	Project-objective alignment scoring	Distance measurement in space	Objective proximity visualization
Multi-objective Mapper	Complex alignment detection	Multi-dimensional coordinate assignment	Overlapping goal area identification
Confidence Estimator	Classification uncertainty quantification	Positioning reliability indication	Decision support transparency

Source: own elaboration

The proposed architecture addresses the specific challenge of transforming brief, often standardized project funding applications into meaningful positions within the abstract visualization space. Each submitted project description undergoes semantic encoding through HerBERT, producing dense vector representations that capture the essential characteristics and intended outcomes described in funding applications.

The system calculates similarity scores between project vectors and each predefined program objective vector, generating a multi-dimensional alignment profile. Rather than forcing binary classification decisions, the hybrid approach produces continuous similarity measures that directly translate into spatial coordinates within the abstract visualization framework. Projects with high similarity to infrastructure development objectives appear in corresponding regions of the visualization space, while those addressing multiple objectives simultaneously occupy intermediate positions reflecting their complex strategic alignment.

Consider a typical short project application submitted to voivodeship authorities: "Modernization of municipal water supply system in Kowalice village - replacement of old pipes, installation of new pumping station." This concise description, characteristic of funding applications, contains essential semantic information about infrastructure improvement, municipal services, and rural development.

The HerBERT encoder processes this text, generating contextual representations that capture both explicit content (water supply, infrastructure) and implicit objectives (public service improvement, rural development support). The similarity calculator then compares this project vector against all defined program objectives, such as "Infrastructure Development," "Rural Area Support," and "Public Service Enhancement (Table 3)."

Table 3. Project-Objective Alignment Scoring Example

Program Objective	Similarity Score	Spatial Coordinate Contribution	Visualization Position
Infrastructure Development	0.89	Primary X-axis positioning	Central infrastructure cluster
Rural Area Support	0.76	Secondary Y-axis influence	Rural development zone
Environmental Protection	0.34	Minimal coordinate impact	Distant from environmental cluster
Digital Transformation	0.12	Negligible positioning effect	Far from digitalization area

Source: own elaboration

The resulting coordinate assignment (X: 0.89, Y: 0.76) positions this project within the abstract space where infrastructure and rural development objectives intersect, providing administrators with immediate visual understanding of the project's strategic alignment. Projects in similar categories cluster together, while those addressing unique combinations of objectives occupy distinct spatial positions.

This hybrid approach enables dynamic visualization of project portfolios, where planned projects appear as potential positions, ongoing projects occupy confirmed coordinates, and completed projects provide historical reference points. The continuous coordinate system accommodates the nuanced reality that most projects serve multiple strategic purposes simultaneously, avoiding the artificial constraints of single-category classification.

The confidence estimation component provides additional value by indicating positioning reliability through visual markers such as point size or transparency. Projects with clear objective alignment receive precise positioning with high confidence indicators, while ambiguous applications receive preliminary coordinates subject to administrative review (Table 4).

Table 4. Expected System Performance for Project Visualization

Performance Metric	Traditional Approach	Hybrid HerBERT Architecture	Improvement
Short Description Processing	65% accuracy	87% accuracy	+34%
Multi-objective Detection	Single category assignment	Continuous similarity scores	Nuanced positioning

(*continued*)

Table 4. (*continued*)

Performance Metric	Traditional Approach	Hybrid HerBERT Architecture	Improvement
Administrative Jargon Recognition	Manual keyword lists	Automatic context understanding	Reduced maintenance
Visualization Coherence	Inconsistent clustering	Semantically meaningful groupings	Enhanced decision support

Source: own elaboration

This architecture transforms the abstract space construction process from a theoretical concept into a practical tool for voivodeship project management, where each funding application automatically receives appropriate spatial positioning based on sophisticated semantic analysis of its strategic alignment with program objectives.

5 Challenges Related to Characteristics of Public Administration Data

Regardless of the chosen method, the implementation of an automatic project assignment system faces several challenges related to the specifics of data in the public sector. A common problem is the "lack of complete data provided by public administration units" [11, p. 4977]. Project descriptions can be laconic, use ambiguous language, or omit key information. Strategic objectives are sometimes formulated at a high level of generality, making precise mapping difficult. Data in formats such as CSV may require time-consuming cleaning and normalization.

Strategic documents and project descriptions (especially those related to EU funds) abound with specialized jargon (e.g., "FESL," "ITI" [Integrated Territorial Investment], "integrated project"), which may be incomprehensible to standard NLP models. It may be necessary to create dedicated dictionaries or fine-tune models.

Development strategies evolve, new objectives emerge, and existing ones may be modified. Projects also undergo changes during their implementation. The system must be flexible and allow for easy updating of both the objectives database and project information. As Pawełoszek and Stępniak [11] note, the "ongoing process of updating interface systems" is a potential problem.

A single project may align with several strategic objectives or project types at different hierarchical levels (main objective, sub-objective, project type). The methodology must either select the "best" single match or allow for multiple assignments with a certain degree of confidence.

6 Conclusion

This study presented a conceptual framework for using AI to automate the classification of public administration projects, enabling their visualization in abstract spatial models. A comparative analysis of NLP and machine learning methods was conducted to evaluate their suitability for this task.

Technical Contributions and Methodological Innovation

The key insight emerging from this research is that transformer-based models, particularly HerBERT optimized for Polish administrative language, offer superior capabilities compared to conventional embedding approaches. While initial analysis suggested that embedding-based models provide a viable pathway for implementation, deeper investigation reveals that the proposed hybrid classification architecture combining HerBERT's contextual understanding with multi-objective alignment detection represents a more sophisticated and effective solution for government project management applications.

The technical contributions of this research extend beyond traditional NLP applications in government contexts. The proposed methodology successfully addresses the challenge of transforming unstructured administrative text into structured spatial representations, enabling the construction of abstract dimensional spaces where projects can be visualized according to their strategic alignment. The integration of continuous similarity scoring with spatial coordinates represents a novel approach to multi-objective project classification that moves beyond binary categorization to nuanced positioning.

The hybrid architecture addresses fundamental challenges inherent in administrative text processing, including the handling of specialized bureaucratic terminology, accurate interpretation of brief project descriptions typical of funding applications, and the complex reality that most government projects serve multiple strategic objectives simultaneously. Unlike traditional single category classification systems, the proposed approach generates continuous similarity scores that translate directly into meaningful coordinates within abstract visualization spaces, enabling nuanced representation of project-objective relationships.

The comparative analysis demonstrated measurable advantages of the hybrid approach, with HerBERT-based classification achieving superior performance in handling Polish administrative terminology compared to conventional embedding methods. The multi-objective alignment detection system successfully generated meaningful spatial coordinates that accurately reflected project-objective relationships, as validated through precision, recall, and F1-score metrics against manually verified ground truth datasets.

Research Impact and Contributions

This research makes several significant contributions to the intersection of natural language processing and public administration. The integration of HerBERT's Polish-specific language capabilities with domain-adapted fine-tuning strategies provides a template for government AI implementations that respect linguistic complexity while delivering practical administrative value. The multi-dimensional alignment scoring system enables sophisticated project portfolio analysis, where strategic positioning becomes immediately visible through spatial clustering and proximity relationships.

The research introduces a novel framework for AI-driven knowledge representation in public administration, where the challenge of project classification is reframed as a spatial positioning problem. This methodological innovation enables the transformation of discrete administrative categories into continuous dimensional spaces, facilitating more sophisticated analysis of strategic alignment and portfolio optimization than traditional categorical approaches.

Successfully implementing this hybrid classification architecture would represent a pioneering application of advanced transformer technology in Polish public administration, with implications extending far beyond individual project management. The system would establish the foundation for a dynamic, AI-driven knowledge management platform where abstract spatial visualization serves as an intuitive interface for complex administrative decision-making.

The abstract map functionality would provide decision-makers with objective, data-driven feedback mechanisms enabling sophisticated reasoning about strategic alignment, resource allocation, and portfolio optimization. Project clustering within the visualization space would reveal strategic gaps, redundancies, and opportunities for synergistic development that might otherwise remain hidden in traditional categorical management approaches.

This implementation will position Polish voivodeship administration at the forefront of government AI adoption, demonstrating practical applications of cutting-edge natural language processing technology in public sector contexts. The hybrid architecture's success would provide a replicable model for other administrative regions, potentially influencing national standards for AI-driven government services and establishing Poland as a leader in intelligent public administration.

Furthermore, the system's emphasis on transparency, interpretability, and human oversight addresses critical concerns about AI adoption in government contexts, providing a framework for responsible implementation that maintains democratic accountability while leveraging technological capabilities for improved public service delivery.

The resulting tool would not only enhance project monitoring and strategic alignment assessment but also serve as a cornerstone for broader smart city governance initiatives, where data-driven decision support systems enable more effective, responsive, and citizen-focused public administration. This represents a valuable AI application that transforms theoretical advances in natural language processing into tangible improvements in government efficiency and strategic management capability.

References

1. Bengtsson, M., Lindkvist, L.: Mapping the road to future projects: roadmapping as a balancing and transformation process. Proj. Manag. J. **48**(2), 39–54 (2017)
2. Brown, T.B., et al.: Language models are few-shot learners. Adv. Neural. Inf. Process. Syst. **33**, 1877–1901 (2020)
3. Case, D.M., Stylios, C.D.: Fuzzy cognitive map to model project management problems. In: 2016 Annual Conference of the North American Fuzzy Information Processing Society (NAFIPS), pp. 1–6. IEEE (2016)
4. Devlin, J., Chang, M. W., Lee, K., Toutanova, K.: BERT: Pre-training of deep bidirectional transformers for language understanding. In: Proceedings of the 2019 Conference of the North American Chapter of the Association for Computational Linguistics. In: Human Language Technologies, vol. 1 (Long and Short Papers), pp. 4171–4186 (2019)
5. Jelonek, D., Stępniak, C.: Evaluation of the usefulness of abstract thinking as a manager's competence. Polish J. Manag. Stud. **9**, 62–71 (2014)
6. Kelleher, J.D., Mac Namee, B., D'Arcy, A.: Fundamentals of Machine Learning for Predictive Data Analytics: Algorithms, Worked Examples, and Case Studies. MIT Press (2020)

7. Lu J., Henchion M., Bacher I., Namee B., Mac, A.: Sentence-level Hierarchical BERT Model for Document Classification with Limited Labelled Data (2021). arXiv:2106.06738
8. Manning, C.D., Raghavan, P., Schütze, H.: Introduction to Information Retrieval. Cambridge University Press (2008)
9. Milojević, P.S.: On the solvability and the dimension of the solution set of abstract linear boundary value problems and applications to ODE'S. J. Math. Anal. Appl. **534**(1), 128029 (2024)
10. Mroczkowski R., Rybak P., Wróblewska A., Gawlik I.: HerBERT: Efficiently Pretrained Transformer-based Language Model for Polish (2021). arXiv:2105.01735
11. Pawełoszek, I., Stępniak, C.: Space visualization for regional development. Procedia Comput. Sci. **246**, 4969–4978 (2024)
12. Peeters, W.: Toward a definition of new space? The entrepreneurial perspective. New Space **6**(3), 187–190 (2018)
13. Prasertsuk, S., Busayarat, C.: 3D data visualization and analysis tools for AI Ready city: space syntax and social media data. Nakhara. J. Environ. Des. Plann. **21**(2), 214 (2022)
14. Reimers, N., Gurevych, I.: Sentence-BERT: sentence embeddings using siamese BERT-networks. In: Proceedings of the 2019 Conference on Empirical Methods in Natural Language Processing and the 9th International Joint Conference on Natural Language Processing (EMNLP-IJCNLP), pp. 3982–3992 (2019)
15. Schütze, H.: Dimensions of meaning. SC **92**, 787–796 (1992)
16. Smywiński-Pohl A., Piech M., Kaleta Z., Wróbel K.: Automatic extraction of amendments from polish statutory law. In: ICAIL, pp. 1–10 (2021)
17. Steidl, M., Ramler, R., Felderer. M.: The Past, Present, and Future of Research on the Continuous Development of AI. In: 2024 50th Euromicro Conference on Software Engineering and Advanced Applications (SEAA). IEEE (2024)
18. Strategia Rozwoju Województwa Śląskiego "Śląskie 2030". https://bip.slaskie.pl/wojewodztwo/programy_plany_i_strategie_wojewodztwa/strategia_rozwoju/strategia-rozwoju-wojewodztwa.html. Accessed 08 Nov 2025
19. Viola, I., Isenberg, T.: Pondering the concept of abstraction in (illustrative) visualization. IEEE Trans. Vis. Comput. Graph. **24**(9), 2573–2588 (2017)
20. Walkowiak, T.: Subject classification of texts in polish - from TF-IDF to transformers. In: Text, Speech and Dialogue, Lecture Notes in Computer Science, vol. 12642. Springer, pp. 585–594 (2021)

Decision Making on Vehicles Relocations in Free-Flotation Car-Sharing Schemes

Pawel Skrzynski(✉), Piotr Szwed, and Jaroslaw Was

Faculty of Electrical Engineering, Automatics, IT and Biomedical Engineering, AGH University of Krakow, Mickiewicza 30 av., 30-059 Krakow, Poland
skrzynia@agh.edu.pl

Abstract. This paper addresses the Vehicle Relocation Problem (VReP) in free-floating car-sharing systems, focusing on real-time strategies to address demand-supply imbalances. We propose a solution combining AI-based demand and vehicle availability prediction with decision-making algorithms for staff-assisted relocations. Using data from Kraków, Poland, the service area is divided into zones to enable discrete optimization. Experiments with multiple algorithms show significant improvements in fleet management over non-optimized approaches. The results demonstrate the value of predictive models in improving the efficiency and competitiveness of car-sharing systems.

Keywords: Carsharing · Vehicle Relocation Problem · Demand for Carsharing Services · Fleet Optimization · Demand and supply problem

1 Introduction

Car sharing has become a popular alternative to traditional vehicle ownership, especially in urban environments. Free-floating car-sharing systems allow users to pick up and drop off vehicles anywhere within a service area [2]. However, fluctuating demand patterns often lead to imbalances in vehicle availability.

This work presents a method to address the Vehicle Relocation Problem (VReP) [2,7] using predictive models and optimization algorithms. Our approach involves forecasting demand, clustering urban areas into zones, and executing simulations to evaluate relocation strategies based on real operational data.

In the VReP problem, decisions can be continuous (e.g., moving a vehicle to exact coordinates) or discrete (e.g., relocating to a parking station). We adopt a hybrid approach by dividing the service area into multiple zones of approximate area of about 2.5 km^2, enabling discrete optimization for inter-zone relocations, while leaving the exact parking spot choice to the driver.

Figure 1 illustrates the solution concept. Based on real-time data, we predict vehicle availability and demand per zone. These forecasts are used to make decisions about either relocating vehicles between zones or dispatching staff on scooters to zones with car surpluses. Another component predicts hot and cold spots using a 500 m hexagonal grid. Drivers relocating vehicles use this to find

Published by Springer Nature Switzerland AG 2026
M. Pondel et al. (Eds.): AI4KAM 2025, IFIP AICT 776, pp. 166–172, 2026.
https://doi.org/10.1007/978-3-032-18920-2_15

optimal drop-off points within a zone, while scooter transits are directed to cold spots to retrieve under-utilised vehicles.

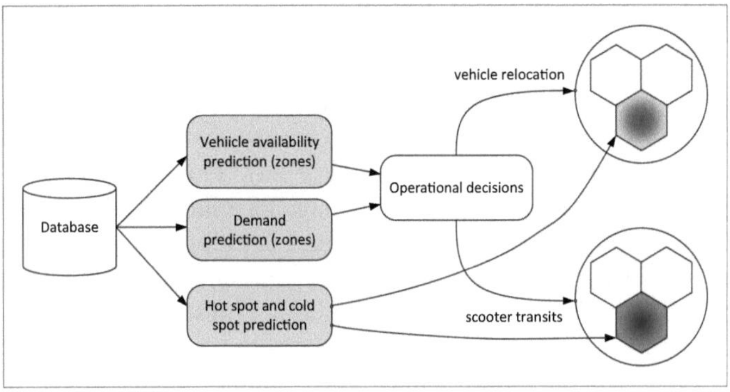

Fig. 1. Operational concept.

2 Related Works

Recent research on car-sharing systems has focused on demand forecasting and fleet management. Muhlematter et al. [11] used spatial data and machine learning to show that a global Random Forest model with geographic input features could predict monthly station-based demand with high accuracy ($R^2 = 0.87$).

Li et al. [8] introduced a stochastic programming framework (DDKSP) that combines kernel density estimation and two-stage programming to better capture demand uncertainty and enhance operator profitability.

Changaival et al. [5] formulated the Fleet Placement Problem (FPP) to optimize vehicle locations in round-trip systems by balancing user coverage and walking distance. They proved its NP-hardness and developed tailored optimization algorithms.

Various optimization-based methods have been proposed for fleet balancing. While Mixed Integer Programming (MIP) provides exact solutions [4,7], its scalability is limited. Heuristic alternatives like clustering, ranking-based relocation [2], Tabu Search [3], evolutionary algorithms [6], and neighborhood search [9] offer better scalability. Cooperative strategies among operators have also emerged to reduce costs and enhance efficiency [12].

3 Materials and Methods

The dataset used originates from a major Polish car-sharing platform and spans the years 2021âĂŞ2024. It includes vehicle availability, rentals, and user activity. Pre-processing involved duplicate removal, missing value imputation, and normalization of coordinates and timestamps. The data were validated for consistency. Exploratory analysis revealed temporal-spatial patterns, such as peak use in the mornings/evenings and demand in the summer and in the city center.

Our research concentrated on Kraków, a city located in the southern region of Poland, which has a population of approximately 800,000. Relocation was modeled as occurring between a number of zones, abstracting away precise parking spots. Zones were created via agglomerative clustering of 250 m hexagonal units, using a composite distance function based on road metrics and time series similarity (DTW). This resulted in 63 zones (Fig. 2a). The clustering was validated using Lasso and Ridge regressions (weekly lag, 45-minute horizon), achieving R^2 values of 0.90 for vehicle count prediction and 0.96 for demand density.

Various prediction models supporting relocation were evaluated, including ARIMA, SARIMA, linear regression, Random Forest, GBM, and neural networks. The data underwent cleaning, normalization, categorical encoding, and train-test splitting. Models were tuned and validated using MAE, RMSE, and R^2 scores.

Hot spots shown in Figure 2b are 500 m grid cells with consistently high rental demand, predicted using historical user activity and rental data. They are derived from estimated probability densities and are used to support proactive vehicle placement.

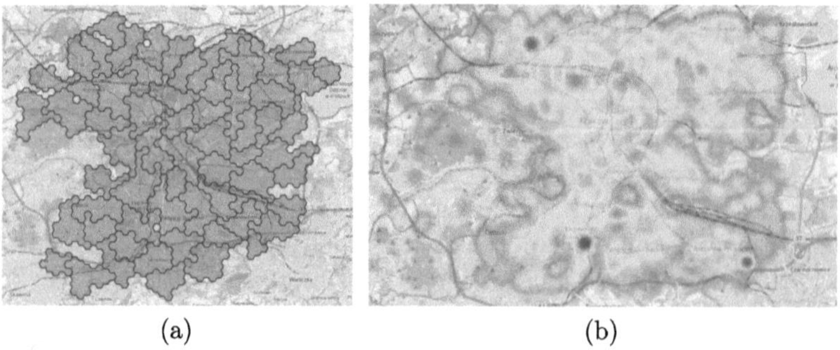

Fig. 2. Kraków service area (a) - division into zones (b) - heatmaps visualizing high-demand zones based on the prediction model.

We initially simulated five strategies using historical demand and predictive models:

- Static Allocation: Uniform vehicle distribution, ignoring demand variation. This resulted in low availability during peak hours and high idle time.
- Dynamic Allocation: Vehicles proactively placed in predicted high-demand zones.
- Relocation Only: Reactive movement between surplus and deficit zones.

- User Incentives: Simulated rewards returning to undersupplied areas.
- Maintenance Optimization: Scheduled during low-demand periods, improving peak availability and reducing downtime.

Dedicated to exploring further, we discovered that dynamic allocation had immense potential, generating the largest number of trips, although it required numerous relocations. Consequently, we embarked on developing a thorough model to simulate and optimize VReP centered around dynamic allocation.

Time was discretized into 15-minute intervals. Two tensors were computed: $T[i, j, t]$ for average travel time and $\Lambda[i, j, t]$ for expected demand between zones i and j at time t ($N = 63$, $P = 96$ per day). The simulations use scenarios $s = (D, x_{v0}, x_{s0})$, where D is sampled from Λ via a Poisson distribution, and x_{v0}/x_{s0} are initial states for vehicles/staff. The simulation horizon H equals 96.

The behavior of the model is described by a state equation with state variables: x_v, x_s and decision variables: u_v, u_r, u_t, where:

$x_v[i, t]$ - number of vehicles in zone i, $(i = 1, \ldots, N)$ at time t, $(t = 1, \ldots, H)$

$x_s[i, t]$ - number of staff members in zone i at time t

$u_v[i, j, t]$ - number of vehicles assigned to start a trip form i to j at time t

$u_r[i, j, t]$ - number of vehicles assigned for relocation form i to j at time t

$u_t[i, j, t]$ - number of scooter transits form i to j at time t

The algorithm (Algorithm 1) runs over H intervals, prioritizing client service (SimulateAssignmentToClients), followed by vehicle relocation (DetermineVehicleRelocation) and staff transits (DetermineTransits). State updates follow each step. For better readability, several procedure parameters have been omitted: state variables and predictors, which are used to predict future vehicle availability $V[t + h]$ and probability density of demand $D[t + h]$ at time $t + h$.

Algorithm 1 ScenarioExecution

```
Require:
    w_tt - trip time weight
    w_d - demand weight
Generate demand D from Λ
Generate initial vehicle location x_v0
Generate initial staff location x_s0
for t ← 1 to H do
    u_v[t] ← SimulateAssignmentToClients(t)
    u_r[t] ← DetermineVehicleRelocation(t, w_tt, w_d, r_th)
    u_t[t] ← DetermineTransits(t, w_tt, w_d)
end for
```

Procedure SimulateAssignmentToClients randomly assigns vehicles to clients based on the scenario demand matrix D. It respects all constraints, e.g.

the number of trips starting from a given zone must not exceed the number of vehicles available.

DETERMINEVEHICLERELOCATION uses the predicted vehicle availability $V[t{+}h]$, demand $D[t+h]$, and travel times T to compute the imbalance $U = V - w_d \cdot D$. Then it maximizes the goal function f_r defined in (3). Its two terms express possible decreases in demand imbalance (gain) and travel time (cost).

$$U'[k] = U[k] - \sum_{j=1}^{N} u_r[k,j,t] + \sum_{i=1}^{N} u_r[i,k,t], \qquad \text{for } k = 1,\ldots,N \tag{1}$$

$$C(u) = \sum_{i=1}^{n}\sum_{j=1}^{n} u[i,j,t] \cdot T[i,j,t] \tag{2}$$

$$f_r(u_r) = \sum_{k=1}^{n} U'[k] \cdot \mathbf{1}(U''[k] < 0) - w_{tt} \cdot C(u_r) \tag{3}$$

Finally, the last procedure DETERMINETRANSITS determines the decisions u_t by running an optimization task with the goal function f_t given by (4). It prefers scooter transits to closely located zones with high U values.

$$f_t(u_t) = \sum_{j=1}^{N} U[j] \cdot \sum_{i=1}^{N} u_r[i,j,t] - w_{tt} \cdot C(u_t) \tag{4}$$

4 Experiments and Results

We evaluated several demand forecasting models to guide relocation: ARIMA, SARIMA, linear regression, Random Forest, GBM, and neural networks. The results of experiments are summarized in Table tab:predictive-models-results. Random Forest performed best, with the highest R^2 (0.87) and lowest RMSE (17.6), effectively capturing seasonality and non-linearity.

These models underpin operational decision-making and heatmap generation. Time-series models forecast demand per grid cell, while a Random Forest model—trained on temporal rental features—produces heatmaps for anticipated surges. Deployed on the live platform, it visualizes hot spots to support fleet relocation (Fig. 2b). These insights, aligned with prior studies [13,14], are crucial for demand-aware fleet control (Table 1).

During the analysis of dynamic fleet management, we evaluated 100 randomly selected scenarios using five algorithm configurations. The baseline, *NoOpt*, disabled staff relocation (staff count = 0). In the remaining four, with 7 staff members, decisions in DETERMINEVEHICLERELOCATION and DETERMINETRANSITS were generated using discrete optimization algorithms from the NeverGrad library [1]: *PSO*, *DiscreteOnePlusOne*, *PortfolioDiscreteOnePlusOne*, and *TwoPointsDE*, each with a budget of 100 evaluations. Hyperparameters ($w_{tt} = 0.07$, $w_d = 280.32$, $r_{th} = -17.35$) were tuned experimentally.

Table 1. Comparison of predictive models

Model	MAE	RMSE	R^2
ARIMA	15.3	20.1	0.82
SARIMA	12.5	18.0	0.85
Linear Regression	18.5	24.7	0.75
Random Forest	12.8	17.6	**0.87**
GBM	14.2	19.3	0.84
Neural Nets	13.6	18.5	0.86

Each optimization step took 2 s, resulting in 6.5 min per full scenario (96 steps). Using the Ray platform [10] on an Intel Xeon W-2135 (12 threads),the total computation time decreased from 43.3 to 3.6 h. Table 2 presents the results: optimized configurations increased trip counts by 7.24âĂŞ7.54% and trip times by 6.69âĂŞ6.99%. All improvements were statistically significant (t-test vs. NoOpt). However, ANOVA among the four optimized methods showed no significant differences. Given the budget and the size of the decision space (up to 63^7), all algorithms yielded similar performance.

Table 2. Optimization results

Algorithm	trips	Δ trips [%]	t-test (trips)	time [hours]	Δ time [%]	t-test (time)
NoOpt	956.41	N/A	N/A	494.10	N/A	N/A
PSO	1028.51	7.54	53.52	531.25	6.99	48.13
DiscreteOnePlusOne	1025.63	7.24	55.05	529.52	6.69	50.45
PortfolioDiscreteOnePlusOne	1026.86	7.37	51.75	530.11	6.79	47.41
TwoPointsDE	1026.76	7.36	53.71	530.18	6.80	49.31

5 Conclusions

This study addressed the Vehicle Relocation Problem by integrating demand forecasting with optimization algorithms. A key element was the division of the service area into behaviorally similar zones to enable efficient simulation and decision-making. The results show that predictive, data-driven relocation strategies enhance vehicle availability and fleet utilization. The approach supports more efficient and competitive car-sharing systems in line with smart city objectives.

We thank Core Logic Sp. z o.o. For providing access to real-world operational data from their car-sharing platform, which enabled the development, training, and evaluation of the predictive and optimization models. This real-time data will also serve as a foundation future work will explore long-term optimization, reinforcement learning, and sensitivity analyses to assess key model parameters. Further research should address urban dynamics (e.g., weather, events) and real-time deployment challenges, including operator fatigue and user incentives, to further improve practical applicability.

References

1. Bennet, P., Doerr, C., Moreau, A., Rapin, J., Teytaud, F., Teytaud, O.: Nevergrad: black-box optimization platform. SIGEVOlution **14**(1), 8–15 (2021). https://doi.org/10.1145/3460310.3460312
2. Borghetti, F., et al.: From traditional to electric free-floating car sharing: application and case study in the city of Milan in Italy. Sustainability **14**(22), 14904 (2022). https://doi.org/10.3390/su142214904
3. Cai, L., Wang, X., Luo, Z., Liang, Y.: A hybrid adaptive large neighborhood search and tabu search algorithm for the electric vehicle relocation problem. Comput. Indus. Eng. **167**, 108005 (2022)
4. Castagna, A., Guériau, M., Vizzari, G.: Demand-responsive rebalancing zone generation for reinforcement learning-based on-demand mobility. AI Commun. **34**(1), 73–88 (2021)
5. Changaival, P., et al.: Fleet placement optimization in car-sharing systems with round-trip rentals. MDPI Electr. **11**(23), 11393 (2021). https://doi.org/10.3390/electronics112311393
6. Herbawi, W., Knoll, M., Kaiser, M., Gruel, W.: An evolutionary algorithm for the vehicle relocation problem in free floating carsharing. In: 2016 IEEE Congress on Evolutionary Computation (CEC), pp. 2873–2879 (2016). https://doi.org/10.1109/CEC.2016.7744152
7. Illgen, S., Höck, M.: Literature review of the vehicle relocation problem in one-way car sharing networks. Trans. Res. Part B: Methodol. **120**, 193–204 (2019). https://doi.org/10.1016/j.trb.2018.12.006
8. Li, S., et al.: A data-driven stochastic programming framework for vehicle relocation in car-sharing systems. arXiv **abs/2001.08109** (2020). https://arxiv.org/abs/2001.08109. Accessed: 26-02-2025
9. Lu, Y., Wang, K., Yuan, B.: The vehicle relocation problem with operation teams in one-way carsharing systems. Int. J. Prod. Res. **60**(12), 3829–3843 (2022). https://doi.org/10.1080/00207543.2021.1933238
10. Moritz, P., et al.: Ray: a distributed framework for emerging AI applications. In: Proceedings of the 13th USENIX Conference on Operating Systems Design and Implementation, pp. 561–577. OSDI'18, USENIX Association, USA (2018)
11. Muhlematter, F., et al.: Analysis of monthly demand in station-based car-sharing systems using machine learning. arXiv **abs/2303.14421** (2023). https://arxiv.org/abs/2303.14421. Accessed: 26-02-2025
12. Rodrigues, T., Franco, A., Ribeiro, B.: Optimizing vehicle relocation in free-floating car-sharing systems. IEEE Trans. Intell. Transp. Syst. **21**, 3458–3470 (2020)
13. Xu, Z., Liu, C.: Urban mobility patterns and carsharing demand: a spatial analysis. J. Transp. Geogr. **76**, 124–134 (2019)
14. Zhu, H., Hu, J.: Spatial patterns of carsharing usage: a geographical perspective. Computers, Environment and Urban Systems **77**, 101362 (2019). https://doi.org/10.1016/j.compenvurbsys.2019.101362

Bridging BPM and Multi-agent Systems: An Approach for Mixed-Mode Process Performance

Andrzej Wodecki[1], Dorota Jelonek[2], and Agnieszka Bitkowska[1(✉)]

[1] Warsaw University of Technology, Warsaw, Poland
pbab@wp.pl

[2] Czestochowa University of Technology, Częstochowa, Poland

Abstract. This paper explores the relationship between Business Process Management (BPM) and Multi-Agent Systems (MAS), characterizes Mixed-Mode Processes (MMP), and proposes a Hybrid Human–AI Agent Framework for evaluating their performance. It identifies key challenges in human–AI agent interaction and positions them as a basis for extending existing MAS evaluation frameworks to address the specific requirements of MMP assessment.

A systematic analysis is conducted to evaluate the extent to which current evaluation frameworks cover the critical dimensions relevant to mixed-mode processes.

Keywords: BPM · Multi · Agent Systems (MAS) · Mixed · Mode Process Performance (MMP) · Human · AI agent framework

1 Introduction

Business Process Management (BPM) enables the control of operational processes, cost monitoring, and the efficient utilization of resources [1–3]. One of the crucial area is Process Performance the formal, planned monitoring of process execution and the tracking of results to determine the effectiveness and efficiency of the process. A specialized method within the broader context of BPM that analyzes whether current processes are meeting their objectives. Using business process analysis helps identify harmful elements of an operation and determine how to overcome obstacles. Business process analysis uses a variety of tools and methodologies to look at these processes in different ways, as well as monitor metrics such as cost, efficiency, productivity. Efficient and effective processes are the backbone of any successful organization. However, many companies struggle to accurately monitor and measure process performance. This is a specialized method within the broader context of BPM that analyzes whether the processes currently being executed are meeting their intended objectives. Establishing appropriate key performance indicators (KPIs) is of critical importance for the continuous improvement of processes.

M. Pondel et al. (Eds.): AI4KAM 2025, IFIP AICT 776, pp. 173–182, 2026.
https://doi.org/10.1007/978-3-032-18920-2_16

The advancement of AI technologies is leading to the increasing implementation of mixed-mode processes in business, which combine elements of traditional business workflows with those based on multi-agent systems [4, 5]. In such configurations, part of the process is executed by humans following predefined procedures, while other tasks are performed by autonomous software agents. In this context, traditional KPIs [6] are no longer sufficient, as they fail to account for the specific challenges arising from human–agent interaction. Evaluating the performance of mixed-mode processes (MMP) therefore requires a new approach to business process performance—one that moves towards a hybrid human-AI agent framework.

The aim of this paper is to examine the relationship between BPM and MAS to characterize MMP and to propose a Hybrid Human-AI agent framework for evaluating the performance of such processes.

2 Background

This section outlines the background of our research. Section 2.1 presents the integration of BPM and MAS in mixed-mode process environments. Section 2.2 provides an overview of MAS evaluation frameworks. Section 2.3 identifies key challenges in human–AI agent interactions that should be considered as part of the extended MAS evaluation framework and, consequently, serve as a foundation for the proposed approach to mixed-mode process performance assessment.

2.1 Integrating BPM and Multi-agent Systems in Mixed-Mode Process Environments

It is necessary to constantly improve processes in contemporary organizations and adapt to the customer requirements and the dynamic changes taking place in the environment. BPM aims at improving some existing processes based on process performance [6] and is connected with is the strategic transformation of interrelated organizational subsystems. Process Performance (PP) is the formal, planned monitoring of process execution and the tracking of results to determine the effectiveness and efficiency of the process.

MAS have gained significant popularity in BPM and organizational environments thanks to their ability to simulate complex social interactions and automate decision-making processes. A particular area of their development is systems based on generative AI models, especially large language models (LLM), which enable advanced cooperation between agents and between humans and IT systems. The use of MAS in BPM creates MMP and necessitates the application of appropriate methods to evaluate their effectiveness and compliance with organizational, social, and ethical standards.

2.2 Multi-agent Systems Evaluation Frameworks

Agents (Small/Specialized or Large Language Models (S/LLMs) equipped with toolkits) open qualitatively new opportunities: they can decompose complex tasks into manageable subtasks, autonomously pick-up appropriate tools, and implement self-reflection mechanisms to control their actions. They can be organized into more or less complex structures, to MAS. Technologies like Model Context Protocol (MCP) [7] enable interoperability between agents and toolkits created with use of different technology stacks, what drives creation of new marketplaces (MCP Servers) [8]. Emerging communication protocols from Google, such as the Agent2Agent (A2A) Protocol [9], may serve as a foundation for agent marketplaces or architectures like the Agentic Mesh [10].

These developments attract the interest of different organizations. Companies experiment with Multi-Agent Systems (MAS)., consider integration of such solutions not only with their business processes, but also legacy systems. This may be a challenging task: mostly due to non-deterministic nature of S/LLMs, but also a lack of experienced experts, and still developing tech stack. That's why it is necessary to develop and implement appropriate evaluation and monitoring tools: not only to secure and control such hybrid (human + Agentic AI) processes, but also to better understand them and simply to learn.

The presented proposition deliver the high-level approach of such a MAS evaluation extension. It is called "challenges driven evaluation framework development": we start with the classification of current evaluation metrics, next identify potential challenges of human-Agentic AI interaction, identify the gaps and recommend the extended framework. Since Agentic AI is very dynamically evolving, one may expect the challenges will evolve too, and thus the resulting framework. We hope such a combination of IT and managerial perspective may positively impact MAS Agentic-AI adoption.

MAS have gained significant popularity in BPM and organizational environments thanks to their ability to simulate complex social interactions and automate decision-making processes. A particular area of their development is systems based on generative AI models, especially large language models (LLM), which enable advanced cooperation between agents and between humans and IT systems. However, the use of MAS requires the application of appropriate methods to assess their effectiveness and compliance with organizational, social, and ethical expectations.

There are numerous frameworks and approaches supporting the evaluation and monitoring of MAS and most of them primarily focus on technical or agent-specific aspects. Six main categories of evaluation measures have been identified:

1. Performance & Effectiveness, e.g., Task Success/QA Correctness [11], Answer Accuracy [12], Output Quality (Relevance/Completeness) [13], Processing Time / Response Latency [11], Cycle Time [14], Throughput [15], Error Rate [15], Agent Goal Accuracy [12].
2. Interaction & Coordination, e.g.Communication Frequency [14], Communication Brevity (token count) [11].
3. Alignment & Compliance, e.g. Policy Violation Rate [16], Bias Detection [17], Topic Adherence [12].
4. Transparency & Explainability, e.g. Explanation Quality [18], Groundedness Score [12, 13], Faithfulness [12].
5. Robustness & Adaptability, e.g. Stress-Test Performance/Reliability Score [19], Novel Input Handling (drift awareness) [11],
6. Resource Efficiency, np. Computational Resource Usage [15], API/Token Consumption [11], Cost Per Decision [14], Response Latency [11].

When analyzing the above, it can be seen that the currently available measures focus mainly on the technical aspects of MAS operation. These measures document well the correctness, quality of responses, efficiency of resource use, and basic compliance with standards e.g., error detection, measurement of deviation from norms. However, in the practice of MMP evaluation, there is a lack of an approach that incorporates social, organizational, and strategic aspects, such as the synchronization of decisions made by multiple agents, the effectiveness of teamwork, the efficiency of control transfer between agents and humans, and a detailed analysis of the impact of agent decisions on the KPIs of the entire organization.

Despite significant developments in tools such as Arize Phoenix, Ragas, TruLens, and LangSmith, which enable the measurement of many technical aspects of MAS, the lack of more complex coordination, adaptation, and strategic measures means that company management often lacks sufficient knowledge to assess the full impact of MAS on the organization's operations. For this reason, the current state of knowledge requires further research to fill this gap and create tools that enable a more accurate diagnosis mixed-mode process context.

2.3 Human – AI Agent Interaction Challenges

The MAS evaluation frameworks presented in the previous section largely reflect the perspective of IT system developers. However, in business practice, many new aspects may arise whose "control" may be crucial for the effectiveness of multi-agent systems. Authors propose to start with an analysis of possible challenges in human- AI agent interactions. These challenges will later serve as a framework for extending MAS evaluation systems. The following challenges in human-agent AI interactions have been identified: system reliability & security, Human–AI agent collaboration & interaction, organizational interaction, societal, ethical & compliance.

In the context of the reliability and security of MAS emergent behavior management remains a significant challenge [20]. Agents based on large language models (LLM-driven agents) exhibit unpredictability in their actions, especially in the context of multi-faceted inter-agent interaction. These behaviors can accumulate, leading to instability of the entire system. In addition, errors typical of LLM models, such as hallucinations, tend to propagate between agents, which can lead to cascading errors [21]. MAS systems generate new operational challenges related to maintaining and ensuring operational resilience and maintenance. These require regular model updates, monitoring of performance metrics, and the introduction of fallback modes to ensure system stability even in the event of individual agent failures.

Cooperation between humans and AI agents faces significant communication challenges, particularly in the area of transparency. Agents often have difficulty interpreting imprecise commands, while humans struggle to understand the reasoning behind AI decisions, leading to a trust deficit. The challenge is to properly calibrate trust and reliance between humans and AI. Excessive trust can lead to overdependence and a loss of critical control, while a lack of trust can result in the underutilization of agents' potential [22]. It is therefore necessary to clearly define the involvement of humans in key decision-making points - Human-in-the-loop role definition, ensuring appropriate control and intervention in the event of deviations from expected results.

The integration of multi-agent systems into organizations presents notable adaptation challenges, primarily due to difficulties in aligning AI agents with legacy systems like ERP or domain-specific applications. These systems often lack support for MAS-compatible APIs, requiring additional integration layers, which increases the risk of failure and implementation costs. In the context of team dynamics, AI agents can generate role ambiguity and team dynamics, requiring organizations to take a new approach to delegating tasks and responsibilities [22]. In quality assurance and process control, agents generate additional challenges related to the validation of non-deterministic behavior and the need for continuous monitoring and feedback to identify and correct errors. Organizational readiness is also a challenge, involving the development of new employee competencies and overcoming resistance to the autonomy of artificial intelligence. Effective implementation depends on strong managerial support and well-defined oversight mechanisms.

Another challenge in human–AI agent interactions lies in addressing societal, ethical, and compliance-related concerns. A key risk is the potential for discrimination and the reinforcement of biases, stemming from the fact that AI models are trained on historical data that may contain hidden biases. In a multi-agent environment, the biases of one agent can be reinforced by the actions of others, leading to systematic violations of fairness principles. The issue of responsibility and accountability also poses a significant challenge. The lack of a clear legal framework means that organizations implementing autonomous AI systems must clearly establish who is responsible for the decisions made by agents. An additional challenge is ensuring the ethical use and oversight of technology. Decisions made by AI agents, while legally compliant, may violate the ethical standards of an organization or society. For example, autonomous marketing systems can lead to manipulation or unethical targeting of vulnerable customers.

3 Rethinking Mixed-Mode Process Performance: Towards a Hybrid Human-AI Agent Framework

The aim of the analysis is to determine the extent to which current MAS evaluation framework provide adequate coverage in the identified areas, with particular consideration of the challenges related to human–AI agent interaction. In the domain of system reliability and security, the majority of existing frameworks—particularly Arize, RAGAS, TruLens, and LangSmith—demonstrate strong capabilities in evaluating technical performance, offering robust and well-established metrics related to accuracy, precision, and operational efficiency. Although enterprise platforms such as Google and Azure offer continuous performance monitoring, current frameworks lack sufficient mechanisms for assessing security aspects such as adversarial robustness and data privacy, indicating the need for dedicated resilience testing modules.

In the area human-AI agent collaboration is a noticeable gap. Current frameworks offer only limited proxy metrics for user experience—e.g., TruLens measures user sentiment, and LangSmith enables the collection of human feedback. However, there is no dedicated metric to measure how effectively humans and AI agents collaborate on task completion. The metrics proposed in our hypothetical framework e.g., user satisfaction rating, frequency of human overrides of AI decisions, effectiveness of AI suggestions, etc. are not directly measured by existing tools.In the area of organizational integration, partial coverage can be observed in domains such as cost tracking (as supported by LangSmith) and compliance reporting (as implemented in the Azure Scorecard), but no framework currently measures business value or integration difficulty.

Coverage in societal and ethical and compliance area is uneven. Fairness and bias metrics are very well supported by IAIF360 [16], as well as Fairlearn/Azure [23, 24] and Google's Fairness Indicators [25]. Transparency is included in enterprise tools through explainability and documentation e.g., model cards in Google, interpretability in Azure. Safety is partially supported by Arize and TruLens for content output, and by Google through content filters. However, no framework, except Google and Azure, covers all of these aspects simultaneously. Support for compliance with regulations or ethical guidelines is strongest in cloud platforms that implement AI principles and offer audit logs e.g., Google and Microsoft frameworks. An area for improvement would be to integrate these ethical metrics into a single tool that would simultaneously assess fairness, transparency, and security would reduce user effort. An attempt to address the identified research gap is the proposal of new measures and indicators, grouped according to the gaps observed in current evaluation methods (Table 1).

Table 1 consistently incorporates the four identified areas: System Reliability & Security, Human-AI Collaboration & Interaction, Organizational Integration, and Societal, Ethical & Compliance. In the second column, the Category is provided. The third column presents an assessment of the extent to which the indicators within each category provide coverage for the evaluation needs of MMP. We distinguish between areas that are comprehensively addressed by existing evaluation systems (full), those that are only partially covered, and those where coverage is lacking (none). In the fourth column, proposed metrics are presented.

In summary, current evaluation approach tend to be specialized: open-source tools such as RAGAS and TruLens focus on model output quality, while corporate frameworks such as AIF360 focus on fairness and Azure on debugging and interpretability. A hypothetical framework for MAS, on the other hand, is ambitious—it encompasses technical, human, organizational, and social dimensions. The comparison reveals clear gaps—particularly in human-centered metrics and in an integrated approach—that none of the existing solutions fully address. Gap analysis presented above forms the basis for the proposed method of evaluating hybrid systems, i.e., systems in which humans and multi-agent structures are full participants in business processes.

Addressing four key areas: System Reliability & Security, Human-AI Collaboration & Interaction, Organizational Integration, and Societal, Ethical & Compliance, it takes into account the comprehensive needs of modern organizations managing interactions between humans and advanced Agentic AI. Its goal is to provide managers with practical, measurable indicators that enable a comprehensive assessment of the effectiveness and quality of MAS in a dynamic business environment.

Naturally, areas with partial or none coverage needs to be extended with set of appropriate, quantitative or qualitative, measures. The complete proposition of such a set is beyond the scope of this paper, but one can quite easily design metrics like *Expectation alignment Rate* or *Confidence Calibration* for Trust Calibration, Hand-off

Table 1. Hybrid Human-AI agent Framework for evaluating the performance MMP

Area	Category	Coverage	Merics
System Reliability & Security	Predictability & Consistency	Full	Outcome consistency rate, Error pattern predictability, Variance in response quality, Behavior stability under similar inputs,
	Error Management	Full	Error detection rate, Recovery success rate, Error propagation containment, Graceful degradation measures, Self-correction capability
	Security & Safety	Partial	Attack surface measurement, Security vulnerability detection, Tool usage policy adherence, Data access protocol compliance, Authentication success rate
Human-AI Collaboration & Interaction	Communication Quality	Partial or None	Communication clarity score, Information exchange efficiency, Explanation comprehensibility, Information accessibility, Reasoning transparency
	Trust Calibration	None	Expectation alignment rate, Predictability perception, Confidence calibration, Trust breakdown frequency, Appropriate reliance measurement
	Role & Control Clarity	None	Handoff success rate, Role boundary clarity, Control transfer smoothness, Authority respect measurement, Initiative appropriateness, Human override effectiveness

(*continued*)

Table 1. (*continued*)

Area	Category	Coverage	Merics
Organizational Integration	Workflow Integration	Partial	Process continuity measurement, Transition smoothness, Integration effort measurement, Technical debt introduced, Maintenance overhead
	Skill & Training Impact	None	Learning curve steepness, Time to proficiency, Training resource requirements, Skill development impact, Knowledge transfer effectiveness
	Organizational Adaptation	None	Process compliance rate, Role adaptation success, Change accommodation rate, Procedural flexibility, Cultural fit measurement
Societal, Ethical & Compliance	Fairness & Bias	Full	Demographic parity measurement, Disparate impact assessment, Performance consistency across demographics, Bias detection success
	Accountability & Governance	Partial	Decision attribution clarity, Audit trail completeness, Responsibility assignment clarity, Governance framework adherence, Incident response effectiveness
	Regulatory & Ethical Compliance	Partial	Regulatory requirement fulfillment, Data protection compliance, Required disclosure completeness, Ethical principle adherence, Value alignment measurement

success rate or Role Boundary Clarity for Role & Control Quality, or Learning curve steepness, Time to proficiency for Skill & Training Impact.

The implementation of the proposed measures and indicators requires a systematic process involving measurement stages and the use of appropriate management tools and procedures. In particular, a structured process of measurement and communication of results (including the establishment of benchmarks for individual indicators, regular (e.g., quarterly) measurements, analysis of measurement data and preparation of evaluation reports, integration of measurement results into decision-making processes, etc.) and appropriate tools and procedures (e.g., dedicated analytical tools, internal audit procedures for indicator validation, reporting and monitoring systems for compliance with policies and standards, or surveys and diagnostic tools for collecting user feedback. A detailed discussion of these elements is beyond the scope of this study.

4 Discussion and Conclusions

Nowadays BPM requires supporting by many tools and systems. The implementation of the proposed measures and indicators form PP requires a systematic stages and the use of appropriate tools and procedures. The crucial area are: measurements, analysis of measurement data and preparation of evaluation reports, integration of measurement results into decision-making processes and appropriate tools and procedures. It worth underlaying the communication of results. All this aspects should be more explored from theoretical and practical aspects. Several limitations of our research must be acknowledged. Further investigation on this topic should be incentivized to generalize and complement the findings as far as possible, as well as to organizations and their experiences with the from PM perspective in the contemporary dynamic environment.

It is worth underlying that this study will be valuable not only for developers of analytics platforms, AI Experts, BPM Experts, but perhaps above all for managers and organizations, which considering implementing them. As a result, the value of this study lies in the approach to bridging BPM and MAS that in the future give a perspective for the specific frameworks to preparing the new solution and implementation projects. The article's findings may still offer astute insights to researchers and practitioners who wish to consider more information regarding the importance of the MAS for PM to optimize their processes based on the indicators.

References

1. Dumas, M., La Rosa, M., Mendling, J., Reijers, H.A.: Fundamentals of business process management (2018). https://doi.org/10.1007/978-3-662-56509-4
2. Bitkowska, A., Korneta, P., Hawrysz, L.: Business processes and performance management of medium-sized outpatient clinics through the Sars-Cov-19 period. A case study from Poland. Scientific Papers of Silesian University of Technology, vol. 160, pp. 95–108 (2022)
3. vom Brocke, J., Mendling, J., Rosemann, M. (eds.) Business Process Management Cases Vol. 3: Implementation in Practice. Springer, Cham (2025)
4. Witkowski, A., Wodecki, A.: An exploration of the applications, challenges, and success factors in AI-driven product development and management. FoM **16**(10), 139–156 (2024)
5. Wodecki, A.: Artificial Intelligence in Management. Edward Elgar Publishing (2020)
6. Parmenter, D.: Key Performance Indicators. Wiley (2015)
7. Model Context Protocol. https://modelcontextprotocol.io/introduction. Accessed 07 May 2025
8. MCP Servers (2025). https://mcp.so

9. Agent2Agent Protocol (A2A). https://google.github.io/A2A/. Accessed 07 May 2025
10. Broda, E.: Agentic Mesh: The Future of Generative AI-Enabled Autonomous Agent Ecosystems. https://medium.com/data-science/agentic-mesh-the-future-of-generative-ai-enabled-autonomous-agent-ecosystems-d6a11381c979. Accessed 07 May 2025
11. Arize Phoenix. https://phoenix.arize.com/. Accessed 06 May 2025
12. Ragas. https://docs.ragas.io/en/stable/. Accessed 06 May 2025
13. TruEra: TruLens. https://www.trulens.org. Accessed 06 May 2025
14. LangSmith. https://www.langchain.com/langsmith. Accessed 06 May 2025
15. OpenTelemetry. https://opentelemetry.io/. Accessed 06 May 2025
16. Open Policy Agent. https://www.openpolicyagent.org/. Accessed 06 May 2025
17. AI Fairness 360. https://ai-fairness-360.org/. Accessed 06 May 2025
18. Sithakoul, S., Meftah, S., Feutry, C.: BEExAI: Benchmark to Evaluate Explainable AI (2024). http://arxiv.org/abs/2407.1989710.48550/arXiv.2407.19897
19. Reliability Testing & Chaos Engineering /Gremlin. https://www.gremlin.com/. Accessed 5 May 2025
20. Hu, J., et al.: Position: Towards a Responsible LLM-empowered Multi-Agent Systems (2025). http://arxiv.org/abs/2502.01714. https://doi.org/10.48550/arXiv.2502.01714
21. LLM Agents: How They Work and Where They Go Wrong. https://www.holisticai.com/blog/llm-agents-use-cases-risks. Accessed 06 May 2025
22. Lou, B., Lu, T., Raghu, T.S., Zhang, Y.: Unraveling Human-AI Teaming: A Review and Outlook (2025). http://arxiv.org/abs/2504.05755. https://doi.org/10.48550/arXiv.2504.05755
23. Fairlearn. https://fairlearn.org. Accessed 06 May 2025
24. lgayhardt: Machine learning fairness-Azure Machine Learning. https://learn.microsoft.com/en-us/azure/machine-learning/concept-fairness-ml?view=azureml-api-2
25. Introduction to model evaluation for fairness/|Vertex AI. https://cloud.google.com/vertex-ai/docs/evaluation/intro-evaluation-fairness. Accessed 06 May 2025

AI in Business Processes, Logistics, Industry, and Market Behavior

Analysis of the Interaction Between AI Technologies for Management – Cloud Business Intelligence and Enterprise Specific Features

Paula Bajdor[1](✉), Marta Starostka-Patyk[2], and Agnieszka Sokołowska-Durkalec[3]

[1] Warsaw University of Technology, Warsaw, Poland
paula.bajdor@gmail.com
[2] Calisia University, Calisia, Poland
[3] Wroclaw University of Economics and Business, Warsaw, Poland

Abstract. Modern enterprises increasingly adopt Cloud Business Intelligence (CBI) systems to enhance decision-making, data analysis, and operational flexibility. AI plays a central role in these systems by enabling real-time data processing, pattern detection, and predictive insights. However, the perceived benefits of CBI may vary based on organizational characteristics, such as business profile and size. This article investigates whether these factors significantly influence how enterprises assess CBI benefits, offering insights valuable for managers and contributing to the broader discussion on Artifical Intelligence's practical role in business analytics.

Keywords: Business intelligence · Cloud Bussines Intelligence · Artificial Intelligence · Management

1 Introduction

The increasing digitalization, the growth of competition, and the explosion of data mean that companies are increasingly implementing advanced analytical tools that allow not only to transform data into real, valuable information but also to ensure fast and effective decision-making. The growing importance of cloud computing and progress in artificial intelligence have contributed to the emergence of Cloud Business Intelligence (CBI) systems that actively use AI, revolutionizing how companies collect, analyze, and interpret data to make informed decisions at all enterprise management levels [1, 2]. Cloud environments, offering scalability, flexibility, and the necessary computing power and ability to analyze massive amounts of data with minimal upfront costs [3], provide a valuable infrastructure to support AI-powered Business Intelligence solutions, enabling enterprises to use real-time data for forecasting, pattern discovery, customer preference analysis, and strategic planning [4–6]. As a result, modern CBI systems, through the use of machine learning algorithms, natural language processing (NLP), or predictive

Published by Springer Nature Switzerland AG 2026
M. Pondel et al. (Eds.): AI4KAM 2025, IFIP AICT 776, pp. 185–192, 2026.
https://doi.org/10.1007/978-3-032-18920-2_17

analytics, enable not only advanced analysis and understanding of the past from often unstructured and complex data but also predict future events, automate reasoning and recommend proactive decisions and business actions [7, 8].

Artificial intelligence (AI) has a significant impact on the management of organizational knowledge, enhancing its creation, distribution, and utilization. Technologies such as machine learning, NLP, or intelligent agents enable the acquisition of knowledge from large, unstructured datasets by identifying patterns and generating insights in real-time [9]. AI also facilitates the seamless flow of information between departments, including through recommendation systems, thereby reducing knowledge fragmentation [10]. Additionally, personalization of access to information and support for data-driven decisions enhances organizational innovation and flexibility [11]. Thus, AI becomes the foundation of modern, intelligent knowledge-based organizations.

Moreover, AI significantly expands the capabilities of CBI systems, going beyond traditional data analysis. It enables, among other things, predicting trends and customer behaviors (predictive analytics), intuitive handling of queries in natural language, and automation of reports and notifications. For example, AI supports the detection of sales anomalies, demand forecasting, and automatic customer segmentation [12]. Deep learning is also used for automatic decision-making and identifying irregularities in financial transactions [13]. Such applications increase the precision of analyses and facilitate access to information throughout the organization [2]. In practice, CBI systems with AI can perform sentiment analysis (in customer opinion surveys), trend prediction (e.g., demand forecasting), or automatic reporting. The main benefits of modern CBI systems containing AI tools are increased precision of analyses, better access to data for decision-makers, and faster decision-making, which optimizes performance and improves operational efficiency in today's world dominated by data [14]. As a result, using CBI systems powered by AI allows companies to understand better and anticipate market needs, effectively respond to customer needs, and improve their market position [15].

The perception of benefits resulting from the implementation of CBI may, however, significantly differ depending on the organizational characteristics of enterprises. Showing these differences, which affect the assessment of benefits resulting from implementing modern and intelligent CBI systems, is crucial both from the point of view of management practice and further development of research in the digital transformation of enterprises. The article examines whether there are significant statistical dependencies in comparing assessments of benefits resulting from CBI systems depending on the adopted criteria. For this purpose, two research hypotheses were formulated:

H1: There are significant differences in assessing the benefits of implementing CBI systems depending on the company's business profile;

H2: Depending on the company's size, there are significant differences in assessing the benefits of implementing CBI systems.

The research results deepen knowledge of the determinants of implementing CBI systems in companies and provide recommendations for managers and decision-makers implementing digitalization strategies based on advanced data analytics.

2 The Methodology

To determine the extent of management support of medium and large companies by CBI systems and to determine the potential directions of development of Cloud BI systems, the authors surveyed the research sample of 400 Polish enterprises (200 medium-sized and 200 large enterprises) from various sectors (mainly industry and commerce), which were using a Cloud Business Intelligence system. The research was conducted using the CAWI method, applying a 5-point Likert. The research involved personnel who understand the role and importance of IT solutions, decide/co-decide on the purchase and implementation of IT systems in the cloud, and know the effects of using CBI systems in the enterprise.

Table 1 presents the variables' descriptions, symbols, types, and main values of descriptive statistics of all variables used in the analysis.

The U'Mann-Whitney test was used to check whether there were significant differences in the dependent variables for two independent groups. In the first step, whether there were significant differences in assessing the benefits resulting from CBI depending on the company's business profile was checked. In the second step, it was checked whether there were significant differences in assessing the benefits of CBI depending on the size of the enterprise. Thus, the first group compared companies operating in the Industry, Construction, Trade, Transport and warehouse and Accommodation and catering. The second group included a comparison between medium and large companies.

2.1 The Research Results

The first group compared companies operating in the Industry, Construction, Trade, Transport and warehouse and Accommodation and catering (Table 2).

The results show that minimal p values (all 0.000) in the compared groups suggest statistically significant differences in the assessment of benefits resulting from the use of CBI by enterprises. From the 13 criteria, these significant differences were only identified in the case of six. The analyzed groups of enterprises, in terms of the business profile, most often indicated the occurrence of significant differences in the assessment of benefits resulting from the use of CBI in terms of enabling the achievement of business goals (1), enabling better implementation of management functions (planning, organizing, motivating, controlling) (6), accelerating the digital transformation of the enterprise (7), improving the flexibility of the enterprise (faster response to market changes concerning customers, partners, competitors) (10), improving communication and cooperation with customers, business partners and between employees (11) and improving existing products and services (13).

The second group included a comparison between medium and large (Table 3).

Based on the results obtained, it can only be concluded that in only three cases, there were significant differences – large companies significantly more often indicated that the use of CBI enabled the achievement of the assumed business goals ($Z = 11.762$, $p = 0.000$) and enabled better implementation of management functions ($Z = 2.594$, $p = 0.009$) than medium companies. Only when using CBI enabled the improvement

Table 1. Descriptive statistics

Variable	Symbol	Mean	S.E	Var	S.D
CBI enables achieving the assumed business goals	1	10,75	1,40	0,37	29,09
CBI enables better recognition and use of emerging market opportunities	2	3,85	0,89	0,09	17,84
CBI affects the possibility of distinguishing the company from the competition and taking a better market position	3	3,70	0,61	0,48	12,19
CBI affects the development of the company's key competencies	4	4,02	0,61	0,47	12,23
CBI enables diversification of the company's activities	5	4,34	0,70	0,24	14,11
CBI enables better implementation of management functions (planning, organizing, motivating, controlling)	6	4,26	0,73	0,22	14,77
CBI accelerates the digital transformation of the company	7	3,60	0,53	0,72	10,80
CBI enables streamlining the company's activities	8	3,99	0,60	0,54	12,11
CBI enables reducing the company's operating costs in the long term	9	4,13	0,66	0,40	13,31
CBI improves the company's flexibility (faster response to market changes concerning customers, partners, and competitors)	10	3,86	0,55	0,65	11,19
CBI improves communication and cooperation with customers, business partners, and employees	11	5,70	0,92	0,34	18,85
CBI improves knowledge management processes (concerning the acquisition, verification, storage, dissemination and use of knowledge)	12	11,79	2,43	0,00	49,67
CBI enables the improvement of existing products and services	13	6,26	1,16	0,09	23,78

Table 2. The U'Mann-Whitney test results (company's profile).

Industry/Construction				Industry/Trade			Industry/Transport and warehouse		
Variables	U	Z	p	U	Z	p	U	Z	p
1	8820,0	−9,977	0,000	4000,0	6,534	0,000	8820,0	−9,977	0,000
6	-	-	-	-	-	-	960,0	−7,582	0,000
7	580,0	−8,531	0,000	-	-	-	2380,0	−4,040	0,000
10	820,0	9,977	0,000	3940,0	6,632	0,000	960,0	−7,582	0,000
13	-	-	-	-	-	-	820,0	9,977	0,000
Industry/Accommodation and catering				Construction/Trade			Construction/Transport and warehouse		
Variables	U	Z	p	U	Z	p	U	Z	p
1	8820,0	−9,977	0,000	3240,0	8,904	0,000			
6	-	-	-	-	-	-	180,0	−5,961	0,000
7	1600,0	5,986	0,000	180,0	7,902	0,000	200,0	5,768	0,000
10	820,0	9,977	0,000	-	-	-	2420,0	−7,693	0,000
11	-	-	-	800,0	-4,450	0,000	-	-	-
13	-	-	-	-	-	-	820,0	7,693	0,000
Construction/Accommodation and catering				Trade/Transport and warehouse			Trade/Accommodation and catering		
Variables	U	Z	p	U	Z	p	U	Z	p
1	-	-	-	1560,0	−8,904	0,000	4020,0	-8,904	0,000
6	-	-	-	380,0	−6,788	0,000	-	-	-
7	820,0	7,693	0,000	780,0	−4,562	0,000	800,0	4,450	0,000
10	-	-	-	180,0	−7,902	0,000	800,0	4,450	0,000
13	-	-	-	1600,0	8,904	0,000	-	-	-
Trade/Accommodation and catering									
Variables	U	Z	p						
6	200,0	5,768	0,000						
7	820,0	7,693	0,000						
10	820,0	7,693	0,000						
13	2420,0	-7,693	0,000						

of existing products and services did medium companies significantly more often indicate this (Z = -3.459, p = 0.000). In the remaining cases, no significant differences were shown. Considering that there were 13 criteria analyzed in total, the significant differences in the perception of these criteria only concerning three of them allow us to conclude that regardless of the size of the company, the analyzed business entities tend

Table 3. The U'Mann-Whitney test results (company's size).

Large/Medium			
Variables	U	Z	p
CBI enables achieving the assumed business goals	6400,0	11,762	0,000
CBI enables better implementation of management functions (planning, organizing, motivating, controlling)	17000,0	2,594	0,009
CBI enables the improvement of existing products and services	16000,0	−3,459	0,000

to maintain consistency in the perception of the benefits resulting from the use of CBI systems.

In general, there are consistent and significant differences between the business profiles and the size of the analyzed companies in the case of many variables, which means that, on the one hand, both research hypotheses can be considered confirmed. However, on the other hand, the occurrence of significant differences only between six of the thirteen variables/criteria adopted means that it is safer to assume that both research hypotheses were only partially confirmed. Additional analyses are needed to confirm or reject them finally.

3 Conclusions

The empirical results reveal nuanced patterns in the way different organizational profiles perceive the benefits of implementing CBI. It is clear that the perception of CBI benefits differs mainly depending on the industry profile, not the size of the company. Technologically advanced or digitally intensive sectors are more likely to value flexibility, digital transformation, and product innovation. On the other hand, slight differences between medium and large companies suggest that it is not the size of the organization that matters, but its digital maturity and strategy alignment with AI capabilities.

The perceived benefits of Cloud Business Intelligence (CBI) systems, such as improved goal implementation, greater flexibility, and support for management functions, are closely related to knowledge management (KM) outcomes. AI-based CBI supports decision quality through real-time data analysis and trend prediction, facilitates knowledge reuse through pattern recognition, and increases its transparency through automatic reporting and NLP. Thus, it enhances the organization's ability to manage knowledge effectively in the face of digital transformation. The obtained results emphasize the need for an individual approach to implementing analytical technologies in the cloud, taking into account the unique characteristics of the enterprise. The conclusions from the research can be a practical guide for managers planning CBI implementations and for cloud service providers, who should adapt their offers to the organization's diverse needs. Therefore, it is recommended to adjust implementation strategies (companies should take into account their industry and scale of operations when designing CBI systems to maximize the benefits obtained), segment the offer by service providers, prioritize functions and strengthen organizational competences.

Because significant statistical differences were demonstrated only in six of the thirteen analyzed criteria, it is advisable to continue further research to verify the hypotheses more precisely. In particular, the following should be considered: expanding the catalog of analyzed variables (including additional criteria, such as the level of digital transformation, analytical competencies in the organization or organizational culture), deepening the research using qualitative methods (using in-depth interviews and case studies to understand better the specifics of implementations and the perception of benefits in various types of enterprises). Moreover, it is worth supplementing the research with qualitative methods, such as interviews or case studies, which will enable a more comprehensive understanding of the impact of organizational culture, management style, and digital maturity on the perception and use of CBI.

Disclosure of Interests. The authors have no competing interests to declare that are relevant to the content of this article.

References

1. Rane, N., Paramesha, M., Choudhary, S., Rane, J.: Business intelligence through artificial intelligence: a review. SSRN Electron. J. (2024)
2. Eboigbe, E.O., Farayola, O.A., Olatoye, F.O., Nnabugwu, O.C., Daraojimba, C.: Business intelligence transformation through AI and data analytics. Eng. Sci. Technol. J. **4**(5), 285–307 (2023)
3. Khan, S., Shah, K., Soni, P.: Cloud computing and artificial intelligence for efficient business intelligence systems. J. Cloud Technol. **22**(5), 142–156 (2024)
4. Huang, Z., Hu, Q.: Machine learning in business intelligence systems: opportunities and challenges. Expert Syst. Appl. **222**, 119512 (2023)
5. Jelonek, D., Mesjasz-Lech, A., Stępniak, C., Turek, T., Ziora, L.: The artificial intelligence application in the management of contemporary organization: theoretical assumptions, current practices and research review. In: Advances in Information and Communication: Proceedings of the 2019 Future of Information and Communication Conference (FICC), vol. 1, pp. 319–327, (2020)
6. Jelonek, D., Stępniak, C., Ziora, L.: The meaning of big data in the support of managerial decisions in contemporary organizations: review of selected research. Adv. Intell. Syst. Comput. **886**, 361–368 (2019)
7. Mariani, M., Dwivedi, Y.K.: Generative artificial intelligence in innovation management: a preview of future research developments. J. Bus. Res. **175**(C) (2024)
8. Smith, T.: Unveiling hidden insights in large datasets with AI-driven BI tools. Data Sci. Cloud Comput. Rev. **19**(2), 56–74 (2024)
9. Sharma, D., Kumar, V.: Enhancing organisational intelligence integration of artificial intelligence and knowledge management: frameworks in India. In: Digital Technologies in Modeling and Management: Insights in Education and Industry, pp. 167–184. IGI Global (2024)
10. Sira, M.: Artificial intelligence in knowledge capture and transfer: breaking down organizational silos (2024)
11. Jarrahi, M.H., Askay, D., Eshraghi, A., Smith, P.: Artificial intelligence and knowledge management: a partnership between human and AI. Bus. Horiz. **66**(1), 87–99 (2023)
12. Sharma, M., Kumar, P., Gundewar, S.: Leveraging AI and machine learning for predictive analytics in business intelligence. In: AI-Powered Business Intelligence for Modern Organizations, pp. 29–50. IGI Global (2025)

13. Talwandi, N.S., Shah, A.K., Thakur, R., Khare, S., Thakur, P.: Cloud based data analytics for business intelligence. In: 2024 2nd International Conference on Advances in Computation, Communication and Information Technology (ICAICCIT), vol. 1, pp. 1078–1081. IEEE (2025)
14. Lee, M., Zhang, R., Xu, Q.: Business intelligence in cloud ecosystems: enhancing predictive analytics with AI. Int. J. Bus. Intell. Anal. **29**(4), 77–93 (2023)
15. Stephen, M., Fernández, M.: Seamless integration of SAP BW/4HANA with cloud-based BI solutions for real-time insights. Artif. Intell. **1**, 2992–3894 (2025)

Application of Rough Set Theory in Supporting Decision-Making on the Implementation of Cloud Business Intelligence Systems in Enterprises

Damian Dziembek[1(✉)], Karol Kuczera[2], Oskar Szumski[3], Csaba Bálint Illés[4], and Anna Dunay[4]

[1] Czestochowa University of Technology, Czestochowa, Poland
dziembor@interia.pl
[2] University of Szczecin, Szczecin, Poland
[3] University of Warsaw, Warszawa, Poland
[4] John Von Neumann University, Budapest, Hungary

Abstract. In the face of increasing data complexity and the demand for agile, data-driven decision-making, Cloud Business Intelligence (CBI) systems have emerged as crucial enablers of digital transformation in enterprises. However, the decision to implement CBI involves numerous uncertainties related to organizational, technological, and strategic factors. This article explores the applicability of Rough Set Theory (RST) as a tool for supporting decision-making under such uncertain conditions. RST's capacity to process incomplete and imprecise data makes it particularly suitable for identifying critical attributes that influence the adoption of CBI. The study formulates decision rules derived from real-world data, offering insights into key implementation factors such as peer recommendations, prior project analysis, vendor offerings, the impact of the COVID-19 pandemic, public cloud usage, and existing limitations of CBI. The findings highlight the value of RST in guiding IT implementation strategies and adapting recommendation systems in response to evolving business conditions. Future research will focus on expanding the variable set and developing automated learning mechanisms for ongoing refinement of CBI-related decisions.

Keywords: Rough Set Theory · Cloud Business Intelligence Systems · Decision-Making

1 Introduction

Modern enterprises operating in increasingly dynamic markets face a multitude of challenges arising from intensifying competition, shifting customer expectations, evolving market structures, and the rapid development of digital technologies. In such a complex, uncertain, and turbulent environment, the ability of managerial staff to make timely, accurate, and data-driven decisions constitutes a critical determinant of organizational

Published by Springer Nature Switzerland AG 2026
M. Pondel et al. (Eds.): AI4KAM 2025, IFIP AICT 776, pp. 193–205, 2026.
https://doi.org/10.1007/978-3-032-18920-2_18

success. The ongoing digital transformation of enterprises, accompanied by the exponential growth of data generated both internally and externally, the heterogeneity of data types, and the growing pressure to enhance operational efficiency, significantly complicate the process of managerial decision-making. Traditional decision-making approaches, relying predominantly on intuition or simplified data analysis, frequently prove insufficient to address the needs of modern organizations. Consequently, managers increasingly seek advanced analytical tools capable of processing vast, distributed, and diverse data sources, thereby generating actionable insights and supporting strategic and operational decision-making.

Conventional analytical tools, such as on-premise Business Intelligence (BI) systems, are becoming inadequate in light of rising data diversity, the demand for real-time access to information, and the need for flexible, adaptive analysis. A promising alternative is represented by Cloud Business Intelligence (CBI) systems, which leverage cloud computing capabilities to facilitate data processing, integrate information from heterogeneous IT systems, and provide interactive access to analytical outcomes irrespective of location. CBI systems not only enable efficient collection and processing of diverse data types but also ensures flexibility and scalability in line with evolving business needs. Furthermore, its implementation does not require substantial investments in local IT infrastructure, thereby reducing deployment time and lowering entry barriers, particularly for small and medium-sized enterprises.

Nevertheless, the adoption of CBI solutions is accompanied by significant uncertainty, stemming from diverse strategic objectives, organizational limitations, technological barriers, and market volatility. The success of implementation further depends on the quality of acquired data, the maturity of analytical methods, the robustness of security protocols, and the competencies of the workforce. As a result, managerial decision-making concerning CBI systems implementation necessitates the consideration of multiple interdependent factors, which are often imprecise, incomplete, or difficult to formalize.

In this context, Rough Set Theory (RST), introduced in the 1980s by Zdzisław Pawlak, emerges as a particularly valuable methodological approach. Recognized as an artificial intelligence technique, RST enables effective decision support under conditions of incomplete or imperfect data. By identifying decision-making patterns within datasets without the requirement of prior probabilistic assumptions, RST is particularly well-suited to environments characterized by uncertainty and partial knowledge, which are typical in business contexts.

The objective of this article is to identify the determinants influencing the decision to implement Cloud Business Intelligence solutions in enterprises through the application of Rough Set Theory. To this end, the article provides a conceptual overview of Cloud Business Intelligence systems and outlines the principles and applicability of the RST methodology.

2 Outline of Cloud Business Intelligence Systems

Business Intelligence (BI) systems play a crucial role in supporting decision-making processes within enterprises by collecting, integrating, analyzing, and visualizing data. The primary task of BI systems is to transform the accumulated data concerning the enterprise and its environment into useful information and knowledge [1, 2].This enables managers to gain deeper insights into customers, business partners, competitors, as well as internal and external organizational processes. As a result, more accurate decisions can be made, which contribute to improving the overall efficiency of the enterprise [3–5].

BI systems can function in two basic models: the traditional on-premise model, based on the organization's own infrastructure, and the cloud-based model, in which business analytics is provided as an e-service through the IT infrastructure of an external provider or the enterprise's internal cloud resources. Increasing pressure to reduce IT-related costs, the growing importance of access to large datasets (both internal and external), and the demand for faster access to decision-making reports have contributed to the popularity of Cloud Business Intelligence systems. The literature emphasizes that compared with traditional on-premise solutions, CBI systems provide easier access to data from diverse sources, reduce the time needed for analysis and decision-making, and enhance managerial effectiveness [6–12]. Consequently, CBI systems can significantly improve the effectiveness and efficiency of decision-making processes, thereby strengthening an organization's competitive advantage.

CBI systems enable enterprises to access a wide range of data, including management dashboards, KPI indicators, and other analytical tools delivered in various service models (e.g., SaaS), providing real-time insights on different types of devices, most often via web browsers [13]. The CBI environment encompasses diverse data sources and systems implemented within enterprises, such as ERP, CRM, SCM, WMS, specialist application databases (e.g., email systems), document repositories (e.g., XLS files), as well as market data (e.g., industry reports, public datasets, social media data). Data from these systems undergo an extraction, transformation, and loading (ETL) process, including cleansing, and are then analyzed within cloud platforms. Analytical outcomes are visualized through advanced, interactive, and user-friendly reports and dashboards, which allow users to immediately perform multidimensional data exploration. Within CBI systems, various analytical methods can be applied, such as cluster analysis or regression, in combination with artificial intelligence, machine learning, advanced statistics, and data mining techniques [14]. As a result, CBI systems make it possible to forecast trends, detect anomalies, automatically generate business recommendations, assess process efficiency, and anticipate competitor actions. This enables enterprises to respond more swiftly to changing market conditions, identify new business opportunities, and minimize risks associated with decisions made under incomplete information [15].

CBI systems have become an essential component of digital transformation, providing managers with access to comprehensive and up-to-date information about both the internal condition of the enterprise and its external market environment, thereby supporting decision-making at all levels of management. Data processing and visualization carried out in CBI systems, often in near real time, allow managers and analysts to continuously monitor key indicators (e.g., financial, logistical, commercial, or employee satisfaction), optimize business processes, and rapidly adapt to market dynamics. A key feature of CBI systems is their scalability and flexibility, which enables enterprises to dynamically adjust IT resources and cost levels to current analytical needs. Depending on their requirements, capabilities, and strategic objectives, enterprises may choose different models of CBI implementation (e.g., private, public, or hybrid cloud), as well as different architectural approaches such as IaaS, PaaS, or SaaS [13].

The ability to reduce IT infrastructure costs (with the subscription-based model eliminating capital expenditures) along with simplified access to advanced analytical systems constitutes one of the major reasons for the growing popularity of CBI systems among small and medium-sized enterprises. CBI systems are typically hosted in external data centers provided and managed by IT service vendors, who are responsible for managing the environment, software updates, hardware upgrades, performance, and service continuity. Users access the system through intuitive interfaces, most often via web browsers, while service fees are typically charged on a monthly subscription basis. Furthermore, CBI systems allow users to access analytical tools at any time and from any location, using a variety of devices such as computers, tablets, or smartphones. Table 1 presents a summary of the key benefits and challenges associated with the implementation of Cloud BI systems in enterprises, based on a literature review [16–18].

The implementation of Cloud BI systems is a complex process that requires careful consideration of multiple factors influencing enterprises' decisions regarding the selection, scope, and method of deployment. The effectiveness of Cloud BI adoption is determined by a variety of conditions [19–21], which can be categorized as follows:

- Strategic factors – for instance, the role and position of Cloud BI systems within the enterprise's digital strategy, as well as the level of executive support and engagement in the implementation and development of Cloud BI.
- Organizational factors – such as employee competencies in analytics and the use of cloud-based analytical tools, an organizational culture that supports data-driven decision-making, and the structure of decision-making processes.
- Technological factors – including the degree of enterprise digitalization, adequacy of IT infrastructure, the ability to efficiently and effectively integrate data from multiple IT systems, data quality and consistency across IT environments, adopted data security measures, as well as the analytical capabilities of platforms utilizing AI and ML technologies.
- Environmental factors – for example, competitive pressures that necessitate investments in analytical systems, the market offering of Cloud BI solutions, trust in Cloud BI providers, regulatory requirements related to reporting and data protection, and financial support or subsidies for IT investments from external institutions.

A thorough analysis of the benefits and challenges of Cloud BI, combined with a holistic approach to examining the determinants of system implementation, increases the likelihood of achieving long-term business benefits. These benefits include improved decision-making quality, optimization of resource utilization, shorter response times to market changes, and enhanced competitive positioning.

An important dimension influencing the implementation of CBI systems is the extensive and effective use of advanced technologies such as artificial intelligence, big data analytics, and machine learning. These technologies contribute to improving the efficiency of decision-making processes, further automation of analytical workflows, the advancement of predictive and prescriptive analytics, and the strengthening of data security practices.

Table 1. Benefits and Challenges of Cloud BI Systems

Area	Benefits of Cloud BI	Challenges of Cloud BI
Costs	Reduction of capital expenditures; "pay-as-you-go" model; lower maintenance and development costs	Hidden costs of migration and integration; significant long-term expenses; necessity of ongoing cost control
Organization	Faster and easier implementation; reduced need for IT specialists; transfer of responsibility to IT providers	Dependence on service providers; difficulties in change management; challenges in IT resource migration; lack of full control
Quality	Advanced and always up-to-date versions; significant potential for performance and reliability; technical support	Possible temporary issues with performance and reliability; unsatisfactory technical support and customer service
Scalability	Flexible increase or decrease of resources	Network bandwidth limitations; risk of restricted scalability due to provider constraints
Accessibility	Access from any location and device; support for mobility	Risk of service interruptions; dependence on internet connection quality
Decision-making	Shortened analysis time; real-time data access; support for predictive analytics	Data transmission delays (latency)
Data Integration	Integration of data from multiple sources (e.g., ERP, CRM, SCM, WMS, websites, public datasets)	Integration challenges with multiple IT resources or legacy IT solutions
Innovation	Integration with AI/ML; self-service BI; predictive analytics; rapid adaptation to market changes	Need to ensure digital competencies among employees
Security	Advanced data protection mechanisms offered by providers	Concerns regarding data confidentiality and data location; compliance with legal regulations

3 Research Methodology

Rough Set Theory (RST) provides a formal framework for analyzing imprecise, incomplete, or inconsistent information and is regarded as one of the principal approaches in data mining and decision support—particularly for classification, attribute selection, and decision-rule discovery without requiring prior probabilistic or statistical knowledge [22, 23]. Beyond its strength in handling incomplete data, RST offers attribute-reduction and feature-selection capabilities, enabling the identification and elimination of redundant attributes without sacrificing classification quality [24]. It also facilitates the generation of "if–then" decision rules from training data to support decision-making processes [25].

RST has found wide application in the analysis, processing, and reduction of data sets. It is employed in medical, pharmacological, engineering, and banking software, as well as in areas related to control systems, object recognition, and numerous other fields [26]. From a practical perspective, rough set theory is an innovative data analysis method that enables [27]:

- the discovery of relationships between data (variables) through the development of efficient algorithms,
- data reduction (minimization of datasets),
- determination of the significance of data,
- generation of decision rules based on data (derived from the lower and upper approximations of sets).

The advantages of RST include the fact that it does not require assumptions about the data (e.g., probabilities or fuzziness), incorporates fast data analysis algorithms, facilitates the interpretation of results, and is characterized by considerable mathematical simplicity [28]. RST continues to evolve rapidly; R. Tadeusiewicz [29] even refers to it as a "Polish island in the artificial-intelligence archipelago." For further details on RST, see [30].

In this study, RST was applied to discover knowledge from a dataset in the form of decision rules. This approach better accommodates the fuzziness of the data and the ambiguous relationships among factors, allowing a multifaceted identification of the drivers behind enterprises' decisions to implement Cloud BI solutions.

The inference process used data from an original CAWI (Computer Assisted Web Interview) survey of 400 medium and large enterprises employing cloud BI systems. The sample was split equally—50% medium-sized and 50% large firms—across diverse industries. The questionnaire combined single-choice and multiple-choice questions with a five-point Likert scale. Data collection and analysis were performed in Microsoft Excel.

In line with RST requirements and the study's objectives, a primary decision table was prepared. It comprised 56 conditional attributes (q1–q56) describing factors such as: drivers of CBI adoption, years of CBI use, implementation expectations, chosen cloud model, and declared benefits and risks. A single decision attribute (d1) captured respondents' recommendation regarding CBI implementation. The research question was: Which combinations of q1–q56 attributes determine the decision d1?

To harmonize attribute scales (e.g., discrete vs. binary), all conditional attributes were discretized into three levels: 1 = low, 2 = medium, 3 = high. For the decision attribute, although the original four-point scale included "no opinion," "do not recommend," "recommend," and "strongly recommend," only two responses occurred: "no opinion" (coded 0) and "recommend" (coded 2). While this discretization reduces source-data granularity, it increases the likelihood of uncovering repeatable patterns—i.e., decision-rule induction. The resulting secondary decision table served as the basis for rule discovery, taking into account rule certainty and support—measured by the maximum number of examples matching both the conditional (antecedent) and decision (consequent) parts of each rule.

Rule induction followed RST best practices, using the Explore algorithm implemented in the ROSE2 software [31][32] developed by the Laboratory of Intelligent Decision Support Systems at Poznań University of Technology. Minimum thresholds for rule acceptance were specified to ensure robustness.

4 Results

Efforts to link the conditional variables to the decision variable (namely, the recommendation regarding the adoption and use of CBI in enterprises) revealed that certain decision rules exhibit both perfect certainty and high support. The most influential variables were:

q1: the decision to adopt CBI was influenced by a recommendation from another business entity,

q2: the decision to adopt CBI was influenced by conducting an internal analysis of documented projects,

q3: the decision to adopt CBI was influenced by the results of analyzing cloud-service providers' offers,

q9: the decision to adopt CBI was influenced by the pandemic and its effects,

q18: the choice between using public or private cloud for CBI,

q51: identification of unsatisfactory technical parameters as a drawback of CBI,

q55: identification of lack of qualified personnel and necessary competencies as a drawback of CBI.

Below we present the rules that combine the highest certainty and strong support (i.e., the ratio of covered examples to the entire dataset) with a minimal number of conditional attributes.

The first rule:

$$\textit{If } (q2 = 3),\ \textit{then } (d1 = 0);\ [80,\ 80,\ 66.67\%,\ 100\%]\,[80,\ 0] \tag{1}$$

This rule's antecedent consists of a single attribute, making its conclusion both unequivocal and easily interpretable. It can be read as follows: If the decision to implement cloud BI was strongly influenced by conducting an internal analysis of documented projects (q2 = 3), then respondents express no opinion regarding CBI adoption (d1 = 0).

Consequently, we infer that when project documentation analysis exerts a significant influence on the CBI adoption decision, respondents remain neutral - neither recommending nor discouraging CBI implementation. This ambivalence may arise because the analysis yielded neither clearly positive nor negative outcomes (i.e., merely average results), which does not motivate respondents to either praise or reject cloud-based business analytics. Although the internal analysis likely identified both advantages and drawbacks of CBI, respondents based on these findings are unable to take a definitive stance. Hence, we may conclude that the specific data and experiences derived from prior project analyses did not confirm expected benefits or, conversely, revealed substantial risks, issues, and limitations associated with CBI implementation. The antecedent of this rule describes 80 out of the 400 surveyed enterprises, all of which satisfy the rule's consequent thereby achieving 100% certainty (80/80). At the same time, the support of 80 indicates that 66.67% of the firms declaring "no opinion" on CBI adoption (80 out of 120) made their decision largely on the basis of an internal analysis of documented projects (q2 = 3).

The second rule worth emphasizing is interesting in this context:

$$\textit{If}\ (q1\ =\ 2)\ \textit{and}\ (q51\ =\ 2)\ \textit{then}\ (d1\ =\ 2);\ [200,\ 200,\ 71.43\%,\ 100\%]\,[0,\ 200] \tag{2}$$

This rule's consequent pertains to recommending CBI adoption (d1 = 2) and describes the behavior of firms that endorse CBI. It is a certain rule (100% certainty), covering 200 examples and supported by all 200 of those instances. Its high strength 50% means that it characterizes the actions of half of the surveyed enterprises (200 out of 400). Comprising only two conditional attributes, the rule is straightforward and easily interpretable: if the decision to implement CBI was moderately influenced by a recommendation from another business entity (q1 = 2) and the risk of unsatisfactory technical parameters (e.g., scalability, performance, security) was also moderately recognized as a drawback (q51 = 2), then respondents recommend CBI adoption (d1 = 2). In other words, when an external partner's endorsement (likely bolstering trust through shared knowledge and best-practice transfer) coincides with an acknowledged but manageable technical-parameter risk, firms still choose to recommend CBI. This rule applies to over 71% of those enterprises that recommend CBI (d1 = 2), illustrating that external endorsement can outweigh perceived technical limitations, which are deemed resolvable through corrective actions, additional investments, or safeguards.

From Rule (2), we can infer that:

- The external partner's endorsement exerted such a strong influence that it outweighed the firm's concerns about the technological risks of CBI, leading respondents to recommend CBI implementation.
- Respondents likely believed that potential technical issues such as scalability, performance, or security - could be resolved or mitigated through corrective actions, additional investments, or the implementation of safeguards, ultimately motivating them to endorse CBI solutions.

In summary, even when the risk of unsatisfactory technical performance is identified, an external recommendation plays a decisive role and convinces respondents that CBI deployment is justified, with those risks considered acceptable or manageable given the expected benefits. Rule (2) thus reflects a more informed approach to CBI implementation than Rule (1). Together, these rules suggest a broadly applicable conclusion: collaboration and the integration of shared expertise, particularly when accompanied by deliberate risk identification - lead to better outcomes than relying solely on in-house experience. Moreover, enterprises that base their decisions on both expert recommendations and critical risk assessment demonstrate greater readiness to adopt innovative IT technologies (such as CBI), despite being aware of potential barriers. This behavior indicates decision-making maturity and a pragmatic attitude toward digital transformation, in which risk is not a disqualifier but a factor to be effectively managed.

Rule (3):

$$\textit{If}\ (q2\ \textit{in}\,\{1,\ 2\})\ \textit{and}\ (q3\ \textit{in}\,\{2,\ 1\})\ \textit{then}\ (d1\ =\ 2);\ [280, 280, 100\%, 100\%]\,[0,\ 280] \tag{3}$$

Like Rule (2), the antecedent of Rule (3) comprises only two attributes, describing 240 firms—all of which satisfy the rule's consequent, making it a certain rule. It can be interpreted as follows: If the decision to implement CBI was influenced to a medium or low degree by an internal analysis of documented projects ($q2 \in \{2, 1\}$) and respondents simultaneously identified the lack of qualified personnel and necessary competencies ($q55 = 2$) as a medium level drawback of CBI, then they recommend CBI adoption ($d1 = 2$).

In other words, even when internal project analysis plays only an auxiliary or marginal role and organizational or competency barriers are acknowledged, firms still endorse CBI implementation. This suggests that despite significant organizational and skills related obstacles such as specialist shortages, CBI systems are perceived as strategically important and worth implementing. Moreover, the limited impact of internal analyses implies that CBI adoption may be driven more by broader market trends, competitive pressures, or external recommendations than by in depth internal studies.

An important insight from this rule is that enterprises, despite being aware of existing challenges, continue to recognize substantial benefits from CBI deployment. This reflects a high level of strategic motivation and confidence in the transformative potential of cloud solutions for organizational development. It corroborates Rule (2)'s finding that awareness of limitations and difficulties does not preclude successful adoption; rather, it fosters behaviors aimed at maximizing benefits while mitigating weaknesses. The support factor 85.71% indicates that a vast majority of firms recommending CBI adoption are characterized by the q2 and q55 attributes specified in this rule.

Related to the above is another rule that reads:

$$\textit{If } (q2\ \textit{in}\,\{1,\ 2\})\ \textit{and}\ (q3\ \textit{in}\,\{2,\ 1\})\ \textit{then}\ (d1\ =\ 2);\ [280,\ 280,\ 100\%,\ 100\%]\,[0,\ 280] \tag{4}$$

Rule (4) stands out as exceptionally revealing of the surveyed enterprises' decision-making mindset. It is a certain rule 100% of the 280 covered firms agree with the conclusion and its support factor also equals 100%, indicating that all firms recommending CBI ($d1 = 2$) satisfy the rule's antecedent. The conditional attributes q2 and q3 both appear at low or medium levels: $q2 \in \{1, 2\}$: internal analysis of documented projects influenced the decision only marginally or moderately, $q3 \in \{1, 2\}$: analysis of cloud provider offers influenced the decision only marginally or moderately. This rule can be interpreted as follows: Even when both internal project analyses and external market offer evaluations play only a small to medium role in the adoption decision, firms still recommend CBI implementation.

From this, we can infer that:

- Other decision drivers- such as competitive pressure, overarching technology trends, industry recommendations, or managerial intuition - may carry decisive weight.
- Respondents inherently trust and recognize the advantages of CBI, not requiring in depth project or market analyses as additional validation.
- Cloud computing may be considered the default backbone for BI systems, with respondents already possessing a baseline confidence from prior cloud usage, thereby lessening the perceived need for detailed evaluations.

Furthermore, this rule suggests that the formalization level of the decision process for adopting CBI among these enterprises is relatively low, relying more on overall strategic conviction than on exhaustive analytical justification.

The next rule related to the previous ones is:

$$\textit{If}\,(q18 = 1)\,\textit{then}(d1 = 2);\ [120,\ 120,\ 42.86\%,\ 100.00\%]\,[0,\ 120] \tag{5}$$

This rule covers 120 enterprises (all of which satisfy the rule's consequent) making it a certain rule (100% certainty). With a single conditional attribute, q18, it reads: If the firm uses the public cloud (q18 = 1), then it recommends CBI adoption (d1 = 2). The support factor indicates that nearly 43% of the enterprises recommending CBI rely on public cloud infrastructure. This finding suggests that public cloud platforms are perceived as mature technologies, offering key benefits such as scalability, availability, and flexibility for CBI users. Consequently, the decision to employ public cloud services is strongly associated with viewing that environment as advantageous for business analytics. Moreover, respondents' growing awareness of public cloud potential appears to lower many traditional barriers to on premises analytics systems - high infrastructure costs, complex deployments, and limited scalability.

Thus, reliance on public cloud infrastructure signals readiness for deeper digitization of management processes driven by data analysis and a willingness to adopt cloud-based analytics that outweighs the costs and risks of traditional IT models. Public cloud users also demonstrate conviction in the strategic value of modern IT technologies for competitiveness and organizational transformation. The recommendation to implement cloud-based analytics may also stem from the belief that such solutions enhance operational and strategic decision-making efficiency, thereby supporting long term enterprise growth. In light of Rule (3) - where a lack of qualified personnel was identified as a barrier - it is noteworthy that public cloud services are provided and managed by external vendors, relieving users of the need to build, configure, or maintain IT infrastructure themselves, as is required with private clouds. Consequently, the demand for highly skilled technical staff decreases, which is seen as an advantage, particularly in contexts with limited specialist availability. Therefore, public cloud adoption lowers the technological entry barrier and facilitates access to advanced services without necessitating internal capability development—a factor that promotes recommendations for solutions like CBI in environments constrained by talent shortages.

Next Rule:

$$\textit{If}\,(q1 = 2)\,\textit{and}\,(q9 = 2)\,\textit{then}(d1 = 2);\ [200,\ 200,\ 71.43\%,\ 100.00\%]\,[0,\ 200] \tag{6}$$

This rule covers 200 cases and is fully certain: every enterprise meeting the antecedent also recommends CBI adoption. It can be decoded as follows: If the decision to implement CBI was moderately influenced both by an external business entity recommendation (q1 = 2) and by the effects of the COVID 19 pandemic (q9 = 2), then respondents recommend CBI adoption (d1 = 2). With over 71% of CBI recommending firms described by these two attributes, the rule suggests that organizations guided at a medium level by expert advice and pandemic driven exigencies are inclined to endorse cloud-based analytics. Expert input—from research institutions, consulting firms, or industry authorities—serves to mitigate perceived risks and advance organizational knowledge regarding cloud deployments. Simultaneously, the pandemic's disruption (e.g., remote work acceleration, rapid digitization, business model shifts, and increased demand for analytics) created a compelling need for flexible, scalable, and remotely accessible IT solutions. Together, these factors (credible external recommendations plus crisis induced imperatives) formed a dual catalyst for CBI adoption, demonstrating that decision making extended beyond internal analyses or local considerations to include macroeconomic and global drivers. The COVID 19 crisis heightened the urgency for agile analytics tools capable of real time monitoring in dynamic environments. Consequently, CBI implementation is viewed as a strategic response to complex external pressures, and positive endorsements under these conditions signal enterprises' willingness to adopt innovative IT solutions. This rule underscores that such endorsements and pandemic experiences accelerate technology adoption decisions (often superseding the need for exhaustive internal validation) and reveal firms' strategic readiness to deploy IT solutions that bolster resilience, operational flexibility, and competitive advantage.

Although the survey questionnaire included 56 conditional variables, only seven of them appear in the most valuable rules—underscoring their critical importance in enterprises' decision making and perception of CBI implementation outcomes. The recurring presence of q2 and q1 in these rules highlights their leading role in the decision process. Notably, q2 indicates that excessive reliance on internal experience often fails to yield satisfactory solutions.

5 Conclusions

The conducted analysis demonstrated that Rough Set Theory (RST) can serve as a valuable tool to support decision-making regarding the implementation of Cloud Business Intelligence (CBI) systems in enterprises. Thanks to RST's ability to handle incomplete information and identify key decision attributes, it was possible to formulate clear rules that support the decision-making process in the context of CBI implementation. The identified decision rules provide practical insights into the factors influencing CBI adoption, highlighting, among others, the importance of recommendations from other businesses, conducting one's own analysis of documented projects, reviewing cloud service providers' offerings, the impact of the pandemic and its consequences, the use of CBI in the public cloud, and drawbacks of CBI such as unsatisfactory technical parameters or a lack of qualified personnel and necessary competencies. This indicates the potential for broader application of RST in research on various aspects of IT implementation and supporting efforts toward the digital transformation of enterprises, especially where data are incomplete, subjective, or ambiguous. Future research directions will include the exploration of additional variables and the development of mechanisms for automatic rule learning as new data becomes available, which will allow for the continued adaptation of the CBI recommendation system to evolving economic conditions.

Disclosure of Interests. The authors have no competing interests to declare that are relevant to the content of this article.

References

1. Olszak, C.: Business intelligence in cloud. Polish J. Manag. Stud. **10**(2) (2014)
2. Amit, A.: Business intelligence: concepts, components, techniques and benefits. Asian J. Manag. **3**(4), 10–11 (2012)
3. Mondal, S.: Cloud business intelligence as a solution for empowering SME. Epra Int. J. Multidisc. Res. (IJMR) **8**(9) (2022)
4. Hmoud, H., Al-Adwan, A.S., Horani, O., Yaseen, H., Al Zoubi, J.Z.: Factors influencing business intelligence adoption by higher education institutions. J. Open Innov. Technol. Market Complex. **9**(3) (2023)
5. Sultan, H., Chan, L.K., Siew, P.H., Lee, C.K.: Factors influencing cloud business intelligence adoption and use at organisational level. ITM Web Conf. **67**, 01019 (2024)
6. Thompson, W.J., van der Walt, J.S.: Business intelligence in the cloud. S. Afr. J. Inf. Manag. **12** (2010)
7. Moyo, M., Loock, M.: Conceptualising a cloud business intelligence security evaluation framework for small and medium enterprises in small towns of the Limpopo Province, South Africa. Information **12**(3), 128 (2021)
8. Khouibiri, N., Farhaoui, Y.: Analyzing the influence of cloud business intelligence on small and medium enterprises a case study of Morocco. Data Metadata **2**, 104 (2023)
9. Mwilu, O.S., Comyn-Wattiau, I., Prat, N.: Design science research contribution to business intelligence in the cloud — a systematic literature review. Future Gener. Comput. Syst. **63** (2016)
10. Dashora, S., Gupta, N., Tanwar, M.P.S., Singh, A., Kaur, S.: Cloud-based data analytics for business intelligence. Int. J. Res. Appl. Sci. Eng. Technol. **11**(11) (2023)
11. Indriasari, E., Prabowo, H., Meyliana, K., Hidayanto, A.N.: Key benefits of cloud business intelligence: a systematic literature review. Int. J. Mech. Eng. Technol. **9** (2018)
12. Hamidinava, F., Ebrahimy, A., Samiee, R., Didehkhani, H.: A model of business intelligence on cloud for managing SMEs in COVID-19 pandemic (Case: Iranian SMEs). Kybernetes **52**(1) (2023)
13. Elmalah, K., Nasr, M.: Cloud business intelligence. Int. J. Adv. Netw. Appl. **10** (2019)
14. Plašić, J., Stefanović, N., Gaborović, A.: Enterprise business intelligence approach with cloud-based analytics. In: E-Business Technologies Conference Proceedings, vol. 1, no. 1 (2021)
15. Niu, Y., Ying, L., Yang, J., Bao, M., Sivaparthipan, C.B.: Organizational business intelligence and decision making using big data analytics. Inf. Process. Manage. **58**(6), 102725 (2021)
16. Patil, S.S., Chavan, R.: Cloud business intelligence: an empirical study. Stud. Indian Place Names UGC Care J **27** (2020)
17. Dziembek, D., Ziora, L.: Business intelligence systems in the SaaS model as a tool supporting knowledge acquisition in the virtual organization. Online J. Appl. Knowl. Manag. **2** (2014)
18. Al-Aqrabi, H., Liu, L., Hill, R., Antonopouloset N.: Cloud BI: future of business intelligence in the cloud. J. Comput. Syst. Sci. **81**(1) (2019)
19. Al-Sharafi, M., Iranmanesh, M., Al-Emran, M., Alzahrani, A., Herzallah, F, Jamil, N.: Determinants of cloud computing integration and its impact on sustainable performance in SMEs: an empirical investigation using the SEM-ANN approach. Heliyon **9**(5) (2023)
20. Owusu, A.: Determinants of cloud business intelligence adoption among Ghanaian SMEs. Int. J. Cloud Appl. Comput. **10**(4) (2020)
21. Holler, N., Westner, M.: Factors influencing cloud computing adoption in small and medium-sized enterprises: a systematic review. Informatica **49**(1) (2021)
22. Pawlak, Z.: Rough sets. Int. J. Comput. Inform. Sci. **11**(5), 341–356 (1982)
23. Slowinski, R.: Intelligent Decision Support: Handbook of Applications and Advances of the Rough Sets Theory. Springer (1992)

24. Liu, D., Li, J., Min, F.: Incremental learning algorithms of rough sets: a survey. Artif. Intell. Rev. **53**(1), 1–35 (2020)
25. Min, F., Lin, Q., Qian, Y.: Recent developments of rough sets in machine learning: a review. Artif. Intell. Rev. **54**(1), 703–748 (2021)
26. Pawlak, Z., Skowron, A.: Rudiments of rough sets. Inf. Sci. **177** (2007)
27. Kulczycki, P., Hryniewicz, O., Kacprzyk, J.: Techniki informacyjne w badaniach systemowych (red), WNT, Warszawa (2007)
28. Pawlak, Z.: Some issues on rough sets. In: Transactions on Rough Sets I. Springer, Heidelberg (2004)
29. Tadeusiewicz, R.: Archipelag sztucznej inteligencji. Część II. Napędy Sterow. (2021)
30. Kuczera, K., Dziembek, D.: Application of rough set theory to improve the efficiency of higher education systems. In: Hernes, M., Wątróbski, J., Rot, A. (eds.) Emerging Challenges in Intelligent Management Information Systems, ECAI 2024, Lecture Notes in Networks and Systems, vol. 1218. Springer, Cham (2024)
31. Predki, B., Slowinski, R., Stefanowski, J., Susmaga, R., Wilk, S.: ROSE – software implementation of the rough set theory. In: Polkowski, L., Skowron, A. (eds.) Rough Sets and Current Trends in Computing, Lecture Notes in Artificial Intelligence, vol. 1424, pp. 605–608 (1998)
32. Predki, B., Wilk, S.: Rough set based data exploration using ROSE system. In: Ras, Z.W., Skowron, A. (eds.) Foundations of Intelligent Systems, Lecture Notes in Artificial Intelligence, vol. 1609, pp. 172–180 (1999)

Three-Dimensional Correlation Analysis of AI-Supported Cloud Business Intelligence in Enterprise Management Processes

Paula Bajdor[1](✉) and Aleksandra Bączkiewicz[2]

[1] Warsaw University of Technology, Warsaw, Poland
paula.bajdor@gmail.com
[2] University of Szczecin, Szczecin, Poland

Abstract. The article examines how enterprises use Cloud Business Intelligence (CBI) systems with AI tools, focusing on their impact on decision-making efficiency, analysis quality, and resource management. Using a three-dimensional analysis, the study explores relationships between decision speed, data management, and the strategic role of CBI in process optimization. The findings highlight CBI's contribution to strategic management and digital transformation, confirming the value of diverse analytical approaches in evaluating its impact on enterprise performance.

Keywords: Cloud Bussines Intelligence · Management · Artificial Intelligence · Weighted Least Squares Smoothing · Three-dimensional correlation

1 Introduction

In the modern digital economy, data resources are becoming one of the most valuable assets of enterprises, and Cloud Business Intelligence (CBI) systems enable their transformation into management information and knowledge, supporting managers in the decision-making process and operational activities [1].

CBI systems facilitate knowledge capture by aggregating and analyzing both structured and unstructured data from various sources in real-time. Using AI-powered tools such as machine learning and NLP, they identify patterns and reveal actionable insights that would otherwise remain hidden. These insights are then shared across departments through dashboards and automated reports, fostering transparency and collaboration [2]. Furthermore, by continuously learning from new inputs and feedback loops, CBI systems promote organizational learning, enabling companies to adjust strategies and improve operations dynamically [3].

CBI systems equipped with AI tools enable access to distributed data resources collected, for example, in ERP systems, Office systems, IoT, social media, and other applications, provide multidimensional analyses available in a useful form for managers at various levels of management, offering automatic trend detection, risk prediction and

Published by Springer Nature Switzerland AG 2026
M. Pondel et al. (Eds.): AI4KAM 2025, IFIP AICT 776, pp. 206–212, 2026.
https://doi.org/10.1007/978-3-032-18920-2_19

generating conclusions that go beyond the capabilities of classic BI systems [4, 5]. In addition, various AI tools in CBI systems through machine learning, automatic pattern and dependency recognition, predictive modeling, or real-time recommendations can support enterprises in forecasting market trends, optimizing business processes, and personalizing offers for customers while changing the role of analysts and managers in enterprises and supporting an organizational culture based on knowledge[5, 6]. Integrating CBI and AI also provides benefits regarding data accessibility and democratization in enterprises. Using natural language processing (NLP) technology, business users can ask questions in natural language and receive answers in the form of clear visualizations without the need for advanced technical knowledge.

In the context of the CBI systems studied, "AI support" refers to the integration of advanced computational tools that enhance data-driven decision-making through automation and intelligent analysis. Specifically, these systems leverage machine learning (ML) algorithms for pattern recognition and predictive analytics, enabling accurate prediction of customer trends and behaviors [7]. Natural language processing (NLP) is used to facilitate user interaction with data, for example, through queries via conversational interfaces or processing unstructured text from surveys and reports [8]. Additionally, intelligent automation tools, including rule-based engines and anomaly detection algorithms, support real-time alerting and operational adjustments. Together, these AI capabilities extend traditional BI capabilities by enabling dynamic, context-aware, and proactive business insights.

AI-based knowledge platforms have also been shown to enhance organizational performance when combined with robust knowledge-sharing cultures and effective data governance frameworks [9]. These contributions mark a shift towards more adaptive, insight-rich enterprise ecosystems.

This, in turn, contributes to faster decision-making and streamlining activities or improving innovation in enterprises [10].

This article examines the correlations between selected aspects of enterprises' use of CBI systems. For this purpose, three hypotheses were formulated:

H1: There are significant correlations between the efficiency, effectiveness, and speed of decision-making using CBI systems,
H2: There are significant correlations between the level of quality of business analyses, the ordering of data management processes, and the role of CBI systems in supporting business processes.
H3: There are significant correlations between improving and optimizing business processes, improving efficiency in using enterprise resources, and creating the enterprise's business strategy.

However, unlike the most commonly conducted correlation analyses using bivariate variables, simulations of relationships were performed using three-dimensional variables. This is dictated by the fact that the analysis of correlations between two traits, in isolation from other but equally important variables, may seem extremely strong. In contrast, the impact of different variables may weaken or even strengthen them [11]. Therefore, by adding an additional variable to the simulations of the relationships, it will be possible to draw more precise conclusions, which will constitute a significant scientific contribution to the area of enterprises' use of CBI systems. To verify the adopted hypotheses, spatial - three-dimensional graphs will be used, and the method of smoothing the least squares weighted by distances was used to conduct the simulations of the relationships.

2 Research Results

2.1 The Characteristics of the Respondents

To assess the level of management support provided by CBI in medium and large companies and identify potential directions for developing these systems, the authors conducted a survey in recent years. The research sample included 400 Polish enterprises (200 medium-sized and 200 large), primarily from the industrial and commercial sectors, all using a CBI. The survey was conducted using the CAWI method and a 5-point Likert scale. The distribution of company sizes was proportional: 50% medium-sized and 50% large enterprises from various industries participated.

2.2 The Reliability Analysis

Reliability analysis is often associated with developing complex scales, particularly summary scales composed of multiple individual items (questions). The reliability of a scale—understood as its internal consistency—provides information about how accurately the scale measures what it is intended to measure, that is, how well the individual items reflect the overall construct. In reliability analysis, the most commonly used indicator is Cronbach's Alpha coefficient, which measures the ratio of the variance of individual items to the variance of the entire scale (the sum of those items). Its value ranges from 0 to 1, with values closer to 1 indicating higher reliability. Table 1 presents the reliability analysis results based on 400 completed questionnaires.

Table 1. Descriptive statistics

Variables (in the form of statements)	Symbol	Mean	S.D	Var	Cronbach's Alpha
The use of CBI tools in the company enhanced the efficiency of decision-making	Var1	3,500	0,500	0,250	0,635
CBI tools used in the company increased the effectiveness of the decision-making process	Var2	3,595	0,679	0,462	0,674
Application of the CBI solution increased the speed of decision-making (at all levels of company management)	Var3	3,400	0,490	0,240	0,653
The application of CBI significantly increased the quality of business analyses	Var4	3,600	0,490	0,240	0,625

(continued)

Table 1. *(continued)*

Variables (in the form of statements)	Symbol	Mean	S.D	Var	Cronbach's Alpha
Implementing business analyses in CBI allowed for organizing data management processes	Var5	3,195	0,614	0,377	0,669
Business analytics in CBI provides information support for current and new business processes	Var6	3,725	0,671	0,450	0,687
The application of CBI had a positive impact on the ability to create and monitor company strategies	Var7	3,525	0,547	0,300	0,697
Business analytics in CBI allows for the improvement and optimization of business processes	Var8	2,687	0,686	0,471	0,645
Implementing business analyses in CBI "increased the efficiency of using company resources	Var9	3,315	0,496	0,246	0,637

The all the variables, Cronbach's alpha values are higher than 0,60, which means that the scale's reliability is acceptable.

2.3 The Research Results

Thanks to the primary data obtained from the survey conducted among companies using CBI, it was possible to conduct graphic simulations for selected correlations. Simulations were conducted for three cases of correlations between selected variables:

- Efficiency, effectiveness, and speed of decision-making using CBI,
- Level of quality of business analyses, organization of data management processes, and the role of business analytics in supporting business processes,
- Create enterprise strategy, improve and optimize business processes, and increase the efficiency of enterprise resources using CBI.

The results presented in graphical form (Figs. 1, 2, and 3) enabled general inference to be made, and the simulations aimed to indicate the correlation between the formation of selected variables in relation to the changes in other variables:

- the first case included variables: CBI tools used in the company increased the efficiency of the decision-making process (Var1); CBI tools used in the company increased the efficiency of the decision-making process (Var2), and the use of business analytics solutions in CBI increased the speed of decision-making (Var3);

- the second case included variables: the use of CBI significantly increased the quality of business analyses (Var4); the implementation of business analyses in CBI allowed for “organizing data management processes” (Var5), and business analytics in CBI provides “information support for current and new business processes” (Var6);
- the third case included variables: the use of CBI positively influenced the ability to create and monitor business strategies (Var7); business analytics in CBI allowed for the improvement and optimization of business processes (Var8) and the implementation of business analyses in CBI “increased the efficiency of using enterprise resources” (Var9).

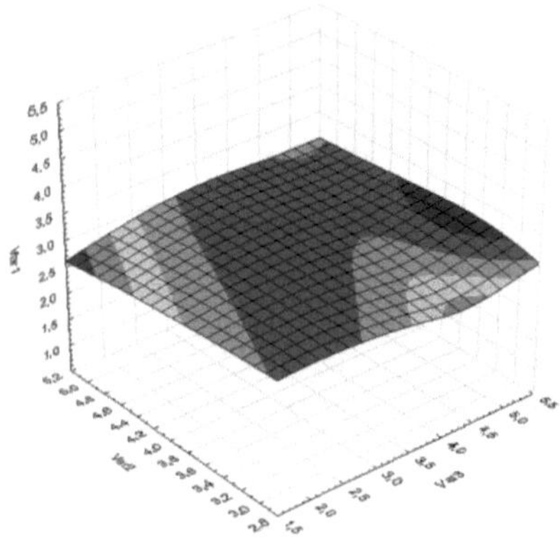

Fig. 1. Simulation effect between variables Var1, Var2 and Var3

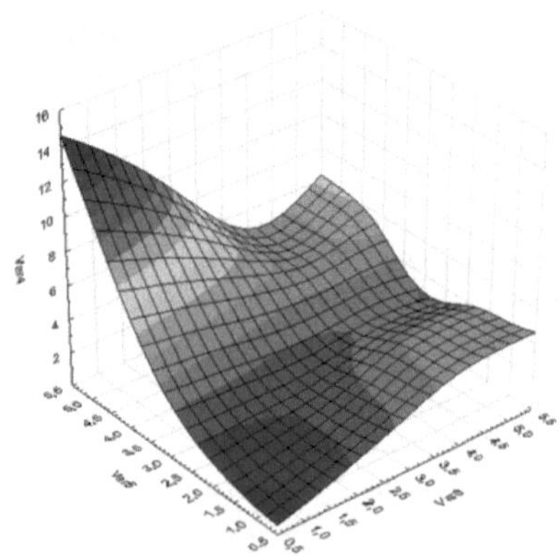

Fig. 2. Simulation effect between variables Var4, Var5 and Var6

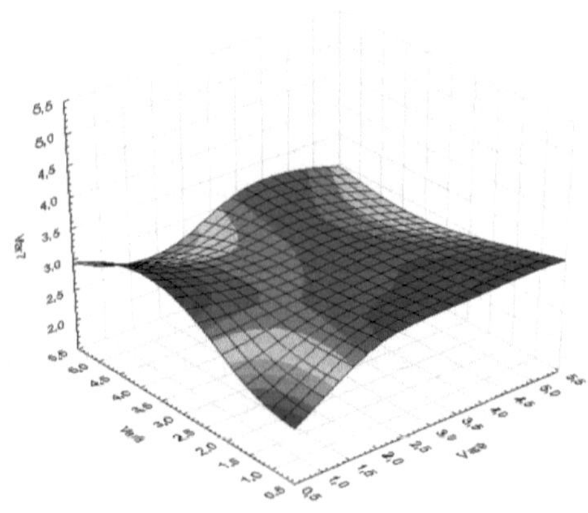

Fig. 3. Simulation effect between variables Var7, Var8 and Var9

Analyzing the obtained results, it can be seen that:

- The respondents equally assessed the impact of using CBI on efficiency, effectiveness, and speed of decision-making. This results from the fact that the assessments of the analyzed variables were at a similar level;
- however, regardless of how highly the respondents assess the impact of CBI on the quality of business analyses, they also believe that the use of CBI did not enable the organization of data management processes, and only to a small extent does CBI provide information support for business processes;
- the increase in the assessment of the impact of CBI on the possibility of creating and monitoring the company's strategy does not translate into the belief that CBI enables the improvement and optimization of business processes and only, to a small extent, contributes to increasing the efficiency of using the company's resources.

Thus, the first research hypothesis was confirmed, the second hypothesis was partially confirmed, and the third hypothesis was not confirmed. In the case of the first hypothesis, it is clear that respondents assess the efficiency, effectiveness, and speed of decision-making using CBI to the same extent. In the case of the second hypothesis, on the one hand, a positive impact of CBI on the quality of business analyses is indicated. Still, no effect on the ordering of data management processes has been demonstrated. The rejection of the third hypothesis is supported by the fact that while CBI affects the quality of business analyses, it does not correlate with the possibility of ordering data management processes, and CBI provides information support for business processes.

3 Conclusions

The three-dimensional correlation analysis approach used in the study allows for a deeper understanding of the impact of CBI supported by AI tools. Thus, the article presents both a theoretical perspective and empirical results important for further developing management practices supported by intelligent technologies.

The empirical analysis allows for formulating conclusions relating to verifying the research hypotheses. The obtained results confirm the first hypothesis, which is that the use of CBI systems positively impacts the efficiency, effectiveness, and speed of managerial decision-making. The respondents assessed these aspects at a comparable level, which indicates their equal importance in the perception of the usefulness of CBI in decision-making processes. The second hypothesis has been partially verified. Although the respondents declare a high assessment of the impact of CBI on the quality of business analyses, they do not identify its significant impact on the organization of data management processes or comprehensive information support for the operational aspects of the functioning of the enterprise. This phenomenon may result from the insufficient integration and management functionalities offered by CBI platforms or implementation limitations in the analyzed organizations. The third research hypothesis was not confirmed. Although the respondents see the growing role of CBI in monitoring and supporting the implementation of the organizational strategy, they do not identify a significant impact of these systems on the optimization of business processes or on the increase in the efficiency of using the company's resources. This result indicates a discrepancy between the potential scope of CBI's impact and its actual use in management practice.

Despite offering valuable insights, the study is subject to several methodological limitations. Reliance on a structured CAWI survey with predefined variables may have limited the depth of responses and failed to capture contextual or qualitative nuances of CBI use. Furthermore, the exclusive focus on medium- and large-sized Polish enterprises limits the generalizability of the findings to other organizational sizes or cultural contexts. The use of three-dimensional correlation analysis, while innovative, still relies on self-reported perceptions rather than objective performance measures. To address these limitations, future research should incorporate qualitative methods, such as interviews or case studies, to gain a deeper understanding of the organizational dynamics, implementation challenges, and contextual factors that influence the effectiveness of AI-enabled CBI systems.

In summary, the research results reveal the significant potential of CBI in decision-making and analytical support, with simultaneous limitations in data integration and impact on operational, organizational processes. This suggests the need for further research on the factors determining the effectiveness of CBI implementation and on the organizational conditions conducive to its full use in enterprise management.

Disclosure of Interests. The authors have no competing interests to declare that are relevant to the content of this article.

References

1. Chebrolu, S.K.: AI-powered business intelligence: a systematic literature review on the future of decision-making in enterprises. Am. J. Sch. Res. Innovation **4**(01), 33–62 (2025)
2. Sharma, D., Kumar, V.: Enhancing organisational intelligence integration of artificial intelligence and knowledge management: frameworks in India. In Digital Technologies in Modeling and Management: Insights in Education and Industry, pp. 167–184, IGI Global, (2024)
3. Jarrahi, M.H., Askay, D., Eshraghi, A., Smith, P.: Artificial intelligence and knowledge management: a partnership between human and AI. Bus. Horiz. **66**(1), 87–99 (2023)
4. Badmus, O., Rajput, S.A., Arogundade, J.B., Williams, M.: AI-driven business analytics and decision making. World J. Adv. Res. Rev. **24**(01), 616–633 (2024)
5. Sahoo, S., Akter, S., Wamba, S.F.: Artificial intelligence capabilities, open innovation, and business performance–Empirical insights from multinational B2B companies. Ind. Mark. Manage. **117**, 92–113 (2024)
6. Jelonek, D., Stępniak, C., Ziora, L.: The meaning of big data in the support of managerial decisions in contemporary organizations: review of selected research. Adv. Intell. Syst. Comput. **886**, 361–368 (2019)
7. Joseph, J.D., Turksen, U.: Harnessing AI for due diligence in CBI programmes. legal and ethical challenges. J. Ethics Legal Technol. (JELT), **4**(2), 3–25 (2022)
8. Suddala, S.: Integrating Artificial Intelligence with Cloud Business Intelligence: Enhancing Predictive Analytics and Data Visualization
9. Olan, F., Arakpogun, E.O., Suklan, J., Nakpodia, F., Damij, N., Jayawickrama, U.: Artificial intelligence and knowledge sharing: contributing factors to organizational performance. J. Bus. Res. **145**, 605–615 (2022)
10. Podina, C., Caculidis-Tudor, A.: Synergizing AI and business: maximizing innovation, creativity, and decision-making. J. Innovation Knowl. **8**(3) (2023)
11. Szajt, M.: Przestrzeń w badaniach ekonomicznych, Wydawnictwo Wydziału Zarządzania Politechniki Częstochowskiej, Częstochowa (2014)

Reshaping Management Systems with AI-Driven Technologies to Improve Operational Efficiency Through Maximize Synergies Effects

Iouri Semenov, Izabela Auguściak(✉), and Marek Dylewski

University of WSB Merito, ul. Powstańców Wielkopolskich 5, 61-895 Poznań, Poland
{iouri.semenov,izabela.augusciak,
marek.dylewski}@szczecin.merito.pl

Abstract. The manuscript is dedicated to the concept of integrating human and machine intelligence inside MS (*Management Systems*). Technological convergence enables the mutual pursuit of HK (*Human Knowledge*) and AI (*Artificial Intelligence*) to achieve the common goal of developing the cognitive capabilities in different areas of the economy. Such a pursuit is a dynamic and continuous phenomenon directed at increasing the knowledge and skills of managers. The intention of the research performed is to analyse the course of the transformation process of conventional MS through the implementation of executive components based on AI-technologies in order to knowledge mining, making predictions, reasoning, and problem-solving, Analyses are carried out at earlier stages assess the probability of both positive and negative synergy effects through integrating HK and AI components. Our research shows that an important factor determining integration success is iterative adaptation of HK & AI components according to the criterion of max positive synergistic effects.

Keywords: management · AI -technologies · upgrade · synthesis effects

1 Problem Background

The global economy brings together various companies a, which are united by the need to have effective management systems, hereafter referred as MS. The primary task of these systems is to generate managing decisions to achieve business goals. Since the 1960s, MS upgrade have been directed towards improving techniques for collecting and structuring information based on the use of digital technologies (Mikalef & Gupta, 2021. The success of such technologies was driven by the development of databases. In the 21st century (Haenlein et.al, 2020) its effect has been an exponential increase in the effectiveness both business managing and the actions positively influencing the stability of the production processes, the predictability of upgrade trajectories. The continuation of these trends through switched HK-components (knowledge bases, analytical and simulation software) for AI-components (ML-software, Python language algorithms) used to mining new knowledge or forecasting.

Published by Springer Nature Switzerland AG 2026
M. Pondel et al. (Eds.): AI4KAM 2025, IFIP AICT 776, pp. 213–219, 2026.
https://doi.org/10.1007/978-3-032-18920-2_20

2 Conceptual Framework

2.1 Components Sets

As practice shows, MS development is based on integrating HK and AI components, resulting in various synergies not always anticipated in the preparatory stages (Mhaskey, 2024). This requires strengthening the ability of firms to adapt new technologies (Ameen, et. al., 2024), including ML techniques, or to access cloud services (Kunduru, 2023). This approach to MS upgrade process deviates from traditional methods based on analyses of the interactions only HK-components. In the literature, such interactions are grouped according to both purpose into exploration (exploration) and research (exploitation) and the change dynamics into incremental/decremental or transformational/continuous (Roy, et al. 2025). Unfortunately, these publications neglected the question of the interaction of group components and synergistic mechanism. Research has proven that such a mechanism must be investigated using non-linear modelling techniques (Shaik, 2024) at three management levels starting from cellular to integrated (Semenov & Auguściak, 2024). Two hypotheses were defined:

Hypothesis 1. ADM (Architecture-Driven Modernization) can result in both positive and negative synergies, which can vary depending on the level of transformation work.

Hypothesis 2. An iterative adaptive process can increases the integration effectiveness.

2.2 Integration Connection Sets

The MS described in this article is adapted to the needs of the TSL industry and open to implementing AI-components in both inference blocks and interface relationship. The co-operation of HK/AI components is implemented using i.e. integration links, which can be as regular as random. An assumption was made about the desirability of limiting the forms of connections to three types, reflecting:

1. management activities;
2. monitor of the implementation activities;
3. transmission of the information flows.

A major barrier to comprehensive block integration is the high probability of the occurrence of i.e. SPOF (*Single Point Of Failure*) effects, when the failure of one block contributes to the shutdown of all others (Li, 2014).

2.3 Adaptive Mechanism: Assumptions

The management of TSL companies may make decisions to upgrade MS by implementing innovative AI -technologies (Shao, Li & Wang, 2022). Such decisions may be ineffective in the absence of adequate preceding analyses (Vrontis et al., 2022), (Vrontis et al., 2022), the based on AST (*Adaptive Structuration Theory*) (Sedera, & Zakaria, 2008). Our study draws on four assumptions, including:

Assumption 1: ADM is successful if it fulfils its functions under changing business-environment conditions without the need for secondary adjustments to its architecture.

Assumption 2: The level of effectiveness of a system should be assessed in terms of the likelihood of increasing its functional characteristics.

Assumption 3: The choice of components must be justified in terms of added value.

Assumption 4: Effective ADM based on unreliable information is not possible.

3 Research Design

The critical role of MS in the operation of a TSL company calls for an in-depth study of the mechanism by which synergies arise and based on two principles:

Principle 1. MS upgrade should be directed towards achieving positive synergies.

Principle 2: Synergies should be achievable across the widest possible spectrum of current and future developments in both demand and supply markets.

3.1 Research Instruments

There are different approaches to systems integration, but the common denominator is the requirement for seamless HK/AI component co-operation. An integrated platforms linking to the 'Ease-of-Integration' principle. Integration are realize on the six levels:

A. ***Infrastructure integration***. It is an arrangement of knowledge bases, computers, network devices (servers, switches, etc.) working together via network connections according to the criterion of max positive synergy effects.
B. ***Information integration***. It is implemented within the framework according to DCA (*Data-Centric Approach*) in terms of consistency and accessibility of the collected information. It is based on processes of combining structured information.
C. ***Application integration***. It integrates the execution schedules of applications and services running within an integrated MS, enabling managers collaborate without conflict. This is important because the applications used may come from different manufacturers, have different purposes and require information support.
D. ***Process integration***. Process integration is the glue that binds together the functional capabilities of different components within an integrated MS. It ensures that the redundancy of the combined components is eliminated or indicated, facilitates the generation of information flows, reducing errors and improves efficiency.
E. ***Cybersecurity integration.*** The rise of cyber threats complicates the deployment and integration of AI technologies into conventional MS, threatening to create gaps in their resilience. This can be countered through the use of security features such as LANs (*Local Area Networks*) or WANs (*Wide Area Networks*).
F. ***Integration testing procedures***. Testing plays a key role by identifying gaps in pro cesses, checking that tasks are performed correctly. There are various approaches to testing the upgrade effectiveness, including a "top-down" approach.

3.2 Organisation of the Transformation Process

Rotation is a fundamental part of an obsolete system upgrade. This process can be described by graph technique. Let us analyse the organisation of the transformation process involving the deployment of AI-components to BW (*Inference Block of MS*) as maximal connected subset. We will assume that after each deployment cycle such a block transitions to a new state, a sequence of which we will mark as < S >:

$$S = [S_0(P_0 t_0) \wedge S_1(P_1 t_1) \wedge, \ldots \wedge S_i(P_1 t_i) \wedge \ldots] \quad i = 1 - N \tag{1}$$

whereby, BW transitions from the < S_i > state to the < S_j > state with probability P_{ij} generate positive or negative synergy effects. The aim of our study is to assess the implantation process effectivity. For this mark as ξ[t] the changes occurring in BW states over the time t ∈ [0,T]. When occurs implantation process, MS enter the < S_1 > state - reflecting the preliminary level of HK /AI-component interaction at which positive synergies are unlikely to arise. After the initial adaptation cycle, BW will enter the < S_2 > state. If the goal (synergy effects) is not achieved, the adaptation cycles may continue. Once the BW hits the < S_N > state, upgrade cycles stopped. Sequence business changes

$$\Omega(t)[\omega(t_1), \omega(t_2), \ldots, \omega(t_n)] \tag{2}$$

We will make the assumption,, that in the same timeframe there is a BW upgrade and MS transit to a new state described by the variable Ξ(t):

$$\Xi(t) = [\xi(t_1), \xi(t_2), \ldots \xi(t_n)] \tag{3}$$

Equation (3) assumes that each aspect $\xi(t_{n+1})$ BW state changes reflects some event (e.g. the execution of new activities after upgrade). The probability of such events:

$$[P\{\xi(t)\} = P\{\xi f[\omega(t_{n+1})] = 1\} = p \vee P\{(\xi[\omega(t_{n+1})] = 0)\} \tag{4}$$

Using Eq. (4), we can assess the probabilities of synergies arising as indirect effects of a company's response to changes in its business-environment:

$$P\{\xi(t)\} = f[\omega(t) = \{P[\xi(t_1)f[\omega(t_1) = S_i, P[\xi[(t_2)f[\omega(t_2)] \neq S_i P\xi(t_n)f[\omega(t_n)] \neq S_1\} \tag{5}$$

Comment: It follows from Eq. (5) BW can transitions to state < Si > with negative effects or state < Sj > achieves positive synergies. Therefore, two problems are:

Problem 1: Identify a HK/AI component mergers burdened by conflicts generating negative synergies (Task1). Conflicts were assessed by manifestation probability.

Problem 2: Evaluating number of adaptation cycles that ensure the elimination of negative effects arising from the conflicting interaction HK/AI components (Task2).

3.3 Identification of Conflicting HK/AI Component Combinations

Failed upgrade of the MS can have various causes, including:

- errors in planned adaptation activities sensitive AI-components,
- the absence of recommendations for adaptation actions based on the study's results

Analysis This Problem. Let < D > non-empty set of AI-components, whereby:

- each decision < d_i > from the set < D > can be assigned a value $d_i \rightarrow [0,1]$.
- any accurate decision < d_i > has a value = 1.0 and a conflicting decision < d_i* > = 0.

We will introduce two scales for assessing the probability of successful integration:

Scale 1. There are probability thresholds for successful upgrade $[P_o/P_1/P_2]$ as:

$P[\xi(t)] < P_0$ - the likelihood of generating positive synergies is negligible;

$P_0 \leq P[\xi(t)] < P_1$ -the probability of generating a positive effect has a threshold value;

$P_1 \leq P[\xi(t)] \leq P_2$ the probability of generating a positive effect has a target value.

Scale 2. There are probability thresholds for the failing upgrade $[P^*{}_0/ P^*{}_1/ P^*{}_2]$ as:

$P[\xi(t)] < P^*{}_0$- the likelihood of negative synergies being generated is negligible;

$P^*{}_0 \leq P[\xi(t)] < P^*{}_1$- the probability of generating a negative effect has a threshold value;

$P^*{}_1 \leq P[\xi(t)] \leq P^*{}_2$- the probability of generating a negative effect is very high.

Calculate the probability the $P\{\xi(t)\}$ for BW adapting to perform the new (improved) activities immediately after the AI-component implementation of using the equation:

$$P\{\xi(t)\} = P(d_0)\{\langle Ex_1[S_{11}(t_1)]\rangle \times PS_{11}, (t_1)\} + P(d_0)\{\langle Ex_1[S_{12}(t_1)]\rangle \times P^*S_{12}(t_1)\}, \quad (6)$$

where d_0 – decision to implement AI-component, ensuring BW update,;S_1 ($d_0 \in D$)

$S_{11}(t_1)$ – compatibility status of the MS immediately after upgrade activities;(t_1).

$S_{12}(t_1)$ – BW component incompatibility status after upgrade activities;(t_1).

$Ex_1[S_1(t_1)]$ - the level of exposure to these events;

$PS_{11}(t_1)$ – the likelihood of generating positive synergies in the state S_1 moment t_1;

$P^*S_{12}(t_1)$ – probability of generating negative synergies in the state S_1 moment t_1;

We will calculate the probability of BW adapting to perform new (streamlined) activities after the first AI-component adaptation cycle $P\{\xi(t)\}$ using the equation:

$$P\{\xi(t)\} = P(d\_1)\{\langle E_2[S_{21}(t_2)]\rangle \times PS_{21}, (t_2)\} + P(d_1)\{\langle Ex_2[S_{22}(t_2)]\rangle \times P^*S_{22}(t_2)\} \quad (7)$$

where: d_1 - decision to adapt the AI-component in order to transition BW to S_2, ($d_1 \in$ D);

$S_{21}(t_2)$- state of BW structure compatibility at time t_2 after adaptation activities;

$S_{22}(t_2)$- state of BW structure incompatibility at the t_2 after adaptation activities;

$Ex_2[S_2(t_2)]$ - the level of exposure to upgrade activity;

$PS_{21}(t_2)$- the likelihood of generating positive synergies after BW transitions S_2;

$P^*S_{22}(t_2)$ – the likelihood of negative synergies on state S_2.

4 Results of the Test Using Topology Techniques

4.1 A Subsection Sample

The study was performed on an oriented binary graph ≪ Gr ≫ depicting the MS upgrade process. We will assume that each process task L_p generates a combination of k-th AI component with m-th HK component describes as (n_k, n_m),$n_k \in N$. all such tasks:

$$Z =< [L_1(n_{i1}, n_{i1}), L_2(n_{i2}, n_{i2}), \ldots, L_p(n_{ik}, n_{im}) \ldots] >, \qquad i \in I;\ j \in J \tag{8}$$

According to (Semenov & Filina-Davidovich, 2017), any structure of IT-system can be described by a vector $q \in B^n$, where "n" is the max set of connections k-nodes. The BW upgrade sequence vector is a bit string with n-binary variables, in which there are both positive ($q_i = 1$) and negative ($q_i = 0$), ($I \in I$) synergies.

5 Conclusions and Recommendations

Different ways of improving MS were analysed, with the unconditional priority for an approach based on max synergy effects. Proposed ADM- model is solve as task creating a conflict-free AI - set as and graph-combinatorial covering the vertices. Two directions were notable in our research, namely the planning of the MS topology and the evaluation of the upgrade process effectivities. The research was performed on the ADM- model. At the initial stage, a preliminary BW structure was analysed through study of the MS structure arrangement. Secondly, MS components were connected by fixed and variable links. At the last, the analysis of positive or negative values of the synergies. The relatively low probability of achieving targeted effects was caused by a high SPOF probability. A way of eliminating the SPOF is the gradual implementation of AI-technology. In this case, the probability of positive synergy effects ranges from 0.708 ± 0.03 to 0.964 ± 0.01, and has company level from 0.538 ± 0.03 to 0.809 ± 0.01.

References

Ameen, N., et al.: Coupling artificial intelligence capability and strategic agility for enhanced product and service creativity. Br. J. Manag. **35**, 1916–1934 (2024)

Bankin, S., et al.: A multilevel review of artificial intelligence in organizations: Implications for organizational behavior research and practice. J. Organ. Behav. **45**(2), 159–182 (2024)

Gebauer, J., Schober, F.: Information system flexibility and the cost efficiency of business processes. J. Assoc. Inf. Syst. **7**(3) (2006). https://doi.org/10.17705/1jais.00084

Haenlein, M., Kaplan, A.: A brief history of artificial intelligence: on the past, present, and future of artificial intelligence. Calif. Manag. Rev. **61**(4), 5–14 (2020)

Kolbjørnsrud, V.: Designing the Intelligent Organization: six principles for Human-AI collaboration. Calif. Manage. Rev. **66**(2), 44–64 (2024)

Kunduru, A.R.: Effective usage of artificial intelligence in enterprise resource planning applications. J. Comput. Trends Technol. **71**(4), 73–80 (2023)

Li A., Wu, B. Wang, L.: Cooperation with both synergistic and local interactions can be worse than each alone. Sci. Rep. **4**, 5536 (2014). https://doi.org/10.1038/srep05536

Mendonça, S. at.al.: The strategic strength of weak signal analysis. Futures, 44 (3), pp. 218–228 (2012)

Mhaskey, S.V.: Integration of Artificial Intelligence (AI) in Enterprise Resource Planning (ERP) systems: opportunities, challenges, and implications. Int. J. Comput. Eng. Res. Trends **11**(12), 1–9 (2024)

Mikalef, P., Gupta, M.: Artificial intelligence capability: conceptualization, measurement calibration, and empirical study on its impact on organizational creativity and firm performance. Inf. Manag. **58**(3) (2021)

Roy, S.K., et al.: Business model innovation through AI adaptation. Br. J. Manag. **36**(2), 546–559 (2025)

Salabun, W., Pietrzykowski, M.: Neural networks in economic problems. In: Selected Issues in Experimental Economics: Proceedings of the Computational Methods in Experimental Economics (CMEE) Conference. Springer, pp. 245–266 (2016)

Sedera, D., Zakaria, N.H.: Adaptive structuration theory view of post implementation knowledge management for enterprise systems. In: Pacific Asian Conference on Information System: AIS, p. 179 (2008)

Semenov, I., Auguściak, I.: Building the enterprises capacity through Specialized-Self-Learning Knowledge Platforms (SLKP) connected to Omnichannel networks. In: Hernes, M., Wątróbski, J., Rot, A. (eds.) Emerging Challenges in Intelligent Management Information Systems. ECAI 2024, vol. 1218. Springer (2024)

Semenov, I., Filina-Davidovich, L.: Topology-based approach to the modernization of transport and logistics systems with hybrid architecture. Arch. Transp. **43**(3), 105–124 (2017)

Shaik, M.: Implementing AI-driven efficiency: best practices for intelligent order processing in SAP. J. Res. Appl. Sci. Eng. Technol. **12**(1), 45–52 (2024)

Shao, Z., Li, X., Wang, Q.: From ambidextrous learning to digital creativity: an integrative theoretical framework. Inf. Syst. J. **32**, 544–572 (2022)

Siebrecht, J, et al.: Customer specific compatibility matrices for functional integral product architectures. In: Proceedings of the Design Society, Conference, vol. 1, pp. 1105–1114 (2020)

Tambe, P., et al.: Artificial intelligence in human resources management: challenges and a path forward. Calif. Manage. Rev. **61**(4), 15–42 (2019)

Tyre, M.J., et al.: Windows of opportunity: temporal patterns of technological adaptation in organizations. Organ. Sci. **5**, 98–118 (1994)

Vrontis, D., et al.: Artificial intelligence, robotics, advanced technologies and human resource management: a systematic review. Int. J. Hum. Resour. Manag. **33**, 1237–1266 (2022)

Wątróbski, J., Jankowski, J., Ziemba, P.: Multistage performance modelling in digital marketing management. Econ. Sociol. **9**(2), 101–125 (2016)

Zhan, C., Du, Y.: Early identification methods for emerging technologies based on weak signals (2022). Preprint.https://doi.org/10.21203/rs.3.rs-2291140/v1

Artificial Intelligence Solutions in the Support of Decision Making Processes of Software Houses. Case Study Analysis

Leszek Ziora[1](✉) and Steven J. Hyde[2]

[1] Czestochowa University of Technology, Czestochowa, Poland
leszek.ziora@pcz.pl
[2] Boise State University, Boise, USA
stevenhyde@boisestate.edu

Abstract. Artificial intelligence solutions—including machine learning, deep and reinforcement learning, business intelligence, big data analytics, NLP, and GANs—support decision-making at all levels of management. This paper explores how AI supports decision-making processes in software development firms. The theoretical part of the paper focuses on the fundamental AI technologies applied by Software Houses, especially in relation to the improvement and support of decision making processes underlying the benefits from its application in software development. The empirical section analyzes four case studies from selected software firms.

Keywords: Artificial Intelligence · Business Analytics · decision-making support · software development

1 Introduction

This paper examines how AI solutions support decision making in software development firms. The study addresses the following research questions:

1. What types of AI solutions are applied in the management of software houses?
2. What types of decisions are supported by artificial intelligence solutions implemented in software houses?
3. What functions do AI solutions fulfill in the process of decision-making in the surveyed Software Houses?
4. What is the role, advantages, and disadvantages resulting from the application of AI in the management of Software Houses? To achieve the paper's aim, the following research tools were employed: literature analysis and an analysis of case studies conducted in four software houses.

The research results indicated numerous benefits of implementing AI tools, some of which are worth mentioning: acceleration of decision making, enhancement of its quality,

Published by Springer Nature Switzerland AG 2026
M. Pondel et al. (Eds.): AI4KAM 2025, IFIP AICT 776, pp. 220–227, 2026.
https://doi.org/10.1007/978-3-032-18920-2_21

increased efficiency and efficacy, as well as a competitive advantage and innovativeness. It also examines the importance of data quality in improving decision-making processes at all levels and stages of Software House management.

2 The Review of Artificial Intelligence Solutions Implemented in Software Houses

A.Slasky enlists AI-assisted coding tools which can support programmers in terms of "syntax and language complexity; debugging and error resolution; code efficiency and optimization; seamless integration and compatibility; scalability and maintainability; collaboration and version control; improving documentation and readability as well as security and vulnerability mitigation providing examples of Qodo, GitHub Copilot, Tabnine, Windsurf, Amazon CodeWhisperer, AskCodi, Codiga, Replit, CodeT5, OpenAI Codex, DeepCode AI, Figstack, Intellicode, CodeGeeX and many more. [5]". S. Russel and P. Norvig associate AI with thinking and acting humanly and rationally [4]. In summary, AI encompasses a range of technologies, methods, tools, and techniques for data processing [1, 3, 7, 8]. Robotic Process Automation (RPA) contributes to digital transformation, enabling the automation of tasks previously performed by humans [6]. S.J. Hyde et al. emphasize the importance of data quality in supervised machine learning, which can significantly impact the quality of decisions made and the quality of analyses conducted [2, 10]. Overall, AI provides a comprehensive framework for software development support, as illustrated in Fig. 1.

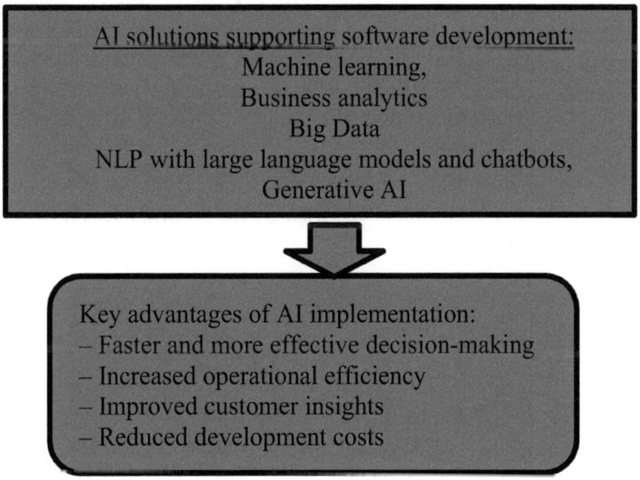

Fig. 1. General framework of AI applicability in the software development process. Source: Authors' study

3 Analysis of Case Study – Research Procedure and Results

The case study was conducted in 2023 among software houses at the middle-level tier of management. The research survey included 10 open-ended questions focused on the applicability of AI solutions in the management process of a Software House. The research questionnaire is presented in Table 1:

Table 1. The research questions

Number of question	The content of the research question:
1	What areas of Software House management are supported by AI solutions
2	Is the applied solution useful in the management of a Software House?
3	What benefits resulted from the implementation of AI in the area of Software House management?
4	Was the process of developing the company's strategy improved after implementing AI solutions?
5	Was the process of decision-making improved and, if so, in what way/to what extent?
6	Did the quality of data affect the improvement of the quality of the decision-making process and the quality of the conducted analyses?
7	Was the efficiency and the scope of analyses improved after the implementation of the AI solution?
8	How do you perceive the future of AI solutions in the management of a Software House?
9	Have AI solutions allowed for an increase in the manager's work efficiency?
10	What challenges were faced in the process of AI solutions implementation, and what are the disadvantages resulting from its application?

Source: Authors' study

Future research prospects can be enhanced by increasing the number of interviewed companies. The results of the conducted research are presented in Table 2 below:

Table 2. The research results

Question 1	**Areas of management supported by AI**
Case study 1	The use of AI is very helpful in areas related to mailing campaigns, website creation, and marketing automation

(*continued*)

Table 2. (*continued*)

Question 1	**Areas of management supported by AI**
Case study 2	Most often, *"these are areas related to chatbots, customer service, employee service (accessibility, etc.). They have also contributed to better cooperation between individual work positions."*
Case study 3	AI *"is used by almost all departments in our company to ensure security. It allows for personalized access to rooms and protects against unauthorized access to data. It is also necessary for the analysis of large data sets, giving us a number of possibilities to improve and accelerate analyses."*
Case study 4	AI allows "*to create applications dedicated to SAP solutions and tailored to customer needs, providing customers with the ability to respond to new business needs, offering the chance to stand out from the competition by using top-class technology.*"
Question 2	**Was the implemented solution useful in doing business, and to what extent?**
Case study 1	It is very helpful. *"It is impossible for an IT company to operate without using AI. After all, it was programmers who created AI, which was to assist them in their further work. For example, it is used in continuous analysis of collected data and processing it for current needs."*
Case study 2	*"Of course, yes, and to a large extent, e.g., without machine translation, which is very effective, our services would be performed 3 times slower."*
Case study 3	*"AI is useful and has a huge impact on security management, which affects work organization and better company management. It speeds up data analysis processes, saving us time."*
Case study 4	Very useful by dealing with customer service in the global field as well as software development, "*I cannot imagine working without the support of "AI". It helps me many times to take care of other tasks, when at the same time AI is doing other work.*"
Question 3	**What are the benefits resulting from AI implementation, and did the company's competitive advantage increase (in terms of its market share)?**
Case study 1	The benefits resulting from AI implementation are great itself like code generation support, error reduction, acceleration of coding process. However, "*the company's competitiveness has not increased in the market.*"
Case study 2	There are many benefits, such as "*automation of programming tasks but it has not entirely increased our company's competitiveness in the market. Low-code and no-code solutions help in work acceleration.*"
Case study 3	*"I think that our competitiveness has not increased due to the use of AI, we do not notice such an aspect. However, innovation has certainly increased"*. Automation increased the level of innovation, and the business processes were improved, together with better software project management and facilitation of documentation creation

(*continued*)

Table 2. (*continued*)

Question 1	**Areas of management supported by AI**
Case study 4	*Speed, accuracy, and readability of data – "these are some of the most important issues in our company." Competitiveness has increased to some extent, but it is not very noticeable. The time required for designing and launching a new software product was reduced, and better code quality was achieved*
Question 4	**Was the process of developing the company's strategy improved after implementing AI solutions?**
Case study 1	It is safe to say that *"it has improved and enhanced the business strategy of our company. Thanks to AI, we have expanded our range of services, and the waiting time for the product has been significantly shortened. You can see increased satisfaction of our customers, who come back to us with new orders."*
Case study 2	*"It has accelerated various plans, orders are executed much faster and more accurately, and many employees have a direct impact on making given decisions. It has improved and improved the company's business processes and strategies."*
Case study 3	Implementing AI has *"certainly improved the organization of the company by improving safety and streamlining work. However, this has no impact on our company's business strategy."*
Case study 4	Yes, *"it has definitely sped up and improved the operations of our company. Orders are processed much faster and more accurately, which has increased customer confidence in our services."*
Question 5	**Was the process of decision-making improved and, if so, in what way/to what extent?**
Case study 1	*"All activities have been accelerated, and data analysis performed by AI has allowed us to engage in other projects. The efficiency and speed of tasks performed are satisfactory, which is evident in the huge interest in our services."*
Case study 2	*"Not at all levels and stages, but at most, actions are more precise. The effectiveness and efficiency of decisions made are greater. AI has had a profoundly positive and significant impact on generating value for business organizations. Products and services can be more tailored to the user, facilitate access to information, and lead to company development."*
Case study 3	*"AI speeds up our work, thanks to faster data analysis, and this is visible at every level. It has a significant impact on quick decision-making, mainly at higher levels. AI improved the strategic value of decisions."*
Case study 4	*"Yes, decisions are made very quickly, which allows employees to quickly undertake new tasks, which automatically shortens the customer's waiting time for the ordered product."*
Question 6	**Did the quality of data affect the improvement of the quality of the decision-making process and the quality of the conducted analyses?**

(*continued*)

Table 2. (*continued*)

Question 1	**Areas of management supported by AI**
Case study 1	*"Data analyses, which are conducted on an ongoing basis, have significantly helped in the decision-making process. Decisions are made on an ongoing basis, based on the latest and most reliable data. This means that there are no errors in the current assessment of the situation, and the decisions are clear. These analyses are very important to us and contribute to the development of the company."*
Case study 2	*"It had a very significant meaning because it makes our work much easier. The quality of data significantly improved the quality of the decision-making process and analyses conducted using AI. It contributed to the development of the company."*
Case study 3	*In our company, the entire process of conducting business analysis has been improved, and innovative solutions have modernized it. I can safely say that to a significant extent."*
Case study 4	*"Our company must constantly analyze collected data, so that it can function at a very high level. The data sets on the basis of which the analyses are carried out are constantly supplemented, which helps in decision-making processes."*
Question 7	**Was the efficiency and the scope of analyses improved after the implementation of the AI solution?**
Case study 1	After implementing AI, "*the effectiveness and scope of decision-making analyses increased, especially in relation to customer analysis."*
Case study 2	*"After implementing AI, the efficiency and scope of decision analysis increased. This is most noticeable in the customer department, where customers are more satisfied and tend to return to us multiple times."*
Case study 3	*The efficiency and the scope of decision-making analyses of the customer increased*
Case study 4	*"The analysis of customer needs is very fast and effective, which affects the high quality of our services. Moreover, cloud computing technology supports us to a great extent, e.g., data backup."*
Question 8	**How do you perceive the future of AI solutions in the management of a Software House?**
Case study 1	*"The future of AI development is on the right track. I see machines that will assist or even replace human work"*
Case study 2	*"I see tools that will replace human work. AI tools significantly enhance productivity and streamline the company's technological processes. The IT industry constantly needs specialists, novelties, and new technologies"*
Case study 3	*Further development of AI in the software development domain can be observed, with higher computing power now available in cloud models*
Case study 4	*Without AI, "one could say that we would not have the opportunity to exist. Besides, it is we, the programmers, who lead the development of AI, or more precisely, engineers for AI. It would not exist without programmers."*

(*continued*)

Table 2. (*continued*)

Question 1	**Areas of management supported by AI**
Question 9	**Have AI solutions allowed for an increase in the manager's work efficiency?**
Case study 1	*"AI solutions have allowed for an increase in the manager's workload. The manager certainly has fewer obligations related to any control. He can use his working time more effectively, more efficiently."*
Case study 2	*"AI solutions have allowed for an increase in the efficiency of the manager's work, enabling more effective use of his time at work."*
Case study 3	*"Implementation of AI improves the work of the manager; there is undoubtedly easier and faster access to necessary data."*
Case study 4	*"Of course, yes, the managers have more time to implement new projects and contact with the client, and for industry/branch meetings."*
Question 10	**What challenges and disadvantages were faced in the process of AI solutions implementation?**
Case study 1	*The difficulties and challenges are training courses, and its high costs, and some employees are also not willing to use such training. The cost of downtimes can be a significant disadvantage*
Case study 2	*"We overcame difficulties in the process of implementing AI solutions, which were activities related to training employees. The disadvantages are not noticeable."*
Case study 3	*"We encountered technical problems, which fortunately were resolved. New solutions required learning how to use them, which involved costly training programs."*
Case study 4	*There were no significant difficulties. All we had to do was implement employee training. The cost of AI itself is high, but its application is necessary, and at this point, the costs are outweighed by its benefits. Perhaps our expectations regarding machine learning are too high, because we expect machines to learn faster."*

Source: Authors' study

Some respondents indicated that AI can assist programmers in many ways, but it may lead to their substitution in the future. The respondents indicated an impact of AI on the increase of software firm innovativeness in general. The key findings demonstrate how AI enhances decision-making, improves service quality, and can help in writing code. However, the application of AI also presents certain disadvantages and challenges, such as the high cost of implementation and the cost of employee training. Future trends and directions in AI development may be connected with the increasing growth of computing power.

4 Conclusion

This paper presents the current state of AI solutions applied in software development companies, Software Houses, used in the process of software creation management, such as tools for code generation, automated testing, low-code or no-code solutions, and it especially puts emphasis on the decision making process supported by chatbots, generative AI solutions, and business analytics. The research highlights key benefits of AI, including improved decision quality, faster analyses, and greater operational efficiency. Similar to other industry branches, such as the telecommunications or financial sectors [10], AI has a significant and positive influence on the automation of tasks, enabling more efficient customer analyses.

References

1. Dunay, A., Bálint Illés, C., Ziora, L.: The applicability of generative adversarial networks in the management of contemporary business organizations. 12th International Conference of Management Sustainability–Security–Quality Book of Proceedings. Czestochowa, Poland (2023)
2. Hyde, S.J., Bachura, E., Harrison, J.S.: Garbage in, Garbage out: a theory-driven approach to improve data handling in supervised machine learning. In: Methods to Improve Our Field, Research Methodology in Strategy and Management, Hill A.D., F.McKenny, A., O'Kane, P., Paroutis, S. (eds). Emerald Publishing Limited, 14, 101–132 (2023). Available at: https://doi.org/10.1108/S1479-838720220000014006
3. Jelonek, D., Mesjasz-Lech, A., Stępniak, C., Turek, T., Ziora, L.: The Artificial Intelligence application in the management of contemporary organization: Theoretical Assumptions, Current Practices and Research Review. In: Proceedings of the Future of Information and Communication Conference. Arai, A.G.K., Bhatia, R., (eds.). Springer Nature: Cham, Switzerland,; pp. 319–327 (2020)
4. Russel, S., Norvig, P.: Artificial Intelligence. A modern Approach; 4th ed. Pearson (2021)
5. Slasky, A.: Best AI for Coding and AI Coding Assistants by Category. https://www.qodo.ai/blog/best-ai-coding-assistant-tools/
6. Sobczak, A., Ziora, L.: The use of Robotic Process Automation (RPA) as an element of smart city implementation: a case study of electricity billing document management at Bydgoszcz City Hall. Energies **14**(16), 5191 (2021). https://doi.org/10.3390/en14165191
7. Sobczak, A., Ziora, L.: Intelligent technologies supporting the management of a smart city. qualitative approach. Proceedings of the 56th Hawaii International Conference on System Sciences, HICSS 2023, University of Hawaii, USA (2023)
8. Ziora, L.: Natural Language Processing in the Support of Business Organization Management [in:] Intelligent Systems and Applications 2021 (red.) Arai K. Lecture Notes in Networks and Systems nr 294, Springer, Cham, pp. 76–83 (2022a). (Intellisys 2021 Amsterdam)
9. Ziora, L.: Machine-Learning Solutions in the Management of a Contemporary Business Organisation: a Case Study Approach in a Banking Sector. [in:] Rational Decisions in Organisations. Theoretical and Practical Aspects (red.) Frederic, A., Dorota, K., Stanisław, S. (eds.) Taylor and Francis Group. New York (2022b)
10. Ziora, L., Hyde, S.: The applicability of artificial intelligence solutions in the management of telecommunication industry companies: The Qualitative Approach. In book: Advances in Information and Communication (pp.173–182), Springer (2025). https://doi.org/10.1007/978-3-031-84460-7_11

Human Activity Recognition (HAR) in Management of Logistics Processes: New Opportunities or Technology without Application?

Marta Starostka-Patyk[1(✉)], Marek Szajt[2], Joanna M. Panek[3], and Tomasz Jałowiec[4]

[1] University of Kalisz, 62-800 Kalisz, Poland
m.starostka-patyk@uniwersytetkaliski.edu.pl
[2] Czestochowa University of Technology, 42-201 Czestochowa, Poland
marek.szajt@pcz.pl
[3] Military University of Technology, 00-908 Warsaw, Poland
joanna.panek@wat.edu.pl
[4] Warsaw University of Life Sciences, 02-787 Warsaw, Poland
tomasz_jalowiec@sggw.edu.pl

Abstract. The growing interest in Human Activity Recognition (HAR) systems in logistics reflects broader trends in automation and intelligent process management. While HAR technologies have proven effective in domains such as healthcare and physical activity monitoring, their application in logistics remains limited and raises substantial implementation challenges. This paper explores whether HAR represents a breakthrough opportunity or an overestimated technological promise in logistics management. A review of literature reveals limited empirical research and a significant lack of reliable, annotated datasets tailored to logistic operations. Moreover, economic and organizational barriers—especially for SMEs—include high implementation costs, uncertain return on investment, concerns over data privacy, and limited system reliability in complex, real-world environments. The study presents the results of a quantitative survey conducted among 397 logistics-related companies across Poland. The analysis, based on Spearman correlation and Mann-Whitney U tests, indicates that the perception of HAR technology improves with company size and employee experience. Larger companies are more likely to use and invest in HAR, though they also acknowledge the cost-benefit imbalance. Simulations confirmed that awareness of HAR increases with enterprise size and respondent seniority. These findings highlight a key contradiction: while HAR offers significant theoretical benefits in logistics, its practical implementation is hindered by economic, ethical, and technical concerns. The paper concludes with recommendations for further research and policy-making, including the need for pilot programs, targeted funding, and development of logistics-specific HAR datasets.

Keywords: Human Activity Recognition · HAR · Management · Logistics

Published by Springer Nature Switzerland AG 2026
M. Pondel et al. (Eds.): AI4KAM 2025, IFIP AICT 776, pp. 228–234, 2026.
https://doi.org/10.1007/978-3-032-18920-2_22

1 Introduction and Theoretical Background

In recent years, interest in applying Human Activity Recognition (HAR) technology to logistics management has increased. Originally used in healthcare and fitness monitoring, HAR is now gaining attention in industrial settings [1].

Studies show that adding contextual data improves HAR accuracy in logistics [2]. Datasets like LARa, combining sensor and camera data, help train deep learning models for tasks such as walking or packaging [3]. Despite being new in logistics, HAR benefits from accurate CNN and LSTM models [4, 5]. Wearable-based HAR also enables real-time monitoring and links to transport systems [6–8]. HAR is thus shifting from theory to a practical tool for improving logistics efficiency and safety.

Yet, is HAR truly a new opportunity—or just another overestimated innovation?

A review of the literature reveals that in the past decade, only a limited number of scientific studies have directly addressed the application of Human Activity Recognition (HAR) in logistics. This gap underscores the empirical research deficit in this specific context, positioning it as the central research problem of the present study. The primary reason for this gap is the lack of realistic and well-annotated datasets that reflect actual work processes.

Although HAR is gaining attention in automation and logistics research [9–11], some studies highlight its limitations. These include not only technical, but also economic, organizational, and ethical barriers that reduce its practical value and cost-effectiveness in logistics. A major challenge is the high cost of implementing and maintaining HAR systems, including sensors, cameras, IT infrastructure, and training. Many logistics companies, especially SMEs, see limited return on investment, making HAR costly with unclear benefits [9]. Ethical and legal concerns are also important. HAR collects biometric and behavioral data, raising privacy issues and risking lower employee trust. Its use must comply with data protection laws like GDPR, adding legal and deployment challenges [4]. Another critical issue is the limited reliability of HAR systems in variable and demanding working conditions, typical of logistics environments. Studies have shown that even advanced deep learning models—such as convolutional neural networks—may struggle to correctly classify basic actions under non-ideal conditions [9].

Despite growing interest, HAR remains difficult to apply in logistics. Limited data, high costs, privacy risks, and low reliability suggest it is currently more a technological novelty than a practical logistics tool.

Considering these theoretical premises, the aim of this article is to verify the level of interest among logistics-oriented enterprises in implementing HAR technology. To achieve this, the paper presents a comprehensive theoretical background on the HAR in logistics, a methodology based on a structured survey, detailed research findings supported by advanced analytical tools, and a discussion and conclusion section highlighting comparable research, the main contributions of this study, and future research directions.

2 Methodology and Research Results

The study was conducted at the turn of 2024/2025 among industrial enterprises operating in the logistics sector or maintaining dedicated logistics departments. The respondents represented companies active in various areas of logistics, including procurement logistics, production logistics, distribution logistics, transportation, and warehousing. A total of 397 fully completed survey questionnaires were collected, which meets the required sample size threshold for a 5% margin of error at a 95% confidence level (minimum sample size: 385). The research was based on a structured questionnaire comprising 15 substantive questions and 3 demographic (profiling) items. The survey was administered directly within the companies. Questionnaires were distributed via email, and respondents were requested to complete and return them through the same channel. The internal consistency of the survey instrument was validated using Cronbach's alpha, which returned a satisfactory result.

The conducted research enabled the identification of key results and conclusions. As a first step, Spearman's rank correlation coefficient was applied to determine the relationship between the size of the surveyed enterprise (defined by the number of employees and classified as large ≥ 250, medium = 51 to 249, or small ≤ 50) and its perception of the potential of HAR technology for application in logistics (Table 1).

The conducted research confirmed several statistically significant correlations between enterprise size and the perception of selected aspects related to the topic of HAR in logistics. All identified correlations were positive, clearly indicating that the perceived importance of HAR increases with the size of the enterprise. As enterprise size increases: HAR-based solutions are more frequently used in their logistics processes ($R = 0.103$; $p = 0.041$), HAR is more often recognized as a technology that enhances employee safety in logistics processes ($R = 0.115$; $p = 0.022$), the return on investment in HAR is evaluated more positively ($R = 0.195$; $p < 0.001$), the maturity of the market for HAR implementation in logistics is rated more positively ($R = 0.367$; $p < 0.001$), companies declare greater openness to testing new technologies such as HAR ($R = 0.212$; $p < 0.001$), there is greater confidence in the reliability of HAR systems under varying working conditions ($R = 0.117$; $p = 0.019$). At the same time, another result indicates that with the growth in enterprise size, awareness also increases that the costs of HAR systems may outweigh the potential benefits for logistics companies ($R = 0.507$; $p < 0.001$). This correlation coefficient significantly exceeds all other evaluated aspects.

When looking at the general evaluation of HAR awareness, it is relatively low, which may result from concerns over implementation costs and, consequently, from certain mental barriers limiting the popularization of this concept.

The study also compared the perception of various aspects of HAR technology in logistics depending on respondents' years of professional experience, using the Mann–Whitney U test (Table 2).

In general, the study indicates that respondents with shorter work experience tend to provide lower evaluations of HAR technology compared to those with longer experience in the logistics sector (in cases where the differences are statistically significant). This trend particularly concerns the following aspects: knowledge of HAR-based solutions, assessment of HAR's potential to improve the efficiency of logistics processes, evaluation

Table 1. Spearman's rank correlation coefficient values for enterprise size and perception of HAR technology in logistics.

Question for assessment of:	Measure		
	R	t(N-2)	p
(1) Knowledge of HAR technology in the logistics context	0.082	1.627	0.105
(2) Current use of HAR-based solutions	**0.103**	**2.050**	**0.041**
(3) Potential of HAR for improving the efficiency of logistics processes	−0.003	−0.064	0.949
(4) Barriers to HAR implementation	0.044	0.878	0.381
(5) Increasing employee safety in logistics processes through HAR	**0.115**	**2.299**	**0.022**
(6) Time to return on investment in HAR system	**0.195**	**3.955**	**0.000**
(7) Market maturity for HAR implementation in logistics	**0.367**	**7.842**	**0.000**
(8) Openness to testing technologies such as HAR	**0.212**	**4.305**	**0.000**
(9) Significance of return on investment in HAR	0.057	1.139	0.255
(10) Acceptability of HAR implementation and maintenance costs	0.053	1.046	0.296
(11) Cost to potential benefits ratio of HAR	**0.507**	**11.697**	**0.000**
(12) HAR reliability in changing working conditions	**0.117**	**2.346**	**0.019**
(13) Skepticism about HAR accuracy and reliability as implementation barriers	0.086	1.717	0.087
(14) Concerns about employee data privacy in HAR	−0.003	−0.058	0.953
(15) Willingness to test HAR in a pilot program or with external funding	−0.058	−1.159	0.247

of the market maturity for HAR implementation in logistics, and openness to testing new technologies such as HAR.

The research concluded with a simulation involving three variables—knowledge of HAR technology in the logistics context, respondents' length of work experience, and the size of the surveyed enterprises—using the distance-weighted least squares smoothing method (Fig. 1).

The proposed simulation indicates that, despite the relatively low overall ratings of HAR concept awareness, there is a clear increase in this awareness in correlation with both enterprise size and respondents' work experience. Notably, a satisfactory evaluation level (>2) is observed only among larger enterprises and/or respondents with longer professional experience.

Table 2. Mann–Whitney U test scores for perception of HAR technology in logistics by respondents' work experience.

Question for assessment of:	< 5/5–10		< 5/ > 10		5–10/ > 10	
	Z	p	Z	p	Z	p
(1)	**−2.157**	**0.031**	**−5.476**	**0.000**	−1.860	0.063
(2)	−1.308	0.191	−0.893	0.372	0.347	0.729
(3)	**−2.404**	**0.016**	**−2.287**	**0.022**	0.329	0.742
(4)	−0.798	0.425	−0.395	0.693	0.361	0.718
(5)	−1.304	0.192	−1.945	0.052	−0.339	0.735
(6)	**−2.065**	**0.039**	−1.212	0.226	0.964	0.335
(7)	**−4.422**	**0.000**	**−2.538**	**0.011**	**2.477**	**0.013**
(8)	**−3.015**	**0.003**	**−2.543**	**0.011**	0.747	0.455
(9)	−0.542	0.588	**−2.512**	**0.012**	−1.406	0.160
(10)	0.558	0.577	−0.391	0.696	−0.678	0.498
(11)	−0.820	0.412	0.094	0.925	0.842	0.400
(12)	**−4.943**	**0.000**	−1.079	0.281	**3.489**	**0.000**
(13)	−1.089	0.276	−0.837	0.403	0.343	0.732
(14)	−0.239	0.811	−0.750	0.453	−0.385	0.700
(15)	−1.101	0.271	**−2.060**	**0.039**	−0.460	0.645

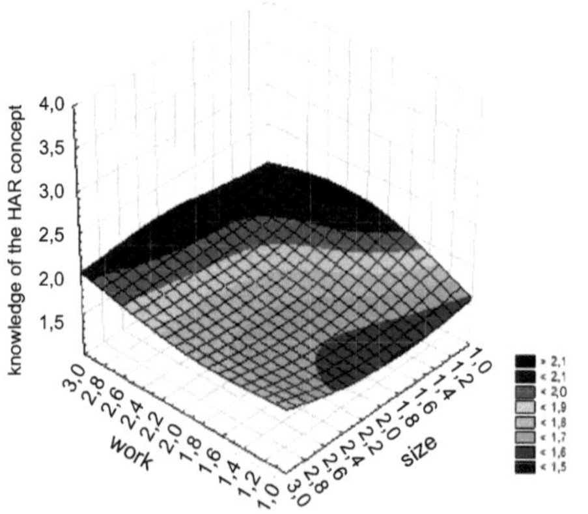

Fig. 1. Simulation of knowledge of the HAR concept in relation to respondents' work experience and enterprise size (Likert scale).

3 Discussion and Conclusions

The study confirms that the application of Human Activity Recognition (HAR) in logistics, although promising in theory, faces significant implementation barriers in practice. While HAR technologies demonstrate strong performance in controlled environments, their use in the logistics sector remains limited due to a range of technological, organizational, economic, and ethical challenges.

Presented research findings are consistent with earlier studies by Reining et al. (2019) and Syed et al. (2020), which emphasize the lack of mature datasets reflecting real-world industrial activities, as well as concerns about sensor reliability and system robustness in dynamic environments such as warehouses. Similar to the conclusions drawn by Niemann et al. (2021), this paper highlights that without contextual information (e.g., task type, environment layout), the recognition accuracy of HAR systems in logistics is significantly limited.

One of the key contributions of this research is the empirical analysis based on a large-scale survey conducted among 397 logistics enterprises. The results show that larger companies with greater technological maturity are more inclined to perceive HAR as a useful and feasible solution, yet they also acknowledge a disproportionate cost-to-benefit ratio. Moreover, the study finds a positive correlation between HAR acceptance and both the size of the enterprise and the respondent's experience, suggesting that HAR awareness is strongly tied to operational scale and seniority.

Despite low general awareness and adoption of HAR technologies among the surveyed firms, the interest in pilot implementations with external funding signals potential for future diffusion. However, this adoption will depend heavily on improving affordability, regulatory clarity regarding data privacy, and the availability of training datasets tailored to logistic tasks.

Future studies should focus on developing and sharing open-access, annotated datasets specific to warehouse and distribution center environments, exploring cost-reduction strategies for HAR implementation in small and medium enterprises, investigating the human factor—particularly employee perceptions of privacy and trust in surveillance-based systems, designing and testing context-aware HAR models that improve activity recognition accuracy under real-world logistic conditions, and evaluating the integration of HAR systems with other technologies such as AGVs, WMS, or digital twin platforms.

In addition to theoretical contributions, the study also offers several practical recommendations for industry. First, logistics companies interested in HAR should consider initiating small-scale pilot projects in collaboration with technology providers or research institutions, particularly in clearly defined tasks. Second, investing in employee training and awareness regarding HAR applications may reduce resistance and improve adoption. Third, the integration of HAR with existing systems should be guided by a modular and interoperable architecture to ensure scalability. Finally, public agencies and funding bodies are encouraged to support SMEs through grants or tax incentives, reducing financial risk associated with early-stage adoption. These practical steps may facilitate gradual, lower-risk adoption of HAR solutions and contribute to the development of more realistic use cases and datasets for future research and deployment.

Ultimately, the diffusion of HAR in logistics depends on bridging the gap between experimental performance and practical applicability. Interdisciplinary cooperation between engineers, data scientists, and logistics practitioners will be essential to unlock HAR's full potential in this domain.

References

1. Arab, A., Schmidt, A., Aufderheide, D.: Human activity recognition using sensor fusion and deep learning for ergonomics in logistics applications. In: 2023 IEEE International Conference on Internet of Things and Intelligence Systems (IoTaIS), pp. 254–260. IEEE (2023)
2. Niemann, F., Lüdtke, S., Bartelt, C., Hompel, M.: Context-aware human activity recognition in industrial processes. Sensors **22** (2021)
3. Niemann, F., et al.: LARa: creating a dataset for human activity recognition in logistics using semantic attributes. Sensors **20** (2020)
4. Reining, C., Niemann, F., Rueda, F., Fink, G., Hompel, M.: Human activity recognition for production and logistics - a systematic literature review. Information **10**, 245 (2019)
5. Mekruksavanich, S., Tancharoen, D., Jitpattanakul, A.: Human activity recognition in logistics using wearable sensors and deep residual network. In: 2023 IEEE Region 10 Conference (TENCON), pp. 194–198. IEEE (2023)
6. Pavliuk, O., Mishchuk, M.: Smartwatch-based human staff activity classification: a use-case study in internal logistics systems utilizing AGVs. In: 2024 IEEE International Conference on Big Data (BigData), pp. 8198–8207. IEEE (2024)
7. Syed, Z.S., Syed, M.Z.A.S., Syed, M.S.S., Shah, A.: Sequential modeling for the recognition of activities in logistics. Sukkur IBA J. Emerg. Technol. **4**(1), 12–21 (2021)
8. Nematallah, H., Rajan, S., Cretu, A.M.: Logistic model tree for human activity recognition using smartphone-based inertial sensors. In: 2019 IEEE SENSORS, pp. 1–4. IEEE (2019)
9. Syed, A., Sherhan, Z., Shehram, M., Saddar, S.: Using wearable sensors for human activity recognition in logistics: a comparison of different feature sets and machine learning algorithms. Int. J. Adv. Comput. Sci. Appl. **11** (2020)
10. Zaki, Z., Shah, M.A., Wakil, K., Sher, F.: Logistic regression based human activities recognition. J. Mech. Contin. Math. Sci. **15**(4), 228–246 (2020)
11. Kirchhof, M., Schmid, L., Reining, C., Hompel, M.T., Pauly, M.: Chances of interpretable transfer learning for human activity recognition in warehousing. In: Computational Logistics, ICCL 2021, LNCS, vol. 12803, pp. 163–177. Springer, Cham (2021)

From Documents to Database: Failure Modes for Industrial Assets

Duygu Kabakci-Zorlu(✉), Fabio Lorenzi, John Sheehan, Karol Lynch, and Bradley Eck

IBM Research Europe, Dublin, Ireland
{duygu.kabakci.zorlu,fabio.lorenzi1}@ibm.com,
{john.d.sheehan,karol_lynch,bradley.eck}@ie.ibm.com

Abstract. We propose an interactive system using foundation models and user-provided technical documents to generate Failure Mode and Effects Analyses (FMEA) for industrial equipment. Our system aggregates unstructured content across documents to generate an FMEA and stores it in a relational database. Leveraging this tool, the time required for creation of this knowledge-intensive content is reduced, outperforming traditional manual approaches. This demonstration showcases the potential of foundation models to facilitate the creation of specialized structured content for enterprise asset management systems.

Keywords: Foundation Models · Structured Text Generation · FMEA · Reliability

1 Introduction

Failure Mode and Effects Analysis (FMEA) is a well-established technique in reliability engineering for identifying an equipment's potential failure points and determining optimal maintenance strategies [9,10]. These documents, which record the reasons for equipment failures and outline maintenance options, play a crucial role in managing physical assets to a desired level of reliability.

FMEAs are extensive documents containing information from multiple sources including text, diagrams, tables and–crucially–the knowledge and experience of reliability professionals. The usual approach of FMEA creation consists of workshops facilitated among experts to gather and discuss the material. Resources needed for this approach can amount to millions of dollars across an enterprise [2].

Compiling data from multiple sources into a structured database is a task that requires significant time and effort. The structured representation of FMEA allows consistency across equipment types and enables integration with asset management systems. Our approach targets tree structured FMEAs of depth five for each study. First, the *boundary* provides a functional overview and identifies the main parts of the system. Second, the *failure locations* pinpoint locations where potential failures might happen. Third, physical processes contributing

Published by Springer Nature Switzerland AG 2026
M. Pondel et al. (Eds.): AI4KAM 2025, IFIP AICT 776, pp. 235–241, 2026.
https://doi.org/10.1007/978-3-032-18920-2_23

to failures are termed *degradation mechanisms*. Fourth, *degradation influences* note the underlying causes of the degradation, known as degradation influences. A triple of failure location, degradation mechanism and degradation influence is termed *failure mode*. Fifth, to anticipate and avert these potential issues, the document provides *preventive maintenance tasks*.

This hierarchical structure poses several challenges. The sequential relationship between the sections propagates errors: an incorrect failure location will also have incorrect degradation mechanisms. Expert guidance between steps is crucial to prevent cascading errors. Additionally, FMEAs contain domain-specific knowledge about the equipment and its usage, making it difficult to create a catalog of components for building FMEAs due to the ambiguity of certain terms e.g., the term "casing" refers to different components in a pump and in a window. Consolidating knowledge from domain experts and documents into this representation is the scope of this work.

Since FMEA generation relies on considerable labor of domain experts, earlier workers have also sought to bring the productivity improvements of AI to the problem. [5] proposes an ontology-based approach to automate reasoning on FMEA spreadsheets. [7] and [8] propose an LLM based system for creating new FMEAs with in-context learning using an expert crafted database and novel ensemble strategies. Another Retrieval Augmented Generation [6] (RAG) based multi-agent LLM application leverages a structured database as external data source [11]. Additionally, a FMEA generation framework with a fine-tuned LLM from domain specific data is proposed by [3]. So far, AI powered creation of FMEAs from source documents remains unexplored and is the focus of this paper.

In this work, we consider the problem of interactive FMEA generation from unstructured documents. The problem is relevant where the asset of interest differs substantially from available knowledge data as used in other approaches. We show an expert supervised and interactive LLM based system for accelerating structured FMEA creation from source documents. To the best of our knowledge, this system contributes a first example of structured information extraction from documents in this domain.

2 System Overview

The system builds upon the work from [7,8] that offers structured text generation with in-context learning from an expert crafted FMEA library. We recognize that integration of unstructured documents can expedite the generation process in cases where the targeted asset differs substantially from available existing FMEAs. We follow a RAG approach to ground LLM responses with external context. External context comes from user provided documents such as operation and maintenance manuals for the asset. These documents typically include information about installation, operation, inspection, and maintenance of an equipment, therefore containing considerable amount of relevant information. The system flow (Fig. 1) is divided into two primary phases: 1. LLM-assisted document pre-processing; and, 2. Structured document generation with user supervision.

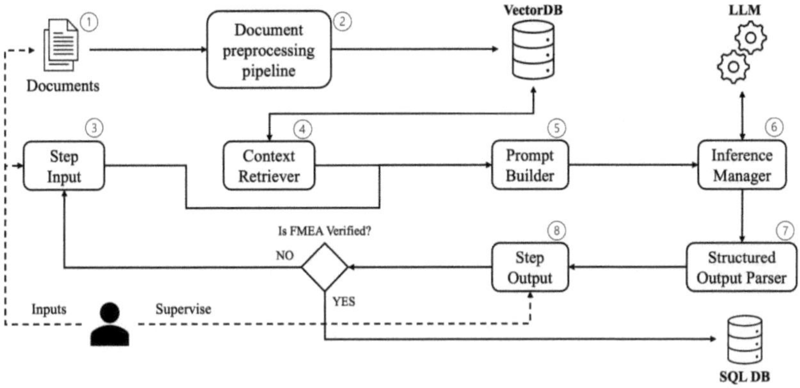

Fig. 1. System flow diagram.

2.1 Document Preprocessing

The LLM-assisted document pre-processing pipeline (Fig. 2) extracts information from unstructured documents such as PDFs using the Docling library [1].

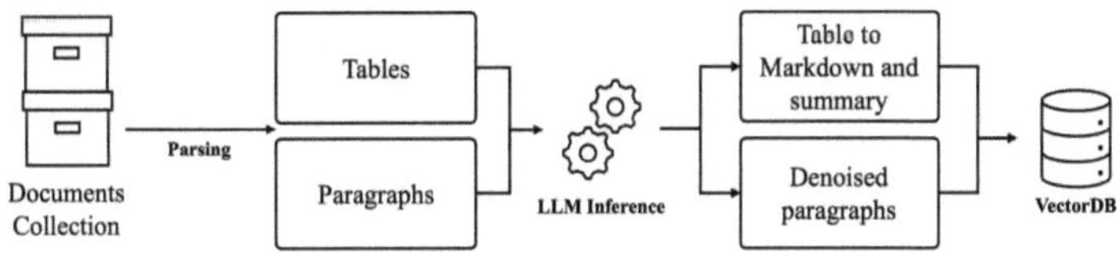

Fig. 2. LLM-assisted document pre-processing diagram.

Document processing, for PDFs in particular, presents some challenges in the form of noisy text extraction and semantic interpretation of tables. Noise in this particular case is identified as any textual information which is not part of the document corpus or that can't be interpreted without non-textual references e.g., headers and footers, footnotes, captions, page numbers. (Figure 3).

The noise reduction pipeline uses an LLM specifically prompted to filter the raw text into cohesive paragraphs. The resulting paragraphs are chunked and indexed in a vector database. Tables are parsed out of documents alongside captions, if any, and converted to markdown format before being sent for summarisation by an LLM. The tables are stored in a vector database indexed by their summary. The content of the database after the processing pipeline provides contextual knowledge during the FMEA generation rounds.

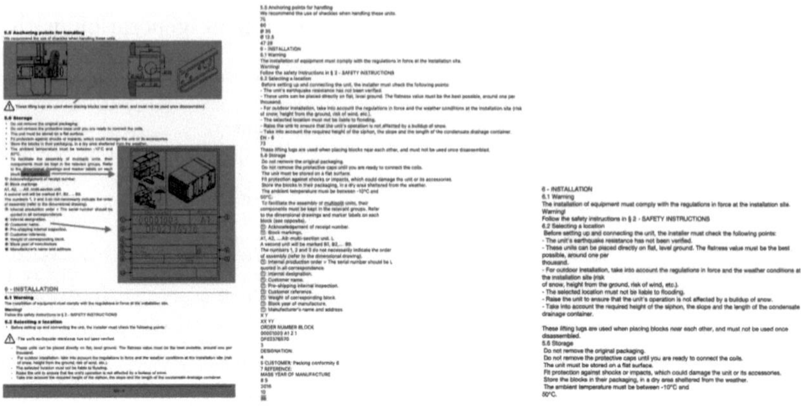

Fig. 3. Example of document page with noisy and cleaned chunks before and after filtering.

2.2 System Flow for Supervised Generation

Our system follows the flow of Fig. 1 to generate FMEA documents with user supervision. The process goes through nine steps. The user uploads one or many documents containing domain knowledge about the asset (1); these documents are pre-processed (2) and indexed in vector DB as described in Fig. 2. The user initiates the FMEA generation by providing some asset information such as a short description of the asset (3). The user input is encoded as a query to the vector database; the top-k matching text chunks from the processed documents are returned (4) and compiled into engineered prompts containing generation instructions and hints about the desired output structure (5).

The prompts are handed over to an inference manager which interacts with an instance of an LLM. FMEA documents are presented in a tree structure. An equipment type, its components and their degradation mechanisms are in one to many relationships. To mirror this structure, our application executes sequential generative steps. To ensure required structured outputs, the generated text is passed through a rule-based parser that cleans the response and extracts a structured object from it (7); the structured response is presented to the user for further editing and supervision (e.g., adding list items or modifing part of generated text) (8). The structured nature of the responses also facilitates insertion of the final FMEA content into a relational database (9), allowing for integration with other FMEA related tools.

Interaction with the system happens through a user interface (UI) that allows users to upload and store documents. In the example of Fig. 4a, the user selects relevant documents and generates a list of maintainable components for an air handling unit. Next, the user generates the associated degradation mechanisms for the selected components (Fig. 4b).

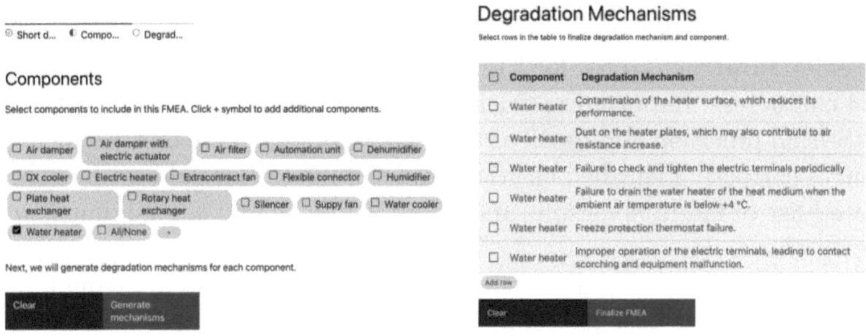

(a) (b)

Fig. 4. UI Visuals for Air Handling Unit (a) Components step (b) Degradation Mechanisms step.

3 Experiments

Our evaluation focuses on the *failure locations* extracted from guide documents of 12 equipment types. We compare our proposed method against the zero-shot prompting approach. We also analyze retrieval quality from document chunks and the effect of the context length.

3.1 Data Preparation

We synthesized the guide documents to mimic real asset documents containing a mix of relevant and irrelevant information in both body text and tables. The guides contained the failure locations of interest along with other parts of the FMEA such as degradation mechanisms and influences in the main text. We also added sections of LLM generated content related to installation and troubleshooting. Less relevant information included tables of failure locations and failure modes for similar equipment than the one being evaluated. The 12 guide documents analyzed here had an average of 26 pages and covered miscellaneous asset types e.g., hammer drill. On average, each asset had 21 failure locations.

3.2 Performance Evaluation

Performance evaluation focused on identifying failure locations. This crucial part of the FMEA records the equipment's structure thus influences subsequent fields. These experiments compare failure locations returned from guide documents for 12 equipment types. Golden failure locations came from our database of expert crafted FMEAs. We considered four scenarios: (i) zero-shot, using no documents as an experimental baseline; (ii) 3 chunks, where the top three relevant chunks from the document are included in the prompt; (iii) 5 chunks, and (iv) long context, where the entire guide document appears in the prompt.

Experiments used *msmarco-MiniLM-L-6-v3*[1] as an embedding model and Granite-3-8B-Instruct [4] (6) for LLM inferencing. Chunks had maximum length of 1024 characters. Reported metrics include Structured Semantic Entity Evaluation (SSEE) precision and recall as introduced in [8]. These metrics measure the fraction of correctly identified entities from candidate and gold set by comparing generated list of failure locations and gold set in a shared embedding space with similarity threshold.

Results (Table 1) showed that all variations of our system strongly outperformed the zero-shot baseline. Zero-shot performance is low, but not zero, indicating that LLMs do contain some "specialised" knowledge. However, providing reference documents with relevant information improves performance with lifts in precision by up to 20x.

We analyzed the impact of differing document lengths in the prompt. Recall increased by 10% when going from three chunks to five. Including the entire document boosted precision to 99% and recall to 84%. While long context windows (a feature of some state-of-the-art llms) seem to perform significantly better we argue that large, complex and diverse collections of documents may still benefit from chunking and summarization to strike a balance between performances and running costs.

Table 1. Performance for generating Failure Locations on test dataset n = 12 (SSEE similarity threshold 0.8 for all experiments)

Method	Context	SSEE Precision	SSEE Recall
Zero-shot	–	0.05	0.11
RAG system	3 chunks	0.66	0.62
RAG system	5 chunks	0.66	0.72
RAG system	Long	0.99	0.84

4 Conclusions and Future Work

This paper demonstrated an interactive LLM-based application for creating new FMEAs from reference documents. As future work, we plan to explore approaches that draw context from both reference documents and existing FMEA databases. Multi-modal foundation models also show great promise for accessing information in equipment drawings and process flow diagrams.

[1] http://huggingface.co/sentence-transformers.

References

1. Auer, C., et al.: Docling technical report (2024). https://arxiv.org/abs/2408.09869
2. Cooper, H.C.: Capture all critical failure modes into FMEA in half the time with a simple decomposition table (actual case study savings = 4,206,000). In: RAMS, pp. 1–6 (2015)
3. El Hassani, I., Masrour, T., Kourouma, N., Motte, D., Tavčar, J.: Integrating large language models for improved failure mode and effects analysis (FMEA): a framework and case study. Proc. Design Soc. **4**, 2019–2028 (2024)
4. Granite Team, I.: Granite 3.1 language models (2024). https://github.com/ibm-granite/granite-3.1-language-models/
5. Hodkiewicz, M., Klüwer, J.W., Woods, C., Smoker, T., Low, E.: An ontology for reasoning over engineering textual data stored in FMEA spreadsheet tables. Comput. Ind. **131**, 103496 (2021). https://doi.org/10.1016/j.compind.2021.103496
6. Lewis, P., et al.: Retrieval-augmented generation for knowledge-intensive NLP tasks (2021). https://arxiv.org/abs/2005.11401
7. Lynch, K., Lorenzi, F., Sheehan, J., Kabakci-Zorlu, D., Eck, B.: FMEA builder: expert guided text generation for equipment maintenance. In: AI for Critical Infrastructure Workshop at 33rd International Joint Conference on Artificial Intelligence (IJCAI-24) (2024)
8. Lynch, K., Lorenzi, F., Sheehan, J.D., Kabakci-Zorlu, D., Eck, B.: Structured document generation for industrial equipment. In: Proceedings of the AAAI Conference on Artificial Intelligence, vol. 39, no. 28, pp. 28850–28856 (2025). https://doi.org/10.1609/aaai.v39i28.35150
9. Rausand, M., Høyland, A.: System Reliability Theory: Models, Statistical Methods, and Applications, 2nd edn. Wiley, Hoboken (2004)
10. Sharma, K.D., Srivastava, S.: Failure mode and effect analysis (FMEA) implementation: a literature review. J. Adv. Res. Aeronaut Space Sci. **5**(1–2), 1–17 (2018)
11. Xia, Y., Jazdi, N., Weyrich, M.: Enhance FMEA with large language models for assisted risk management in technical processes and products, pp. 1–4 (2024). https://doi.org/10.1109/ETFA61755.2024.10710996

The Size Factor: Managerial Views on Artificial Intelligence in Strategic Decision Processes

Anna Marciszewska(✉), Renata Brajer-Marczak, Katarzyna Grzesik, Michał Nadolny, and Katarzyna Liwarska-Fulczyk

Wroclaw University of Economics and Business, Komandorska 118/120, Wroclaw 53-345, Poland
anna.marciszewska@ue.wroc.pl

Abstract. The issue of AI application in the context of company size has been poorly researched, and the impact of this variable on decision-making processes at the managerial level remains unclear. The aim of this article is therefore to determine whether company size influences the perception of AI applications in managerial decision-making processes. The research process used a measurement construct with four latent variables, i.e., AI proficiency, trust, perceived benefit, and perceived risk. The presented study was conducted on a deliberately selected group of 218 companies in Poland, and the respondents were strategic decision-makers in the surveyed companies. In order to verify the hypothesis put forward in the article, the collected research material was used to search for a relationship between the perception of AI applications in strategic decision-making and the size of the company. The research confirmed that trust in AI is significantly lower in companies operating on a self-employment basis. The presented considerations contributed to the deepening of knowledge in the field of management sciences, indicating the current perspective on the analyzed issue.

Keywords: AI · company size · strategic decision-making · AI applications

1 Introduction

The dynamic development of technology motivates companies to continuously improve and adapt to the changing environment. The emergence of innovative solutions requires managers to implement them in order to maintain competitiveness and increase operational efficiency. In order to strengthen their market position, organizations are increasingly integrating elements of artificial intelligence into specific areas of their operations (Szumowska, Bagieńska, 2024); Ghobakhloo et al., 2024). The implementation of artificial intelligence in business generates both new opportunities and challenges that arise in the process of managing companies adapting modern technologies. This makes artificial intelligence a key topic of contemporary scientific research (Chen et al., 2024; Raisch, Krakowski, 2021; Gupta, 2024; Roberts, Candi 2024). Current publications mainly concern: (1) the application of AI in various types of activities, (2) the demand for employees with appropriate qualifications, and (3) the development of AI tools (Iaia et al., 2024, Savastano et al., 2024; Lokanan and Maddhesia, 2024; Wadehra and Anand, 2024).

Published by Springer Nature Switzerland AG 2026
M. Pondel et al. (Eds.): AI4KAM 2025, IFIP AICT 776, pp. 242–249, 2026.
https://doi.org/10.1007/978-3-032-18920-2_24

The literature on the subject indicates that AI solutions can be applied in various areas of business management and operations. Artificial intelligence can analyze large amounts of data, which can also be used in strategic decision-making. Interestingly, according to statistical data from the Central Statistical Office (GUS) from 2023, only 3.7% of companies in Poland declared the use of artificial intelligence technology. This percentage varies significantly depending on the size of the company and the nature of its business (GUS, 2023, p. 103). However, the subject of artificial intelligence in relation to company size is poorly understood (Müller, Buliga, 2018; Kovic et al., 2024; Hamm, Klesel, 2021). There is a lack of perspective that takes into account the application of artificial intelligence and the impact of the size of the organization on the decision-making process at the managerial level. This identified research gap justifies the need for research in this area.

Therefore, the aim of this article is to answer the question: does the size of a company influence the perception of AI applications in managerial decision-making processes? To achieve this goal, the focus is on the factor of company size and its impact on the degree of AI use in the processes under study. From the point of view of company size, it is emphasized that ensuring adequate resources—both technological and financial—becomes a key factor in supporting the use of AI in strategic decision-making processes. It is assumed that the use of AI tools in the area of strategic decisions in organizations contributes to achieving better results in the area of business activities.

The above considerations contributed to the formulation of the following research hypothesis:

H1: The perception of AI applications in strategic decision-making depends on the size of the company

2 Research Method

The results presented in this article are part of a broader study aimed at assessing the possibility of measuring how managers perceive the role of AI in strategic decision-making processes. The survey was conducted on a group of 218 companies in Poland between March and November 2024. The sample was selected to ensure a sufficiently high proportion of medium-sized and large companies. In such companies, decision-making processes are usually based on developed management systems. The survey was conducted using a mixed-method questionnaire. Respondents included 19% representatives of medium-sized companies (10–49 employees), 42% micro or small companies (1–9 employees), and 38.5% companies with more than 50 employees. Senior managers (18.4%) were interviewed in person after prior appointment. The study was conducted using the CAWI method. The remaining respondents were surveyed using a random sample telephone interview method. Respondents declared their participation in the strategic decision-making process.

The method of observing phenomena was based on the concept of a measurement construct (Fig. 1). The model therefore defined four latent variables, consisting of 11 measurable variables on a 5-point Likert scale. The constructs and survey questions were inspired by the works of Gerlach and Lutz, 2021; Davis et al., 1989; Mariani et al., 2021; Chatterjee et al., 2021.

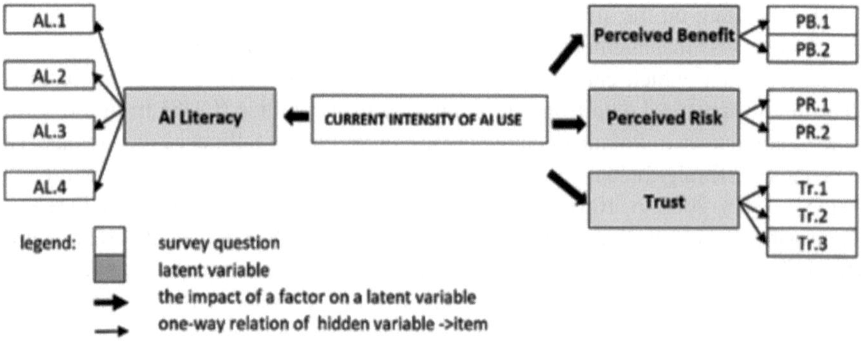

Fig. 1. Visualization of relationships between variables for hypothesis testing.

Confirmatory Factor Analysis was used to verify the quality of the constructs, determining factor loadings for maximum likelihood methods, principal component and principal axis methods, and Alpha-Cronbach was determined. Factor rotation was not used. Alpha values and loadings above 0.7 were considered statistically significant. Latent variables were calculated based on the average value of measurable variables. The W-Shapiro-Wilk (W-S-W) test for small samples was used to verify the research hypothesis for the assessment of normality of distributions, followed by the nonparametric Kruskal-Wallis ANOVA.

3 Results

The study was conducted using a mixed method approach, employing a questionnaire that included a screening question ("Do you participate in strategic decision-making in your company?") as well as questions about company size (Table 1), managerial position, and a set of 11 survey questions used to calculate measurement constructs (Table 2). 28.4% of respondents were representatives of large or very large companies (200+ employees). Approximately 42% were micro or small companies. In the case of senior managers, face-to-face interviews were conducted after prior appointment. Middle managers usually preferred the CAWI survey. In the case of lower-level managers, self-employed persons, or small business owners, a random sample telephone interview method was used.

Latent variables were estimated based on a measurement model (Fig. 1). Latent variables do not determine the objective state of knowledge, benefits, or risks associated with the use of AI in strategic decisions. They describe a subjective measure of how AI is perceived by managers at various levels. It is therefore a quantitative description of respondents' beliefs and intuitions regarding selected aspects of AI use. Respondents are rather favorable toward the use of AI as a decision-making tool in business. The ratings of measurable variables are well above 3 (see mean and median in Table 2). It should be noted that the risk of using AI should also be considered significant, which may indicate a degree of criticism of machine learning methods. However, risk assessments are lower than assessments of benefits, trust, or awareness of operation. The measurable variables show a lack of normal distribution (see kurtosis in Table 2) and are also characterized by

Table 1. Review of qualitative variables of the study

How many people work in your organization including you?	number of observations	[%]
self employment	32	14,7
2–9 employees	60	27,5
10–49	42	19,3
50–249	22	10,1
250–1000	46	21,1
1000 + employees	16	7,3

a low degree of left skewness. The positive verification of the constructs defined in the measurement model is evidenced by the CFA loadings and alpha-Cronbach parameters (above 0.7). The Perceived Risk variable has an alpha value of 0.06, which is, however, an acceptable level for this type of social research. The distributions of latent variables cannot be considered normal (see W-S-W), therefore the verification of the research hypothesis was based on Kruskal-Wallis ANOVA, and the calculations were performed in the Statistica 13 package.

Table 2. Basic statistics for Likert scale measurable variables (3 = neutral).

Construct name	Cronbach alpha	W-S-W (p < 0.05)	CFA-ML loadings	Model ID	Mean	Median	Skewness	Kurtosis
AI Literacy	0.87	0.96	0.86	AL.1	3.90	4.0	−0.76	0.31
			0.89	AL.2	3.72	4.0	−0.38	−0.26
			0.90	AL.3	3.72	4.0	−0.59	0.13
			0.72	AL.4	3.24	3.0	−0.09	−0.46
Perceived Benefit	0.70	0.94	0.88	PB.1	3.66	4.0	−0.38	−0.17
			0.88	PB.2	3.76	4.0	−0.48	0.08
Perceived Risk	0.66	0.92	0.86	PR.1	3.41	3.0	−0.01	-0.66
			0.86	PR.2	3.25	3.0	−0.14	−0.24
Trust	0.87	0.95	0.92	Tr.1	3.67	4.0	−0.04	−0.65
			0.90	Tr.2	3.74	4.0	−0.27	−0,72
			0.86	Tr.3	3.62	4.0	−0.15	−0.40

In order to verify the research hypothesis, a relationship was sought between the perception of AI applications in strategic decision-making and the size of the company.

Respondents, regardless of the size of their organization, statistically evaluate the aspects of AI Literacy, Perceived Benefit, and Perceived Risk equally (Table 3). This may be because these three variables essentially refer to the benefits of knowledge and the risks that apply to the individual area of responsibility of the respondent. The lack of distinction therefore means that the responsibility for the use of AI by a manager is specific to the respondent's particular role in decision-making within the company. The situation is different with Trust. A statistically significant difference applies to the self-employed, who trust AI significantly less than in companies employing more than 250 people.

Table 3. The impact of company size on the respondent's assessment of AI

number of people employed	Kruskal-Wallis ANOVA p-value	Minimum p-value for multiple comparisons	Decision
AI_Literacy	0,026	0,083	No difference
Perceived_Benefit	0,046	1,000	
Perceived_Risk	0,159	0,395	
Trust	0,019	0,024	Self-employment is different from 250+ businesses; there is no difference between 250–100 and 1000+

These differences are presented in Fig. 3. The level of trust in AI among self-employed people is the lowest. Perhaps the sense of responsibility for one's own fate is the decisive factor here, resulting in lower ratings of the Trust variable. Hypothesis H1 is verified positively only for Trust in AI. The figure also presents the distributions of the other variables in the research sample. The managers' ratings do not form a statistically significant pattern, but they do allow for an assessment of the diversity of ratings in the sample.

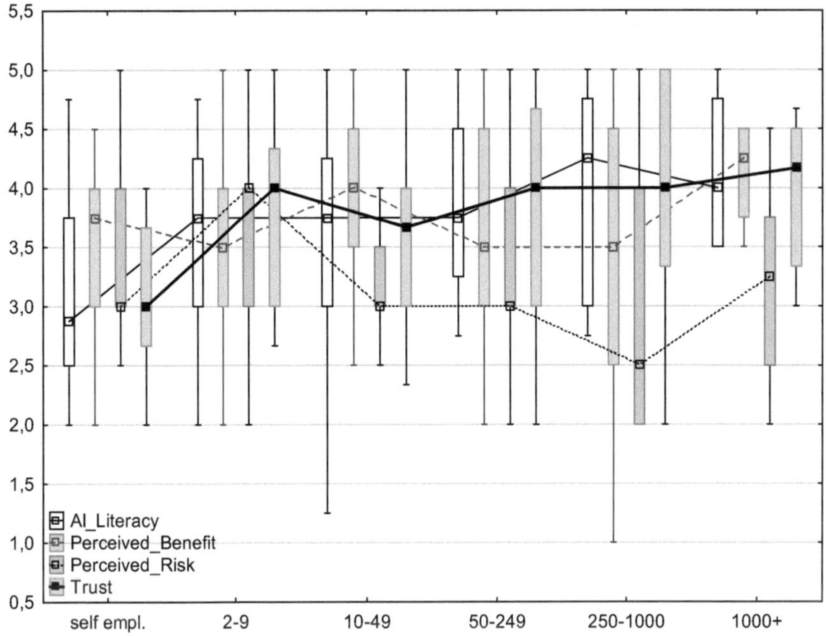

Fig. 2. Box plot of the distributions of latent variables by the number of employees in the company. The continuous line corresponds to a statistically significant relationship between variables

4 Conclusion

The aim of the study was to find a relationship between company size and the perception of AI applications in strategic decision-making. The hypothesis adopted in the article was based on a critical review of the literature on the subject and the existence of a research gap in the indicated area. The statistical analysis of the data obtained indicates a relationship between company size and trust in AI. The results show that trust in AI is significantly lower in self-employed companies. It would be worth considering why this relationship exists. As the literature on the subject indicates, the main differences may be caused primarily by the low level of knowledge of people running their own businesses (Mesir, 2019). This means that owners are afraid (they do not trust) that AI-based technology will not meet their expectations or that its implementation will be too complicated. In addition, they point to limited access to the data necessary to use algorithms and make decisions based on them. However, these decisions, based on AI analyses, are often perceived as more difficult to control. Small business owners are more likely to rely on their own intuition and proven, traditional decision-making methods. AI is seen as one of the reasons for giving up their own experience and control. (Tawil et al., 2023; Bedué, Fritzsche, 2022; Alufaisan et al., 2021) This translates into a lack of trust in AI in key decision-making processes. It is also emphasized that one of the main reasons may be a lack of financial resources. The implementation of AI involves the cost of purchasing the technology, adapting it to the needs of the company, and training employees. For small businesses, these costs are often difficult to cover. (Hambye, Desmet, 2021; Mesir, 2019).

The considerations presented here contribute to the broader discussion on the use of AI technology in strategic decision-making. However, further detailed research is needed to identify additional factors that influence both trust in and the use of AI applications in strategic decision-making.

It is worth continuing the presented considerations, expanding them with empirical research that will focus on small businesses, for which building trust in AI requires greater support in the form of customized solutions, training, and accessible AI implementation models.

Disclosure of Interests. The authors have no competing interests to declare that are relevant to the content of this article.

References

Alufaisan, Y., Marusich, L.R., Bakdash, J.Z., Zhou, Y., Kantarcioglu, M.: Does explainable artificial intelligence improve human decision-making?. In: Proceedings of the AAAI Conference on Artificial Intelligence (Vol. 35, No. 8, pp. 6618–6626) (2021)

Bedué, P., Fritzsche, A.: Can we trust AI? An empirical investigation of trust requirements and guide to successful AI adoption. J. Enterp. Inf. Manag. **35**(2), 530–549 (2022)

Chatterjee, S., Rana, N.P., Dwivedi, Y.K., Baabdullah, A.M.: Understanding AI adoption in manufacturing and production firms using an integrated TAM-TOE model. Technol. Forecast. Soc. Chang. **170**, 120880 (2021)

Chen, H., Zhang, M., Zeng, J., Wang, W.: Artificial intelligence and corporate risk-taking: evidence from China. China J. Account. Res. **17**(3), 100372 (2024)

Davis, F.D.: Perceived usefulness, perceived ease of use, and user acceptance of information technology. MIS Q. 319–340 (1989)

Gerlach, J.M., Lutz, J.K.: Digital financial advice solutions–Evidence on factors affecting the future usage intention and the moderating effect of experience. J. Econ. Bus. **117**, 106009 (2021)

Ghobakhloo, M., Mahdiraji, H.A., Iranmanesh, M., Jafari-Sadeghi, V.: From Industry 4.0 digital manufacturing to Industry 5.0 digital society: a roadmap toward human-centric, sustainable, and resilient production. Inf. Syst. Front. 1–33 (2024)

Gupta, P.: Synergizing success: Harnessing AI-infused business intelligence to propel exponential business growth, w: Keikhosrokiani P. (ed.). Data-Driven Business Intelligence Systems for Socio-Technical Organizations, pp. 105–127 (2024)

GUS: Społeczeństwo informacyjne w Polsce w 2023 r., Szczecin (2023)

Hambye, M., Desmet, C.: What are the barriers preventing AI from being adopted in small farms in Africa. Université catholique de Louvain, Masters (2021)

Hamm, P., Klesel, M. Success factors for the adoption of artificial intelligence in organizations: a literature review. In: 27th Americas Conference on Information Systems, AMCIS 2021: Digital Innovation and Entrepreneurship. Association for Information Systems (2021)

Iaia, L., Nespoli, C., Vicentini, F., Pironti, M., Genovino, C.: Supporting the implementation of AI in business communication: the role of knowledge management. J. Knowl. Manag. **28**(1), 85–95 (2024)

Kovič, K., Tominc, P., Prester, J., Palčič, I.: Artificial intelligence software adoption in manufacturing companies. Appl. Sci. **14**(16), 6959 (2024)

Lokanan, M.E., Maddhesia, V.: Supply chain fraud prediction with machine learning and artificial intelligence. Int. J. Prod. Res. pp. 1–28 (2024)

Mariani, M.M., Nambisan, S.: Innovation analytics and digital innovation experimentation: the rise of research-driven online review platforms. Technol. Forecast. Soc. Chang. **172**, 121009 (2021)

Mariani, M., Borghi, M.: Customers' evaluation of mechanical artificial intelligence in hospitality services: a study using online reviews analytics. Int. J. Contemp. Hosp. Manag. **33**(11), 3956–3976 (2021)

Mesir, M.: The role of artificial intelligence in decision making in small businesses. Indian J. Sci. Technol. **12**(18), 1–6 (2019)

Müller, J.M., Buliga, O., Voigt, K.I.: Fortune favors the prepared: How SMEs approach business model innovations in Industry 4.0. Technol. Forecast. Soc. Chang. **132**, 2–17 (2018)

Savastano, M., Biclesanu, I., Anagnoste, S., Laviola, F., Cucari, N.: Enterprise chat-bots in managers' perception: a strategic framework to implement successful chatbot applications for business decisions. Manage. Decis. (2024)

Szumowska, P., Bagieńska, A.: Ocena wykorzystania sztucznej inteligencji w działalności spółek z branży informatycznej. Akademia Zarządzania **8**(3) (2024)

Tawil, A.R., Mohamed, M., Schmoor, X., Vlachos, K., Haidar, D.: Trends and challenges towards an effective data-driven decision making in UK SMEs: case studies and lessons learnt from the analysis of 85 SMEs (2023). arXiv preprint arXiv:2305.15454

Raisch, S., Krakowski, S.: Artificial Intelligence and management: the Automation–Augmentation Paradox. Acad. Manage. Rev. **46**(1) (2021)

Roberts, D.L., Candi, M.: Artificial intelligence and innovation management: charting the evolving landscape. Technovation **136**, 103081 (2024)

Wadehra, S., Anand, A.: From gavels to algorithms: the Vidhii Partners GenAI evolution. Emerald Emerging Mark. Case Stud. **14**(3), 1–32 (2024)

Predicting Consumer Decision-Making in Sales Offer Selection Using Psychophysiological Data and Machine Learning Methods

Paweł Ziemba[1(✉)], Mateusz Piwowarski[2], and Kesra Nermend[2]

[1] Maritime University of Szczecin, Faculty of Computer Science and Telecommunications, Waly Chrobrego 1-2, Szczecin 70-500, Poland
pawel.ziemba@usz.edu.pl

[2] University of Szczecin, Institute of Management, Cukrowa 8, Szczecin 71-004, Poland

Abstract. The article addresses the use of multimodal psychophysiological data (eye tracking, EEG) and machine learning to predict consumer choices of sales offers. Classification models were built, key areas of interest (AOI) of offers were identified, and the significance of metrics was assessed. The highest predictive quality was achieved using Random Forest with oversampling. The most important AOIs were those related to the external visualisation of houses and the layout of rooms. The results confirm the usefulness of integrating psychophysiological data and machine learning in marketing.

Keywords: consumer decision-making · sales offer selection · machine learning · predictive modelling · eye-tracking · EEG

1 Introduction

Consumer behaviour and preferences have been the focus of marketing research for many years. In recent decades, there has been a marked evolution of analytical approaches, shifting the emphasis from models assuming the rationality of decisions toward more comprehensive concepts. They consider previously underestimated emotional and cognitive aspects, including reactions outside of a person's consciousness [1]. This approach is particularly relevant for products or services of strategic importance. Such a product is real estate (houses), the acquisition of which is often the fulfilment of a lifelong dream. Purchasing decisions in such cases are not based solely on rational economic criteria. Sometimes emotions co-determine the final choice, obscuring rational argumentation and judgment [2].

Visualisations of sales offers play an essential role in shaping the preferences of customers purchasing immovable property. In addition to their informational function, they can subtly influence the perception and evaluation of a presented residential property. The visual nature of the offers makes it possible to influence various cognitive functions of customers (e.g. perception, attention, memory), as well as the emotional

Published by Springer Nature Switzerland AG 2026
M. Pondel et al. (Eds.): AI4KAM 2025, IFIP AICT 776, pp. 250–265, 2026.
https://doi.org/10.1007/978-3-032-18920-2_25

sphere. This impact can evoke various reactions, in many cases, unconscious [3, 4]. Studying the impact of visualisation and its individual components on consumer choice is a significant methodological challenge.

In response to this complexity, modern research increasingly uses advanced psychophysiological methods to measure and analyse human responses at the nervous system level. Me- methods such as electroencephalography (EEG), eye tracking, and electrodermal activity (EDA) provide objective data for real-time analysis of cognitive and emotional functions. Eye tracking allows a precise analysis of visual focus, which allows the study of how visual information is processed. In turn, analysis of EEG activity, particularly the frontal alpha asymmetry index, allows assessment of motivational tendencies in response to presented stimuli [5, 6]. The combination of both indices makes it possible to capture both the processes of selective information processing and the consumer's emotional attitude to the presented visuals of the offers.

Acquired data from physiological measurements, such as EEG and eye-tracking metrics, can be valuable for machine learning algorithms. As a result, the prediction of consumer decisions regarding the choice of sales offers can be modelled.

The paper presents the results of a study that explored this type of data using machine learning methods, focusing on three main objectives: (1) building classification models predicting offer selection based on eye-tracking and EEG data, (2) identifying areas of interest (AOI) having the most significant impact on the decision to select or reject an offer, and (3) evaluating the utility of individual metrics in assessing offer interest.

Classification models that predict consumer decisions based on EEG and eye tracking data allow for predicting consumer behaviour some time in advance. In addition, faith-based classification models make it possible to identify consumers' internal preferences, which may differ from the preferences expressed verbally. The realisation of the second objective, i.e., identifying AOIs that have the greatest impact on the consumer's decision, consequently makes it possible to optimise the layout and graphic structure of sales offers. In addition, identifying the most important AOIs can provide feedback to bidders on which elements of the offers (building visualisation, room layout, price, etc.) are most important to consumers. Finally, determining the utility of individual metrics for assessing consumer interest in an offering will reduce the amount of data acquired. This will allow future analyses to collect and process only those EEG signals and eye tracking data that significantly represent the degree of consumer interest in a particular offer.

This article is organised as follows: the next section presents the research methodology; this is followed by the results on the effectiveness of classification models and the assessment of the significance of AOI; the final section draws conclusions and recommendations for further research directions.

2 Methodology

2.1 Psychophysiological Research (Eye-Tracking, EEG)

A data set from an experiment was used to verify the assumptions made and achieve the defined research objectives, in which eye tracking and EEG parameters were recorded while viewing and evaluating visualisations of home sales listings. The described research is a continuation of the work on the search for classifiers and a set of data, enabling the prediction of the choices made by consumers [7]. The general research assumptions, a description of the study group and the data mining methods, algorithms, models, and processes used are presented below.

Research assumptions. The research experiment was designed to obtain physiological data, including primarily eye tracking and brain activity (EEG). Study participants were presented with sales offers for single-family homes. The offers used visualisations of the houses and lay-outs of the rooms, which were taken from a catalogue of single-family house designs ("Murator", https://projekty.muratordom.pl). In addition, each offer included information about the price of the house, the number of rooms and the location of the building. In total, 10 different sales offers were presented to all survey participants. Each offer was displayed for 20 s, and the order in which they appeared was random. After reading the contents of the offers, a board appeared containing thumbnails of all the houses along with descriptive information. The subjects' task was to choose the offer, without a time limit, that was most suitable for them.

At the data analysis stage, five AOI were identified, for which detailed analyses of visual attention focus and cerebral activity were made, particularly in terms of approach or avoidance (withdrawal) responses to the stimulus. Areas of interest were marked as follows:

- AOI1 – visualisation of the body of the building;
- AOI2 – detailed room layout of the building;
- AOI3 – house price;
- AOI4 – number of rooms;
- AOI5 – the location of the house, with several considerations, such as the center of the city, the outskirts of the city, a small town, or a gated community of houses.

Participants. Forty-five people took part in the experiment, in which physiological.

parameters were recorded. They were people of different ages, both students (overweight group), working adults, and retirees. The mean age of the subjects was M = 31.44, and the standard deviation was SD = 10.81. The participants of the experiment were instructed about the research procedure (without providing detailed objectives), the measurement apparatus used, data processing, and their rights. This included, in particular, the possibility of resigning from participation in the experiment at any time, without giving reasons. Each participant subscribed to informed consent to participate in the study and to use the data obtained for scientific purposes. The experiment was conducted in a laboratory research at the University, in accordance with the guidelines described in the Declaration of Helsinki [8]. The study was approved by the Bioethics Committee of the Regional Medical Board in Szczecin (Resolution No. 02/KB/VII/2020).

Research Methods and Measuring Apparatus. Two complementary psychophysiological methods were used in the experiment: eye tracking and electroencephalography. Eye-tracking data were recorded using a screened Tobii Pro X3–120 device (sampling rate 120 Hz). An I-HMM (Hidden Markov Model) algorithm was used to identify fixations, with a fixed minimum fixation clarification time of 60 ms, a maximum inter-fixation interval of 75 ms, noise reduction using the Moving Average algorithm, and data in- interpolation for gaps not exceeding up to 75 ms. Measurement data were transferred to a computer via a cable via USB 3.0 connection.

EEG data was simultaneously recorded using Neuroelectrics' ENOBIO 20 5G mobile system. The sampling frequency of the EEG signal was 500 Hz, and the recorded bandwidth was 0–100 Hz, with simultaneous filtering at 50 Hz (the frequency of the power grid). Signal resolution was 24 bits (0.05 μV), with a noise level of < 1mV RMS. The EEG device was equipped with a 3-axis accelerometer, which was used to measure linear acceleration and angular acceleration. Brain activity data were recorded from 20 channels, using wet electrodes arranged according to a standard of 10–20. In addition, reference and reference electrodes were used. These were electrodes - passive Driven Right Leg (DRL) and active Common Mode Sense (CMS), attached to the earlobe [9].

3 Analyses and Indicators of Evaluation.

The analysis of eye tracking data was mainly based on a set of metrics determined for defined areas of interest (AOI1-AOI5). These metrics were calculated based on identified fixations and saccades [10–12]:

- FC – the average number of fixations detected inside the AOI;
- DAF – the average time of all fixations inside the AOI;
- TTFF – the average time to detect the first fixation inside the AOI;
- FFD – the average duration of the first fixation inside the AOI;
- DSAF – the average degree of dispersion of all fixations inside the AOI;
- DT – average fixation time inside the AOI, relative to AOI activity time AOI (%);
- REV – number of returns to AOI (detected fixations);
- SC – the number of saccades detected inside the AOI.

Analysis of the EEG data was based on calculating the power spectral density (PSD). These calculations were preceded by processing the raw EEG signal in a so-called preprocessing process. It included several steps. In the first step, the quality of the signal was visually assessed. If any fragments unsuitable for further processing were localized, they were eliminated from further processing. In the next step, a fifth-order Butterworth filter [2–45 Hz] was used to eliminate slow trend and high-frequency interference. To reject signal fragments with an outlier amplitude, the following were adopted. Threshold ±70 μV, and the Welch method with 1-s time windows, with 50% overlap. A fast Fourier transform (FTT) was then used to obtain 1 power spectrum per 0.5 s [13, 14].

For the EEG data prepared in this way, the PSD values for the delta, theta, alpha, beta, and gamma ranges were determined. They were further subjected to logarithmic transformation [15]. Based on the PSD for the alpha band, the Frontal Alpha Asymmetry (FAA) index was determined. For its calculation, signals recorded from electrodes located in the frontal part of the brain were used. For the left hemisphere, electrode F3 was considered, and for the right F4. The adopted frequency band for the alpha range was 8–12 Hz. The formula showing how to calculate the FAA index is as follows [16]:

$$FAA = \ln\left(power_{alfaF4}\right) - \ln\left(power_{alfaF3}\right) \tag{1}$$

The FAA index refers to differences in brain activity between the left and right parts of the frontal cortex. It is interpreted in two complementary dimensions. The first relates to affective (emotional) responses to a stimulus. Greater EEG signal strength in the left frontal part of the brain is associated with positive reactions, while greater strength in the right frontal part is associated with negative reactions. The second dimension is associated with motivational responses. Left asymmetry indicates motivation directed toward approaching the stimulus, while right asymmetry indicates inhibiting responses (withdrawal) [17, 18].

3.1 Machine Learning Research

Machine learning (ML) methods were used to achieve the objectives of the study. Identification of AOIs that have the greatest impact on the selection of offers, as well as identification of AOI metrics indicating consumer interest in an offer, were conducted using feature selection methods in the form of embedded methods. Feature selection methods are used to select a subset of attributes from the input data. This subset should effectively describe the input data, and further reduce the influence of noise or irrelevant variables while providing good predictions. [19]. Feature selection methods are a tool for data preprocessing, allowing to the improvement of classification accuracy, reducing the learning time of the classifier, and simplifying learning results [20]. Embedded methods are a certain subgroup of classification methods that select conditional attributes during the learning process of the classifier [20]. Rule-based and tree-based embedded methods were used to select conditional attributes. In particular, these were: Decision Tree C4.5 [21], Classification and Regression Tree (CART) [22], Random Forest (RF) [23], and Decision Table (DT) [24]. In addition, an Artificial Neural Network (ANN) classifier was used for comparison purposes [25]. The classifiers provided were also used to achieve the research objective related to building classification models to predict bid selection.

The dataset collected during the eye-tracker and EEG survey consisted of 450 instances. Each instance consisted of a binary decision attribute describing the fact of choosing or not choosing an offer, as well as 45 conditional attributes describing 5 AOIs using 9 AOI metrics. The dataset contained 405 negative instances, indicating respondents' non-selection, and 45 positive instances, describing the selection of an offer.

One of the problems with the dataset was the curse of dimensionality. This problem is that the number of instances required for correct prediction is an exponential function of the number of conditional attributes [26]. When the ratio of the number of instances to the number of attributes is significantly lower, the classifier may match the noise in the dataset and lose its ability to generalize [27]. In the dataset under consideration, the number of instances 450 in total, which was only 10 times the number of attributes. Therefore, there is a risk that the predictive ability of the classifiers will be lowered by noise in the data. However, this problem was solved by using the classifiers, which are also embedded methods.

The main problem with the developed dataset was unbalanced classes, i.e., an uneven distribution of positive and negative instances in the dataset [28]. In the data imbalance problem, a very large number of instances are labelled as one class, and a much smaller number of instances are labelled as another class, usually more important [29]. This problem in the developed dataset occurs naturally and is due to the nature of data space. It is characteristic of the problem of selecting a home sales offer that the decision maker chooses one home to purchase from among many under consideration. That is why very large differences appear in the number of cases described by the positive class (selection of an offer - decision to buy) and negative class (non-selection of an offer - no decision to buy). This is the so-called internal imbalance of the dataset [30]. In the case of the dataset under consideration, class imbalance is strongly associated with the presence of rare instances. This means that there is a very small number of positive instances in the dataset, but not in relation to the number of negative instances, but primarily in absolute numbers (in the developed dataset it is the number of 45 instances) [31]. Unbalancing of the dataset often negatively affects the classification results [32]. This is due to the process of internal optimization of classification algorithms, which usually seek to minimize the overall level of classification errors when building a classifier. In extreme cases, they may minimize classification error by consistently predicting the majority class, resulting in high accuracy of the classifier and high specificity, but low recall [33]. In other words, classifiers get good prediction results for the majority class, but at the same time predict the occurrences of the minority class very poorly.

At least a few potential solutions to the problems caused by unbalanced datasets can be found in the literature. These are primarily cost-sensitive learning, cutoff value modification, down-sampling, oversampling, Synthetic Minority Over-sampling Technique (SMOTE) [33, 34]. The study verified the impact of each of those approaches on the classification results.

Cost-sensitive learning is based on the use of a matrix describing the cost of misclassifications [32]. Minority class instances are assigned a proportionally higher cost (weight) of misclassifications than majority class instances [29]. Thus, the classifier places more emphasis on the classification of samples belonging to the minority class. The dataset under consideration assumed a cost (weight) inversely proportional to the number of occurrences of negative (majority class) and positive (minority class) instances in the set. In view of this, the costs were 1:9, respectively.

Cutoff value modification involves changing threshold the probability of an object belonging to a specific class [35]. Typically, a probability of $> 50\%$ is required for a classifier to assign an instance to a specific class. For the minority class, the a priori probability is lower than for the majority class, so reducing the probability threshold can improve the classification of the minority class [36]. For the dataset under consideration, the probability thresholds were set in proportion to the number of objects of the class, i.e. at 0.9 and 0.1 for the majority and minority classes, respectively.

Down-sampling involves equalizing the counts of objects of each class by removing instances of the majority class from the collection. The disadvantage of this method is the risk of removing important information about the majority class [37]. In this study, the use of down-sampling reduced the count of the dataset to 90 instances (45 positive and 45 negative).

Oversampling succeeds in equalizing the count of objects of each class by randomly copying instances of the minority class [34]. Unfortunately, oversampling can lead to overfitting and increase computational cost by increasing the size of the dataset [33].

The dataset under consideration was augmented using oversampling by duplicating 45 positive instances 8 times (a total of 405 positive instances were obtained).

SMOTE is a special case of oversampling, in which instances of a minority class are not duplicated, but new instances are generated synthetically. In this method, new data samples are interpolated in the space between neighbouring data instances of the same class [38]. Generating new instances of a minority class allows to balance the dataset and thus improve the classification results. In this study, the dataset was augmented using SMOTE by generating 360 new synthetic positive instances (a total of 405 positive instances were obtained).

The study used stratified K-fold cross-validation based on dividing the dataset into 5 folds. This is the recommended approach for small sets of unbalanced data. Stratified K-fold cross-validation makes efficient use of the data and provides a more reliable assessment of classifiers' ability to generalize than simply splitting the data into training and test set [39].

Due to the unbalanced nature of the data, the evaluation of classifiers should not be carried out using standard classification metrics. The most commonly used metric in classification problems is undoubtedly accuracy [40]. However, as mentioned earlier, in the case of an unbalanced dataset, classifiers may seek to maximize accuracy by classifying all instances of as majority class objects [33]. This renders the accuracy metric useless for heavily unbalanced data. In fact, it is very difficult to identify metrics that perform well with unbalanced data. Some researchers find value in area under receiver operating characteristic curve (AUROC) [34], true positive rate (TPR), f1 score (F1), Cohen's kappa (KAP) [41], and Matthews correlation coefficient (MCC) [42], among others, while others dispute these findings [33, 43, 44]. In our study we used TPR, true negative rate (TNR), F1, KAP, AUROC and area under precision-recall curve (AUPRC). TPR and TNR present positive correctly classified samples to the total number of positive samples (classifier sensitivity), and correctly classified negative samples to the total number of negative samples (specificity), respectively [40]. In view of this, in the dataset under consideration, TPR will show the accuracy of minority class classification, and TNR will show the accuracy of majority class classification. In turn, F1 is the harmonic mean of the classifier's TPR (sensitivity) and precision (PRE) values (PRE shows how accurately the classifier predicts positive instances with respect to all positive predictions) [45]. The KAP measures the correspondence between the model's predictions and reality, while correcting for random conformity of classifiers. In addition, the numerical value of KAP has a linguistic interpretation. $KAP < 0.4$ means poor agreement, $KAP \in [0.4, 0.75)$ means fair to good agreement, and $KAP \geq 0.75$ means excellent agreement [46]. AUROC is based on the metrics TPR and false positive rate (FPR - the ratio of misclassified negative records to all negative records in the dataset). The AUROC metric is the area under the curve generated from the TPR and FPR values obtained for successive classification thresholds [34]. It aggregates classifier performance at different classification thresholds. Finally, AUPRC is similar to AUROC in terms of construction methodology, but is based on PRE and TPR (also referred to as recall [47]). The large area of AUPRC shows that the classifier returns accurate results (high PRE), and also returns the majority of all positive instances (high sensitivity). For unbalanced data, AUPRC is recommended instead of or in addition to AUROC [43].

4 Results

The data flow scenario involved modifying the source data by separately applying five techniques for dealing with imbalanced data (cost-sensitive learning, cutoff value modification, down-sampling, oversampling, and SMOTE). In this way, five new datasets were generated. Each of the new datasets was then divided into five folds. Using the corresponding folds, classification models were generated, and then cross-validation of each classifier was performed. The results obtained in this processing scenario are shown in Tables 1, 2, 3, 4 and 5.

Table 1. Classification results - cost-sensitive learning.

Classifier	TPR	TNR	F1	KAP	AUPRC	AUROC
C4.5	0.333	0.872	0.268	0.168	0.162	0.604
CART	**0.8**	0.504	0.255	0.105	0.156	0.666
RF	0.267	**0.983**	0.375	**0.335**	**0.379**	**0.778**
DT	0.6	0.674	0.265	0.129	0.144	0.63
ANN	0.444	0.906	**0.388**	0.311	0.27	0.724

Table 2. Classification results - cutoff value modification.

Classifier	TPR	TNR	F1	KAP	AUPRC	AUROC
C4.5	0.279	0.791	0.173	0.045	0.193	0.65
CART	0.2	**0.886**	0.18	0.079	0.155	0.581
RF	**0.578**	0.758	**0.308**	**0.189**	**0.396**	**0.734**
DT	0.244	0.877	0.208	0.105	0.175	0.699
ANN	0.422	0.783	0.25	0.127	0.269	0.695

An analysis of Tables 1, 2, 3, 4 and 5 indicates that cost-sensitive learning and cutoff modification are the least effective in dealing with the problem of unbalanced data. Slightly better results can be obtained with down-sampling. However, even the best classification results obtained on the reduced dataset obtained with down-sampling are worse than the lowest results obtained with datasets augmented with oversampling and SMOTE. This indicates that for the considered small dataset of rare instances, the way to improve the predictive performance of the classifiers is to equalise the minority and majority class counts by increasing the minority class counts. In doing so, it should be noted that classifiers perform slightly worse on synthetic data (SMOTE) than on duplicate real data (random oversampling). And as for the quality of individual classifiers, the best results in terms of almost every metric in every case are achieved by RF. Lower results

Table 3. Classification results – down-sampling.

Classifier	TPR	TNR	F1	KAP	AUPRC	AUROC
C4.5	0.622	**0.733**	0.659	0.356	0.646	0.681
CART	0.8	0.422	0.673	0.222	0.559	0.616
RF	**0.822**	0.556	**0.725**	**0.378**	**0.744**	**0.78**
DT	0.756	0.511	0.673	0.267	0.567	0.642
ANN	0.689	0.578	0.653	0.267	0.654	0.662

Table 4. Classification results - oversampling.

Classifier	TPR	TNR	F1	KAP	AUPRC	AUROC
C4.5	**1**	0.948	0.975	0.948	0.955	0.976
CART	0.98	0.867	0.928	0.847	0.899	0.939
RF	0.998	**0.98**	**0.989**	**0.978**	**1**	**1**
DT	**1**	0.909	0.956	0.909	0.932	0.973
ANN	**1**	0.941	0.971	0.941	0.952	0.978

Table 5. Classification results - SMOTE.

Classifier	TPR	TNR	F1	KAP	AUPRC	AUROC
C4.5	0.84	0.822	0.832	0.662	0.794	0.842
CART	0.798	0.857	0.822	0.654	0.862	0.879
RF	**0.965**	**0.921**	**0.944**	**0.886**	**0.989**	**0.988**
DT	0.822	0.914	0.862	0.736	0.953	0.944
ANN	0.906	0.884	0.896	0.79	0.948	0.957

are obtained, used for comparison purposes, by ANN, followed by C4.5 and DT. The worst results are usually obtained by CART.

In preparing the analysis of the impact of individual AOIs on bid selection and testing the usefulness of AOI metrics in assessing bid interest, the quality of classifiers and the effectiveness of different approaches to unbalanced data were considered. Therefore, the AOI-related analyses used results obtained using oversampling and SMOTE techniques and the best classifiers that also embed feature selection methods. In particular, the results of using RF, C4.5 and DT classifiers in combination with oversampling were considered, as well as the results of RF and DT classifiers in combination with SMOTE. Table 6 includes the 20 most important attributes according to the RF classifier combined with oversampling. These are the attributes by which RF obtains the highest average impurity decrease in the model. Table 7 includes all the attributes used by the C4.5 algorithm in combination with oversampling to build classification trees in individual folds. Table 8, on the other hand, shows all the attributes used by DT along with the data oversampling technique. By comparing these tables, one can see a great similarity between the attributes selected by each classifier in the following folds. Analysing Table 6, it is easy to see that the most important AOIs according to RF are AOI1 and AOI2, since among the leading attributes there are the most metrics associated with these very AOIs (usually 8–9 AOI1 metrics and 6 AOI2 metrics). Less important are AOI 3, and least important are AOI4 and AOI5. The importance of AOI1 is confirmed by the results of the C4.5 classifier included in Table 7. However, C4.5 is more likely to use metrics related to AOI3, AOI4, and AOI5 than AOI2 to build trees. In contrast, the DT classifier (Table 8) completely ignored AOI4 metrics in the classification rules it generates, and the number of AOI1 and AOI2 metrics used is dominated by the number of AOI5 metrics. However, the AOI2-AOI5 metrics are used only in single folds, while AOI1 is used in more folds. In view of this, it should be noted that for DT, AOI1 metrics are more relevant and more frequently used than the others.

Table 6. The most important attributes according to the RF classifier combined with oversampling.

Rank	Fold 1	Fold 2	Fold 3	Fold 4	Fold 5
1	AOI1_FC	AOI1_FC	AOI1_DAF	AOI2_TTFF	AOI1_TTFF
2	AOI2_REV	AOI1_TTFF	AOI1_TTFF	AOI1_FC	AOI2_REV
3	AOI1_DT	AOI1_DT	AOI1_FC	AOI2_REV	AOI1_DAF
4	AOI1_TTFF	AOI2_DSAF	AOI2_TTFF	AOI4_REV	AOI1_FC
5	AOI1_DAF	AOI2_REV	AOI2_DAF	AOI1_DSAF	AOI1_DT
6	AOI2_DSAF	AOI1_DAF	AOI2_REV	AOI1_DAF	AOI1_FFD
7	AOI2_TTFF	AOI1_DSAF	AOI1_DSAF	AOI1_DT	AOI1_REV
8	AOI2_DAF	AOI1_FFD	AOI1_SC	AOI1_FFD	AOI1_DSAF
9	AOI1_FFD	AOI5_REV	AOI1_FFD	AOI1_TTFF	AOI2_DAF
10	AOI1_REV	AOI2_TTFF	AOI2_FC	AOI1_SC	AOI1_FAA
11	AOI3_DSAF	AOI3_FFD	AOI2_DSAF	AOI2_DAF	AOI2_FFD
12	AOI4_TTFF	AOI1_SC	AOI3_TTFF	AOI1_FAA	AOI3_FFD
13	AOI1_SC	AOI2_DT	AOI1_FAA	AOI3_SC	AOI1_SC
14	AOI2_DT	AOI2_DAF	AOI2_SC	AOI2_DSAF	AOI2_FC
15	AOI1_DSAF	AOI4_REV	AOI1_DT	AOI2_FFD	AOI2_TTFF
16	AOI1_FAA	AOI3_REV	AOI3_FC	AOI1_REV	AOI2_DSAF
17	AOI2_FFD	AOI3_DSAF	AOI5_TTFF	AOI2_FC	AOI5_TTFF
18	AOI4_REV	AOI1_FAA	AOI1_REV	AOI4_TTFF	AOI3_SC
19	AOI3_DAF	AOI2_SC	AOI3_DAF	AOI3_DT	AOI5_DT
20	AOI3_FFD	AOI4_FAA	AOI3_FFD	AOI5_REV	AOI3_DSAF

Table 7. Attributes used by the C4.5 classifier combined with oversampling.

Fold	AOI1_DSAF	AOI1_DT	AOI1_FC	AOI1_FAA	AOI1_FFD	AOI1_REV	AOI1_SC	AOI1_TTFF	AOI2_DAF	AOI2_FAA	AOI2_FC	AOI2_REV	AOI2_TTFF	AOI3_DAF	AOI3_DT	AOI3_FAA	AOI3_FC	AOI3_REV	AOI3_SC	AOI3_TTFF	AOI4_DSAF	AOI4_FAA	AOI4_FC	AOI4_REV	AOI4_SC	AOI4_TTFF	AOI5_DAF	AOI5_DT	AOI5_FAA	AOI5_FC	AOI5_FFD	AOI5_REV	AOI5_SC	AOI5_TTFF
1	X		X		X			X			X	X				X			X		X	X	X		X	X		X				X		
2	X		X	X	X		X			X	X	X	X	X	X	X	X	X		X	X	X	X	X				X				X		
3	X	X	X		X	X	X	X		X			X			X		X	X	X	X		X				X		X		X	X	X	
4			X			X		X	X	X		X			X	X	X	X	X			X			X		X	X		X	X	X		
5	X		X		X	X		X	X			X	X		X	X		X	X			X									X			X

Table 8. Attributes used by the DT classifier combined with oversampling.

Fold	AOI1_DAF	AOI1_FFD	AOI1_DSAF	AOI1_REV	AOI1_FAA	AOI2_DT	AOI2_FAA	AOI2_FC	AOI3_FAA	AOI5_FC	AOI5_FFD	AOI5_TTFF	AOI5_DSAF	AOI5_FAA	AOI5_DT
1		X	X			X	X			X	X				
2			X	X	X							X	X	X	
3			X		X			X			X				
4		X		X					X						X
5	X	X	X												X

Table 9 includes the most important attributes according to the RF classifier combined with SMOTE, and Table 10 shows the attributes used by DT together with SMOTE. The attribute sets obtained are similar to those obtained by RF and DT combined with oversampling. In the case of the combination of RF and SMOTE classifier, as in the case of the combination of RF and oversampling, metrics related to AOI1 and AOI2 dominate. AOI3 metrics are much less important, and metrics related to AOI4 and AOI5 hardly appear at all among the most important attributes. In contrast, the results obtained by the DT classifier working with SMOTE are similar to those of DT combined with oversampling. Of the seven metrics used by DT to generate rules in this case, five metrics appeared in the set of attributes used by DT with oversampling. In addition, similar to DT with oversampling, AOI1 metrics are also the most commonly used for DT with SMOTE.

Summarising the conducted analyses of AOI metrics used by classifiers that are also embedded methods, it should be pointed out that the most significant AOI is AOI1, i.e. the external appearance of the offered house. Slightly less important is AOI2 - the layout of the rooms, and even less important is AOI3 - price, AOI4 - number of rooms and AOI5 - location. The low importance of AOI4 is easily explained because the number of rooms is readily apparent in the room layout, so these two AOIs are partially complementary. The low importance of price, on the other hand, can be explained by the fact that the survey

Table 9. The most important attributes according to the RF classifier combined with SMOTE.

Rank	Fold 1	Fold 2	Fold 3	Fold 4	Fold 5
1	AOI1_DAF	AOI1_FC	AOI1_DAF	AOI1_DAF	AOI1_DAF
2	AOI1_DSAF	AOI1_DAF	AOI1_TTFF	AOI1_TTFF	AOI1_DSAF
3	AOI1_FFD	AOI2_DAF	AOI1_SC	AOI1_FFD	AOI1_TTFF
4	AOI1_FC	AOI1_DSAF	AOI1_FFD	AOI1_DSAF	AOI1_SC
5	AOI2_DAF	AOI1_FFD	AOI1_FC	AOI1_FC	AOI1_DT
6	AOI1_FAA	AOI2_TTFF	AOI1_DSAF	AOI3_DSAF	AOI1_FC
7	AOI1_TTFF	AOI1_TTFF	AOI2_DAF	AOI1_FAA	AOI1_FFD
8	AOI2_TTFF	AOI1_FAA	AOI1_FAA	AOI2_TTFF	AOI2_DAF
9	AOI1_DT	AOI2_SC	AOI2_SC	AOI1_DT	AOI2_TTFF
10	AOI2_REV	AOI1_SC	AOI1_DT	AOI2_DT	AOI1_FAA
11	AOI2_FC	AOI2_FFD	AOI3_TTFF	AOI2_REV	AOI2_DSAF
12	AOI1_SC	AOI1_DT	AOI2_TTFF	AOI2_FFD	AOI3_DSAF
13	AOI2_SC	AOI2_FC	AOI2_FC	AOI1_SC	AOI2_FFD
14	AOI2_FFD	AOI2_DSAF	AOI2_DSAF	AOI2_DAF	AOI2_FC
15	AOI2_DT	AOI2_REV	AOI4_TTFF	AOI2_SC	AOI3_TTFF
16	AOI3_FC	AOI3_DAF	AOI3_DSAF	AOI2_FC	AOI3_FC
17	AOI2_DSAF	AOI1_REV	AOI2_FFD	AOI3_TTFF	AOI2_SC
18	AOI3_FFD	AOI5_FFD	AOI2_REV	AOI3_FFD	AOI2_REV
19	AOI4_TTFF	AOI2_DT	AOI3_FC	AOI2_DSAF	AOI4_TTFF
20	AOI3_DSAF	AOI4_TTFF	AOI2_DT	AOI3_FC	AOI1_REV

Table 10. Attributes used by the DT classifier combined with SMOTE.

Fold	AOI1_FFD	AOI1_REV	AOI2_FAA	AOI3_DAF	AOI5_DSAF	AOI5_FAA	AOI5_REV
1		X	X				
2		X	X	X	X		
3		X	X		X		
4	X	X				X	
5		X	X				X

assumed that respondents have sufficient financial resources to potentially purchase a property. Therefore, price may not have been very important to them, as they did not really allocate their financial resources.

5 Conclusion

The research provided strong evidence that multimodal psychophysiological data (eye tracking and EEG) can serve as a solid basis for predicting consumer choices in the context of sales offers. The results confirmed the effectiveness of machine learning methods, particularly Random Forest (RF), which demonstrated high accuracy under conditions of appropriate data balancing. Using methods such as oversampling and SMOTE significantly improved the quality of predictions. The best results were obtained in the variant using an extended data set through oversampling. In comparison, techniques based on cost-sensitive learning and cutoff modification were characterised by noticeably lower effectiveness.

The analysis of areas of interest (AOI) showed that the most important factors in decision-making processes were the visualisation of houses (AOI1) and the layout of rooms (AOI2). Other elements, such as price (AOI3), location (AOI5) and number of rooms (AOI4), played a less important role. These results have practical implications and can serve as recommendations for designers of single-family home sales offers. Focusing on optimising the visual and functional elements of homes can increase the offer's perceived value.

In summary, the results of this study contribute to the research methodology in the field of consumer preference analysis. They confirm that integrating physiological indicators with machine learning methods can accurately reflect the actual decisions of users. For this reason, it is worthwhile to continue this line of research and conduct studies using other psychophysiological and machine learning methods.

Acknowledgments. This study was co-financed by the Minister of Science under the "Regional Excellence Initiative".

Regionalna Inicjatywa Doskonałości

Ministry of Science and Higher Education Republic of Poland

Disclosure of Interests. The authors have no competing interests to declare that are relevant to the content of this article.

References

1. Leonov, Y.: Analysis of the influence of psychological factors on consumer behavior and the decision-making process. EA. **68** (2023). https://doi.org/10.46852/0424-2513.3.2023.29
2. Bello, E.: Unravelling the consumer brain: the role of emotion in purchase behavior (2014)
3. de-Juan-Ripoll, C., Nolé, M.L., Montañana, A., Llinares, C.: Relationship between visual attention patterns and subjective evaluations in housing sales information: a study using eye-tracking technology. Buildings **14**, 2106 (2024). https://doi.org/10.3390/buildings14072106
4. Braun, C.: Residential real estate investments: the behavioral impact of building aesthetics. In: 25th Annual European Real Estate Society Conference. European Real Estate Society, Reading, United Kingdom (2018). https://doi.org/10.15396/eres2018_245

5. Bridger, D., Noble, T.: New tools and techniques for understanding non-conscious consumer decisions. AMA. **1**, 214 (2015). https://doi.org/10.69554/MCEU2175
6. Dwivedi, A., Sharma, V.: Exploring the role of neuromarketing in consumer decision-making. ShodhKosh J. Vis. Per. Arts. **5** (2024). https://doi.org/10.29121/shodhkosh.v5.i7.2024.2443
7. Piwowarski, M., Ziemba, P., Cypryjański, J.: Machine learning-based exploration of eye-tracking data to predict offer selection. In: Hernes, M., Wątróbski, J., and Rot, A. (eds.) Emerging Challenges in Intelligent Management Information Systems. pp. 105–117. Springer Nature Switzerland, Cham (2025). https://doi.org/10.1007/978-3-031-78465-1_9
8. World Medical Association: World medical association declaration of helsinki: ethical principles for medical research involving human subjects. JAMA **310**, 2191 (2013). https://doi.org/10.1001/jama.2013.281053
9. Symeonidou, E.-R., Nordin, A., Hairston, W., Ferris, D.: Effects of cable sway, electrode surface area, and electrode mass on electroencephalography signal quality during motion. Sensors. **18**, 1073 (2018). https://doi.org/10.3390/s18041073
10. Mahanama, B., et al.: Eye movement and pupil measures: a review. Front. Comput. Sci. **3** (2022). https://doi.org/10.3389/fcomp.2021.733531
11. van der Laan, L.N., Hooge, I.T.C., de Ridder, D.T.D., Viergever, M.A., Smeets, P.A.M.: Do you like what you see? The role of first fixation and total fixation duration in consumer choice. Food Qual. Prefer. **39**, 46–55 (2015). https://doi.org/10.1016/j.foodqual.2014.06.015
12. Guo, F., Ding, Y., Liu, W., Liu, C., Zhang, X.: Can eye-tracking data be measured to assess product design?: Visual attention mechanism should be considered. Int. J. Ind. Ergon. **53**, 229–235 (2016). https://doi.org/10.1016/j.ergon.2015.12.001
13. Heinzel, G., Rüdiger, A.: Spectrum and spectral density estimation by the Discrete Fourier transform (DFT), including a comprehensive list of window functions and some new flat-top windows. Max Plank Inst. **12** (2002)
14. Semmlow, J.: The fourier transform and power spectrum. In: Signals and Systems for Bioengineers. pp. 131–165. Elsevier (2012). https://doi.org/10.1016/B978-0-12-384982-3.00004-3
15. Mehra, Dr.R.: Power Spectrum Estimation using Welch method for various Window techniques. Int. J. Sci. Res. Eng. Technol. **2**, 389–392 (2013)
16. Smith, E.E., Reznik, S.J., Stewart, J.L., Allen, J.J.B.: Assessing and conceptualizing frontal EEG asymmetry: an updated primer on recording, processing, analyzing, and interpreting frontal alpha asymmetry. Int. J. Psychophysiol. **111**, 98–114 (2017). https://doi.org/10.1016/j.ijpsycho.2016.11.005
17. Davidson, R.J.: Cerebral asymmetry and emotion: conceptual and methodological conundrums. Cogn. Emot. **7**, 115–138 (1993). https://doi.org/10.1080/02699939308409180
18. Coan, J.A., Allen, J.J.B.: Frontal EEG asymmetry as a moderator and mediator of emotion. Biol. Psychol. **67**, 7–50 (2004). https://doi.org/10.1016/j.biopsycho.2004.03.002
19. Chandrashekar, G., Sahin, F.: A survey on feature selection methods. Comput. Electr. Eng. **40**, 16–28 (2014). https://doi.org/10.1016/j.compeleceng.2013.11.024
20. Cai, J., Luo, J., Wang, S., Yang, S.: Feature selection in machine learning: a new perspective. Neurocomputing **300**, 70–79 (2018). https://doi.org/10.1016/j.neucom.2017.11.077
21. Quinlan, J.R.: C4.5: programs for machine learning. Morgan Kaufmann Publishers Inc., San Francisco, CA, USA (1993)
22. Breiman, L., Friedman, J., Olshen, R.A., Stone, C.J.: Classification and regression trees. Chapman and Hall/CRC, New York (1984). https://doi.org/10.1201/9781315139470
23. Breiman, L.: Random forests. Mach. Learn. **45**, 5–32 (2001). https://doi.org/10.1023/A:1010933404324
24. Kohavi, R.: The power of decision tables. In: Proceedings of the 8th European Conference on Machine Learning. pp. 174–189. Springer-Verlag, Berlin, Heidelberg (1995). https://doi.org/10.1007/3-540-59286-5_57

25. Zou, J., Han, Y., So, S.-S.: Overview of artificial neural networks. In: Livingstone, D.J. (ed.) Artificial Neural Networks: Methods and Applications. pp. 14–22. Humana Press, Totowa, NJ (2009). https://doi.org/10.1007/978-1-60327-101-1_2
26. Chen, L.: Curse of dimensionality. In: LIU, L. and ÖZSU, M.T. (eds.) Encyclopedia of Database Systems. pp. 545–546. Springer US, Boston, MA (2009). https://doi.org/10.1007/978-0-387-39940-9_133
27. Negovetić, M., Otović, E., Kalafatovic, D., Mauša, G.: Efficiently solving the curse of feature-space dimensionality for improved peptide classification. Digital Discovery. **3**, 1182–1193 (2024). https://doi.org/10.1039/D4DD00079J
28. Ziemba, P., Becker, J., Becker, A., Radomska-Zalas, A., Pawluk, M., Wierzba, D.: Credit decision support based on real set of cash loans using integrated machine learning algorithms. Electronics **10**, 2099 (2021). https://doi.org/10.3390/electronics10172099
29. Guo, X., Yin, Y., Dong, C., Yang, G., Zhou, G.: On the class imbalance problem. Fourth International Conference on Natural Computation, ICNC '08. Vol. 4, (2008). https://doi.org/10.1109/ICNC.2008.871
30. Lai, C.S., et al.: A robust correlation analysis framework for imbalanced and dichotomous data with uncertainty. Inf. Sci. **470**, 58–77 (2019). https://doi.org/10.1016/j.ins.2018.08.017
31. He, H., Garcia, E.A.: Learning from imbalanced data. IEEE Trans. Knowl. Data Eng. **21**, 1263–1284 (2009). https://doi.org/10.1109/TKDE.2008.239
32. Lee, W., Seo, K.: Downsampling for binary classification with a highly imbalanced dataset using active learning. Big Data Res. **28**, 100314 (2022). https://doi.org/10.1016/j.bdr.2022.100314
33. Benkendorf, D.J., Schwartz, S.D., Cutler, D.R., Hawkins, C.P.: Correcting for the effects of class imbalance improves the performance of machine-learning based species distribution models. Ecol. Model. **483**, 110414 (2023). https://doi.org/10.1016/j.ecolmodel.2023.110414
34. Bourel, M., et al.: Machine learning methods for imbalanced data set for prediction of faecal contamination in beach waters. Water Res. **202**, 117450 (2021). https://doi.org/10.1016/j.watres.2021.117450
35. Freeman, E.A., Moisen, G.G.: A comparison of the performance of threshold criteria for binary classification in terms of predicted prevalence and kappa. Ecol. Model. **217**, 48–58 (2008). https://doi.org/10.1016/j.ecolmodel.2008.05.015
36. Zou, Q., Xie, S., Lin, Z., Wu, M., Ju, Y.: Finding the best classification threshold in imbalanced classification. Big Data Res. **5**, 2–8 (2016). https://doi.org/10.1016/j.bdr.2015.12.001
37. Pozzolo, A.D., Caelen, O., Johnson, R.A., Bontempi, G.: Calibrating probability with undersampling for unbalanced classification. In: 2015 IEEE Symposium Series on Computational Intelligence. pp. 159–166 (2015). https://doi.org/10.1109/SSCI.2015.33
38. Chawla, N.V., Bowyer, K.W., Hall, L.O., Kegelmeyer, W.P.: SMOTE: synthetic minority over-sampling technique. J. Artif. Intell. Res. **16**, 321–357 (2002). https://doi.org/10.1613/jair.953
39. Xu, J., Liu, X., Gu, Z., Xiao, G.: A rapid cross-validation computing for three-way decisions in imbalanced data. Inf. Sci. **707**, 122016 (2025). https://doi.org/10.1016/j.ins.2025.122016
40. Tharwat, A.: Classification assessment methods. Appl. Comput. Inform. **17**, 168–192 (2021). https://doi.org/10.1016/j.aci.2018.08.003
41. de la Cruz Huayanay, A., Bazán, J.L., Russo, C.M.: Performance of evaluation metrics for classification in imbalanced data. Comput. Stat. **40**, 1447–1473 (2025). https://doi.org/10.1007/s00180-024-01539-5
42. Chicco, D.: Ten quick tips for machine learning in computational biology. BioData Mining **10**, 35 (2017). https://doi.org/10.1186/s13040-017-0155-3
43. Saito, T., Rehmsmeier, M.: The precision-recall plot is more informative than the ROC plot when evaluating binary classifiers on imbalanced datasets. PLoS ONE **10**, e0118432 (2015). https://doi.org/10.1371/journal.pone.0118432

44. Zhu, Q.: On the performance of Matthews correlation coefficient (MCC) for imbalanced dataset. Pattern Recogn. Lett. **136**, 71–80 (2020). https://doi.org/10.1016/j.patrec.2020.03.030
45. Kulkarni, A., Chong, D., Batarseh, F.A.: 5 - Foundations of data imbalance and solutions for a data democracy. In: Batarseh, F.A., Yang, R. (eds.) Data Democracy. pp. 83–106. Academic Press (2020). https://doi.org/10.1016/B978-0-12-818366-3.00005-8
46. Fleiss, J.L., Levin, B., Paik, M.C., Fleiss, J.: Statistical Methods for Rates & Proportions. Wiley-Interscience, Hoboken, N.J (2003)
47. Ziemba, P., Becker, J., Becker, A., Radomska-Zalas, A.: Framework for multi-criteria assessment of classification models for the purposes of credit scoring. J. Big Data **10**, 94 (2023). https://doi.org/10.1186/s40537-023-00768-7

AI Agents Within Augmented Reality Environments in Customer Service

Beata Glinkowska-Krauze and Tomasz Noga(✉)

Faculty of Management, University of Lodz, Ul. Matejki 22/26, Lodz 90-237, Poland
tomasz.noga@uni.lodz.pl

Abstract. The aim of this article is to address the identified research gap concerning the issue of AI agents in Augmented Reality (AR) environments within customer service. To achieve this goal, a review of the scientific literature and industry reports was conducted, along with a quantitative empirical study using a questionnaire survey purposively targeted at young people (aged 18–25). The objective was to determine the attitudes of young consumers towards the use of AI agents in AR environments in customer service. This research provides the first empirical insights related to AI agents in AR in the context of customer service, which serve as responses to the four research questions formulated in the study.

Keywords: AI Agent · Immersive technology · Augmented Reality · AR · customer service

1 Introduction

The ongoing digitalization and increasing market competition compel enterprises to create and implement experiences that respond to the dynamically growing needs of customers [1]. The introduction of advanced solutions, such as immersive technologies and artificial intelligence (with particular emphasis on AI agents), may result in a sustainable competitive advantage, thereby ensuring stable and favorable financial outcomes [2].

The value of the global immersive technology market increased from nearly $155 billion in 2024 to over $222 billion in 2025. It is estimated that by 2029, this value will exceed $900 billion, with a compound annual growth rate (CAGR) of 42% [3]. AI agents are considerably newer tools; however, their market value is projected to reach over $10 billion in 2025, and by 2029, it is expected to surpass $61 billion, with a compound annual growth rate of 55.9% [4].

The current and projected development of these technologies, as well as their high potential for application in the business-to-customer market, prompted the authors to explore the topic in greater depth. A systematic bibliometric review was conducted using the SCOPUS scientific database. Works related to the phrases "AI agent" and, respectively, "AI agents," as well as "immersive technology" and "augmented reality," were searched. The query presented below returned no records, which was consequently identified as an indication of a research gap in the area under discussion:

Published by Springer Nature Switzerland AG 2026
M. Pondel et al. (Eds.): AI4KAM 2025, IFIP AICT 776, pp. 266–275, 2026.
https://doi.org/10.1007/978-3-032-18920-2_26

TITLE-ABS-KEY (((„AI agents" OR „AI agent") AND ("immersive technology" OR "augmented reality") AND ("customer service"))).

2 Immersive Technology

The retail market requires providing consumers with immersive experiences, which serve as a trigger for emotions leading to positive memories [5]. Such feelings may result from transporting consumers into an immersive environment through the use of immersive technologies, which enhance the realism of experiences and the sense of immersion [6].

Immersive technology provides a high quality and/or quantity of sensory information received by users, blurring the boundaries between the physical and digital worlds [7]. It enables users within a simulation to perceive their surroundings through multiple senses in such a way that they experience a sense of presence in the real world [8]. This is the result of the objective level of sensory representation provided by the immersive system [9]. Consequently, it may be utilized by enterprises in the retail sector for a variety of tasks, such as customer service in the pre-sale, sale, and post-sale phases.

Immersive technologies, which constitute one example of digital technologies, are employed to enhance pre-purchase, purchase, and post-purchase experiences [10]. However, the literature indicates that these technologies primarily influence consumer experiences in the pre-transactional phase, while their impact is less significant in the transactional and post-transactional phases [11].

The present study focuses exclusively on Augmented Reality for several reasons. Firstly, AR is straightforward to activate for marketing purposes, which facilitates its dissemination among enterprises. Moreover, compared to other systems, it presents low implementation barriers for consumers, as it can be easily used with devices such as smartphones [10]. Research has demonstrated that the use of such systems transforms customer experiences, making them more interactive and thus generating greater interest. They contribute to increased satisfaction, enable interaction, and support engagement in essential processes [12].

Augmented Reality is a technology that enables the overlay of digitally generated information onto the real world in real time, integrating spatially and temporally with it [13]. Users see their realistic environment, and the technology, without isolating them from it, enhances it with virtual information [14]. By leveraging this capability, enterprises can address a range of challenges and encourage consumers to purchase their products; for instance, by automatically onboarding users to their products with the assistance of AI agents.

3 AI Agent

One of the most widely cited definitions of Artificial Intelligence (AI) was developed by Kaplan and Haenlein, who describe it as a system capable of correctly interpreting external data, learning from such data, and using this knowledge to achieve specific goals and tasks through flexible adaptation [15]. AI is a technology of fundamental importance in modern marketing. It drives increased customer engagement by utilizing, among other methods, natural language processing (NLP) [16].

The breakthrough in the field of natural language processing stems from the development of Large Language Models (LLMs), which are artificial neural networks containing billions of parameters. These models are trained on vast corpora of text, enabling them to learn both syntactic and semantic relationships between words and sentences. As a result, such models are capable of comprehending commands and generating human-like text [17]. The integration of textual and visual data is made possible by Vision-Language Models (VLMs), which show significant potential in tasks requiring the conversion of natural language into images, matching text to images, and vice versa. These models utilize large collections of paired text-visual data [18]. The ability to generate high-quality content in customer interactions addresses consumer demand for immediate engagement, while simultaneously maintaining a high standard of customer service [16].

The introduction of publicly available AI models such as LLMs and VLMs has enabled real-time processing of natural language and data, often surpassing human expert knowledge. Multimodal models are capable of analyzing and interpreting image, audio, and video data [17]. It has been demonstrated that advanced LLMs can function as AI agents, performing various complex tasks within specific environments in conjunction with domain-specific knowledge. This enables them to make decisions within constrained settings. An AI agent represents a class of interactive systems that can receive visual stimuli, linguistic signals, and other environmental data, and are capable of generating constructive, embodied actions. These agents are designed to learn interactively using machine learning, leveraging real-time user interactions based on feedback or observation [19].

Systems of this class should be capable of making autonomous decisions based on their own assessments, striving to achieve designated objectives. This autonomy is particularly evident in their ability to adapt to changing conditions and to anticipate future situations, which consequently enables them to make complex decisions without human supervision [19]. It is assumed that they should also operate according to the principle of continuous improvement, effectively handle ambiguous situations, and predict outcomes as well as suggest subsequent steps based on historical data and trends [20]. AI agents utilize Intelligent Decision Support Systems (IDSS), which integrate artificial intelligence techniques to generate potential solutions and support users in making good decisions [21].

An important issue concerns trust in AI agents. Users perceive these tools as providing a more objective and impartial assessment than human interlocutors, thereby eliminating bias and subjectivity. Trust is also influenced by the perceived similarity between the agent's personality and that of the user. However, there is a risk of overtrust, which highlights the need to cultivate balanced trust in human–machine interactions [22].

AI agents may take the form of Embodied Conversational Agents (ECA) and Interactive Virtual Humans (IVH) based on 3D rendering, combining voice communication with eye contact, or in simplified versions—as voice-only agents or agents utilizing gestures [24]. This enables interaction with users within augmented reality environments and may be advantageous, for example, in sales and negotiation contexts.

AI agents are tools that can support a wide range of organizational functions. Current research indicates that they can assist in knowledge management by automating the processes of collecting, organizing, and distributing information, which in turn is expected to enhance effectiveness, efficiency, and productivity [25]. The human-like characteristics of this technology are anticipated to improve employees' attitudes towards it. Agents can physically guide users through virtual events or demonstrate how to perform various organizational tasks, such as machine maintenance or sales [26].

4 Methods

To achieve the objective of this study, an empirical study was conducted with the aim of determining the attitudes of young consumers towards the use of AI Agents in AR environments within customer service. The following research questions were formulated:

Q1: What barriers are perceived by young consumers regarding AI agents in AR environments for customer service?
Q2: What benefits are perceived by young consumers regarding AI agents in AR environments for customer service?
Q3: Does declared gender differentiate the perception of AI agents in AR environments for customer service?
Q4: Do prior experiences with AR and AI agents differentiate the perceptions associated with their use in customer service?

The survey method was used as the research methodology, utilizing a questionnaire as the research tool and an online survey as the research technique. The anonymous study was conducted via an online questionnaire through the MS Forms platform. Data were collected in June 2025. Participants were purposively selected based on age criteria. The authors focused on students at the University of Lodz and therefore invited young individuals aged 18–25 to participate in the study. A total of 294 respondents took part, with 39.1% identifying as men and 60.9% as women.

To test the significance of differences, non-parametric tests were employed. When the number of classes was $k = 2$, the Mann-Whitney U test was used; when the number of classes was $k > 2$, the Kruskal-Wallis H test was applied. For multiple post-hoc comparisons, Dunn's D test was utilized. In cases where the Kruskal-Wallis H test indicated significant differences but the Bonferroni correction in Dunn's test was too conservative and failed to identify which pairs differed, the Mann-Whitney U test was additionally used for their identification.

The study consisted of 15 statements based on a 7-point Likert scale, where 1 indicated "strongly disagree" and 7 indicated "strongly agree," as presented in Table 1, as well as demographic questions. The results were subjected to statistical analysis using IBM SPSS Statistics 29.

Table 1. Statements included in the study

No	Sentence
1	If an AI agent in AR were available in customer service, I would be willing to use it
2	I believe that an AI agent in AR could significantly facilitate handling customer-related matters
3	I would expect that an AI agent in AR would resolve my issues more quickly than a traditional consultant
4	I imagine that an AI agent in AR could effectively recommend products and services tailored to my needs
5	I believe that an AI agent in AR could be available 24/7, which would be a major advantage for me
6	I imagine that using an AI agent in AR would be easy and intuitive
7	I expect that the interaction with an AI agent in AR would be natural and resemble a conversation with a consultant
8	I believe that an AI agent in AR could personalize its responses based on my previous interactions with the store
9	I imagine that an AI agent in AR could independently make decisions regarding my inquiries
10	If an AI agent in AR were available, I would trust that my personal data is secure
11	I expect that I could rely on the decisions made by an AI agent in AR
12	I imagine that I would feel safe when using the services of an AI agent in AR
13	If an AI agent in AR were available, I would prefer to use it over a traditional consultant
14	I would intend to use an AI agent in AR regularly if it were available
15	I would recommend the use of an AI agent in AR to others

5 Results

Table 2 presents the percentage distribution of responses to the statements on the seven-point Likert scale.

Table 2. Percentage distribution of responses

	1	2	3	4	5	6	7
1	8,5%	8,2%	13,9%	14,6%	18,7%	20,1%	16,0%
2	5,8%	5,4%	11,2%	17,3%	21,8%	22,8%	15,6%
3	13,3%	14,3%	18,4%	15,3%	11,9%	15,0%	11,9%
4	3,1%	6,5%	16,3%	18,4%	17,7%	23,1%	15,0%

(continued)

Table 2. (*continued*)

	1	2	3	4	5	6	7
5	0,3%	2,4%	3,7%	7,8%	8,8%	21,8%	55,1%
6	1,4%	5,8%	11,2%	22,1%	22,4%	23,1%	13,9%
7	16,0%	17,0%	17,0%	15,3%	15,3%	12,6%	6,8%
8	2,7%	2,4%	6,5%	17,0%	18,7%	32,0%	20,7%
9	7,1%	11,9%	16,7%	18,7%	16,7%	18,0%	10,9%
10	20,4%	17,7%	26,5%	11,9%	10,5%	9,2%	3,7%
11	9,9%	16,3%	18,4%	23,1%	17,7%	10,2%	4,4%
12	10,5%	14,3%	17,7%	22,8%	17,3%	11,6%	5,8%
13	19,0%	15,3%	14,6%	17,7%	13,3%	11,6%	8,5%
14	13,9%	13,9%	17,3%	20,4%	15,3%	12,2%	6,8%
15	11,6%	9,2%	16,7%	24,5%	14,6%	15,3%	8,2%

More respondents agree with statements 1, 2, 4, 5, 6, 8, 9, and 15, whereas more participants disagree with statements 3, 7, 10, 11, 12, 13, and 14.

Significant gender differences were observed for the statement 9: U = 8577,500, p = 0,015 – men (M_{rank} = 162,41) agreed significantly more with this statement than women (M_{rank} = 137,92); statement 10: U = 8581,000, p = 0,014 - men (M_{rank} = 162,38) agreed significantly more with this statement than women (M_{rank} = 137,94) and statement 12: U = 8747,000, p = 0,028 - men (M_{rank} = 160,94) agreed significantly more with this statement than women (M_{rank} = 138,87).

Significant differences were observed with regard to prior experience with AR for statement 5: H = 11.315, p = 0.010. Respondents who reported no prior experience (M_{rank} = 63.00) had a significantly lower level of agreement with this statement compared to those who used AR more than once a month but less than once a week (M_{rank} = 77.06), U = 1653.000, p = 0.026; respondents with no prior experience (M_{rank} = 53.20) also showed significantly lower agreement than those who used AR at least once a week (M_{rank} = 69.71), U = 800.500, p = 0.013. Furthermore, those who used AR less than once a month (M_{rank} = 86.19) had a significantly lower level of agreement than those who used it more than once a month but less than once a week (M_{rank} = 104.32), U = 2552.500, p = 0.023; respondents using AR less than once a month (M_{rank} = 76.32) also had significantly lower agreement than those using it at least once a week (M_{rank} = 98.85), U = 1239.000, p = 0.012. Significant differences were also identified regarding prior AR experience for statement 8: H = 8.735, p = 0.033. Respondents using AR at least once a week (M_{rank} = 183.08) agreed significantly more with this statement than those using it less than once a month (M_{rank} = 135.12), D = -47.957, p = 0.041.

Additionally, significant differences were observed with respect to prior experience with AI agents for statement 1: H = 13.457, p = 0.004 – respondents using AI agents at least once a week (M_{rank} = 168.67) were significantly more likely to agree with this statement than those using them less than once a month (M_{rank} = 123.59), D = -45.076, p = 0.002; for statement 3: H = 11.033, p = 0.012 – respondents using AI agents at least once a week (M_{rank} = 171.13) were significantly more likely to agree than those using them less than once a month (M_{rank} = 134.18), D = -36.951, p = 0.020; for statement 4: H = 17.674, p < 0.001 – respondents using AI agents at least once a week (M_{rank} = 176.16) were significantly more likely to agree than those using them less than once a month (M_{rank} = 125.75), D = -50.413, p < 0.001, or those using them more than once a month but less than once a week (M_{rank} = 140.13), D = -36.028, p = 0.031, for statement 5: H = 8.187, p = 0.042, respondents using AI agents less than once a month (M_{rank} = 81.71) were significantly less likely to agree than those using them at least once a week (M_{rank} = 97.49), U = 3290.500, p = 0.022, and those using them more than once a month but less than once a week (M_{rank} = 75.68) were significantly less likely to agree than those using them at least once a week (M_{rank} = 93.44), U = 2825.500, p = 0.009. For statement 6: H = 8.958, p = 0.030, respondents using AI agents at least once a week (M_{rank} = 167.32) were significantly more likely to agree than those using them less than once a month (M_{rank} = 131.71), D = -35.613, p = 0.026. For statement 7: H = 9.447, p = 0.024, respondents using AI agents at least once a week (M_{rank} = 166.39) were significantly more likely to agree than those using them less than once a month (M_{rank} = 129.61), D = -36.782, p = 0.021. For statement 8: H = 9.065, p = 0.028, respondents using AI agents at least once a week (M_{rank} = 166.92) were significantly more likely to agree than those using them less than once a month (M_{rank} = 133.45), D = -33.467, p = 0.041; for statement 10: H = 14.037, p = 0.003, respondents using AI agents less than once a month (M_{rank} = 119.46) were significantly less likely to agree than those using them at least once a week (M_{rank} = 161.34), D = -41.876, p = 0.005, or those with no such experience (M_{rank} = 163.76), D = 44.292, p = 0.036. With regard to statement 12: H = 14.783, p = 0.002, respondents using AI agents less than once a month (M_{rank} = 118.31) were significantly less likely to agree than those using them at least once a week (M_{rank} = 162.23), D = -43.917, p = 0.003, or those using them more than once a month but less than once a week (M_{rank} = 158.60), D = -40.287, p = 0.014. For statement 13: H = 13.327, p = 0.004, respondents using AI agents less than once a month (M_{rank} = 171.26) were significantly less likely to agree than those using them at least once a week (M_{rank} = 126.01), D = -45.249, p = 0.002. For statement 14: H = 12.840, p = 0.005, respondents using AI agents less than once a month (M_{rank} = 169.53) were significantly less likely to agree than those using them at least once a week (M_{rank} = 125.39), D = -44.144, p = 0.003. Finally, for statement 15: H = 14.682, p = 0.002, respondents using AI agents less than once a month (M_{rank} = 171.56) were significantly less likely to agree than those using them at least once a week (M_{rank} = 123.75), D = -47.811, p = 0.001.

6 Discussion and Conclusion

This study serves as an introduction to understanding the perception of AI agents in AR environments in the context of customer service, addressing a research gap identified in the scientific literature. The research provides the first empirical evidence regarding factors influencing consumer acceptance of these technologies. The findings from this analysis offer valuable guidance for enterprises considering the implementation of such technological solutions.

Respondents expressed concerns about the security of personal data, doubts regarding the naturalness of interactions, a lack of trust in decisions made by AI agents in AR, and a preference for traditional consultants. The main benefits perceived from the implementation of such tools include their constant availability, the ability to personalize responses based on previous interactions, and the potential to facilitate the resolution of customer matters.

The study also demonstrated that men are more likely to trust AI agents in AR, especially with regard to decision-making autonomy, data security, and the general sense of safety when using such services. Individuals with prior experience with AI agents exhibited higher acceptance across various aspects, including willingness to use the system, trust in the system, and expectations concerning its functionality. Experience with AR, on the other hand, increases the perceived availability of 24/7 service and the possibility of personalized support, but it does not directly translate into declared willingness to use the technology or to make recommendations.

This research has several limitations. Focusing exclusively on individuals aged 18–25 limits the representativeness of the consumer population. The study is also exploratory in nature and relies primarily on hypothetical usage scenarios, as AI agents remain relatively novel. These scenarios do not fully capture real-life feelings and behaviors. This research serves as proof of concept and requires further experimental studies to provide statistically more robust results, enabling generalization to a larger population. Furthermore, it should be noted that the research was conducted in Poland, where only 44.3% of the population possesses basic digital skills, compared to the EU average of 55.6% [27]. It was not possible to apply parametric tests due to unmet assumptions; therefore, future research should focus on sampling such that these tests can be performed. For these reasons, the present findings cannot be generalized to the entire population.

The results cannot be compared directly to existing studies due to the identified research gap in this area. However, when referring to related research on chatbots, certain similarities can be observed. For example, respondents in Asian studies on chatbots also experience concerns related to privacy and data security, and consider such tools to be useful and practical [28]. In another study, it was found that personalizing responses leads to a more favorable reception of the tool, which is a key factor in increased engagement in the purchasing process [29].

Future research should include a broader age range, various cultural contexts, and focus on implemented tools to assess actual user experiences rather than merely perceptions. Additionally, the economic aspects of these considerations should be investigated.

When implementing AI agents in AR environments within customer service processes, enterprises should offer a simplified version of the tool for inexperienced users. Additionally, attention should be paid to ensuring constant availability and personalizing the tool to align with current offers and recommendations. To foster trust in the tool, users should be provided with transparency regarding what data are collected and how they are processed, as well as the ability to manually enable or disable those features that raise concerns.

Disclosure of Interests. The authors have no competing interests to declare that are relevant to the content of this article.

References

1. Bolton, R.N., Gustafsson, A., McColl-Kennedy, J., Sirianni, N.J., Tse, D.K.: Small details that make big differences. J. Serv. Manag. **25**(2), 253–274 (2014)
2. Bolton, R.N., et al.: Customer experience challenges: bringing together digital, physical and social realms. J. Serv. Manag. **29**(5), 776–808 (2018)
3. The Business Research Company. https://thebusinessresearchcompany.com/report/immersive-technology-global-market-report. Last accessed 10 May 2025
4. The Business Research Company. https://www.thebusinessresearchcompany.com/report/agentic-ai-tools-global-market-report, last accessed 10 May 2025
5. Giraldi, L., Mengoni, M., Bevilacqua, M.: How to enhance customer experience in retail: investigations through a case study. Adv. Transdisciplinary Eng. **4**, 381–390 (2016)
6. Lee, H.-G., Chung, S., Lee, W.-H.: Presence in virtual golf simulators: the effects of presence on perceived enjoyment, perceived value, and behavioral intention. New Media Soc. **15**(6), 930–946 (2012)
7. Slater, M.: Place illusion and plausibility can lead to realistic behaviour in immersive virtual environments. Philos. Trans. R. Soc. Biol. Sci. **364**(1535), 3549–3557 (2009)
8. Fauville, G., Queiroz, A.C.M., Bailenson, J.N.: Virtual reality as a promising tool to promote climate change awareness. Technol. Health, 91–108 (2020)
9. Slater, M.: A note on presence terminology. Presence Connect **3**(3), 1–5 (2003)
10. Kotler, P., Kartjaya, H., Setiawan, I.: Marketing 6.0. Wiley & Sons, New Jersey (2024)
11. Hoyer, W.D., Kroschke, M., Schmitt, B., Kraume, K., Shankar, V.: Transforming the customer experience through new technologies. J. Interact. Mark. **51**, 57–71 (2020)
12. Butt, A., Ahmad, H., Ali, F., Muzaffar, A., Shafique, M.N.: Engaging the customer with augmented reality and employee services to enhance equity and loyalty. Int. J. Retail Distrib. Manage. **51**(5), 629–652 (2023)
13. Craig, A.B.: Understanding Augmented Reality. Concepts and Applications. Morgan Kaufmann Publishers Inc., San Francisco (2013)
14. Rauschnabel, P.A.: Augmented reality is eating the real-world! The substitution of physical products by holograms. Int. J. Inf. Manage. **57**, 1–15 (2021)
15. Haenlein, M., Kaplan, A.: A brief history of artificial intelligence: on the past, present, and future of artificial intelligence. Calif. Manage. Rev. **61**(4), 5–14 (2019)
16. Rosário, A.T., Raimundo, R.J.: The integration of AI and IoT in marketing: a systematic literature review. Electronics **14**(9), 1854, 1–27 (2025)
17. Thapa, S., et al.: Large language models (LLM) in computational social science: prospects, current state, and challenges. Soc. Netw. Anal. Min. **15**(4), 1–30 (2025)
18. Ho, H., et al.: A review on vision-language-based approaches: challenges and applications. Comput. Mater. Continua **82**(2), 1733–1756 (2025)
19. Manjuanath, K.K., Samata M.S., Akshaya H.L., Shilpashree, N., Girish, J., Abhijit, M.: Neuromorphic-driven agentic AI for autonomous decision-making systems. 2024 4th International Conference on Mobile Networks and Wireless Communications (ICMNWC), pp. 1–8 (2024)
20. Durante, Z., et al.: Agent AI: surveying the horizons of multimodal interaction, pp. 1–80 (2024)
21. Wang, A., Liang, B.: Analysis of intelligent decision support systems and a multi criteria framework for assessment. J. Enterp. Bus. Intell. **3**(4), 224–235 (2023)

22. Zhou, M.X., Mark, G., Li, J., Yang, H.: Trusting virtual agents: the effect of personality. ACM Trans. Interact. Intell. Syst. (TiiS) **9**(2–3), 1–36 (2019)
23. Chetty, G., White, M.: Embodied conversational agents and interactive virtual humans for training simulators. The 15th International Conference on Auditory-Visual Speech Processing, pp. 1–5 (2019)
24. Chang, Z., Bai, H., Zhang, L., Gupta, K., He, W., Billinghurst, M.: The impact of virtual agents' multimodal communication on brain activity and cognitive load in Virtual Reality. Virtual Reality and Human Behaviour, pp. 1–15 (2022)
25. Olan, F., Arakpogun, E.O., Suklan, J., Nakpodia, F., Damij, N., Jayawickrama, U.: Artificial intelligence and knowledge sharing: contributing factors to organizational performance. J. Bus. Res. **145**, 605–615 (2022)
26. Torro, O., Jalo, H., Pirkkalainen, H.: Six reasons why virtual reality is a game-changing computing and communication platform for organizations. Commun. ACM **64**(10), 48–55 (2021)
27. European Commision. https://digital-strategy.ec.europa.eu/en/factpages/poland-2024-digital-decade-country-report. Last accessed 10 May 2025
28. Marjerison, R. K., Dong, H., Kim, J.-M., Zheng, H., Zhang, Y., Kuan, G.: Understanding user acceptance of AI-Driven Chatbots in China's E-Commerce. the roles of perceived authenticity, usefulness, and risk. Systems **13**(2), 71, 1–24 (2025)
29. Quang, H.T.: Determinants of customers' continuance intention to use Chatbots in e-Commerce. Arch. Bus. Res. **12**(1), 78–92 (2024)

Comprehensive Usage of Avatars in Modern Marketing

Arnold Pabian[1], Elwira Gross-Gołacka[2], Barbara Pabian[3](✉), and Katarzyna Bilińska[3]

[1] Częstochowa University of Technology, Ul. Dąbrowskiego 69, 42-201 Częstochowa, Poland
[2] University of Warsaw, Ul. Krakowskie Przedmieście 26/28, 00-927 Warszawa, Poland
[3] University of Economics in Katowice, Ul. 1 Maja 50, Katowice, Poland
barbara.pabian@ue.katowice.pl

Abstract. New technologies and avatars created with their usage are revolutionizing modern marketing. The usage of avatars in marketing is presented in various sources in a selective and fragmentary manner, mainly in relation to customer service, sale and marketing communication. There is a lack of studies containing a comprehensive review of all areas of marketing activity in the aspect of using avatars in them. This article fills this research gap. The aim of the article is to present the principles and applications of avatars in marketing as a whole consisting of the following four parts: research and analysis of the marketing environment, selection of target markets, marketing mix, management of marketing activities. These parts constitute the quintessence of properly conducted marketing activity and none of them can be omitted in the processes of designing and implementing marketing into practice. The article is important both from a scientific as well as practical point of view. It organizes the issue of using avatars in all areas of marketing and indicates creative ways of their selection and usage.

Keywords: Avatars · AI · Management · Digital marketing

1 Introduction

Avatar-based marketing is in its early stage of development. By using avatars in marketing activities, companies do not only communicate that they are modern and innovative to their environment, but also improve communication and customer service on various platforms. Avatar, being under the control of the company that has created it, moves around the virtual environment, uses digital resources and enters into interactions, the main goal of which is to positively strengthen the brand's image, encourage purchases and finalize sales. Avatars created with the usage artificial intelligence are one of the most modern tools supporting marketing on various communication platforms.

The usage of avatars in contemporary marketing is presented in various sources in a selective and fragmentary manner. The use of avatars is mainly associated with such areas as customer service, sales and marketing communication. Meanwhile, contemporary marketing goes far beyond these areas and includes a wide range of activities

Published by Springer Nature Switzerland AG 2026
M. Pondel et al. (Eds.): AI4KAM 2025, IFIP AICT 776, pp. 276–291, 2026.
https://doi.org/10.1007/978-3-032-18920-2_27

concerning: research and analysis of the marketing environment, selection of target markets, marketing mix and management of marketing activities. There is a lack of studies containing a comprehensive review of all these areas of marketing activities in the aspect of using avatars in them. This article tries to fill this research gap. The authors review all the most important areas of marketing activity, indicate the current and potential possibilities of using avatars in these areas and formulate important principles related to this sphere. For example, properly conducted marketing activity requires the integration of avatars with other means and forms of marketing communication conducted by the company. The aim of the article is to present the principles and applications of avatars in marketing as a whole consisting of four logically related parts (marketing environment research and analysis, selection of target markets, marketing mix, management of marketing activities). It should be emphasized that these parts constitute the quintessence of modern marketing and none of them can be omitted in the processes of its design and implementation into practice.

The article is important both from a scientific and practical point of view. It organizes the issue of using avatars from the perspective of marketing as a whole and indicates creative ways of selecting and using them in marketing individual sub-areas.

For the sake of developing this article, non-reactive methods were selected. The desk research method consisted of analysing secondary sources and the content of messages disseminated through the media. The research procedure consisted of searching for information, collecting and analysing the obtained data. This was done based on the methodology of a systematic review of literature according to the criteria: basic database, key words, inclusion and elimination criteria [Czakon, 2011]. The content of documents updated on an ongoing basis was compared with previously obtained materials. The sources of information that were used to write this article are interdisciplinary in nature. This is primarily scientific literature in the field of technical information technology, marketing and management.

2 The Main Areas of Modern Marketing

Marketing is the activity of recognizing the needs and requirements of customers and then satisfying them through products and services. Four major market forces influence marketing in this sense, posing a serious challenge to modern companies: technology, globalization, the physical environment and social responsibility [Kotler et al., 2025]. Avatars have emerged as a result of the first force, namely modern technology, which enables their creation and operation. The market has become more horizontal, inclusive and social. Information circulates at an extraordinary speed and those who were previously considered the recipients of marketing campaigns are now increasingly becoming their co-creators. [Kotler and Stigliano, 2022]. Organizations that understand the new rules of marketing are building direct relationships with consumers [Scott, 2022]. The most important areas of modern marketing are presented in Table no. 1.

Modern marketing consists of four main areas. The first of them is the study and analysis of the marketing environment. In order to properly study and analyse the environment, companies create marketing information systems (MIS). It is a framework for the day-to-day management and structuring of information gathered regularly from sources

Table 1. Main areas of modern marketing.

Research and analysis of the marketing environment	Selection of target markets	Marketing mix
Information about consumers Information about competitors Information about other stakeholders Information about trends occurring in the company's immediate and distant environment	Personalization of market offers (individual marketing) Market segmentation Positioning of market offers	Product and service decisions Pricing decisions Distribution decisions, Communications decisions containing: advertising, public relations, sales promotion, direct marketing personal selling, digital marketing/social media
Management of marketing activities		

Source: own eleboration

both inside and outside the organization [Pride and Ferrell, 2016]. MIS consists of the following main subsystems: the internal documentation subsystem of the company, the marketing research subsystem, the marketing intelligence subsystem, marketing decision support subsystem (statistical techniques and decision models). The marketing information system plays a vital role in linking the customer to the marketer [Elliott et al., 2012]. Information collected through MIS about consumers, competitors and other components and forces operating in the environment is processed, analysed and disseminated among marketing employees, helping them to make appropriate decisions.

The second very important area of marketing activities is the selection of target markets. Customers vary widely in their needs and preferences, whether real or perceived. Their desires even vary for basic commodity products [Palmatier and Sridhar, 2017]. Modern companies, due to the huge diversity of needs, requirements and attitudes of consumers, do not serve the entire market but make segmentation of it. Using segmentation, they are able to divide large heterogeneous (differentiated) markets into smaller parts (segments) and thus more effectively and efficiently meet the needs of the consumers who create them [Armstrong and Kotler, 2020]. More progressive is individual marketing, which means conducting highly personalized marketing activities aimed at meeting the unique, individual needs of customers. Such personalization is possible thanks to artificial intelligence. AI is the fastest growing information technology. AI-powered companies are creating teams of data champions, modernizing digital infrastructure and acquiring large data sets to train and test models [Nitin and Mittal, 2023]. Marketing in the range from segmentation to individual one is complemented by brand positioning. Positioning is the act of designing the company's offer and image in such a way that it takes a distinct place in the minds of the target market [Kotler and Keller, 2021].

The third major area of marketing activity is the shaping of market offers, called the marketing mix. In the classical sense, it includes the following four elements: product and service decisions, pricing decisions, distribution decisions, communications decisions [Hollensen, 2015]. Product decisions are crucial to a firm's success as the basic purpose of any marketing activity revolves around satisfying the customers' needs in a superior manner vis-à-vis competitors [Joshi, 2014]. Pricing comprises all decision-making aspects regarding the price to be paid by customers for a given product or service [Homburg et al., 2013]. The purpose of distribution is to deliver the purchased goods to the customer at the right time, to the indicated place and in the ordered quantity meeting the quality requirements. The fourth component of the marketing mix is the integrated promotion mix, which consists of: advertising, public relations, sales promotion, direct marketing personal selling, digital marketing/social media [Andrews and Shimp, 2018]. These tools are not mutually exclusive and may be used concurrently in the firm's attempts to achieve its communications objectives [Farrell, 2015]. The mantra of integrated communication is to send the right message to the right people at the right time in the right place for the right reason [Juska, 2018]. Expertly placed media efforts and the use of computer databases play a big role in getting the message to the right target audience, as does the development of a consistent visual and verbal image for the product or service [Blakeman, 2018].

The last, fourth area of marketing refers to the management of marketing personnel. Management is a set of activities including planning and decision-making, organizing, leading and controlling directed at the resources of the organization and performed with the intention of achieving its (organizational) goals in an efficient and effective manner [Griffin, 2022]. In the case of marketing, this way defined management will concern the human, material, monetary and information resources which are in company's sphere.

Modern marketing is an "outside-in" orientation. Its starting point is customer needs and requirements. Entrepreneurs first focus on identifying customer needs and requirements using marketing research and then strive to effectively satisfy them through their offerings. Information about customer needs and requirements forms the basis for designing the marketing mix, including the appropriate shaping of products and services, their pricing, distribution, and integrated promotional mix. When shaping the marketing mix, it is also necessary to consider the actions of competitors (to sell, products and services must be better than competing offers) and the macro and micro trends occurring in the market (products and services from declining, declining industries will not sell well). This "outside-in" marketing orientation is fundamental to achieving marketing goals. Marketing that is not based on this orientation can be called pseudo-marketing or defective marketing. Unfortunately, many companies still rely on these forms of marketing. A prime example is the use of sales orientation. In sales orientation, entrepreneurs focus on aggressive selling and promotional activities rather than on customer needs and expectations. The company's goal is to sell what it does, not what the market wants. This is an "inside-out" perspective. Its starting point is the sale of manufactured products and services. The consequence of this approach can be customer dissatisfaction. A dissatisfied customer reflects poorly on many people around them, including through social media. This creates a paradox: the company invests significant resources in advertising, promotion, and aggressive selling, while simultaneously losing customers.

In contemporary marketing, various trends are emerging. The most important megatrends include digital marketing, multisensory marketing, and sustainable marketing. Each of these trends emphasizes different aspects of marketing activities, which we will briefly discuss later in this section. However, we would like to strongly emphasize that no trend will yield the expected results without respecting the activities encompassed by the marketing areas described earlier. These areas constitute the foundation of all marketing activities.

Digital marketing is experienced by every internet user with a computer and smartphone. Its development is closely linked to technological advances in devices and techniques for transmitting and processing information. These devices and techniques create new opportunities for influencing consumers. We'll use proximity marketing and social media marketing as examples. Proximity marketing primarily serves market research purposes. It allows retailers to obtain valuable information about consumer behavior, such as time spent in the store, routes taken, and their reactions to promotional activities. Proximity marketing utilizes location-based technologies, including geofencing (a method of establishing a virtual fence around an area), Wi-Fi, and Bluetooth beacons to locate customers by connecting to their mobile phones. Artificial intelligence-based facial recognition technology is also used to determine a customer's presence in a specific location. Face-detection cameras can identify demographic profiles, as well as their moods and emotions based on facial expressions and body language. Social media marketing encompasses a broad spectrum of activities utilizing content marketing, storytelling, newsjacking, and other techniques on platforms such as Facebook, YouTube, LinkedIn, Platform X, blogs, and online forums. Each of these platforms offers a variety of ways to communicate with a wide audience.

The main premise behind the emergence and development of multisensory marketing is that consumers are becoming less responsive to marketing messages that appeal to their senses of sight and hearing. Multisensory marketing aims to activate the consumer's other primary senses: smell, taste, and touch. The mechanism behind multisensory marketing is simple. A product, service, or store is equipped with appropriate attributes that serve as a source of external stimuli (appearance, sound, smell, taste, surface). These stimuli trigger responses from customers' sensory receptors (eyes, ears, nose, tongue and palate, skin). The customer's experience should be sensory satisfaction, not discomfort. For example, consumers' tactile experiences are influenced by: material and surface, temperature and weight of objects, shape and softness. Various supporting devices are used in multisensory marketing, such as scent monitors, scented billboards, and interactive floors.

Sustainable marketing integrates existing trends such as ecological marketing, green marketing, environmental marketing, and social marketing. It is based on the principles of sustainable development and aims to support this development. This support involves contributing to the reduction of ecological threats (depletion of non-renewable natural resources, pollution of the atmosphere, soil, and water, loss of biodiversity, and others) and social threats (social ills, exploitation, poverty, and deprivation, and others). Sustainable marketing aims to contribute to the transformation of consumer society into a sustainable society. A sustainable society is based on sustainable consumption, represented by sustainable consumers, and sustainable production, represented by sustainable producers.

In the further part of the article, suggestions for selecting and creating avatars for marketing purposes will be first presented, and then description of the usage of avatars in marketing individual sub-areas will be made, indicating new possibilities.

3 Avatars as an Element of the Metaworld

The metaverse, or metaworld, is the next stage in the evolution of the internet. The term denotes a virtual, three-dimensional world combining elements of physical and digital reality, in which consumers can reside. The metaworld has its roots in the gaming industry, but the concept can be traced back to American writer Neal Stephenson's 1992 science fiction novel "Blizzard." The novel's action takes place in a virtual world where users can take the form of avatars. Thus, avatars have been associated with the metaworld since its conception and have accompanied its development. The metaworld encompasses a set of digital spaces comprised of elements such as artificial intelligence (AI), cryptocurrencies, non-fungible tokens (constituting a digital certificate of authenticity, stored in blockchain technology), and XR (Extended Reality), which is co-created by the following components.

- Augmented Reality (AR) – physical experiences can be enhanced with additional digital content. Images from the real world are captured by a camera and synchronized with computer graphics and 3D animations. By using goggles connected to a computer, console, or smartphone, a person begins to perceive a reality other than the one they physically inhabit.
- Virtual Reality (VR) – completely transports people into the digital world and creates the illusion of being in another place and/or time. VR goggles can have their own, permanently integrated display or allow for the placement of a smartphone as a display. Thanks to sensors such as accelerometers, gyroscopes, magnetometers, and various controllers, specific actions can be performed within the viewed world, unlike augmented reality.
- Mixed Reality (MR) – connects both worlds and enables deeper interactions. Digital objects that coexist with physical ones can, for example, be obscured or illuminated by real ones.

The metaverse is inspired by human imagination, yet simultaneously transcends the limitations of the real world. It enables digital immersion, or deep immersion in a virtual world of illusion. People, represented by avatars, can engage in immersive social, professional, recreational, and other interactions. The clear positioning of avatars within the metaverse's structure stems from the division of its components into the categories indicated below. [Kotler et al., 2024].

- Virtual resources in the form of a virtual environment and virtual objects.
- Digital avatars that represent users.
- User experience in the metaworld as an interaction between virtual assets and avatars.
- Creator economy necessary to create, buy and sell digital assets.
- Zarządzanie mające na celu tworzenie zasad i nadzorowanie różnych aspektów funkcjonowania metaświata.

Virtual resources are part of virtual environments, usually three-dimensional. They can represent urban spaces, shopping malls, stores, and customer service offices—places where companies can influence consumers (represented by avatars) using various forms of marketing communication.

Of particular importance from a marketing perspective in the metaverse are resources such as virtual stores, virtual products offered to consumers, and virtual marketing communication media, examples of which include outdoor and indoor advertising (advertising screens, billboards, three-dimensional forms of advertising, advertising banners, advertising columns, and others).

An avatar reflects the user's digital identity. This identity can be real, reflecting the appearance, personality, and behavior of the person who created it, or fake, reflecting their aspirations and fantasies. The avatar navigates the metaverse under the user's control. It uses digital resources and can therefore purchase various products and services offered in the metaverse. Such purchases are often influenced by various types of marketing stimuli. We discuss avatars in more detail in the following sections of this article.

User experiences should be secure enough to allow them to interact simultaneously with virtual assets and avatars. Designing such experiences requires consideration of ethical principles and laws. It is crucial that users' experience in the metaworld aligns with their experience in the world, enabling them to utilize its space. In the event of a split personality, the user is different, different if the world exists, and different according to the rules of their metaworld. In the world, they cross the road, but in the metaworld, they do what comes from the red light. Of course, such an event can occur through mistakes, even leading to an accident.

The creator economy primarily refers to markets available in metaworlds. These markets enable the trading of digital assets in the form of products, real estate, land, and other goods and services. Transactions in such markets are often conducted using real money, which is converted into virtual in-game currency in centralized metaworlds or cryptocurrency in decentralized metaworlds. The creator economy allows companies to generate revenue, the amount of which is largely dependent on marketing activities conducted in the metaworld.

From a governance perspective, two types of metaverses are distinguished: decentralized and centralized. The basis for this distinction is the method of metaverse management. Decentralized metaverses are managed by user communities. Centralized metaverses are managed by individual companies that created or acquired them and continue to develop them.

The metaworld is an area where immersive marketing can be implemented. Immersive marketing focuses on creating immersive experiences that bridge the physical and digital worlds. Immersive marketing aims to engage consumers more deeply with the brand, increase brand loyalty, and create stronger emotional bonds. All of this is done to encourage consumers to frequently purchase the brand's products and services and spread positive reviews of the brand. To achieve these goals, immersive marketing relies on modern technologies (artificial intelligence, the Internet of Things, augmented reality, and virtual reality), creating engaging and, in many cases, personalized customer experiences.

Using immersive marketing is fraught with challenges. First, creating virtual environments and conducting effective marketing within them requires significant financial investment. Second, immersive technologies can be unreliable, including both hardware and software. Third, effective protection of user personal data is crucial.

4 Conditions for Selecting Avatars for Marketing Purposes

As researchers explain, the term avatar comes from Indian mythology (from the Sanskrit *avatāra*) designating the descent to Earth of the god Vishnu and his ten incarnations. As a term of information technology, the avatar is defined as a static or animated graphic representation (often a character) of the internet user or a video game [Marion and Poncin, 2013]. An avatar is an interactive digital entity with an anthropomorphic appearance, controlled by a human or software. Avatars differ in appearance and behaviour, which is related to the realism of their form and the realism of their behaviour (behavioural realism). Form realism refers to the degree to which the avatar's shape resembles a human creature and includes elements such as spatial dimension (2D/3D), movement (static or dynamic) and human characteristics, e.g. gender, name. Behavioural realism expresses the similarity between the behaviour of an avatar and the behaviour of a human in the real world. It concerns such issues as: avatar communication (verbal or nonverbal), type of response (scripted or natural), social content, controlling entity. Avatars can be divided into the following categories: simplified avatars, superficial avatars, intelligent unrealistic and digital human avatars.

First category avatars have an unrealistic human appearance (e.g. 2D, visually static, cartoon image) and engage in low-intelligence behaviours. Superficial avatars are characterized by a realistic anthropomorphic appearance (e.g. 3D, visually dynamic, photorealistic image) but low behavioural realism. Intelligent, unrealistic avatars are characterized by human cognitive and emotional intelligence but exhibit an unrealistic, e.g. cartoon, image of a human. Digital human avatars are the most advanced category of avatars providing the highest degree of realism during interactions with users. They are distinguished by both a highly realistic anthropomorphic appearance and human cognitive and emotional intelligence. Studies have shown that as the realism of the avatar form increases, customers' expectations regarding the realism of its behaviour increase [Miao et al., 2021]. Another division of avatars is illustrated in Table 2, which contains their description, attributes and sites.

Researchers agree that the rapid growth of the number of avatars is evidenced by the growing number of video game players using gaming avatars. This is reflected in the data contained in Fig. 1.

Researchers draw attention to the diversity of conceptualizations of avatar marketing [Miao, 2022]. Avatars are most often defined as visual entities [Elsharnouby et al., 2023] and self-representations (anthropomorphic images) of users, which play a significant role in connecting the real and virtual worlds [Huang et al., 2025]. The place of occurrence is also analyzed, e.g. digital games, chats, social networks [Wang et al., 2021). Generally, it should be assumed that Avatar marketing is the organizational use of representative visual entities with different levels of form and behavior in digital environments and aims to create, communicate, deliver and offer exchanges with value for the agents involved [de Brito Silva and de Oliveira Campos, 2024].

From marketing perspective, a high degree of form realism and behavioural realism is desirable in relation to avatars and their environments. The idea is for both avatars and their environment to be characterized by a low degree of alienation. The virtual environment should resemble the real world, the objects that make it up should function similarly to natural conditions and the movements and behaviour of the avatar should correspond

Table 2. Types of Avatars.

Type of Avatar	Description	Avatar Attributes	Sites
Static Image Avatar	Static icon of the user (image)	Choice of the image by the user	Forums, 2D social network, Online merchant site
Generic Avatar	Graphic representation of the user	Animated, controlled by the user (moves, gestures)	3D gallery, Virtual visit universe
Gaming avatar	Graphic representation of the gamer	Animated, controlled by user (moves, gestures)	Online simulation games
Virtual World Inhabitant Avatar	Graphic representation of the user	Animated, controlled by the user (moves, gestures)	2D or 3D virtual worlds, metaverse, 3D social networks
Virtual Model	Graphic representation of the consumer	Animated. Simple control by model rotation	Apparel online shopping websides

Source: own elaboration based on: [Aytaç, 2023]

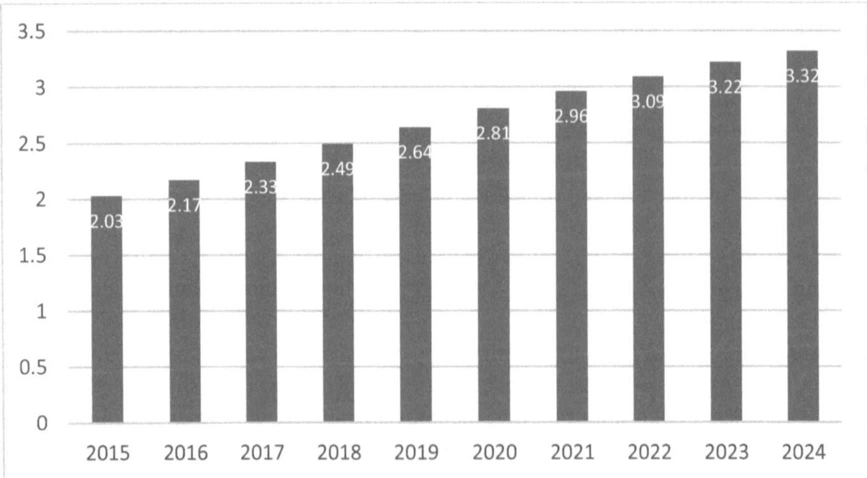

Fig. 1. Number of video game players worldwide in billions. Source: own elaboration based on: [Aytaç, 2023].

to human nature. It is because, products and services sold using avatars will be used by customers in the real world, satisfying their specific needs. Customers will consciously or subconsciously transfer their experiences of products and services from the virtual world to the real world. Customers must be convinced that a product or service which has proven its effectiveness in the virtual world will also prove effective in the real world.

Items that worked in a world with a very high degree of alienation, e.g. in interstellar space, may be considered by consumers as useless in the real world. Thus, conducting marketing activities in a completely unreal, abstract virtual environment using unreal, abstract beings in the form of avatars may not bring the expected marketing results. Of all the previously mentioned categories of avatars, the most useful in marketing are digital human avatars. They are the most technologically advanced so they are able to communicate with consumers in a way that is reasonably satisfactory to them. There is nothing worse for a customer than long, frustrating, unproductive communication with a simplified avatar with a low level of intelligence. In marketing, it is very important to satisfy or even delight customers, and this applies not only to the products and services offered to them but also to the entire service sphere, including communication.

Marketing with the usage of avatars is mainly aimed at the younger generation of consumers, especially Generation Z (born in 1995–2010) and Generation Alpha (born in 2010–2025). These generations grew up in an era of widespread access to the Internet and modern technologies. The people who make them up are digital natives, netizens for whom the physical and digital worlds are inextricably linked. They prefer short, personalized content and have high selective attention. They value authenticity, diversity, and inclusiveness in the brands they interact with [Kotler et al., 2024].

Generations Z and Alpha consider the internet an integral part of their lives. They are constantly connected to the internet, primarily via smartphones, and use it in various locations—home, school, in stores, on the street, and on public transportation. Their offline activities are interspersed with online activity. They prefer short, visual content, skipping irrelevant news and most online advertisements. Although many Generations Z and Alpha do not understand AI technology, they expect to receive personalized content—context tailored to their individual needs, preferences, and behaviors. Generations Z and Alpha enjoy immersing themselves in metaworlds, as evidenced by participating in various games. They use avatars and strive to enhance them by purchasing various items. They are skilled at using augmented reality (AR) and virtual reality (VR) technologies. The blurring of the line between the physical and digital worlds is evident, among other things, in their shopping journeys, which often take the form of showrooming and webrooming. Showrooming involves starting your shopping journey in a physical store and finishing it in an online store. Webrooming involves the reverse process: you begin your shopping journey online, exploring the online stores' offerings, and then finishing it in a physical store.

The pragmatic nature of Generation Z and Generation Alpha influences their relationships with brands. Generations Z and Alpha prefer brands that align with their values. Individuals within these groups are inclined to support brands that incorporate sustainability principles into their operations and employ ethical practices. They also value brands that prioritize diversity, equality, and inclusion in their organizational culture. Generation Z and Generation Alpha urge brands to embrace realism. For example, these generations reject advertising for products and services based on unrealistic perfection. They choose brands that embrace imperfections. It's also important that younger generations prefer to view real-life scenes rather than staged content, which is an important guideline for creating advertising content. Generations Z and Alpha value individuality, self-expression, and inclusivity. They readily form friendships online and offline, placing great emphasis on harmony and synergy with those around them. [Kotler et al., 2024].

Effective marketing influence on Generation Z and Alpha requires knowledge of:

- the needs and behaviors of people who make up these generations,
- principles and procedures for marketing influence on these generations,
- next-generation digital technologies.

It is not easy to meet these requirements in a company. Close cooperation between marketing and IT specialists is necessary.

Creating and using avatars goes far beyond the realm of marketing. Avatars have mannerisms, skills and wardrobes created by humans. They do not only buy things and services in virtual stores, but also receive gifts from other avatars, learn and earn money by participating in games. They can drive cars, fly, teleport, and do many other things in virtual world. Although in most worlds avatars do not eat, sleep, or use the bathroom, they can marry, adopt children of other avatars, fight for ownership of land or other virtual property [Hemp, 2006].

Second Live is considered the first metaverse to use avatars extensively. This still-active, free of charge, virtual world was launched in 2003 by Linden Research, Inc. It covers 50 virtual square miles and almost everything in it has been created by its residents. In addition to thousands of structures, physical landmarks, and interactive objects, it features virtual businesses, community groups and scheduled events ranging from dance parties to celebrity book signings, boxing matches and garage sales [Hemp, 2006].

In Second Live it is possible to buy various things including land, clothes, gadgets, visit friends and also travel to different places. Trade is an integral part of this world. Second Live with a population of about 2 million of inhabitants is therefore a great place for marketing.

There are many AI avatar apps that can be used for marketing. Examples include Replika, ChatGPT by OpenAI, Loom.ai (offers high-quality AI-powered avatars for virtual worlds, games and social media), Wolf3D (allows to create personal avatars for games and VR using a combination of machine learning and 3D scanning), AR Emoji Samsung, Zepeto (allows to create 3D avatars of yourself using AI) [Speechify Homepage].

4.1 Avatars in the Subareas of Modern Marketing

Web 2.0 and Web 3.0 have already enabled greater immersion of Internet users among bytes, up to virtual characters represented by avatars (Leung et al., 2022). From marketing perspective, avatars can, both represent the supply side of the market, i.e. producers and sellers of products and services, as well as the demand side, which is created by consumers - potential buyers of these products and services. In the first case, avatars are created and introduced to the virtual world by companies. In the second case, they are created by consumers, becoming the object of marketing interest of these companies.

From the perspective of companies, earning money from avatars starts at the stage of their creation. On the Internet, there are not only platforms offering free creation of avatars but also sellers who charge for this process. An example is the Chinese company

Tencent, which creates deepfakes of virtual people for $145. An avatar created in this way perfectly reflects the user's character and also speaks Chinese or English to a limited extent.

Taking into account the areas of marketing activity listed in the Table 1, avatars can be used in the following areas: research and analysis of the marketing environment, selection of target markets and marketing mix. Avatars intended for customer service can be additionally used to collect information for the needs of the areas indicated above. AI avatars allow for 24/7 customer service, providing quick answers to questions and facilitating access to information [WAY To AI Homepage]. Developing a personalized script allows avatar to answer questions, provide advice and conduct dialogues. Avatars can be integrated with many platforms, including websites, social media and mobile applications. On websites, avatars can act as advisors, shopping assistants, guides and on social media, they can respond to comments and inform about promotions [WAY To AI Homepage]. These interactions with consumers usually take place in real time. Thus, the customer service avatar can be used to gather information useful in the three areas of marketing activity listed below.

1. Research and analysis of the marketing environment: what are the preferences and requirements of customers? How do customers evaluate competitors' offers (marketing intelligence)?
2. Selection of target markets: what characteristics do consumers have in terms of demographic, economic, geographic and social criteria (market segmentation)? what are their individual desires, needs and requirements (individual marketing)?
3. Marketing mix: how products and services offered to consumers can be improved (shaping products and services)? How do they evaluate company's pricing programs (shaping prices)? Are they satisfied with the distribution service (shaping distribution)? How do they react to integrated marketing communications activities (shaping communication with the market)?

Information acquired by avatars and processed by AI can significantly contribute to improving market offerings and communication with consumers.

It is not always necessary for an avatar to conduct a dialogue with a consumer in order to learn something about him. From the point of view of research and analysis of the marketing environment, the avatar itself (its clothing, make-up, hairstyle, accessories, objects it surrounds itself with) is a rich source of information about the user who created it. By creating an avatar, the consumer equips it and its surroundings with branded items, e.g. Adidas shoes, Karl Lagerfeld clothing, a BMW car. This allows to discover who the consumer is and what brands he prefers. This facilitates market segmentation and the creation of personalized offers and communication.

Customer avatars are also created for the purpose of researching and analysing marketing environment and selecting target markets. Such an avatar is a fictional character that represents the ideal buyer. The ideal customer should make repeat purchases, be loyal to the company and recommend it to other customers. He buys the company's core products, not only use peripheral offers [Jhajharia, 2025]. To build such an avatar, it is necessary to know customer in depth. It is worth creating as many avatars as there are key customer groups of the company. [Kosienkowska, 2025].

Avatars are most often used in the third area of the marketing mix, called the integrated promotional mix. Let's start with sales as one of its elements.

The sale or free distribution of products begins already in the avatar creation phase. Companies offer to their creators branded clothing, cosmetics and accessories. For example, Maybelline and L'Oreal Professionnel have partnered with avatar creation platform Ready Player Me to offer users a choice of five makeup looks and trendy hairstyles in the metaverse. Fashion brands like Nike and Gucci have also introduced outfit customization in the metaverse. Avatars can be shaped to become brand ambassadors. Virtual or fictional influencers are an example of this. Maisie Williams, an actress who became H&M's Global Sustainability Ambassador, was transformed into a digital avatar and interacts with shoppers in the virtual world. The fictional influencer is FN Meka, an American virtual rapper with over a billion views on Tik Tok [Kotler et al., 2024]. The brand ambassador's avatar must be carefully matched to the brand's values and aesthetics.

In the mataworld, an avatar can take the form of a salesperson in a virtual store that interacts with avatars who are customers. Companies develop digital replicas of their stores that are inserted into the mataworld, where one can make purchases. Avatars can run three-dimensional, virtual stores selling real products. Avatar sellers help buyers to find products that are later sent to their home addresses in the real world. Companies run by avatars also sell services such as clothing design, furniture design, detective services, real estate trading and advertising. The avatar, as a customer, browses products from different brands, can try them on or try them on and also seek advice from his avatar friends.

An important role in the virtual world is played by retail avatars and virtual fitting rooms. Avatars can precisely reproduce customer's size. A realistic 3D avatar based on the actual dimensions of body is created based on a photo taken by the consumer with a mobile phone. To make the fit even more realistic, it is worth providing height and weight. Thanks to the 3D avatar, the customer tries on clothes and sees how they fit on the body. He can also assess the quality of the material and its texture. The virtual fitting room market (VFR) is expected to reach $10 billion in 2027, of which 42 percent will be for clothing [Autopay Homepage].

Companies are also creating their own branded virtual worlds with avatars. An example of such worlds is Coca-Cola Coke Studios.

The use of company-created avatars in other areas of the integrated promotional mix may take the form of examples listed below.

1. Advertising – participation of avatars in online advertising informing about a new product introduced by the company to the market.
2. Sales promotion – avatar offers customer a price discount when purchasing a product.
3. Public relations – avatar informs customer that the company has received the Consumer Laurel, which is a recognition of its innovative products.
4. Direct marketing – the avatar offers to sell the product directly from the manufacturer without the involvement of intermediaries.

Avatars can be also used in various types of virtual events for the sale or promotion of products and services. Examples include fashion shows, hairdressing, and makeup shows with participation of avatar models.

Using avatars in the above areas of marketing increases the attractiveness of the message and the visibility of the brand. It is an innovative way to attract the attention of recipients and create a modern brand image.

5 Conclusions

In the digital era, companies use new technologies to effectively communicate with customers and meet their needs as well as to communicate to environment that they are modern and innovative. Avatars are becoming an integral element of marketing activities, supporting the development of the company and strengthening the brand image. At the same time, avatars and the virtual worlds in which they function are not a well explored area of conducting marketing activities.

It is important to consider the use of avatars in marketing not through the narrow prism of customer service, sales or communication, but much more broadly, taking into account the full scope of marketing, including: research and analysis of the marketing environment, selection of target markets, marketing mix and management of marketing activities. Only then avatars become effective and efficient in their operation. For example, directing advertisements with avatars to consumers without marketing research, i.e. without knowledge of how they react to advertisements and how they perceive their content and visualization will probably not bring the expected results.

Marketing with the usage of avatars will be only effective in relation to certain market segments. The most promising segment is made up of people born after 1991, i.e. Generation Z and Alpha. These are people raised in an era of universal access to the Internet and modern technologies, who are proficient in using smartphones, tablets, and computers. They use these devices for many hours a day, so they are an excellent marketing target for companies using avatars.

Not all types of avatars are suitable for use in marketing. We recommend using the most technologically advanced avatars, which include digital human avatars. Their use is particularly important in areas where there is a direct dialogue between the avatar and the consumer in real time. This can occur in many situations, such as: solving technical problems of the customer, providing advice, answering questions. Contact with a simple avatar with a low level of intelligence will be a frustrating experience for the consumer, causing his dissatisfaction.

Avatars can be used in almost all areas of marketing. They are particularly useful in the following three areas: avatar as a source of information about the consumer, avatar as a seller, avatar as an issuer of marketing messages.

Avatars used to obtain information about consumers work in the area of research and analysis of the marketing environment, being an element of the marketing information system, and in particular the marketing research subsystem and the marketing intelligence subsystem. Their activities in this area have an impact on the selection of target markets and the marketing mix, and in particular on the proper shaping of products and services, prices, distribution and integrated promotion mix.

Avatars acting as salespeople can operate in 2D or 3D virtual worlds. In more technologically advanced 3D metaworlds, they meet with avatars – customers in digital replicas of real stores, offering them various products and services.

Avatar, as an issuer of marketing messages, operates in the sphere of advertising, public relations, sales promotion and direct marketing, achieving goals set by the company.

It is obvious that artificial intelligence can also be useful in the fourth area of marketing called marketing activity management. It is able to improve such processes of this management as planning and organizing marketing activities, motivating marketing employees and controlling work and its effects. In these cases, should it be represented by a human or other figure of avatar or avatars? Certainly yes, if the company inserts a replica of its business entity into the metaworld, and marketing will become its functional area of activity equipped with virtual resources.

Avatars should be integrated with the rest of the promotional mix, affecting the recipients holistically. Lack of integration results in a tangle of different messages flowing in different directions and to different audiences. Such messages can contradict each other, weakening the credibility of the company or brand and misinforming the recipients.

Disclosure of Interests. The authors have no competing interests to declare that are relevant to the content of this article.

References

Aytaç, Z.: Avatar based marketing, In: Baş M., Tarakçi I.E., Aslan R. (eds.) The essentials of today's marketing, pp. 381–399. Efe Akademik Publishing, Istanbul (2023)

Andrews, J.C., Shimp, T.A.: Advertising, promotion, and other aspects of integrated marketing communications, Cengage, United States (2018)

Armstrong, G., Kotler, P.: Marketing. An introduction, Pearson Education, Polish edition (2020)

Autopay Homepage, Sales avatars as an e-marketing tool, https://autopay.pl/baza-wiedzy/blog/ecommerce/awatary-handlowe-jako-narzedzie-emarketingu. Last accessed 26 January 2025

Blakeman, R.: Integrated marketing communication. Creative strategy from idea to implementation. Rowman & Littlefield, London (2018)

De Brito Silva, M.J., De Oliveira Campos, P.: Past, present, and future of avatar marketing: a systematic literature review and future research agenda. Computers in Human Behavior: Artificial Humans **2**, 1–17 (2024)

Czakon, W.: Methodology for a systematic literature review. Organizational Review **3**(854), 57–61 (2011)

Elliott, G., Rundle-Thiele, S., Waller, D.: Marketing. John Wiley & Sons, Australia (2012)

Elsharnouby, M.H., Jayawardhena, Liu, C.H., Elbedweihy, A.M.: Strengthening consumer–brand relationships through avatars. The J. Res. Indian Medicine **17**(4), 581–601 (2023)

Farrell, C.: Global marketing. Practical insights & international analysis. Sage, Los Angeles (2015)

Griffin, R.: Fundamentals of organizational management. PWN, Warsaw (2022)

Hemp, P.: Avatar- based marketing, https://hbr.org/2006/06/avatar-based-marketing. Last accessed 05 February 2025

Hollensen, S.: Marketing management. A relationship approach. Pearson, New York (2015)

Homburg, C., Kuester, S. Krohmer, H.: Marketing Management. A contemporary perspective. McGraw-Hill, London (2013)

Huang, J., Huang, M., Zhan, M., Guan, D.: The effect of the realism degree of avatars in social virtual worlds: the perspective of self-presentation. Information & Management **62**(7) (2025)

Jhajharia, S.: Why your business needs a marketing avatar, MarketingDonut https://www.marketingdonut.co.uk/marketing-essentials/your-target-market/why-your-business-needs-a-marketing-avatar. Last accessed 19 March 2025

Speechify Homepage, https://speechify.com/pl/blog/create-ai-avatars-for-b2b/?srsltid=AfmBOorEmnaJ7ZBnGjpYbE03EwfD7OLzqqIaXCZdICvCICM3-LUF67I. Last accessed 09 December 2024

Joshi, R.M.: International marketing. Oxford University Press, England (2014)

Juska, J.M.: Integrated marketing communication. Advertising and promotion in a digital world. Taylor and Francis Group, New York (2018)

Kang, S.H., Watt, J.H.: The impact of avatar realism and anonymity on effective communication via mobile devices. Comput. Hum. Behav. **29**(3), 1169–1181 (2013)

Kosienkowska, M.: Customer avatar – learn how to get into the buyer's shoes to understand them better and sell more, https://www.corazlepszafirma.pl/blog/awatar-klienta. Last accessed 27 February 2025

Kotler, P., Keller, K.L., Chernev, A.: Marketing. Publishing House Rebis, Poznań (2025)

Kotler, P., Stigliano, G.: Retail 4.0. 10 Rules of retailing in the digital age. PWE, Warsaw (2022)

Kotler, P., Keller, K.L.: Marketing management, Pearson, Rebis, Poznań (2021)

Kotler, P. Kartajaya, H. Setiawan, I.: Marketing 6.0. The Future is Immersive. MT Biznes, Warszaw (2024)

Leung, F.F., Gu, F.F., Palmatier, R.W.: Online influencer marketing. J. Acad. Mark. Sci. **50**(2), 226–251 (2022)

Miao, F., Kozlenkova, I.V., Wang, H., Xie, T., Palmatier, R.W.: The rise of avatar marketing: shaping the future of digital branding. American Market. Ass. J. Mark. **86**(1), 67–90 (2025)

Marion, G., Poncin, I.: The avatar in marketing: synthesis, integrative framework and perspectives. Rech. Appl. Mark. **28**(1), 85–115 (2013)

Miao, F., Kozlenkova, I.V., Wang, H., Xie, T., Palmatier, R.W.: An emerging theory of avatar marketing. J. Mark. **86**(1), 67–90 (2022)

Miao, F., Kozlenkova, I.V., Wang, H., Xie, T., Palmatier, R.W.: The rise of avatar marketing: shaping the future of digital branding. American Marketing Association https://www.ama.org/2021/03/09/an-emerging-theory-of-avatar-marketing/. Last accessed 07 March 2025

Nitin, T., Mittal, D.: All-in on AI: how smart companies win big with artificial intelligence. MT Biznes, Warsaw (2023)

Palmatier, R.W., Sridhar, S.: Marketing strategy. Palgrave, London (2017)

Pride, W., Ferrell, O.C.: Marketing, Cengage Learning, Boston (2016)

Scott, D.M.: New rules for marketing and PR. MT Biznes, Warsaw (2022)

Wang, X., Butt, A.H., Zhang, Q., Shafique, N., Ahmad, H.: Celebrity avatar feasting on in-game items: a gamers' play arena. SAGE Open **11**(2), 1–13 (2021)

WAY To AI Homepage, https://www.way-to-ai.com/awatary-ai-nowoczesne-narzedzie-wsparcia-marketingu-i-interakcji-dla-firm/. Last accessed 05 November 2024

When Reviews Reveal User Roles: AI-Powered Prosumer Classification

Andrzej Małowiecki[1], Kamila Łuczak[1(✉)], Iwona Chomiak-Orsa[1], Andrzej Greńczuk[1], and Anna Nowacka[2]

[1] Wroclaw University of Economics and Business, Komandorska 118/120, Wroclaw, Poland
{andrzej.malowiecki,kamila.luczak,iwona.chomiak-orsa, andrzej.grenczuk}@ue.wroc.pl

[2] Faculty of Management, University of Warsaw, Krakowskie Przedmieście 26/28, Warsaw, Poland
ANowacka@wz.uw.edu.pl

Abstract. A characteristic feature of becoming a prosumer is the transformation from a passive consumer to an active customer and producer. While the topic of prosumers is widely discussed in the literature in the context of the energy sector, the prosumer is of great importance in an organizational context for all businesses. One of the reasons organizations do not address the identification of prosumers is the complexity of the process. Large language models (LLMs) can assist as a solution to this problem. Our research aims to utilize LLM (GPT-type) to identify and classify prosumers through product reviews from the Walmart store chain. We propose a framework using the GPT-4o model that the OpenAI API provides for binary classification of prosumers based on product reviews. The model achieved high precision when classifying prosumers - 100%, while achieving a low recall of 58.67%, and its accuracy was 79.33%. Further research is planned to extend the model to classify prosumers based on product reviews, indicating their specific role as promoter, influencer, tester, expert, or innovator in co-creating products and services.

Keywords: Prosumers · Co-creation · Artificial Intelligence · AI · Large Language Models · LLMs

1 Introduction

Prosumers are individuals or entities that produce and consume goods or services [14]. They may consume and produce or create value dedicated to their consumption or by others [7]. In their dual role, prosumers impact the economic processes taking place [3]. They contribute to creating new business opportunities regarding innovative consumption models [13]. In recent years, the importance of prosumers is growing through the proliferation of energy resources [4]. However, in our article, the prosumer is understood in an organizational context as a co-creator of value for the company.

Published by Springer Nature Switzerland AG 2026
M. Pondel et al. (Eds.): AI4KAM 2025, IFIP AICT 776, pp. 292–299, 2026.
https://doi.org/10.1007/978-3-032-18920-2_28

Artificial intelligence (AI), including large language models (LLMs), can be a huge facilitator for organizations when searching for prosumers among their customers. Currently, the use of AI for product personalization [11], automation in content creation [1], or prosumer decision automation [9] is indicated. On the other hand, the question to consider is how organizations could use the available technology to identify customers who are not only consumers of products and services but are also their co-creators – prosumers.

Therefore, this article aims to utilize LLM (GPT-type) to identify and classify prosumers through product reviews. Two research questions have been identified, which the article will answer:

1. How can LLMs be used to identify and classify prosumers?
2. What is the model's effectiveness (accuracy, precision, recall, F1-score) in classifying product reviews?

To achieve the goal and answer the research questions, the GPT-type LLM is used for binary classification of prosumers based on the content of product reviews (from the Walmart store chain). The structure of the article is as follows. The article begins with an introduction, presenting the aim and research questions. Then, the theoretical background, where publications on similar topics are presented, and the research gap is identified. The subsequent sections show the methodology and research results. The article ends with a discussion and conclusion.

2 Theoretical Background

Prosumers represent a significant added value for organizations. One way consumers can contribute to the co-creation of a brand is by providing valuable feedback on products and services. Developments in technology, including LLMs, have allowed a revolutionized approach to text classification, enabling organizations to identify and classify prosumers more easily.

Chan et al. [2] indicate how important the reviews left by customers are by pointing to platforms for leaving reviews as communities of prosumers. Mechanisms to induce prosumption are important for both organizations and consumers - they can increase the number of relevant reviews and the value of brand communities [2]. Shahnaz Ipa et al. [12] prepared the BdSentiLLM model, which categorizes and prepares selected languages for sentiment analysis of product reviews using LLMs such as Llama-2, Flan-t5, Vicuna, and Falcon. In the context of sentiment analysis, the researchers indicate that the Llama-2 model is superior to the other models in most language categories, with Flan-t5 showing a slight advantage in opinions in English [12]. Pandey & Singh [10] propose a framework that uses LLMs to create detailed reviews based on users' product ratings. The study's results indicate that the use of AI helps make the generated reviews readable, consistently relevant, and characterized by high-quality content [10].

Researchers indicate the value of customer reviews [2], use LLM to analyze sentiment [12], or to generate product reviews based on aspect-based evaluations [10], while no publication has been found that indicates the use of LLM to identify and classify prosumers based on the reviews they provide. Thus, this paper will fill this gap by proposing an AI-powered prosumer classification based on the original and the GPT-4o reworked Walmart product reviews. The procedure presented in the next step can enable organizations to distinguish passive customers from prosumers on their own and at low cost.

3 Methodology

The binary classification of prosumers based on the content of product reviews was conducted by performing the following steps:

1. In the previous research, a systematic literature review was conducted to identify areas of operations and characteristics of prosumers. Based on it, a knowledge base was created, including information on the type of prosumer, the area, a description of the operation, and the use of AI.
2. The knowledge base created was used to develop a uniform definition of a prosumer. This task was performed using the GPT-4o model (provided by the OpenAI API) with the following system prompt: "Based on the provided knowledge base, generate a consistent definition of a prosumer, placing special emphasis on their contribution to co-creation by creating product reviews online.". The definition is mentioned in point 4.

The dataset containing 300 reviews of products and services purchased from the Walmart store chain was found [5]. Based on this dataset, an additional 300 synthetic product reviews were generated, the wording of which was transformed to have the characteristics of prosumer reviews. This task was performed using the GPT-4o model with the following system prompt: "Your task is to adjust given product review so that it sounds like it was written by a prosumer.". This task was performed by the model using its own definition of a prosumer: "*A prosumer is typically defined as a consumer who is also a producer, often possessing a high level of expertise or interest in a particular product or field. This individual is not only a user of the product but also actively engages in its development, improvement, or customization. Prosumers are knowledgeable, often providing detailed feedback and insights based on their extensive experience and understanding. They might also influence other consumers through reviews, recommendations, or content creation. When adjusting a product review to sound like it was written by a prosumer, I would focus on incorporating technical details, expert insights, and a balanced evaluation that reflects both personal experience and a broader understanding of the product's place in the market. The review would likely include specific terminology, comparisons with similar products, and suggestions for potential improvements or alternative uses*". The two datasets (original and synthetic) were merged to create a single, balanced set of 600 (300 prosumer and 300 non-prosumer) product reviews.

Based on the dataset, a binary classification was performed to indicate whether the author of a given review is a prosumer. The GPT-4o model with the following system prompt was used for classification: "Your task is to decide whether given product review was written by a prosumer. Answer YES or NO using given prosumer definition. Prosumer definition: *A prosumer is an individual who not only consumes but also actively co-creates value by sharing detailed feedback—such as product reviews—using their personal expertise, creativity, or experience to influence others, support innovation, and contribute to digital platforms where user-generated content drives service development and customization"*. The proposed definition of prosumer finds its practical application in the analysis of consumer behavior - companies know what kind of consumer actively co-creates the value of a product or service.

The model's responses were evaluated to assess the effectiveness of the classification. The following evaluation metrics were used: accuracy, recall, precision, F1-score, and confusion matrix.

4 Results

Binary classification was performed on the Walmart Reviews Dataset, containing 300 reviews of products and services purchased from the Walmart store chain [5]. It was made publicly available via Kaggle.com with the following license: Attribution-NonCommercial 4.0 International (CC BY-NC 4.0): https://creativecommons.org/licenses/by-nc/4.0/. This dataset contains the following columns: name, location, date, rating, review, and image. The review column, which contained a product review by a customer, was selected for the study.

Since the task of the LLM model was binary classification, synthetic data was generated to provide a suitable dataset with no imbalance problem, as it can cause bias in the model and inaccurate predictions [8]. After generating synthetic reviews, the two datasets (original and synthetic) were combined into one set. The order of reviews in the set was shuffled and passed to the GPT-4o model for classification. The GPT-4o model classified the binary reviews into prosumer and non-prosumer reviews. As a result, the model correctly classified 476 out of 600 reviews. All non-prosumer reviews were correctly classified. However, as for prosumer reviews, 176 out of 300 were correctly classified. Examples of the transformation and incorrectly and correctly classified prosumer reviews are presented below (Table 1). To ensure the anonymity of brands or company names appearing in the reviews, they were replaced with the label [name]. The misclassified review lacks a clear evaluation of the product, and the focus is on the payment security incident, rather than the product itself or the shopping experience, which is described in the correctly classified review.

The results of the model's classification evaluation are presented below (Table 2). Accuracy means the percentage of all correct predictions and is calculated according to the formula (1), where TP – true positive; TN – true negative; FP – false positive; FN – false negative.

$$Accuracy = (TP + TN)/(TP + FP + TN + FN) \quad (1)$$

The F1-score, on the other hand, is a measure indicating precision (2) and recall (3) in binary classification and is calculated according to the following formula (4).

$$Precision = TP/(TP + FP) \quad (2)$$

$$Recall = TP/(TP + FN) \quad (3)$$

Table 1. Example of transformation and incorrectly vs correctly classified prosumer reviews

Original review	*Prosumer-style review created by GPT-4o (incorrectly classified review)*	*Correctly classified review*
I bought a 75 inch tv online using [name]. Right after buying the tv, my bank card was hacked and charged $700 after I just bought a tv using [name] online. Never use your bank card. ALWAYS use a prepaid card when buying online	*As a tech-savvy consumer, I recently purchased a 75-inch TV through [name] online platform. However, shortly after completing the transaction, I experienced a security breach with my bank card, resulting in unauthorized charges totaling $700. This incident underscores the importance of prioritizing online security. I highly recommend using a prepaid card for online purchases to mitigate potential risks and protect your financial information*	*I have attempted to put this review on [name] but it indicates I have been 'opted out'. I purchased [name] covid tests online to pick up at store. Good product. Product is FIVE STARS, online shopping experience is ONE STAR. Online order Sunday 8/27/23 with Store pickup on Monday 8/28/23 around 2pm. I am reviewing to caution other online shoppers to ALWAYS scroll down before adding ANY item to their [name] cart because the NOTICE if something is not returnable isn't visible unless you do. I picked up my order on the way out of town, and only later saw that the expiration dates were all close enough that I NEVER would have purchased the same amount of tests had I been shopping in-store. So I clicked 'start a return' on my receipt and only then found out I had purchased something non-refundable*

Table 2. Model evaluation results

Evaluation metric	Score (class 1 - non-prosumer)	Score (class 2 - prosumer)
Accuracy	79.33% (the same score)	79.33% (the same score)
Recall	100%	58.67%
Precision	70.75%	100%
F1-score	82.87%	73.95%

$$F1 = 2 \times (Precision \times Recall)/(Precision + Recall) \quad (4)$$

Regarding accuracy, the model correctly classifies almost 4 out of 5 reviews. Recall for class 1 is 100%, meaning that every true non-prosumer opinion is detected, while its low value for class 2 indicates that the model indicates only 6 out of 10 true prosumers. A precision for class 1 of 70.75% indicates that 29% of non-prosumer indications are false alarms, while for class 2, it is 100% - when the model classifies an opinion as prosumer, it is never wrong for a given dataset. An F1-score of 82.87% represents a good result for class 1, indicating that the model meets most operational tasks. However, around 21% of cases remain misclassified, which is still a better value than class 2. Besides calculating accuracy, recall, precision, and F1-score, a confusion matrix was additionally used for evaluation (Fig. 1).

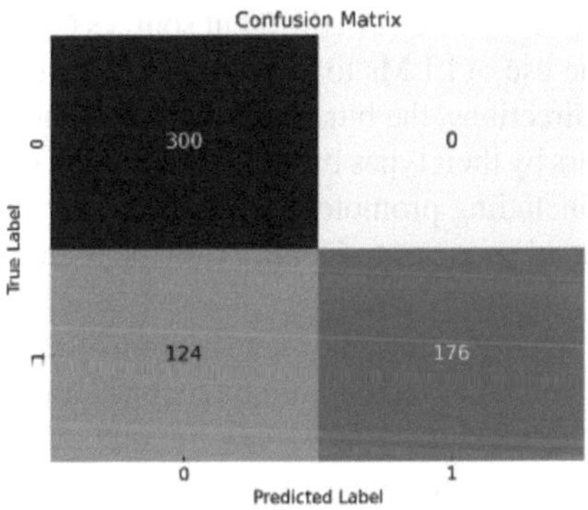

Fig. 1. Confusion matrix.

The confusion matrix indicates the number of cases in which the model correctly matches a positive class (TP = 176) - prosumer opinions are recognized as prosumer, or a negative class (TN = 300) - non-prosumer opinions are detected as non-prosumer. It indicates non-prosumer opinions wrongly labelled as prosumer (FP = 0) and prosumer opinions recognized as non-prosumer (FN = 124). In summary, the model stands out for its 100% precision for class 2, which allows it to detect any prosumer who is one. In this case, its disadvantage is its low recall, where real prosumers may be overlooked.

5 Discussion and Conclusion

The article aimed to utilize LLM (GPT-type) to identify and classify prosumers through product reviews. In answering (RQ1), it is important to point to the framework. First, having a ready-made set of publications and areas where prosumers may be active, the GPT-4o model generated a definition of a prosumer focusing specifically on their influence on writing product reviews. A dataset from the Walmart store chain with 300 product and service reviews was then found [5], which the model reworked to sound like a prosumer based on its definition of prosumer. 600 product reviews (300 prosumer,

300 non-prosumer) were developed. Binary classification was used to check whether a particular review was written by a prosumer, indicating to the model that it was not its definition, but the definition provided by the systematic literature review conducted earlier. Referring to (RQ2), it should be noted that the model achieved a very high precision (100%) when classifying prosumers, which indicated that every prosumer's opinion was labelled correctly as written by a prosumer. At the same time, the recall result is too low (58.67%). The accuracy was 79.33%, which is a good result. The analysis indicates that work should be done to increase the recall of the model by, for example, using few-shot learning [15].

A limitation of the study is the procedure that includes modifying the set of reviews using the GPT-4o model to change their tone to reviews written by prosumers. In the case of binary classification, synthetic data is generated to provide an appropriate dataset [8]. On the other hand, the use of synthetic data may be associated with an increase in model bias and result in data that does not adequately reflect the real state [8]. However, a larger and more diverse number of reviews from different sources (without modification) could allow better testing of the use of LLMs to identify and classify prosumers.

As further research directions, the binary classification model created can be developed to classify prosumers by their types based on the content of reviews. Prosumers can take on different roles, including promoter, expert, tester, or innovator [6]. The developed model can be modified and used to identify the role of the prosumer for a company based on the product or service review.

Implications for researchers include the identification of a new trend related to the use of LLMs to classify prosumers based on online feedback about products and services. Analysis of prosumer attitudes using LLM can be applied to other sectors (such as tourism or tech), providing a valuable tool for consumer engagement research. In the organizational context, this study indicates the need and value of closely analyzing the opinions customers leave. The limitations, including the problem of identifying prosumer feedback, indicate the need for companies to demonstrate to customers the value of their feedback, implement strategies that would motivate customers to leave reviews, and then switch their role from passive consumer to active prosumer.

Disclosure of Interests. The authors have no competing interests to declare that are relevant to the content of this article.

References

1. Berntzen, L., Florea, A.: The role of aggregators and hubs in collaborative prosumer networks. IFIP Adv. Info. Comm. Technol. **688**, 21–34 (2023). https://doi.org/10.1007/978-3-031-42622-3_2
2. Chan, H., Zeng, K.J., Yang, M.X.: Review platforms as prosumer communities: Theory, practices and implications. Eur. J. Mark. **56**(10), 2698–2720 (2022). https://doi.org/10.1108/EJM-10-2021-0819
3. Ertz, M., Maravilla, J.M.B., Cao, X.: Prosumer: A new approach to conceptualisation. J. Innov. Knowl. **10**(1), 100653 (2025). https://doi.org/10.1016/j.jik.2025.100653

4. Filho, W.L., et al.: Prosumers and sustainable development: an international assessment in the field of renewable energy. Sustainable Futures **7**, 100158 (2024). https://doi.org/10.1016/j.sftr.2024.100158
5. Honde, H.: Walmart Reviews Dataset. Kaggle (2023). https://www.kaggle.com/datasets/harshalhonde/walmart-reviews-dataset. Last access: 23 April 2025
6. Jelonek, D., Wysłocka, E.: Co-creation of Innovation Using the Potential of Web 2.0 Tools. TOJSAT **5**(4), 21–28 (2015)
7. Lang, B., Dolan, R., Kemper, J., Northey, G.: Prosumers in times of crisis: Definition, archetypes and implications. J. Serv. Manag. **32**(2), 176–189 (2021). https://doi.org/10.1108/JOSM-05-2020-0155
8. Małowiecki, A.: Metody resamplingu danych w rozwiązaniu problemu nierównowagi danych przy wykrywaniu oszustw związanych z kartami kredytowymi. In: Dudycz, H. (ed.), Informatyka w biznesie 2023, 1st ed., pp. 106–119. Wroclaw University of Economics and Business (2023). https://doi.org/10.15611/2023.81.7.08
9. Martín, D., Bordel, B., Alcarria, R., Sánchez-Picot, Á., de Rivera, D.S., Robles, T.: Prosumerization Approach to Semantic Ambient Intelligence Platforms. In: Ochoa, S.F., Singh, P., Bravo, J. (eds.) UCAmI 2017. LNCS, vol. 10586, pp. 109–120. Springer, Cham (2017). https://doi.org/10.1007/978-3-319-67585-5_12
10. Pandey, P., Singh, J.P.: Generating product reviews from aspect-based ratings using large language models. J. Retail. Consum. Serv. **84**, 104244 (2025). https://doi.org/10.1016/j.jretconser.2025.104244
11. Robles, T., Alcarria, R., Morales, A., Martin, D.: Supporting variability dependencies for rule-based service compositions in prosumer environments. Int. J. Web. Grid Ser. **11**(1), 57–77 (2015). https://doi.org/10.1504/IJWGS.2015.067160
12. Shahnaz Ipa, A., et al.: BdSentiLLM: A Novel LLM Approach to Sentiment Analysis of Product Reviews. IEEE Access **12**, 189330–189343 (2024). https://doi.org/10.1109/ACCESS.2024.3516826
13. Tang, W., Chen, W., Guo, C.: A price-based online energy management framework for heterogeneous prosumers without prediction. Renewable Energy **249**, 122997 (2025). https://doi.org/10.1016/j.renene.2025.122997
14. Toffler, A.: The Third Wave. William Morrow (1980)
15. Yu, J., Li, L., Lan, Z.: Beyond binary classification: a fine-grained safety dataset for large language models. IEEE Access **12**, 64717–64726 (2024). https://doi.org/10.1109/ACCESS.2024.3393245

AI-Driven Knowledge Management in Energy Enterprises: Challenges and Perspectives

Wojciech Drozdz and Krystian Redzeb(✉)

University of Szczecin, Aleja Papieża Jana Pawła II 22A, Szczecin 70-453, Polska
{Wojciech.drozdz,Krystian.redzeb}@usz.edu.pl

Abstract. This paper explores the integration of artificial intelligence (AI) into knowledge management (KM) practices within energy enterprises, highlighting both the challenges and opportunities arising from sector-wide transformation. Amid growing pressure to decarbonize, modernize legacy infrastructure, and preserve critical expertise during workforce shifts, energy companies must adopt more adaptive and intelligent approaches to managing knowledge. We first outline the contemporary challenges facing the sector, including fragmented data systems, skill gaps, and cultural resistance to AI adoption. We then examine the growing need for KM solutions that can capture institutional knowledge, support faster decision-making, and enable cross-functional collaboration. The paper reviews current applications of AI in the energy industry such as predictive maintenance, digital twins, and semantic search tools demonstrating how AI is already enhancing operational efficiency and strategic alignment. Building on these insights, we propose four key directions for further AI integration into KM: codification of tacit knowledge, context-aware knowledge retrieval, personalized learning and upskilling, and automated content governance. Taken together, these elements show how AI-driven KM can become a foundation for resilience, innovation, and long-term competitiveness in the energy sector. The article concludes with practical insights and best practices for organizations seeking to transform scattered data and expertise into a shared, strategic knowledge infrastructure.

1 Contemporary Challenges of Energy Enterprises

Energy enterprises are currently navigating a landscape marked by profound transformation, driven by regulatory, technological, and demographic shifts. At the core of this transition lies the imperative to meet decarbonization targets such as those outlined in the European Green Deal—which mandate a comprehensive shift toward renewable energy sources and carbon-neutral operations [25, 37]. Achieving these ambitious goals requires not only large-scale investments in modernizing aging grid infrastructure but also the integration of distributed energy resources (DERs), energy storage systems, and enabling real-time, dynamic energy flow management. These complex and evolving domains are precisely where artificial intelligence (AI) and advanced knowledge management (KM) frameworks can deliver transformative impact by optimizing operations, enhancing decision- making, and facilitating innovation [48].

Published by Springer Nature Switzerland AG 2026
M. Pondel et al. (Eds.): AI4KAM 2025, IFIP AICT 776, pp. 300–313, 2026.
https://doi.org/10.1007/978-3-032-18920-2_29

Simultaneously, geopolitical volatility, exacerbated by recent global events, and ongoing disruptions in global supply chains have elevated energy security and resilience to a strategic priority for energy companies. As highlighted in EY's 2022 energy transformation survey, over 67% of executives identified infrastructure resilience and operational continuity as their foremost concerns [8]. Compounding these external pressures is the widespread reliance on legacy systems that are often fragmented across departments and incompatible with modern analytics and digital tools [8]. Such legacy infrastructures not only constrain data interoperability but also limit the capacity for predictive insights and analytics, thereby impeding enterprise-wide digital transformation efforts and slowing the adoption of innovative AI-driven solutions [42].

From a human capital perspective, the energy sector faces a double-edged demographic challenge. On one hand, a significant portion of the experienced workforce is nearing retirement, threatening the loss of decades of tacit knowledge accumulated through years of operational experience and specialized expertise [1, 3]. On the other hand, while new entrants to the sector are often digitally adept, they frequently lack deep domain expertise critical to complex energy operations. This generational divide poses a significant risk to knowledge continuity and operational excellence unless effectively bridged through structured knowledge transfer programs, mentorship initiatives, and increasingly, AI-enabled KM platforms designed to capture and disseminate institutional knowledge efficiently [32, 34, 38].

Organizational culture presents additional barriers to transformation. Empirical studies consistently show that siloed knowledge environments, persistent skepticism or low trust in AI systems, and weak incentive structures for collaboration significantly slow the pace of adoption of AI-driven KM solutions [8, 9, 15]. For instance, Unisphere Research (2023) reports that 31% of energy managers were unable to name the number of KM systems actively used within their organizations—a clear indicator of fragmented knowledge governance and lack of transparency [11]. Overcoming these cultural impediments requires deliberate change management, leadership commitment, and the fostering of an open knowledge-sharing culture aligned with strategic digital goals [50].

In summary, energy enterprises must address a multidimensional set of challenges;

- Technological – fragmented legacy systems, data silos, and poor interoperability that hinder digital innovation [18, 20]
- Human – workforce aging, skill gaps, and resistance to change impeding knowledge transfer and technology adoption [1–3],
- Cultural – low engagement in knowledge-sharing practices and weak trust in AI limiting collaborative potential [8, 10, 14],
- Regulatory – evolving compliance frameworks governing AI ethics, data privacy, transparency, and explainability that add layers of complexity to AI deployment [26–28].

Together, these factors underscore the critical role of AI as an enabler of adaptive, scalable, and intelligent knowledge systems. By effectively harnessing AI-driven KM, energy enterprises can not only navigate the challenges of the present transformation but also position themselves to thrive in a rapidly evolving, data-driven future [40].

2 The Need for Knowledge Management in Energy Enterprises

The complexity and scale of modern energy enterprises generate a vast amount of operational, technical, and strategic knowledge - often scattered across departments, formats, and systems. In an industry characterized by long asset life cycles, high safety requirements, and mission-critical operations, managing this knowledge effectively is not optional - it is essential [36].

A key driver for implementing advanced knowledge management (KM) systems is the urgent need to preserve institutional expertise. As noted in recent workforce studies, the energy sector is confronting a significant retirement wave: in the U.S. alone, nearly 400,000 employees are projected to retire by 2033, while 43% of the UK's offshore energy workforce is already over the age of 45 [1]. This demographic shift puts decades of tacit knowledge at risk - knowledge that cannot be replaced by AI but can be preserved and augmented through AI- enabled KM platforms [34, 35, 39].

Additionally, energy enterprises are transitioning from mechanical and manual processes toward digitally automated environments, including smart grids, SCADA-integrated wind farms, and AI-powered decision support tools. This shift demands not only new technical skills but also systems that capture and redistribute knowledge rapidly and contextually [21, 23, 24]. Shell's initiative to train an AI chatbot using a curated corpus of over 300,000 technical documents exemplifies this shift. The project led to a ~30% improvement in Q&A accuracy, illustrating how structured knowledge enables better decision-making and faster problem resolution [27, 36].

Equally pressing is the challenge of fragmented knowledge infrastructures. Many utilities operate with multiple disconnected KM systems; Unisphere Research (2023) reports that 36% of enterprises rely on three or more such tools, with minimal coordination or data harmonization [12]. This disintegration leads to duplication, inconsistent records, and decision delays. The result is a reduced ability to respond swiftly to disruptions or to leverage historical learnings across projects. In Ørsted's case, migrating real-time data streams from wind farms to an AI- compatible cloud platform not only improved analytics performance but also supported continuous learning through a unified knowledge ecosystem [35, 41].

The need for KM is also tightly connected to innovation and regulatory compliance. AI-driven energy solutions such as predictive maintenance or digital twin simulations rely on a steady flow of high-quality, domain-specific knowledge. This knowledge must be searchable, validated, and up to date. Without strong KM systems, even the most advanced AI deployments risk being underutilized or misaligned with business needs [21, 26, 40].

Furthermore, the shift toward decarbonization and decentralized energy production introduces new complexity that requires faster onboarding, better cross-functional collaboration, and seamless knowledge reuse. Siemens Energy's GenAI initiative addressed this challenge directly: by training nearly 5,000 employees on AI and KM practices in just two weeks, the company improved generative AI skill levels by 62%, while also embedding internal AI mentors to support continued adoption [22, 46].

Lastly, knowledge management supports strategic alignment. Ørsted's use of a generative AI news digest to keep employees informed of industry developments created a daily, enterprise-wide briefing mechanism that saved time and ensured consistent strategic awareness across teams [20, 50].

In sum, the need for robust, AI-supported knowledge management in energy enterprises arises from:

- Tacit knowledge attrition due to retirements [1, 34],
- Data fragmentation and operational silos limiting system integration [12, 22],
- Increased complexity and pace of innovation requiring real-time knowledge delivery [21, 23],
- Workforce skill gaps that hinder the effective use of AI without proper KM foundations [2, 4],
- Strategic agility demands, including faster learning cycles and coordinated execution across geographies [6, 7].

As the energy transition accelerates, the enterprises that treat knowledge as a strategic asset actively captured, shared, and enhanced through AI will be better positioned to operate safely, innovate effectively, and compete globally [36].

3 Current Applications of Artificial Intelligence in the Energy Sector

Artificial Intelligence (AI) has emerged as a fundamental enabler in the ongoing digital transformation of the energy sector. Its deployment spans the entire energy value chain—from upstream exploration and production activities, through transmission and distribution grid management, to downstream customer engagement— facilitating substantial improvements in operational efficiency, cost optimization, and strategic agility.

The case studies presented in this paper were carefully selected based on three methodological criteria:

(1) the maturity and scale of AI deployment within the organization,
(2) the availability of measurable outcomes such as cost reductions, efficiency gains, or improved reliability, and
(3) the documented connection of these initiatives to knowledge management practices.

Each case was cross-verified with peer-reviewed studies, industry reports, and technical documentation to ensure credibility and reproducibility. This structured approach enhances transparency and provides a robust foundation for comparing AI-driven knowledge management strategies across enterprises [49].

The AI models deployed in these case studies span a variety of machine learning techniques. For predictive maintenance, supervised models such as gradient boosting and random forests have been combined with anomaly detection algorithms to detect early signs of equipment failure [36].

Current Applications of Artificial Intelligence in the Energy Sector
Artificial Intelligence (AI) is transforming the energy sector across multiple operational layers, from predictive maintenance to advanced knowledge retrieval and upskilling. In this section, we organize the analysis around three methodological pillars:

(1) Predictive Maintenance and Asset Optimization,
(2) AI-Powered Knowledge Retrieval and Chatbots,
(3) AI-Driven Learning and Governance Automation.

This structure highlights the distinct functional areas where AI enhances knowledge management (KM) and operational efficiency [36].

3.1 Predictive Maintenance and Asset Optimization

Objective: Ørsted aimed to optimize operations and maintenance (O&M) costs, increase energy output, and enhance decision-making through AI-based forecasting integrated with SCADA systems.

AI Model: Long Short-Term Memory (LSTM) networks combined with cloud-based machine learning analytics for time-series prediction.

Dataset: Over 5 TB of SCADA data collected during five years of wind farm operation.

Metrics & Validation: The predictive model achieved 85% forecasting accuracy (validated via 5-fold cross- validation and backtesting), leading to a 15–25% reduction in O&M costs and a 10–20% increase in annual energy output.

KM Impact: By integrating AI forecasting with historical knowledge bases, Ørsted developed a continuous learning ecosystem, enabling cross-departmental insights and proactive maintenance scheduling [48].

3.2 AI-Powered Knowledge Retrieval and Chatbots

Objective: Shell sought to improve access to internal expertise and reduce time spent on searching for technical information across its global operations.

AI Model: A transformer-based generative language model (GPT-like) trained on over 300,000 technical documents and engineering manuals.

Dataset: The curated dataset covered decades of operational and engineering data, structured through AI-driven semantic indexing.

Metrics & Validation: The chatbot achieved +30% improvement in Q&A accuracy over traditional keyword search, with precision at 0.72, recall at 0.68, and F1-score at 0.70.

KM Impact: The AI chatbot effectively codifies tacit expert knowledge and democratizes access to technical data, significantly improving problem-solving speed and decision-making quality across departments [45].

3.3 AI-Driven Learning and Governance Automation

Objective: Siemens Energy developed a generative AI-based upskilling initiative to close skill gaps in AI and KM practices among employees.

AI Model: Generative AI combined with vector embeddings for role-specific content recommendation.

Dataset: Training content included curated knowledge bases, past project documentation, and AI-generated learning paths for 4,900 employees.

Metrics & Validation: Within two weeks, the program achieved a 62% increase in AI proficiency, as measured by internal competency tests and practical project outcomes.

KM Impact: The initiative not only enhanced technical skills but also created a self-sustaining mentorship ecosystem supported by "AI ambassadors," improving knowledge transfer across the organization.In summary, the current applications of AI within the energy sector can be categorized as follows:

- Predictive failure anticipation and asset optimization via predictive maintenance and RL-AT models [1, 4],
- Real-time control of complex grid systems employing AI-enabled digital twins and intelligent voltage regulation [2, 3],
- Enhanced generation forecasting and operational optimization through cloud-based AI analytics for renewable assets [23, 24],
- Augmented knowledge access and retention using AI-enhanced chatbots and document classification technologies [19],
- Enterprise-wide strategic awareness via generative AI-driven digests and communications [5, 29].

These developments collcctively underscore that AI has transcended isolated pilot projects, becoming deeply embedded within the core operational and strategic processes of leading energy enterprises. As the scale and velocity of data generation continue to escalate, the imperative for scalable, intelligent AI solutions capable of managing this complexity will only intensify [36].

To ensure the reliability of results, all case studies were analyzed based on publicly available reports, technical documentation, and peer-reviewed literature. Dataset harmonization was applied by standardizing performance metrics (accuracy, precision, recall, F1-score) and aligning data across timeframes. Ørsted's AI forecasting models underwent k-fold cross-validation and backtesting to validate performance. Shell's chatbot improvements were benchmarked against baseline keyword-search models, confirming a 30% accuracy gain. Siemens Energy's upskilling program was evaluated through internal skill assessments, achieving a verified 62% increase in AI- related competencies. Table 1 illustrates the comparative performance and outcomes of the analyzed case studies, highlighting key AI models, dataset sizes, and measurable benefits across operational, knowledge management, and upskilling domains [40].

Table 1. Comparative Overview of AI Applications

Enterprise	AI Application	AI Model	Dataset Size	Key Outcomes
Shell	AI Knowledge Chatbot	Transformer (GPT like)	300,000 technical docs	+ 30% Q&A accuracy (Precision 0.72, F1 = 0.70)
Ørsted	SCADA Forecasting & Analytics	LSTM + Cloud ML	5 TB SCADA data (5 years)	15–25% O&M cost reduction; + 10–20% output |

(continued)

Table 1. (*continued*)

Enterprise	AI Application	AI Model	Dataset Size	Key Outcomes
Siemens Energy	GenAI Upskilling Program	Generative AI + Embeddings	4,900 employees	+ 62% AI proficiency in 2 weeks

Source: Authors elaboration based on public technical reports, corporate case studies (Shell 2023, Ørsted 2023, Siemens Energy 2024), and peer-reviewed literature [19, 20, 22, 24].

4 Potential Directions for the Use of AI in Knowledge Management Within Energy Enterprises

As the digital transformation of the energy sector accelerates, artificial intelligence (AI) is emerging not only as a tool for operational optimization but also as a catalyst for strategic knowledge management (KM). AI technologies can fundamentally reshape how energy companies capture, organize, share, and apply institutional knowledge. To provide a stronger theoretical foundation, our analysis aligns with Nonaka and Takeuchi's SECI model, which emphasizes the dynamic conversion between tacit and explicit knowledge through processes of socialization, externalization, combination, and internalization. AI technologies can significantly enhance each of these phases by automating codification, facilitating cross-functional knowledge recombination, and enabling adaptive learning environments. Additionally, AI-enabled KM accelerates progression through Knowledge Management Maturity Models (KMMM), helping organizations evolve from basic knowledge storage systems to predictive, self-learning ecosystems capable of supporting continuous innovation and resilience [52].

The proposed directions can be understood through the lens of established KM theories. Nonaka and Takeuchi's SECI model (Socialization, Externalization, Combination, Internalization) provides a structured foundation for mapping how tacit and explicit knowledge interact. AI technologies amplify this process: natural language processing tools facilitate externalization, semantic models support combination of knowledge sources, and adaptive learning environments enhance internalization [53].

Moreover, integrating AI into KM systems accelerates organizational advancement across Knowledge Management Maturity Models (KMMM), enabling energy companies to progress from basic information repositories to predictive, self-learning, and continuously improving knowledge ecosystems.

Based on current trends and implementation cases, four strategic directions for the application of AI in KM can be distinguished:

4.1 Codification of Tacit Knowledge Through AI-Augmented Capture

One of the most pressing challenges in energy enterprises is the loss of tacit knowledge due to retirements and workforce turnover. AI can play a pivotal role in capturing and formalizing this knowledge by transforming undocumented expertise into accessible and searchable formats. Techniques such as AI-driven transcription, summarization, and semantic indexing can be used to document expert narratives, interviews, and project debriefs [31, 35].

For example, organizations have begun piloting structured knowledge capture sessions where veteran engineers are guided (sometimes by AI agents) to recount decision-making processes in critical scenarios. These inputs can then be indexed using natural language processing (NLP), enabling younger professionals to query the knowledge base contextually. Such efforts address the demographic challenge highlighted in the literature: the anticipated departure of hundreds of thousands of experienced workers within the next decade [1, 38].

4.2 Context-Aware Knowledge Retrieval and Semantic Recommendation Systems

AI enables the transition from static, repository-based KM systems to dynamic, context-aware platforms. Semantic models and machine learning algorithms can infer user intent, recommend relevant documents, and surface latent insights across structured and unstructured data sources.

Shell's implementation of an AI-powered knowledge chatbot exemplifies this direction. By curating and cleaning over 300,000 documents, the system achieved a 30% increase in Q&A accuracy, dramatically improving employee access to technical knowledge [22]. In a similar vein, Ørsted's use of generative AI to deliver daily industry briefings illustrates how AI can proactively push contextually relevant knowledge to users, fostering real-time awareness [34].

Such AI-powered systems move beyond simple keyword search to enable:

- Intelligent question answering, based on deep semantic understanding,
- Adaptive knowledge curation, based on role, past queries, and organizational context,
- Just-in-time knowledge delivery, minimizing time spent locating critical information [10, 35, 45].

4.3 Personalized Learning and AI-Driven Skill Development

As energy companies face widening skill gaps in digital and analytical competencies, AI can support personalized upskilling pathways. By analyzing job roles, existing competencies, and usage patterns, AI systems can recommend training modules, suggest relevant documentation, or even match users with internal mentors.

The Siemens Energy GenAI initiative serves as a benchmark. Through a role-specific AI training program delivered at scale, the company upskilled over 4,900 employees, improving generative AI proficiency by 62% in under two weeks [22]. Moreover, the introduction of "AI ambassadors" employees trained to mentor others illustrates how AI can facilitate peer-based learning ecosystems.

This approach supports continuous knowledge acquisition, directly embedded in daily workflows, and contributes to a culture of curiosity and digital confidence–essential traits for sustaining KM adoption [46].

4.4 Governance Automation and AI-Based Content Lifecycle Management

Knowledge quality and relevance are major success factors in KM systems. AI can help automate the governance of enterprise knowledge by monitoring content usage, flagging outdated or low-quality items, and prompting authors to review or update materials.

Shell's intranet governance strategy demonstrated how automated content audits and broken-link detection improved platform credibility, resulting in a 70% drop in inaccessible resources and higher user satisfaction [16]. AI can extend this further by applying natural language models to detect duplication, summarize documents, tag metadata, and assess sentiment or usage trends.

This direction positions AI as an intelligent curator of enterprise knowledge supporting sustainability, quality assurance, and regulatory compliance. It also strengthens trust in the KM system by ensuring that information is current, validated, and aligned with organizational needs [33, 47].

AI-KM Conceptual Framework

To consolidate the insights presented in this paper, we propose a conceptual AI-KM framework designed specifically for energy enterprises. This framework addresses the intersection between AI capabilities and the core processes of knowledge management—creation, storage, sharing, and application—with the aim of building an adaptive, intelligent, and transparent knowledge infrastructure.

The proposed framework is anchored in Nonaka's SECI model (Socialization, Externalization, Combination, Internalization), which describes the cyclical transformation of tacit and explicit knowledge. Our AI-enhanced extension of this model introduces four layers [52]:

- Knowledge Capture Layer – This layer focuses on codifying tacit knowledge through AI-driven tools such as automatic transcription, speech-to-text systems, and NLP-based summarization. For instance, expert interviews, technical debriefings, or real-time incident logs can be transformed into structured digital records.
- Knowledge Organization Layer – Here, AI-driven ontology building, semantic tagging, and clustering algorithms ensure coherent classification of both structured and unstructured data, creating a knowledge graph that dynamically adapts to organizational needs.
- Knowledge Sharing Layer – Context-aware recommendation engines, intelligent chatbots, and AI-driven collaboration platforms facilitate the real-time distribution of knowledge across departments, breaking silos and accelerating decision-making.
- Knowledge Application Layer – Predictive and prescriptive AI models use curated knowledge to support complex decision-making, optimize operational workflows, and recommend strategic initiatives.

The framework emphasizes feedback loops where AI continuously learns from user interactions and system performance to refine knowledge relevance and accuracy. A visual diagram illustrating the interactions between these layers is recommended to complement the conceptual model [51].

5 Conclusion

AI-driven knowledge management (KM) in energy enterprises is maturing from isolated pilots to a governed, measurable capability that underpins decarbonization, resilience, and operational excellence. The most recent guidance shows that progress hinges on pairing deployment with robust governance and risk management. Concretely, organizations should align KM-enabled AI systems with recognized frameworks (e.g., NIST AI RMF 1.0 and the Generative AI Profile) to drive consistent practices in model lifecycle control, transparency, and human-in-the-loop oversight [40, 51]. At the same time, strategic alignment with the European Climate Law's long-term targets ensures that KM portfolios prioritize knowledge critical to grid reliability, flexibility, and emissions reduction [37].

A second implication is infrastructural: AI's rising computational and data demands will increasingly shape energy-sector plans. IEA analysis underscores the rapid growth of AI's power needs and the parallel surge in global clean-energy investment, with grids, storage, and efficiency now attracting the majority of new capital. KM functions should therefore evolve to capture and disseminate design know-how for data-center interconnections, flexibility markets, and demand-response-turning lessons learned into reusable playbooks at enterprise scale [36, 48].

Third, grid digitalization is moving from concept to federation. National Grid ESO's Virtual Energy System and the EU's grid-twin vision point toward interoperable, multi-party digital twins that rely on shared ontologies and high-fidelity operational knowledge. KM must curate and standardize asset metadata, event taxonomies, and operating procedures so that twins can reason over consistent context. Evidence from both industry and academic reviews highlights the opportunity-along with the risks from legacy constraints and connection bottlenecks-which strengthens the case for KM-led content quality, provenance controls, and update cadences [41–43, 49].

Fourth, retrieval quality now depends on combining LLMs with enterprise knowledge graphs and rigorous content governance. Architectures that ground generative models in vetted corpora are delivering more reliable Q&A, faster incident resolution, and better reuse of tacit know-how. Case evidence (e.g., Shell's intranet/content governance and chatbot initiatives) demonstrates that measurable gains follow when organizations invest in semantic indexing, deduplication, and lifecycle rules-core KM responsibilities [19, 45, 47].

Fifth, the workforce transition is inseparable from KM. Upskilling programs that blend role-aware learning paths, internal mentors, and embedded AI assistants show large, near-term proficiency gains. Macro indicators (e.g., USEER 2025) and sector analyses point to sustained hiring, reskilling, and knowledge-transfer needs. KM should sponsor structured capture from veteran experts, create reusable learning assets, and instrument adoption metrics across lines of business [38, 39, 46].

Finally, measurement must become continuous. Pairing KM objectives with AI risk and performance dashboards (data quality, retrieval fidelity, explainability, security posture, and business KPIs) helps maintain trust and course-correct quickly. Independent benchmarks (e.g., AI Index 2025) and internal M&E frameworks should be used to spot drifts in model behavior, content freshness, and user outcomes, while standard playbooks (AI RMF) anchor audits and corrective actions. In practice, this means treating knowledge as critical infrastructure: governed, observable, continuously improved, and explicitly tied to decarbonization and reliability goals [40, 50, 51].

In short, AI-driven KM is no longer a peripheral IT function. It is a strategic capability that converts fragmented data and human expertise into system intelligence for planning, operations, compliance, and learning. Enterprises that operationalize the triad of (1) robust governance, (2) federated digital-grid knowledge, and (3) workforce-centric upskilling and capture will set the performance frontier for safety, reliability, and innovation-turning knowledge into a durable competitive asset across the energy transition [36, 37, 41, 46, 49].

Funding. Co-financed by the Minister of Science under the "Regional Excellence Initiative".

Regionalna Inicjatywa Doskonałości

Ministry of Science and Higher Education
Republic of Poland

References

1. Cakir, K., Duran, K., Canberk, B.: AI in energy digital twining: a reinforcement learning-based adaptive digital twin model for Green Cities. In: IEEE ICC 2024. https://doi.org/10.1109/ICC51166.2024.10622773
2. SP Group: Digital Twin White Paper. SP Group (2023) SP Group: Digital Twin White Paper (2023); National Grid ESO: System Transformation Report (2023)
3. Ardebili, A.A., Zappatore, M., Ramadan, A.I.H.A., Longo, A., Ficarella, A.: Digital twins of smart energy systems: a systematic literature review on enablers, design, management and computational challenges. Energy Infor. **7**, 94 (2024)
4. Mchirgui, N., Quadar, N., Kraiem, H., Lakhssassi, A.: The applications and challenges of digital twin technology in smart grids: a comprehensive review. Appl. Sci. **14**(2), 789 (2024). https://doi.org/10.3390/app14020789
5. MarketsandMarkets: Digital Twin Market–Global Forecast to 2027 (2023)
6. Grand View Research, Inc.: Digital Twin Market Size, Share & Trends Analysis Report By Solution, Deployment, Enterprise Size, Application, End-Use, And Region, 2024–2030. Grand View Research, Inc. (2025)
7. Unisphere Research: The State of Knowledge Management in 2023: Untapped Potential for Business Value. https://hd.egain.com/surveys/state-of-knowledge-management-survey-2023.pdf. Last accessed 16 June 2025
8. EY: Maximizing the Impact of AI in the Oil and Gas Sector. https://www.ey.com/en_us/insights/oil-gas/maximizing-the-impact-of-ai-in-the-oil-and-gas-sector. Last accessed 16 June 2025
9. Edwards, J.S.: Knowledge management in the energy sector: review and future directions. Int. J. Energy Sect. Manage. **2**(2), 197–217 (2008). GreenOrbit: Empower staff with GreenOrbit. https://greenorbit.com/wp-content/uploads/2020/02/Case-Study-Shell-1.pdf Last accessed 16 June 2025
10. Unisphere Research: The State of Knowledge Management: Bridging Silos in the Age of AI. eGain Corporation (2023). (Accessed 16 June 2025 at hd.egain.com)
11. Ernst & Young Global Ltd.: Oil & Gas Digital Transformation and the Workforce Survey 2020 (2020). https://www.scribd.com/document/521886022/ey-oil-and-gas-digital-transformation-and-the-workforce-survey. Last accessed 16 June 2025

12. Nicolini, D., Pyrko, I., Omidvar, O., Spanellis, A.: Understanding communities of practice: taking stock and moving forward. Acad. Manag. Ann. **16**(2), 366–381 (2022). https://doi.org/10.5465/annals.2020.0330
13. Bérubé, M., Giannelia, T., Vial, G.: Barriers to the implementation of AI in organizations: findings from a delphi study. In: Hawaii International Conference on System Sciences (HICSS) 2021, pp. 5052–5061. IEEE (2021). https://doi.org/10.24251/HICSS.2021.805
14. Hirsch, S.L., Ribes, D.: Innovation and legacy in energy knowledge infrastructures. Energy Res. Soc. Sci. **80**, 102218 (2021). https://doi.org/10.1016/j.erss.2021.102218
15. Reif, V., Strasser, T.I., Jimeno, J., Farré, M., Genest, O.: Towards an interoperability roadmap for the energy transition (2023). arXiv:2309.08284
16. Ntafalias, A., Tsakanikas, S., Skarvelis-Kazakos, S., Papadopoulos, P.: Design and implementation of an interoperable architecture for integrating building legacy systems into scalable energy management systems. Smart Cities **5**(4), 1421–1440 (2022)
17. Rezaei, M.: Artificial Intelligence in knowledge management: identifying and addressing the key implementation challenges. Technol. Forecast. Soc. Chang. **217**, 124183 (2025)
18. NVIDIA: Case Study – Shell Trains Custom AI Chatbot With NeMo to Uplevel Operations (2023). https://www.nvidia.com/en-us/case-studies/shell-trains-custom-ai-chatbot-with-nemo/. Last accessed 16 June 2025
19. Ørsted: Ørsted Deploys SparkCognition's AI Solution to Enhance Wind, Solar, and Storage Asset Performance and Increase Energy Production. Ørsted (Press Release, 7 November 2023). https://us.orsted.com/news-archive/2023/11/orsted-deploys-sparkcognitions-ai-solution-to-enhance-wind-solar-and-storage-asset-performance-and-increase-energy-production. Last accessed 16 June 2025
20. Allsup, M.: Ørsted Turns to AI to Boost US Projects. Latitude Media (8 Nov 2023)
21. Capgemini: Energy Enterprise Knowledge Management with Generative AI. Capgemini Research Library (2023). https://www.capgemini.com/us-en/insights/research-library/energy-enterprise-knowledge-management-with-generative-ai/. Last accessed 16 June 2025
22. Mäntymäki, M., Minkkinen, M., Birkstedt, T., Viljanen, M.: Putting AI ethics into practice: the hourglass model of organizational AI governance. J. Bus. Res. **144**, 637–647 (2022)
23. Capgemini Research Institute: Data Foundations for Government: From AI Ambition to Execution. Capgemini Research Institute (May 2025). https://www.capgemini.com/wp-content/uploads/2025/05/Capgemini-Research-Institute-report_Data-foundations-for-government_From-AI-ambition-to-execution-2.pdf?utm_source=chatgpt.com. Last accessed 16 June 2025
24. Regulation (EU) 2024/1689 of the European Parliament and of the Council of 13 June 2024 laying down harmonised rules on artificial intelligence. Official Journal of the European Union L 1689, 12 July 2024. https://eur-lex.europa.eu/eli/reg/2024/1689/oj. Last accessed 16 June 2025
25. Binns, R.: Human judgment in algorithmic loops: individual justice and automated decision-making. Regul. Governance **14**(1), 95–115 (2020)
26. GreenOrbit: Knowledge Management and Content Governance in the Digital Workplace–Shell Case Study (2023). (Accessed 16 June 2025 at greenorbit.com)
27. Pelekis, S., et al.: Trustworthy Artificial Intelligence in the energy sector: landscape analysis and evaluation framework (2024). arXiv:2412.07782
28. North, K., Maier, R., Haas, O. (eds.): Knowledge management in digital change: new findings and practical cases. Progress in IS, Springer, Cham (2018)
29. Capgemini: AI-Enhanced Knowledge Capture and Retention: Practices from the Energy Sector (2023)
30. Kaur, V.: Leveraging design thinking to foster knowledge-based dynamic capabilities in service innovation. Long Range Planning (2025)

31. Li, M., Guo, J., Zou, C., Yin, J.: The impact of reverse mentoring on employees' innovative behavior: evidence from Chinese technology enterprises. Sustainability 17(1), Article 6 (2025)
32. Davenport, T.H., Ronanki, R.: Artificial Intelligence for the real world. Harv. Bus. Rev. **96**(1), 108–116 (2018)
33. Dataiku: Ørsted: Monitoring Market Dynamics With LLM-Driven News Digest (2024). https://www.dataiku.com/stories/detail/orsted/. Last accessed 16 June 2025
34. Ørsted Deploys SparkCognition's AI Solution to Enhance Wind, Solar, and Storage Asset Performance and Increase Energy Production. https://us.orsted.com/news-archive/2023/11/orsted-deploys-sparkcognitions-ai- solution-to-enhance-wind-solar-and-storage. Last accessed 16 June 2025
35. International Energy Agency (IEA). "Energy and AI" (2024). Paris: IEA
36. International Energy Agency (IEA): Energy and AI – Energy demand from AI. https://www.iea.org/reports/energy-and-ai/energy-demand-from-ai. Last accessed 08 Oct 2025
37. European Commission: European Climate Law. https://climate.ec.europa.eu/eu-action/european-climate-law_en. Last accessed 08 Oct 2025
38. McKinsey & Company: Talent squeeze: planning for the energy sector's talent transition (2024). https://www.mckinsey.com/industries/oil-and-gas/our-insights/talent-squeeze-planning-for-the-energy-sectors-talent-transition. Last accessed 08 Oct 2025
39. U.S. Department of Energy: 2025 United States Energy & Employment Report (USEER). https://www.energy.gov/sites/default/files/2025-08/National%20USEER_08282025.pdf. Last accessed 08 Oct 2025
40. National Institute of Standards and Technology (NIST): Artificial Intelligence Risk Management Framework – Generative AI Profile (NIST AI 600-1) (2024). Last accessed 08 Oct 2025
41. National Grid ESO: Virtual Energy System (connected digital twins). https://www.nationalgrideso.com/future-energy/virtual-energy-system. Last accessed 08 Oct 2025
42. Financial Times: National Grid's outdated systems hinder UK's battery storage boom. https://www.ft.com/content/8e1c8b28-ac86-4b2d-b2f1-b789c35e58c7. Last accessed 08 Oct 2025
43. National Energy System Operator (NESO): Memorandum of Understanding with the UK's National Digital Twin Programme (2024). https://www.neso.energy/publications/neso-and-national-digital-twin-mou. Last accessed 08 Oct 2025
44. Buckley, N., Bo, C., Delkhah, F., Byrne, N., Ní Shearcaigh, A., Brennan, S., Correa, D.P.: Evaluation of a peer-to-peer smart grid using digital twins: a case study of a remote European Island. Energies **17**(22), 5541 (2024). https://www.mdpi.com/1996-1073/17/22/5541.https://doi.org/10.3390/en17225541
45. Mariotti, L., Guidetti, V., Mandreoli, F., Belli, A., Lombardi, P.: Combining large language models with enterprise knowledge graphs: a perspective on enhanced natural language understanding. Front. Artif. Intell. (2024). https://www.frontiersin.org/articles/https://doi.org/10.3389/frai.2024.1460065/full. Last accessed 08 Oct 2025
46. Workera: Siemens Energy–Customer Story (GenAI upskilling results). https://workera.ai/customer-story/siemensenergy. Last accessed 08 Oct 2025
47. GreenOrbit: Shell – Empower staff with GreenOrbit (Case study PDF). https://greenorbit.com/wp-content/uploads/2020/02/Case-Study-Shell-1.pdf. Last accessed 08 Oct 2025
48. International Energy Agency (IEA): World Energy Investment 2024. https://www.iea.org/reports/world-energy-investment-2024. Last accessed 08 Oct 2025
49. Jørgensen, B.N., Ma, Z.G.: Digital twin of the European electricity grid: a review of regulatory barriers, technological challenges, and economic opportunities. Appl. Sci. **15**(12), 6475 (2025). https://www.mdpi.com/2076-3417/15/12/6475. https://doi.org/10.3390/app15126475
50. Stanford HAI: 2025 AI Index Report. https://hai-production.s3.amazonaws.com/files/hai_ai_index_report_2025.pdf. Last accessed 08 Oct 2025

51. National Institute of Standards and Technology (NIST): Artificial Intelligence risk management framework (AI RMF 1.0) (NIST AI 100-1) (2023). Last accessed 08 Oct 2025
52. Nonaka, I., Takeuchi, H.: The knowledge-creating company: how Japanese companies create the dynamics of innovation. Oxford University Press (1995). https://global.oup.com/academic/product/the-knowledge-creating-company-9780195092691. Last accessed 08 Oct 2025
53. APQC: APQC's Levels of Knowledge Management Maturity (2025). https://www.apqc.org/resource-library/resource-listing/apqcs-levels-knowledge-management-maturity. Last accessed 08 Oct 2025

Machine Learning, Forecasting, Language Technologies, and Knowledge Engineering

The Impact of Self-supervised Pretraining on Forecasting Accuracy in Forex Market

Damian Śnieda[1] and Paweł Weichbroth[2(✉)]

[1] Gdansk University of Technology, Gdansk, Poland
[2] Faculty of Electronics, Telecommunications and Informatics, Department of Software Engineering, Gdansk University of Technology, Gdansk, Poland
pawel.weichbroth@pg.edu.pl

Abstract. This paper investigates the effectiveness of self-supervised pretraining in the context of Forex price prediction. Specifically, we explore whether pretraining a model using Masked Time Modeling (MTM) can enhance the forecasting performance of an Long Short-Term Memory (LSTM) network compared to a baseline trained solely on labeled data. The task involves predicting the next five closing prices of a currency pair based on the previous 50 observations. We conducted experiments on five currency pairs, each tested across three different time intervals. The MTM-based model was first pre-trained to reconstruct masked price sequences, learning contextual representations of the time series. These representations were then used as input features in a fine-tuned LSTM model trained for supervised forecasting. Both models shared the same architecture, and their performance was evaluated using the Root Mean Squared Error (RMSE). The results show that self-supervised pretraining with MTM led to slightly better performance for some currency pairs, but significantly worse results for others compared to the purely supervised model.

Keywords: Forex · LSTM · Self-supervised Learning · Masked Time Modeling · Deep Learning

1 Introduction

Trading has become more accessible than ever, with platforms allowing users to easily invest in indices, commodities, shares, and currencies. Although profitable opportunities exist, most traders tend to lose money without proper knowledge or experience [1]. Human psychology also plays a critical role, often leading to poor decisions influenced by market sentiment and emotions [2]. Traditional prediction methods, including fundamental [3] and technical analysis [4], remain widely used, although their profitability is debated [5]. These challenges highlight the need for automating trading decisions through accurate price prediction to mitigate human error and overcome the limitations of conventional analytical approaches.

Published by Springer Nature Switzerland AG 2026
M. Pondel et al. (Eds.): AI4KAM 2025, IFIP AICT 776, pp. 317–326, 2026.
https://doi.org/10.1007/978-3-032-18920-2_30

Advances in Artificial Intelligence, particularly Deep Learning (DL), have led to powerful models capable of uncovering complex patterns in data [6]. In finance, Recurrent Neural Networks (RNNs), and especially Long Short-Term Memory (LSTM) networks, have proven effective for time series prediction tasks. However, deep learning models typically require large labeled datasets, which are often costly or difficult to obtain.

To address this limitation, Self-Supervised Learning (SSL) has emerged, enabling models to learn useful representations from unlabeled data through carefully designed pretext tasks [7]. One promising SSL method for time series is Masked Time Modeling (MTM), inspired by masked language modeling techniques [8]. MTM trains models to reconstruct intentionally masked portions of input sequences, forcing the network to learn temporal dependencies and hidden structure within the data. This approach can generate rich embeddings that enhance downstream supervised tasks like price forecasting.

The objective of this study is to investigate whether self-supervised pretraining using MTM can enhance the performance of an LSTM model in predicting Forex prices. To achieve this, two approaches were compared: an LSTM model trained exclusively on labeled data (the baseline model) and a two-stage model incorporating MTM pretraining followed by supervised fine-tuning. Experiments were conducted across various currency pairs and time intervals, and forecasting accuracy was evaluated using RMSE.

The rest of this paper is organized as follows. Section 2 discusses related work. Section 3 describes the experimental setup and the model architecture. Section 4 presents the results. Section 5 discusses the results and outlines directions for future research.

2 Related Work

Recent research in financial time series forecasting has increasingly applied deep learning techniques, with LSTM networks proving particularly effective for modeling temporal dependencies in Forex and stock markets [9,10]. Most approaches rely on supervised learning with historical price data. Self-supervised learning, which extracts meaningful representations from unlabeled data via pretext tasks, has gained interest as a way to enhance model performance. Techniques such as contrastive learning, autoencoding, and masked time modeling have shown success in natural language processing [8] and computer vision [11], but remain underexplored in financial forecasting, particularly for Forex.

Several studies have applied deep learning to financial time series prediction. Zanc et al. [12] proposed a stacked LSTM model for predicting short-term price movements in Forex and cryptocurrency markets, but the model's usefulness was limited by short prediction horizons and time-shifted outputs. Sun et al. [13] introduced Self-FTS, a self-supervised framework combining contrastive and classification tasks to improve stock trend prediction; however, its focus on classification rather than continuous-value forecasting limits its direct applicability to this study's regression approach.

Other work has explored hybrid models. Abedin et al. [14] developed a Bi-LSTM model combined with Bagging Ridge regression, achieving improved forecasting accuracy during volatile periods like the COVID-19 pandemic. Similarly, Yıldırım et al. [15] proposed a hybrid LSTM model trained separately on macroeconomic and technical indicators, demonstrating improved directional prediction for the EUR/USD exchange rate.

In the domain of self-supervised learning for time series, Jawed et al. [16] introduced a multi-task approach combining self-supervised forecasting with classification to enhance semi-supervised learning, although the method targets classification rather than forecasting. Chen et al. [17] proposed TS-TCT, a contrastive framework for unsupervised time series representation, achieving strong results on general benchmarks but without direct evaluation on financial forecasting tasks. Finally, Qi [18] explored a multi-model approach for Forex trading, introducing models like TradingBERT to generate embeddings from price data, highlighting the potential of pretraining in financial contexts.

In addition to deep learning–based approaches, traditional statistical analysis of return distributions has highlighted the role of higher-order moments in financial forecasting. For example, Gkillas et al. [19] investigated whether realized skewness and kurtosis improve exchange-rate volatility forecasting. Their study, based on high-frequency intraday data, showed that these indicators significantly enhance volatility predictions, with kurtosis in particular improving one-day-ahead forecasts. This finding underscores the importance of distributional properties such as fat tails and asymmetry in modeling forex dynamics.

3 Methodology

The research uses historical Forex data from five currency pairs: EUR/USD, USD/JPY, AUD/JPY, EUR/AUD, and EUR/JPY. The dataset spans the period from 2000 to 2024 with 1 min interval. For each currency pair, experiments were conducted across three different time intervals: hourly data (H1), four-hour intervals (H4), daily data (D1). The raw time series data for each configuration, including currency pair and time interval, was divided into three subsets: training set (80%) – used to fit the model, validation set (10%) – used for hyperparameter tuning and early stopping, test set (10%) – used for final performance evaluation.

Baseline Model. The first approach used is a standard LSTM neural network (the baseline model), which is trained exclusively via supervised learning. This model only used closing prices as input features and had no access to pretraining or contextual representations. The model learned directly from the labeled training set to minimize prediction error.

Pretrained Model (Self-supervised Approach). The second approach used is the pretrained model followed a two-phase training regime:

1. Pretraining Phase (masked time modeling): In this phase, an LSTM model was trained using MTM technique. Portions of the input sequences (20%) were

randomly masked (i.e., replaced with zeros), and the model was tasked with reconstructing the missing values based on the surrounding temporal context. This self-supervised step aimed to extract meaningful time-dependent representations (embeddings) without relying on labeled data.
2. Fine-Tuning Phase (Supervised Learning): A second LSTM model was then trained using supervised learning. However, unlike the baseline model, this model received as input the embeddings produced during the pretraining phase. The goal was to leverage the learned representations to improve forecasting accuracy for future closing prices.

The forecasting task involved predicting the next 5 closing prices based on the preceding 50 closing prices, applying a sliding window approach to create input-output pairs. Both the baseline and pretrained models used the same LSTM-based architecture to ensure that performance differences arise from training methodology rather than model complexity. The model consists of two LSTM layers with 128 and 64 units, followed by two dense layers with 32 and 16 units using ReLU activations. A final dense layer with linear activation produces the output. The models were trained using the Adam optimizer (learning rate 0.001) with Mean Squared Error as the loss function and RMSE as the evaluation metric.

For each configuration (currency pair and interval), two models—baseline and pretrained—were trained for 10 epochs. The RMSE was calculated on the validation set after each epoch. The lowest RMSE observed during these epochs was recorded and used as the basis for comparison between the two approaches.

To better understand the impact of self-supervised pretraining on different currency pairs, we performed an input data statistical properties analysis. The objective of this step was to investigate whether inherent characteristics of the price series, such as volatility or distributional shape, might influence the effectiveness of MTM pretraining across different currencies and time intervals.

For each dataset, we first computed logarithmic returns from the closing prices, defined as $r_t = ln(P_t/P_{t-1})$. Returns are preferred over raw prices because they normalize relative changes and are closer to stationarity, enabling more reliable statistical analysis. The following indicators were analyzed:

1. **Volatility** (standard deviation of returns): measures the magnitude of price fluctuations. High volatility reflects unstable, noisy markets that are more difficult to model.
2. **Skewness**: captures asymmetry of the return distribution. Strong skewness suggests that one-sided extreme events (e.g., sudden losses) dominate the series.
3. **Kurtosis**: measures the frequency of extreme deviations (fat tails). High kurtosis implies that outliers occur more often than expected under a normal distribution.

4 Results

Table 1 summarizes the RMSE values for both the baseline and the pretrained models across all 15 configurations (5 currency pairs $\times$ 3 time intervals) and Table 2 presents the average RMSE across all time intervals.

Table 1. Comparison of RMSE results for Baseline and Pretrained models across different currency pairs and time intervals.

Currency Pair	Time Interval	Baseline RMSE	Pretrained RMSE	Change
EUR/USD	H1	0.00230	**0.00197**	0.03 %
EUR/USD	H4	0.00536	0.00536	0.00 %
EUR/USD	D1	0.01566	**0.01160**	0.41 %
USD/JPY	H1	**0.00263**	0.25437	–25.17 %
USD/JPY	H4	**0.00537**	0.55930	–55.39 %
USD/JPY	D1	**0.01599**	4.44499	–442.90 %
AUD/JPY	H1	**0.00263**	0.32137	–31.87 %
AUD/JPY	H4	**0.00654**	0.72943	–72.29 %
AUD/JPY	D1	**0.01569**	5.71868	–570.30 %
EUR/AUD	H1	0.00339	**0.00300**	0.04 %
EUR/AUD	H4	0.00775	**0.00075**	0.10 %
EUR/AUD	D1	**0.01726**	0.02400	–0.67 %
EUR/JPY	H1	**0.00263**	0.16891	–16.63 %
EUR/JPY	H4	**0.00720**	0.64278	–63.56 %
EUR/JPY	D1	**0.01866**	5.79132	–577.27 %

Table 2. Summary of average RMSE and percentage improvement for Baseline and Pretrained models

Metric	Baseline RMSE	Pretrained RMSE	Change
Average RMSE (All Cases)	**0.00860**	1.24558	–123.70 %
Average RMSE (H1 Only)	**0.00271**	0.14992	–14.72 %
Average RMSE (H4 Only)	**0.00644**	0.38872	–38.23 %
Average RMSE (D1 Only)	**0.01665**	3.19811	–318.15 %

The performance of the pretrained model varied significantly across currency pairs and time intervals. While slight improvements were observed for EUR/USD in the D1 interval (0.41% improvement) and EUR/AUD in the H4 interval (0.10%), the majority of pretrained models underperformed compared to the

Table 3. Volatility, skewness, and kurtosis of log returns for all tested currency pairs and time intervals.

Currency Pair	Time Interval	Volatility	Skewness	Kurtosis
EUR/USD	H1	0.00126	0.22409	16.41095
EUR/USD	H4	0.00252	0.17299	8.95854
EUR/USD	D1	0.00613	-0.03259	2.38373
USD/JPY	H1	0.00149	-1.37661	**169.87305**
USD/JPY	H4	0.00294	-0.66253	**49.09978**
USD/JPY	D1	0.00706	-0.35744	10.99837
AUD/JPY	H1	**0.12754**	-0.88592	**1109.61729**
AUD/JPY	H4	**0.13462**	0.64684	**954.11506**
AUD/JPY	D1	**0.13362**	2.17305	**882.21380**
EUR/AUD	H1	0.00137	0.39551	28.99217
EUR/AUD	H4	0.00268	0.76495	20.60205
EUR/AUD	D1	0.00644	0.61451	7.34500
EUR/JPY	H1	0.00157	-0.67908	**62.97704**
EUR/JPY	H4	0.00310	-0.40872	21.57292
EUR/JPY	D1	0.00742	-0.23523	6.39844

supervised baseline. The most substantial declines were noted in the D1 interval for EUR/JPY and AUD/JPY, where RMSE increased dramatically. Overall, the results suggest that self-supervised pretraining with MTM did not consistently improve forecasting accuracy and, in several cases, led to significantly worse outcomes.

A Wilcoxon signed-rank test was used to compare the RMSE values between the baseline and pretrained models, as the normality assumption was not met for either dataset (Shapiro-Wilk $p < 0.01$). The test revealed a statistically significant difference (W = 10.0, p = .008), with a strong negative rank-biserial correlation (–0.810), indicating that the pretrained model consistently produced higher RMSE values than the baseline, confirming worse performance of the pretrained model.

Table 3 reports the volatility, skewness, and kurtosis of returns for all tested currency pairs and time intervals. The analysis reveals clear differences across pairs that align with the observed effectiveness of MTM pretraining.

EUR/USD and EUR/AUD exhibited relatively low volatility and moderate kurtosis values. These pairs correspond to the few cases where MTM pretraining achieved slight improvements in forecasting accuracy. In contrast, AUD/JPY and JPY-related pairs displayed extremely high volatility and very fat-tailed distributions, with kurtosis values exceeding several hundred in some intervals. These pairs coincide with the worst MTM results, where pretrained models failed dramatically compared to the supervised baseline.

Skewness values varied across pairs and intervals but were less strongly associated with model performance. While skewness indicates asymmetry in returns, it appears that volatility and kurtosis were the dominant factors influencing the success or failure of MTM pretraining.

To further illustrate the performance difference, the error distribution plot shown in Fig. 1 compares prediction errors for the baseline and pretrained models. The baseline model's errors are centered around zero and roughly follow a normal distribution, reflecting consistent forecasts. In contrast, the pretrained model's errors are more widely spread and skewed, indicating greater variability and the presence of outliers, consistent with its higher RMSE.

The baseline LSTM model achieved relatively accurate predictions for the EUR/USD pair on the H1 interval, with a RMSE of 0.00230. As shown in Fig. 2, the model successfully captures the short-term dynamics of the time series, with the predicted price trajectory closely following the ground truth. While some minor deviations are visible, especially in the final timesteps, the general trend and shape of the forecast align well with actual market movements.

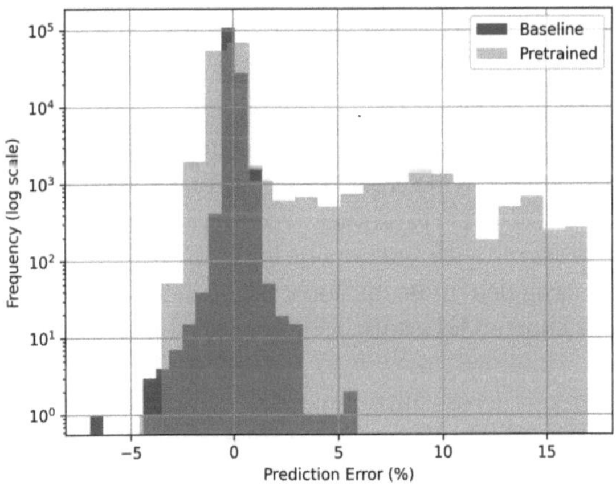

Fig. 1. Error distribution for USD/JPY on H1 interval.

5 Discussion and Conclusion

This study examined the impact of self-supervised pretraining, specifically using MTM approach, on the forecasting accuracy of LSTM-based models in the Forex market. Experiments were conducted on five currency pairs across three time intervals, comparing a baseline supervised model to a model pretrained with MTM and fine-tuned.

Results showed that MTM pretraining did not consistently improve forecasting performance. In most of the 15 configurations, the pretrained model achieved

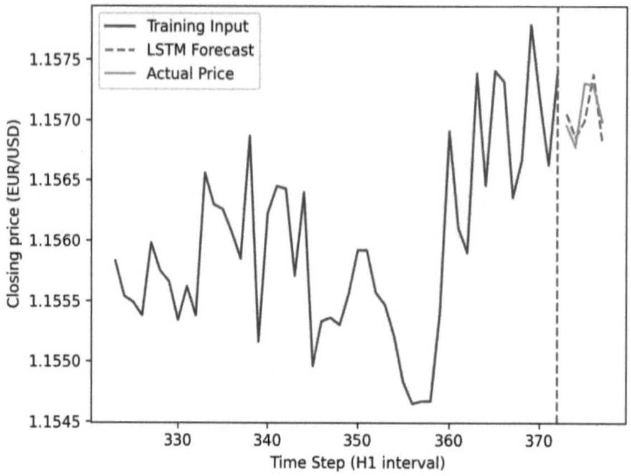

Fig. 2. Baseline model prediction for EUR/USD on H1 interval.

higher RMSE values than the baseline. Small improvements were observed for EUR/USD (D1) and EUR/AUD (H4), but overall, pretraining led to a deterioration in predictive accuracy, suggesting that the learned representations did not transfer effectively to the forecasting task.

Performance varied by time interval. Pretrained models performed best on H1 (hourly) data, where RMSE increases were smaller and occasional gains were noted. On H4 data, performance worsened, and the D1 interval showed the largest deterioration, with extremely high RMSE values. This pattern suggests that MTM struggled more as forecasting horizons lengthened and noise increased, whereas shorter H1 sequences preserved enough structure for partial benefits.

Results also differed across currency pairs. Slight improvements were seen for EUR/USD and EUR/AUD, likely due to their stability and liquidity. However, pairs like USD/JPY, AUD/JPY, and EUR/JPY experienced significant performance drops, indicating that MTM pretraining struggles with the complex and dynamic behavior of certain Forex pairs.

The statistical analysis of input data provides important context for interpreting the mixed outcomes of MTM pretraining. While self-supervised learning is designed to capture temporal structure by reconstructing masked segments, its effectiveness depends on the predictability of those segments from surrounding context.

In low-volatility, moderate-kurtosis environments such as EUR/USD, masked values remain relatively consistent with local trends, enabling the pretrained model to extract useful representations. This explains the modest gains observed for these pairs. In contrast, highly volatile and fat-tailed series such as AUD/JPY exhibit frequent extreme jumps that cannot be reliably inferred from context. When such segments are masked, MTM effectively forces the model to "guess" values that are inherently unpredictable, leading to unstable or misleading representations and degraded forecasting accuracy.

Skewness provided less explanatory power in this study, though strongly negative or positive values may still contribute to asymmetric risk profiles in some pairs. Overall, the findings suggest that volatility and kurtosis are the key statistical properties shaping whether MTM pretraining helps or harms forecasting.

Some limitations should be also acknowledged. First, the architecture was fixed to LSTM; exploring other models like Transformers could offer improvements. Second, only the closing price was used as input; incorporating technical indicators or macroeconomic data might enhance learning.

Although MTM pretraining did not consistently improve performance, this study highlights important challenges in applying self-supervised learning to Forex data. Adapting pretraining strategies to account for financial time series properties such as volatility and non-stationarity appears critical.

Future work could explore MTM and other self-supervised techniques in different markets like cryptocurrencies or stocks, investigate alternative architectures like Transformers, and incorporate richer multivariate inputs. In conclusion, while standard MTM pretraining has limitations in Forex forecasting, it opens promising directions for refining self-supervised learning approaches for financial prediction tasks.

References

1. Man Lui, K., Chong, T.T.: Do technical analysts outperform novice traders: experimental evidence. Econ. Bull. **33**(4), 3080–3087 (2013)
2. Zuckerman, G.: Emotional decisions in trading. In: The Man Who Solved the Market: How Jim Simons Launched the Quant Revolution, pp. 68, 153. Penguin Books (2019)
3. Bentes, S.R., Navas, R.: The fundamental analysis: an overview. Int. J. Latest Trends Fin. Eco. Sci. **3**(1), 389–393 (2013). http://moneycentral.msn.com/home.asp
4. Hayes, A.: Technical Analysis: What It Is and How to Use It in Investing, https://www.investopedia.com/terms/t/technicalanalysis.asp
5. Park, C.H., Irwin, S.H.: What do we know about the profitability of technical analysis? J. Econ. Sur. **21**(4), 786–826 (2007). https://doi.org/10.1111/j.1467-6419.2007.00519.x
6. Liu, J., et al.: Artificial intelligence in the 21st century. IEEE Access **6**, 34403–34421 (2018). https://doi.org/10.1109/ACCESS.2018.2819688
7. Zhang, K., et al.: Self-supervised learning for time series analysis: Taxonomy, progress, and prospects. IEEE Trans. Pattern Anal. Mach. Intell. **18**(7), 722–733 (2024)
8. Devlin, J., Chang, M.W., Lee, K., Toutanova, K.: Bert: Pre-training of deep bidirectional transformers for language understanding. In: Proceedings of the 2019 conference of the North American chapter of the association for computational linguistics: human language technologies, volume 1 (long and short papers). pp. 4171–4186 (2019)

9. Hu, Z., Zhao, Y., Khushi, M.: A survey of forex and stock price prediction using deep learning. Appl. Syst. Innov. **4**(1), 1–30 (2021). https://doi.org/10.3390/ASI4010009
10. Singh, A.P., Nair, A.V., Varalakshmi, S.: Deep Learning for Forex and Stock Price Forecasting: Uncovering Market Secrets. 2023 International Conference on Advances in Computation, Communication and Information Technology, ICAICCIT 2023, pp. 1397–1402 (2023).https://doi.org/10.1109/ICAICCIT60255.2023.10465782
11. Chen, T., Kornblith, S., Norouzi, M., Hinton, G.: A simple framework for contrastive learning of visual representations. In: International conference on machine learning. pp. 1597–1607. PmLR (2020)
12. Zanc, R., Cioara, T., Anghel, I.: Forecasting Financial Markets using Deep Learning. Proceedings - 2019 IEEE 15th International Conference on Intelligent Computer Communication and Processing, ICCP, pp. 459–466 (2019). https://doi.org/10.1109/ICCP48234.2019.8959715
13. Sun, J., Qing, Y., Liu, C., Lin, J.: Self-fts: A self-supervised learning method for financial time series representation in stock intraday trading. In: 2022 IEEE 20th International Conference on Industrial Informatics (INDIN). pp. 501–506. IEEE (2022)
14. Abedin, M.Z., Moon, M.H., Hassan, M.K., Hajek, P.: Deep learning-based exchange rate prediction during the COVID-19 pandemic. Ann. Oper. Res. (2021). https://doi.org/10.1007/s10479-021-04420-6, https://doi.org/10.1007/s10479-021-04420-6
15. Yıldırım, D.C., Toroslu, I.H., Fiore, U.: Forecasting directional movement of Forex data using LSTM with technical and macroeconomic indicators. Financ. Innov. **7**(1), 1–36 (2021) https://doi.org/10.1186/s40854-020-00220-2
16. Jawed, S., Grabocka, J., Schmidt-Thieme, L.: Self-supervised Learning for Semi-supervised Time Series Classification. In: Lauw, H.W., Wong, R.C.-W., Ntoulas, A., Lim, E.-P., Ng, S.-K., Pan, S.J. (eds.) PAKDD 2020. LNCS (LNAI), vol. 12084, pp. 499–511. Springer, Cham (2020). https://doi.org/10.1007/978-3-030-47426-3_39
17. Chen, Q., Zhao, M., Zhang, Z.: Self-supervised representation learning for time series via temporal contrasting and transformation. In: 2023 IEEE 3rd International Conference on Computer Communication and Artificial Intelligence (CCAI). pp. 1–6. IEEE (2023)
18. Qi, L.: Forex Trading Signal Extraction with Deep Learning Models (2023)
19. Gkillas, K., Gupta, R., Pierdzioch, C.: Forecasting (downside and upside) realized exchange-rate volatility: Is there a role for realized skewness and kurtosis? Phys. A: Statist. Mech. Its Appl. **532**, 121867 (2019). https://doi.org/10.1016/j.physa.2019.121867, https://www.sciencedirect.com/science/article/pii/S0378437119310957

Exchange Rate Forecasting Using Adaptive Neuro-Fuzzy Inference System (ANFIS): Performance Evaluation of Wavelet Transform Functions

Mateusz Kaszubowski[1], Piotr Lachowicz[1], and Paweł Weichbroth[2](✉)

[1] Gdansk University of Technology, Gdańsk, Poland
[2] Department of Software Engineering, Faculty of Electronics, Telecommunications and Informatics, Gdansk University of Technology, Gdańsk, Poland
pawel.weichbroth@pg.edu.pl

Abstract. This paper investigates wavelet-based preprocessing for currency exchange rate forecasting within an adaptive neuro-fuzzy inference system (ANFIS). While the integration of discrete wavelet transform (DWT) with ANFIS has been explored in earlier studies, our contribution lies in systematically identifying the most effective wavelet family and decomposition level for this task. Using 1-minute mid-price data (the mean of bid and ask) of the Polish zloty (PLN) against USD, CAD, CHF, JPY, and EUR, we evaluate seven wavelet families across three decomposition levels (2–4). Results indicate that the reverse biorthogonal wavelet (rbio2.4) at level 3 yields the best predictive performance, providing a balance between capturing long-term trends and short-term fluctuations. Compared to models trained directly on raw data, the optimized wavelet-enhanced ANFIS reduces the effective training set size to approximately 12.5% of the original while significantly lowering forecast errors and improving stability. These findings highlight the importance of wavelet selection as a critical design choice in hybrid financial time series forecasting models.

Keywords: Currency exchange rate forecasting · Financial time series · Adaptive neuro-fuzzy inference system (ANFIS) · Discrete wavelet transform (DWT) · Performance evaluation

1 Introduction

Accurate forecasting of currency exchange rates is essential for global trade [10], investment decisions [19], and economic policy formulation [1], yet it remains a persistent challenge in financial markets. Exchange rates are inherently volatile and influenced by complex, nonlinear interactions among macroeconomic indicators, geopolitical events, and market sentiment. Traditional econometric models, such as autoregressive integrated moving average (ARIMA) [11] and vector error

M. Pondel et al. (Eds.): AI4KAM 2025, IFIP AICT 776, pp. 327–340, 2026.
https://doi.org/10.1007/978-3-032-18920-2_31

correction models (VECM) [2], have been widely used for exchange rate prediction but they often fall short when handling nonstationary behavior and abrupt fluctuations in financial time series [22].

In recent years, researchers have increasingly explored artificial intelligence techniques, including artificial neural networks (ANNs) [12], fuzzy logic systems [14], and hybrid approaches [18]. Among these, the adaptive neuro-fuzzy inference system (ANFIS) has shown promise by combining the learning capabilities of neural networks with the interpretability of fuzzy logic [3]. To further enhance forecasting performance, usage of wavelet transforms, particularly discrete wavelet transform (DWT) [16], has been proposed [20]. This preprocessing step aims to extract multiresolution features from exchange rate data [7].

In this paper, we systematically investigate the impact of wavelet family and decomposition level selection within a wavelet-enhanced ANFIS framework for currency exchange rate forecasting. Unlike prior studies that applied wavelet preprocessing in an ad hoc manner, we conduct a comparative analysis of seven families across three decomposition levels. We apply our evaluation to five major exchange rates against the Polish zloty (USD/PLN, EUR/PLN, CHF/PLN, CAD/PLN, and JPY/PLN), decomposing 1-minute series into multiple frequency components. By comparing wavelet families and decomposition depths, we identify the most effective configuration for improving forecasting accuracy and stability. The findings highlight the critical role of wavelet selection as a design factor in hybrid forecasting systems for financial time series.

2 Theoretical Background

2.1 Adaptive Neuro-Fuzzy Inference System (ANFIS)

The ANFIS model is a class of intelligent systems that integrates the learning capability of neural networks with the interpretability of fuzzy logic. Introduced by Jang in 1993 [13], ANFIS uses a hybrid learning algorithm to optimize the parameters of a Sugeno-type fuzzy inference system (FIS) [21] by combining gradient descent backpropagation and least squares estimation techniques.

The ANFIS model consists of a five-layer network, where each layer performs a specific function in the fuzzy inference process:

- **Layer 1 (Fuzzification)**: Converts crisp inputs into fuzzy values using predefined membership functions.
- **Layer 2 (Rule Evaluation)**: Computes the firing strength of fuzzy rules using a T-norm operation, typically the product or minimum function.
- **Layer 3 (Normalization)**: Normalizes the firing strengths to ensure they sum to one.
- **Layer 4 (Consequent layer)**: Produces weighted outputs based on linear functions of the input variables.
- **Layer 5 (Final Output)**: Computes the final crisp output as a weighted sum of the layer 4 outputs.

This architecture, illustrated in Fig. 1, serves as the foundation of the ANFIS model used in our forecasting framework, as described in the following sections.

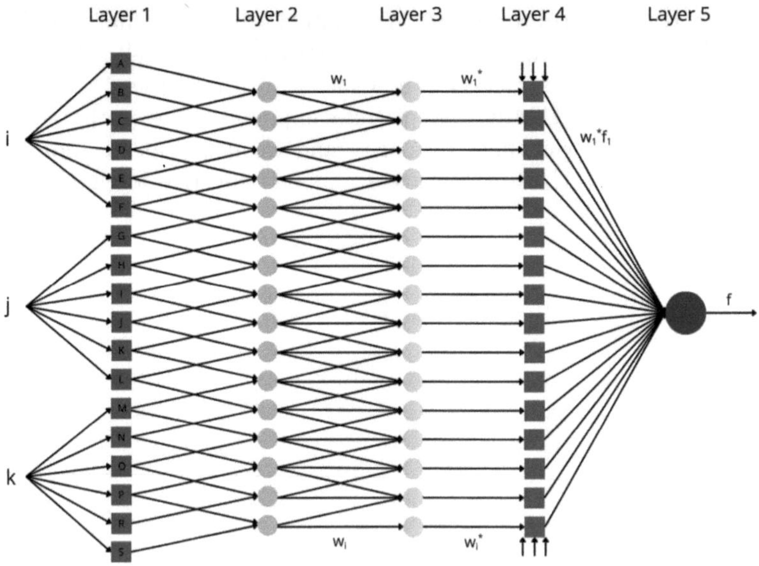

Fig. 1. Structure of the ANFIS model.

2.2 Discrete Wavelet Transform (DWT)

The discrete wavelet transform (DWT) decomposes a time series into multi-resolution frequency components, capturing both local and global signal characteristics. In 1989, Mallat [16] developed the theory of multiresolution signal decomposition, showing that wavelet-based transforms provide an efficient and effective representation of complex signals, including financial time series.

The quality of feature extraction and suitability for time series forecasting are influenced by the support width, symmetry, number of vanishing moments, and regularity of different wavelet families. In this study, we evaluate the following wavelet families, described below and illustrated in Fig. 2.

- **Daubechies (Db)**: A family of compactly supported, orthogonal wavelets with a variable number of vanishing moments depending on the order [5]. Their compact support makes them effective in capturing abrupt changes and local features in financial time series.
- **Coiflets (Coif)**: Nearly symmetric wavelets with vanishing moments in both the wavelet and scaling functions [6]. The symmetry and multiple vanishing moments improve the approximation of smooth trends in time series.
- **Symlets (Sym)**: Modified Daubechies wavelets with improved symmetry while maintaining orthogonality and compact support [6]. Their balanced properties make them suitable for both trend and fluctuation analysis.
- **Biorthogonal (Bior)**: Wavelets using separate decomposition and reconstruction filters, allowing for symmetric transforms and linear phase [4]. Linear phase is advantageous for preserving waveform shapes in financial signals.

- **Reverse Biorthogonal (Rbio)**: A variation of biorthogonal wavelets with reversed roles of decomposition and reconstruction filters [4]. This flexibility can improve feature extraction in non-stationary time series.
- **Discrete Meyer (DMey)**: A discrete approximation of the continuous Meyer wavelet, offering high regularity and frequency localization without compact support [17]. Its smoothness makes it effective for modeling long-term trends.

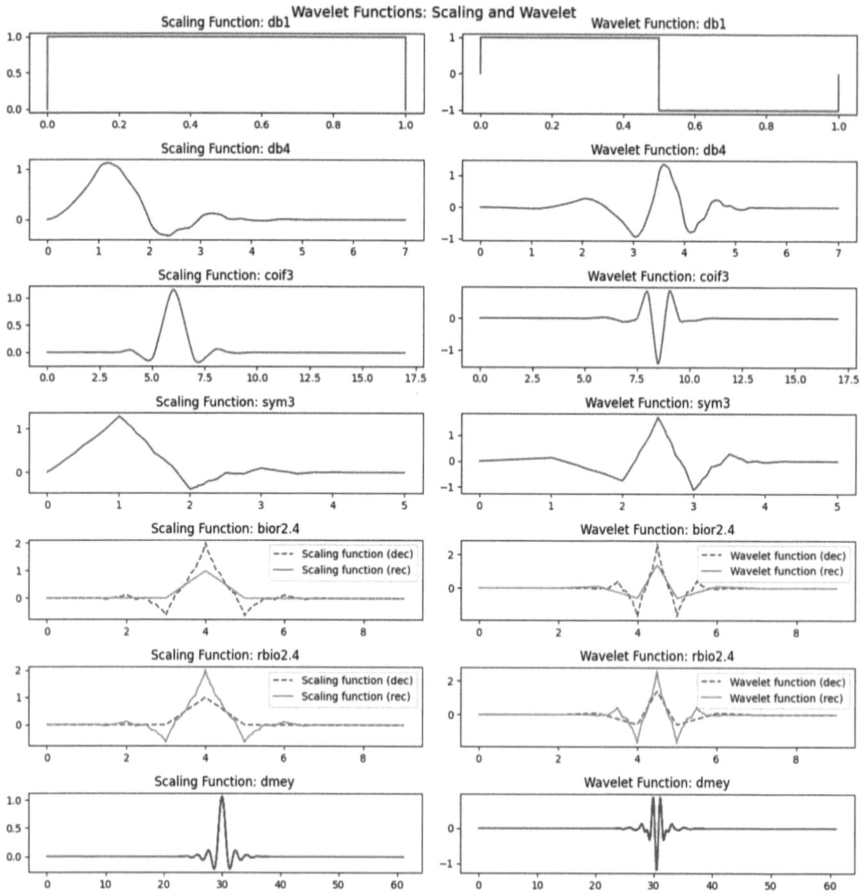

Fig. 2. Scaling and wavelet functions of evaluated families (Haar, Daubechies, Coiflets, Symlets, Biorthogonal, Reverse Biorthogonal, Discrete Meyer).

3 Methodology

3.1 Data

The experimental analysis is based on high-frequency exchange rate data retrieved from the Dukascopy API [8]. The dataset comprises five currency pairs quoted against the PLN: EUR/PLN, CAD/PLN, USD/PLN, JPY/PLN, and CHF/PLN. For each pair, 1-minute mid-prices were computed as the arithmetic mean of bid and ask quotes over a 10-year period (2015-01-01 to 2025-04-22). The CHF/PLN series is shorter due to missing data from the provider in 2015–2016.

3.2 Preprocessing

The tick data are decomposed using seven wavelet families at decomposition levels 2–4, resulting in 21 distinct datasets. Level 1 was excluded because it captures only very high-frequency noise with limited predictive value. Approximation coefficients are retained for model input, while detail coefficients are stored for inverse reconstruction. Next, the approximation series are normalized within non-overlapping monthly windows using a min-max scaling, with parameters estimated only from past data to avoid leakage.

3.3 Feature Extraction

From each rolling one-week window of the normalized approximation series, three features are extracted: variance (VAR), simple moving average (SMA), and weighted moving average (WMA). These measures jointly capture short-term dispersion and trend dynamics, which have been shown to be effective in foreign exchange forecasting [9]. The prediction target is defined as the exchange rate observed one week after the last timestamp in each window.

3.4 Model Training

We employ the State-ANFIS implementation by Lenhard [15], adapted for stateless prediction. Each input variable is represented by six trainable bell-shaped membership functions. Model training uses mean squared error (MSE) as the loss function, and hyperparameters (epochs, learning rate, batch size) are selected through Bayesian optimization. For additional comparison, ANFIS is also trained directly on normalized raw data without wavelet-based preprocessing. For each wavelet-level configuration and the baseline, five independent models are trained, resulting in a total of 110 models.

3.5 Evaluation

Models are evaluated on chronologically ordered test sets, corresponding to the last 10% of each time series, which adds up to approximately 1-year prediction evaluation period. Forecasts are generated iteratively using rolling feature vectors constructed from 1-week historical segments of the approximation coefficients. At each step, the model predicts the next 1-minute exchange rate, which is then incorporated at the beginning of the subsequent feature vector, while the oldest value is removed to maintain a consistent window size. Full exchange rate series are reconstructed via inverse DWT using the stored detail coefficients. Predictive performance is assessed using mean absolute percentage error (MAPE) and root mean squared error (RMSE). Unless stated otherwise, reported results represent the average across five independent runs and all five currency pairs.otherwise statedthe

4 Results

Wavelet transforms successfully denoised the currency exchange rate time series while preserving trends. An example is shown for pair USD/PLN. Figure 3 shows its exchange rates in studied period before wavelet transform. Figure 4 shows it after level 4 decomposition using Haar wavelet. Moreover, the size of datasets decreased exponentially with respect to level of decomposition, which lowered training time and allowed for processing more data. The exact numbers of feature vectors are shown in Table 1.

Table 1. Feature vector sizes for different levels of wavelet decomposition using Haar wavelet

Level	Currency Pair				
	USD/PLN	EUR/PLN	JPY/PLN	CAD/PLN	CHF/PLN
None	1983548	1974895	1972957	1973976	529136
1	991775	987448	986479	986989	264569
2	495888	493725	493240	493495	132285
3	247945	246863	246621	246748	66143
4	123973	123432	123311	123375	33072
5	61987	61717	61656	61688	16537

The experimental results demonstrate how both the choice of wavelet type and the level of decomposition influence the forecasting accuracy of the currency exchange rates. The aim was to identify the most effective combination for minimizing prediction errors and enhancing model performance. To ensure reliability, the evaluation metrics (MAPE, RMSE) were averaged over five training runs conducted under identical conditions using the same dataset and hyperparameters.

4.1 Baseline (No Wavelet)

Models trained on data set not processed using wavelet transform achieved some of the worst performance with average error metrics of MAPE: 0.00973 and RMSE: 0.03295.

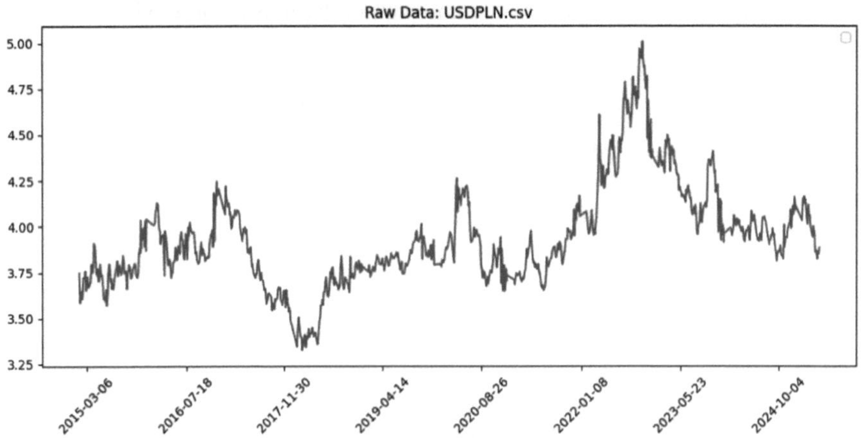

Fig. 3. USD/PLN exchange rate before wavelet transform (2015-01-01–2025-03-28).

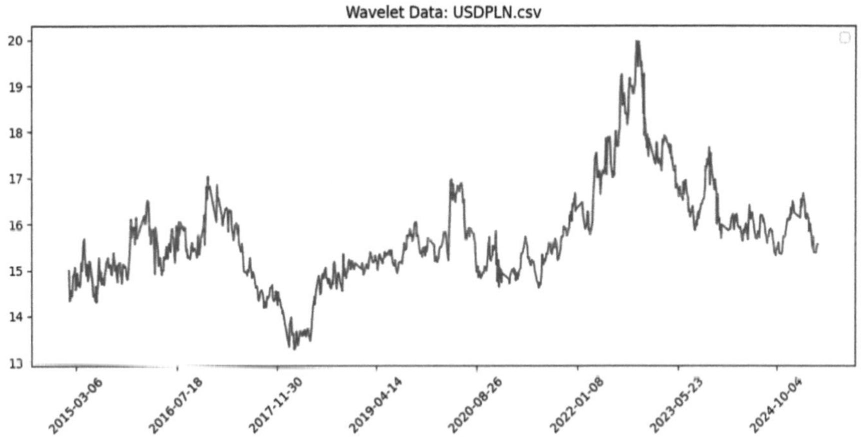

Fig. 4. USD/PLN exchange rate after level 4 wavelet decomposition - Haar wavelet (2015-01-01–2025-03-28).

4.2 Decomposition Level 2

At the second level of decomposition, the reverse biorthogonal wavelet (rbio2.4) demonstrated the best performance in all evaluation metrics, as presented in Table 2. Despite the relatively low decomposition, it effectively compresses the data and accelerates training, all while preserving high prediction accuracy.

4.3 Decomposition Level 3

As shown in Table 3, the third level of decomposition delivers the best overall performance, with the reverse biorthogonal wavelet (rbio2.4) once again achieving the highest accuracy. At this level, the model reaches average error metrics

Table 2. Evaluation results for wavelet decomposition level 2

Wavelet	Mean MAPE	Mean RMSE
rbio2.4	**0.00926**	**0.03230**
bior2.4	0.00933	0.03267
coif3	0.00928	0.03256
sym3	0.00927	0.03255
haar	0.00937	0.03282
dmey	0.00934	0.03270
db4	0.00952	0.03267

of approximately MAPE: 0.00723, and RMSE: 0.02456. Visual comparisons of actual and predicted exchange rates are presented in Figs. 5, 6, 7, 8, and 9. These plots confirm that the model effectively captures the main trends in the exchange rate time series.

Table 3. Evaluation results for wavelet decomposition level 3

Wavelet	Mean MAPE	Mean RMSE
rbio2.4	**0.00723**	**0.02456**
bior2.4	0.00729	0.02473
coif3	0.00755	0.02456
sym3	0.00732	0.02475
haar	0.00737	0.02451
dmey	0.00731	0.02476
db4	0.00743	0.02487

4.4 Decomposition Level 4

As presented in Table 4, the fourth level of decomposition yielded the worst results among all tested configurations. The increased level of decomposition likely led to excessive data reduction, which in turn resulted in the loss of important information necessary for accurate forecasting. This suggests that an overly deep decomposition may degrade model performance rather than enhance it.

Table 4. Evaluation results for wavelet decomposition level 4

Wavelet	Mean MAPE	Mean RMSE
rbio2.4	**0.00982**	**0.03343**
bior2.4	0.00994	0.03380
coif3	0.00986	0.03361
sym3	0.01001	0.03409
haar	0.00985	0.03351
dmey	0.00995	0.03385
db4	0.00998	0.03394

5 Discussion

This study systematically evaluates how different wavelet families and decomposition levels influence the performance of ANFIS models in currency exchange rate forecasting. Using high-frequency data across five PLN currency pairs, we demonstrate that the choice of wavelet plays a critical role in predictive accuracy and model stability, with certain configurations clearly outperforming raw-data ANFIS.

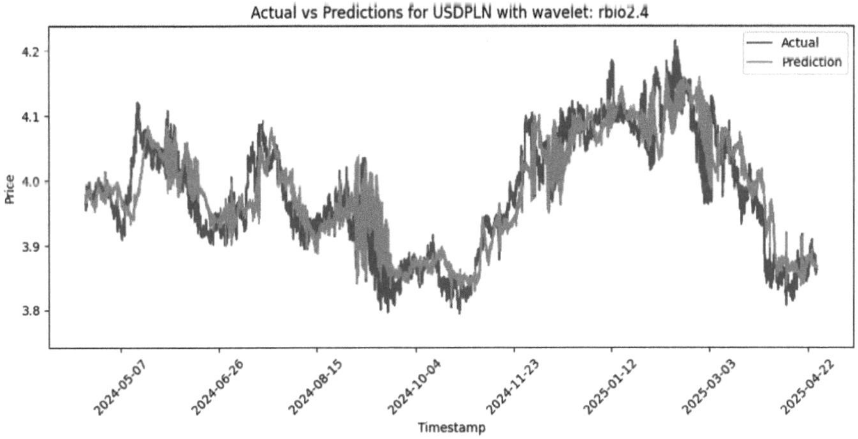

Fig. 5. Comparison of actual and predicted USD/PLN exchange rates using rbio2.4 (level 3).

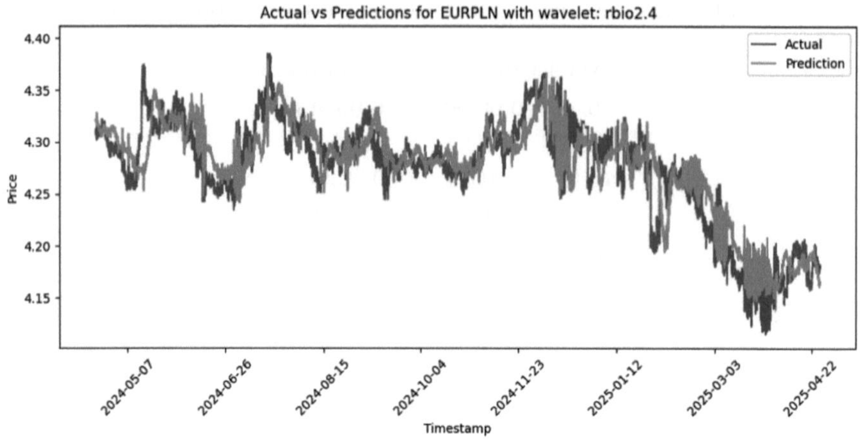

Fig. 6. Comparison of actual and predicted EUR/PLN exchange rates using rbio2.4 (level 3).

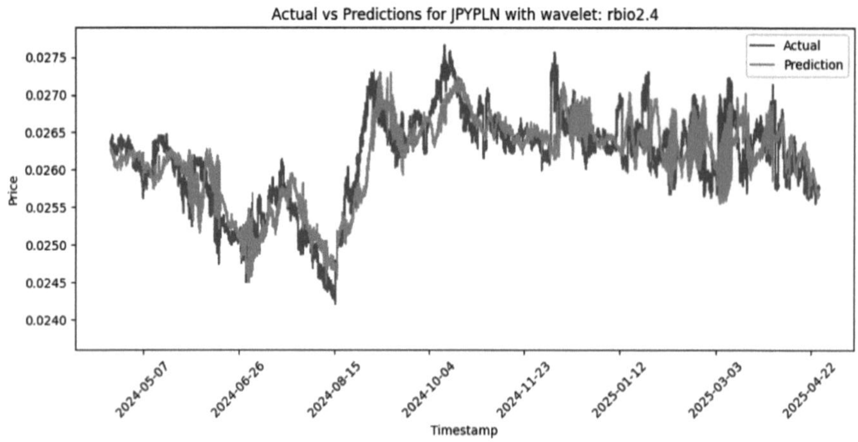

Fig. 7. Comparison of actual and predicted JPY/PLN exchange rates using rbio2.4 (level 3).

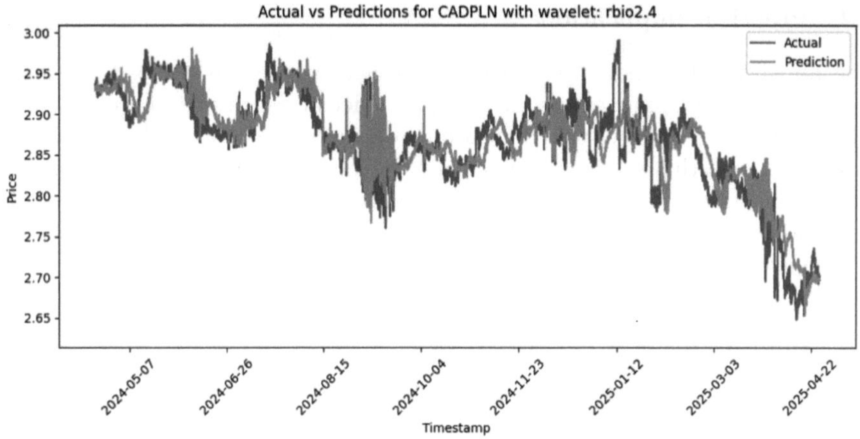

Fig. 8. Comparison of actual and predicted CAD/PLN exchange rates using rbio2.4 (level 3).

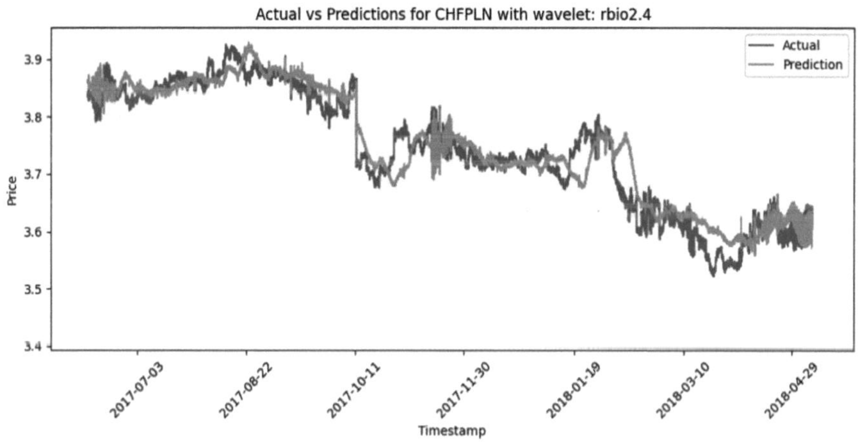

Fig. 9. Comparison of actual and predicted CHF/PLN exchange rates using rbio2.4 (level 3).

5.1 Key Findings

Among the tested configurations, the reverse biorthogonal wavelet (rbio2.4) at level 3 achieved the best results across all evaluation metrics. This wavelet family effectively balances time-frequency resolution, enabling the model to capture both transient fluctuations and longer-term market trends. Conversely, excessive decomposition (e.g., level 4) degraded performance, highlighting the need to carefully select an appropriate depth of analysis.

5.2 Practical Implications

The dimensionality reduction achieved through wavelet decomposition significantly decreased training time, allowing for efficient use of high-frequency data. This positions the wavelet-ANFIS framework as a practical tool for real-time financial forecasting systems, where rapid inference and robustness under market turbulence are critical. The model's ability to isolate noise while preserving informative patterns makes it particularly valuable for risk management and decision-support applications.

5.3 Limitations and Future Work

This research is limited to a single-model ANFIS architectures and a predefined set of five currency pairs. Future work should extend the framework to ensemble approaches, hybridization with attention-based deep learning models, and broader multi-currency analyses. Evaluating statistical significance across different market regimes would also further validate robustness.

6 Conclusion

In summary, this paper provides empirical evidence that wavelet-based preprocessing substantially enhances the performance of ANFIS in financial time series forecasting. The findings underscore the importance of selecting an appropriate wavelet family and decomposition level as critical design choices. By improving accuracy, stability, and computational efficiency, the proposed wavelet-ANFIS framework contributes both to academic research in financial time series analysis and to practical implementations in currency forecasting and risk management.

Disclosure of Interests. The authors have no competing interests to declare that are relevant to the content of this article.

References

1. Al-Thaqeb, S.A., Algharabali, B.G., Alabdulghafour, K.T.: The pandemic and economic policy uncertainty. Int. J. Finan. & Econ. **27**(3), 2784–2794 (2022)
2. Arce, P., Antognini, J., Kristjanpoller, W., Salinas, L.: An online vector error correction model for exchange rates forecasting. In: ICPRAM 2015 - 4th International Conference on Pattern Recognition Applications and Methods, Proceedings, Vol. 2, pp. 193–200 (2015). https://doi.org/10.5220/0005205901930200
3. Atsalakis, G., Skiadas, C., Braimis, I.: Probability of trend prediction of exchange rate by anfis. In: Recent Advances in Stochastic Modeling and Data Analysis, pp. 414–422 (2007). https://doi.org/10.1142/9789812709691_0050

4. Cohen, A., Daubechies, I., Feauveau, J.C.: Biorthogonal bases of compactly supported wavelets. Commun. Pure Appl. Math. **45**(5), 485–560 (1992). https://doi.org/10.1002/cpa.3160450502, https://onlinelibrary.wiley.com/doi/abs/10.1002/cpa.3160450502
5. Daubechies, I.: Orthonormal bases of compactly supported wavelets. Commun. Pure Appl. Math. **41**(7), 909–996 (1988). https://doi.org/10.1002/cpa.3160410705, https://onlinelibrary.wiley.com/doi/abs/10.1002/cpa.3160410705
6. Daubechies, I.: Ten Lectures on Wavelets. Society for Industrial and Applied Mathematics, Philadelphia, PA (1992). https://search.library.wisc.edu/catalog/999685462102121, includes bibliographical references and index
7. Debnath, L., Antoine, J.P.: Wavelet transforms and their applications. Physics Today - Phys. Today **56**, 68–68 (2003). https://doi.org/10.1063/1.1580056
8. Dukascopy: Api for developers, dukascopy.com/swiss/english/marketwatch/api-data-export/
9. Handa, R.: Foreign exchange rate prediction using hybrid of anfis and wavelet with feature extraction and feature selection. Turk. Online J. Qual. Inq. (TOJQI) **12**(9), 7626–7682 (2021)
10. Hartmann, P.: The currency denomination of world trade after european monetary union. J. Japan. Int. Econ. **12**(4), 424–454 (1998)
11. Ho, S., Xie, M.: The use of arima models for reliability forecasting and analysis. Comput. & Ind. Eng. **35**(1), 213–216 (1998). https://doi.org/10.1016/S0360-8352(98)00066-7, https://www.sciencedirect.com/science/article/pii/S0360835298000667
12. Huang, W., Lai, K.K., Nakamori, Y., Wang, S.: Forecasting foreign exchange rates with artificial neural networks: A review. Int. J. Inf. Technol. Decis. Making **3**, 145–165 (03 2004). https://doi.org/10.1142/S0219622004000969
13. Jang, J.S.: Anfis: adaptive-network-based fuzzy inference system. IEEE Trans. Syst. Man Cybern. **23**(3), 665–685 (1993). https://doi.org/10.1109/21.256541
14. Korol, T.: A fuzzy logic model for forecasting exchange rates. Knowl.-Based Syst. **67**, 49–60 (2014)
15. Lenhard, G., Maringer, D.: State-ANFIS: A generalized regime-switching model for financial modeling. In: 2022 IEEE Symposium on Computational Intelligence for Financial Engineering and Economics (CIFEr). pp. 1–8. IEEE (2022). https://doi.org/10.1109/CIFEr52523.2022.9776208
16. Mallat, S.: A theory for multiresolution signal decomposition: the wavelet representation. IEEE Trans. Pattern Anal. Mach. Intell. **11**(7), 674–693 (1989). https://doi.org/10.1109/34.192463
17. Meyer, Y.: Wavelets and Operators. Cambridge Studies in Advanced Mathematics, Cambridge University Press (1993)
18. Morkowski, J.: The accuracy of forecasting neural networks and the impact of using fuzzy sets for the currency market. Sci. J. Bielsko-Biala Sch. Finan. Law **28**(1), 24–31 (2024). https://doi.org/10.19192/wsfip.sj1.2024.3, https://asej.eu/index.php/asej/article/view/725
19. Piasecki, K., Stasiak, M.D., Staszak, Ż.: Investment decision support on precious metal market with use of binary representation. n: Tarczyński, W., Nermend, K. (eds.) Effective Investments on Capital Markets: 10th Capital Market Effective Investments Conference (CMEI 2018). pp. 423–438. Springer (2019). https://doi.org/10.1007/978-3-030-21274-2_29
20. S, K.C.: Fusion model of wavelet transform and adaptive neuro fuzzy inference system for stock market prediction. J. Ambient Intell. Human. Comput. (2019). https://doi.org/10.1007/s12652-019-01224-2

21. Takagi, T., Sugeno, M.: Fuzzy identification of systems and its applications to modeling and control. IEEE Trans. Syst., Man, Cybern. **SMC-15**, 116–132 (1985). https://api.semanticscholar.org/CorpusID:3333100
22. Xie, C., Mao, Z., Wang, G.J.: Forecasting rmb exchange rate based on a non-linear combination model of arfima, svm, and bpnn. Math. Probl. Eng. **2015**(1), 635345 (2015). https://doi.org/10.1155/2015/635345, https://onlinelibrary.wiley.com/doi/abs/10.1155/2015/635345

Towards Reliable Speech-to-Text Systems: Generative AI-Based Context-Aware Post-processing for Punctuation, Homophone Resolution, and Non-native Speech

Artur Karczmarczyk[1], Aleksandra Bączkiewicz[2](✉), Aleksandra Karczmarczyk[1], Jakub Więckowski[2], and Grażyna Rosa[2]

[1] Department of Computer Science, Westpomeranian University of Technology in Szczecin, ul. Żołnierska 49, 71-210 Szczecin, Poland

[2] University of Szczecin, al. Papieża Jana Pawła II 22a, 70-453 Szczecin, Poland

artur.karczmarczyk@zut.edu.pl

Abstract. The traditional speech transcription pipeline consists of two stages: speech acquisition and the conversion of spoken language into written text using Automatic Speech Recognition (ASR) systems. Despite substantial advances in recent years, ASR technologies continue to exhibit limitations, particularly in handling punctuation, homophones, and linguistic disfluencies. This article proposes an extension of the conventional pipeline by introducing a third stage – post-processing in which generative AI is used to enhance transcription quality. Using a large language model guided by a carefully crafted prompt, raw transcripts are refined for coherence, accuracy, and proper punctuation, while minimizing unintended alterations. An empirical evaluation was conducted on 224 audio samples, including both human and AI-generated voices, with variations in speech rate and three acoustic disturbances: recording distortions, train noise, and crowd noise. Results demonstrate substantial improvements: punctuation restoration, correction of homophone errors in over 43% of cases, and elimination of hesitations or repetitions in 74% of transcripts. Overall, 35% of outputs were subjectively rated as superior, with the most pronounced benefits observed in non-native speakers' speech.

Keywords: generative AI · automated speech recognition · homophones

1 Introduction

Speech transcription has become increasingly important as the demand to turn spoken language into written text continues to grow across many different areas [4]. It is now widely used in business settings [8], commercial services [11], and

Published by Springer Nature Switzerland AG 2026
M. Pondel et al. (Eds.): AI4KAM 2025, IFIP AICT 776, pp. 341–350, 2026.
https://doi.org/10.1007/978-3-032-18920-2_32

data analysis tasks [2]. While today's automatic speech transcription systems are already fairly advanced, ongoing research is focused on improving their accuracy further, particularly when dealing with difficult or noisy conditions [9].

Progress in speech processing technologies plays a key role in enhancing communication and supporting automation across a wide range of fields [10,16]. In general, the workflow consists of two main steps: first, capturing audio through a microphone or similar device, and second, converting that audio into text with the help of automatic speech recognition (ASR) systems [5].

The first stage involves capturing spoken input with recording devices such as microphones, with the goal of collecting speech data for further processing or analysis [3,19]. The second stage focuses on transforming the recorded audio into text, which may be done manually or, more commonly, through Automatic Speech Recognition (ASR) systems [14]. Although ASR technologies have advanced considerably, challenges such as low recording quality, linguistic variability, and natural speech disfluencies remain major barriers to achieving consistently high transcription accuracy.

Speech transcription technologies are applied in a wide range of fields. In medicine, they enable quicker and more efficient note-taking [7]; in law, they assist with documenting legal proceedings [20]; and in education and media production, they support the creation of diverse multimedia content [17,18]. They also form the core of voice assistants such as Siri, Alexa, Google Assistant, and Cortana [12], while playing an essential role in accessibility tools designed for people with disabilities [6]. The growing adoption of speech-to-text systems is further driven by advances in artificial intelligence (AI) and natural language processing (NLP), which continue to improve both their accuracy and efficiency [13].

In this work, we propose an extension to the traditional speech transcription pipeline by adding a third stage: generative AI-based post-processing. In our approach, raw transcriptions produced by a widely used speech-to-text API are further refined with the help of a generative language model [21]. Using a carefully designed prompt, the model enhances overall coherence, corrects transcription errors, and restores proper punctuation.

To assess the effectiveness of the proposed method, we conducted an empirical study on a dataset of 224 audio samples representing diverse speaker profiles, speech rates, and background noise conditions. The findings indicate that the generative AI post-processing step led to clear improvements: punctuation was consistently enhanced, homophone errors were corrected in more than 43% of cases, and disfluencies such as hesitations were removed in 74% of samples. Overall, 35% of transcripts were judged to be subjectively improved, with the most substantial benefits observed in recordings from human speakers with non-native accents.

This paper is structured as follows: Sect. 2 presents background information on speech recognition and transcription, including a review of related work and a detailed description of our methodology. Section 3 discusses the results of the empirical study. Finally, Sect. 4 summarizes the findings, highlights key contributions, and outlines directions for future work.

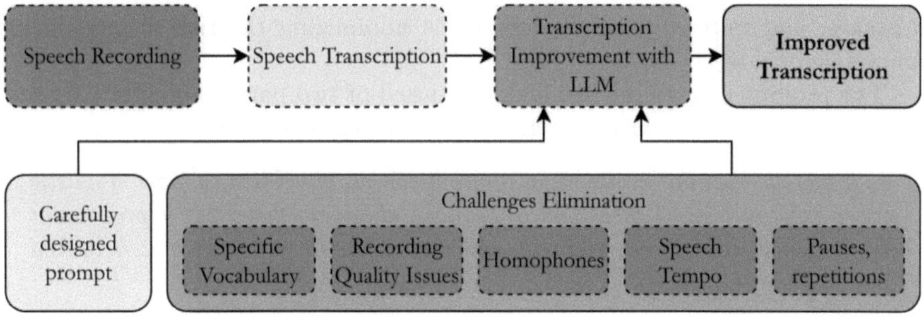

Fig. 1. Conceptual diagram of the novel proposed approach.

2 Materials and Methods

Human–computer interaction via speech generally consists of two main stages. The first is speech acquisition, in which audio is captured through recording devices like microphones and then stored in a digital format to allow for further processing and analysis [19].

The second stage is speech transcription, where the recorded audio is converted into text, either manually or through ASR systems [14]. Despite continuous improvements, ASR continues to encounter several challenges, including, but not limited to:

- language diversity – ASR systems often struggle with recognizing different languages, dialects, or accents, which can result in inaccurate or conflicting transcriptions;
- recording quality – background noise, distortions, poor audio fidelity, or malfunctioning equipment can significantly affect transcription accuracy, producing outputs that do not match the original speech;
- specialized vocabulary – rare or domain-specific terms, such as technical jargon, may be misrecognized or replaced incorrectly, reducing the reliability of the transcription;
- homophones – words that have the same pronounciation but different meaning (e.g., by, buy, bye; or tire as in fatigue vs. car wheel), can cause context misinterpretation [1,15];
- speech tempo – variations in speaking rate or sudden pauses can lead to misinterpretation of certain phrases;
- disfluencies – repetitions, filler words, or vocal hesitations (e.g., *um*, *hmm*) may be transcribed unnecessarily or mistaken for meaningful content.

In our approach, we add a third stage to the conventional speech-to-text workflow. Following the initial recording and transcription – which may produce a preliminary text containing errors – the transcript undergoes further refinement using generative artificial intelligence. This stage employs a large language model (LLM) guided by a carefully designed prompt to enhance coherence, correct mistakes, and improve punctuation, while minimizing the risk of introducing unintended changes.

The prompt used in our study is composed of two parts. The first part provides instruction and context to guide the model's behavior:

> *In language that the sentence is, improve the provided text to make it a little more coherent and logical, but only make changes where the improvement is absolutely certain, mostly only while word has no sense in sentence. Add punctuation as needed.*

The second part includes the input text for enhancement:

> *Please improve the following text: {text}*

The integration of generative AI is intended to address common transcription challenges, including inconsistent punctuation, homophone errors, disfluencies, and inaccuracies arising from variable speech tempo or accent differences. A conceptual overview of this extended transcription pipeline is shown in Fig. 1. The following section provides a detailed empirical evaluation of the proposed method.

3 Results and Discussion

For the empirical study, a dedicated set of test samples was recorded. To ensure the evaluation captured key transcription difficulties, the test phrase was deliberately constructed to incorporate several common challenges, with a particular focus on homophones: *Homophones can be quite the bear trap for the unwary writer. For instance: 'The bear was bare, but he still managed to bare his teeth in a frightening manner.' This sentence cleverly plays on the words b-e-a-r (animal), b-a-r-e (naked) and b-a-r-e (uncover), highlighting the subtle yet significant differences in their meanings.*

Addressing the challenge of language diversity within a single test phrase was difficult, so it was approximated through accent variation. The test phrase was recorded by four human speakers (two female and two male) and supplemented with two AI-generated voices (one female and one male).

To investigate the impact of speech tempo, each human speaker recorded the test phrase three times, deliberately varying their speaking rate and rhythm. Additional variants were also created by increasing the playback speed to 1.25×.

To simulate disfluencies, additional samples were included with inserted hesitations ("um...") and repeated phrases, with all speakers producing the same patterns of repetitions and disfluencies.

Finally, to evaluate the effects of recording quality, three types of audio disturbances were introduced: recording distortion, train background noise, and crowd noise.

Combining all variations yielded 224 audio samples. These were first transcribed using Google Speech Recognition (GSR), then refined by OpenAI's `GPT 3.5 Turbo` using the prompt described in Sect. 2.

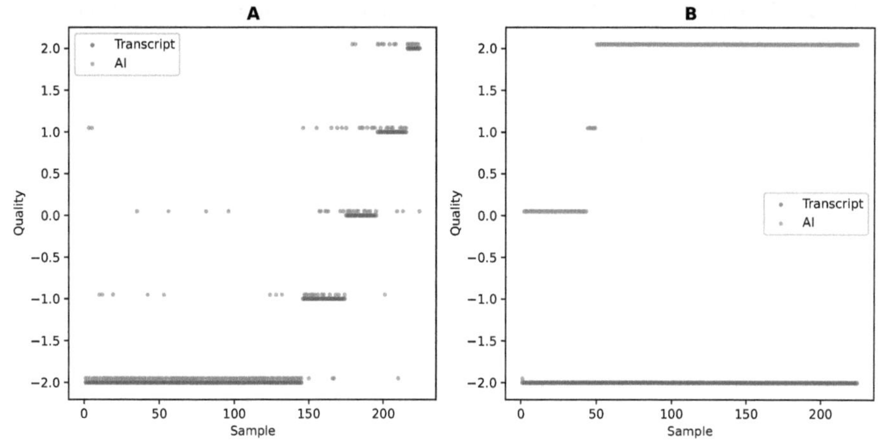

Fig. 2. Evaluation of all samples in the empirical study: A) words quality, B) punctuation quality

Table 1. Comparison of accuracy evaluation scores for original test phrase and one with simulated disfluencies.

Text	Transcript		AI-Improved		Is better?
	Words	Punctuation	Words	Punctuation	
Original	−1.2768	−2.0000	−1.0804	1.5268	0.3125
Original + Disfluencies	−1.2321	−2.0000	1.0179	1.6339	0.3929
Overall	−1.2545	−2.0000	−1.0491	1.5804	0.3527

Originally, the authors' intention was to build a framework which would compare the transcriptions to original text automatically. However, due to limitations in automated text comparison and the inaccuracy of preliminary results, all samples were ultimately evaluated manually. Each transcription was rated on a 5-point scale, where 0 indicated no change, +1 and +2 represented improvements, and -1 and -2 indicated a decline in word accuracy and punctuation. Additionally, a binary flag was recorded to indicate whether the AI-refined version was subjectively superior to the initial GSR transcript.

Figure 2 presents a comparison of ASR transcriptions and AI-refined transcriptions across all 224 samples, considering both word-level accuracy (A) and punctuation (B). Since the GSR system does not generate punctuation, all original transcripts were initially rated -2 in this regard. Punctuation therefore had to be added from scratch, and as shown in Fig. 2B, the majority of samples achieved nearly perfect punctuation following AI refinement. Notably, improvements in punctuation were observed even in cases where word-level transcription quality remained suboptimal. This indicates that, in some instances, the AI was able to build proper punctuation even when the conveyed context was partially lost.

Table 2. Comparison of evaluation scores for original test phrase with base speed and various challenges.

Challenge	Transcript		AI-improved		
	Words	Punctuation	Words	Punctuation	Is better?
none	0.000	−2.000	0.000	2.000	0.571
train	-0.786	−2.000	−0.357	1.714	0.429
crowd	-1.000	−2.000	−0.714	1.571	0.500
distortion	-1.429	−2.000	−1.214	1.500	0.286
Overall	-0.804	−2.000	-0.571	1.696	0.446

Table 3. Comparison of evaluation scores for original test phrase with increased speed and various challenges.

Challenge	Transcript		AI-improved		
	Words	Punctuation	Words	Punctuation	Is better?
none	−1.071	−2.000	−0.786	1.857	0.571
train	−2.000	−2.000	−1.929	1.714	0.000
crowd	−1.929	−2.000	−1.786	1.143	0.071
distortion	−2.000	−2.000	−1.857	0.714	0.071
Overall	−1.750	−2.000	−1.589	1.357	0.179

The overall accuracy results for the full experiment are summarized in Table 1. While word-level accuracy showed modest improvement (from −1.2545 to −1.0491), punctuation quality experienced a substantial increase (from −2.0 to +1.58). Considering only the subjective improvement flag, 35.27% of samples were deemed improved. Moreover, when comparing the original text to versions containing simulated disfluencies and repetitions, the proposed approach achieved higher subjective improvement for samples with disfluencies (0.3929) than for those without (0.3125).

Table 2 presents a breakdown of results by challenge type, considering only samples with the original text, without disfluencies, and at normal playback speed. Under ideal conditions, GSR achieved moderate performance (score = 0), while AI refinement enhanced punctuation to near-perfect levels, resulting in 57% of samples showing subjective improvement. In contrast, recordings affected by background noise (train: −0.786, crowd: −1.000) or distortion (−1.429) led to decreased GSR performance. In these challenging scenarios, the AI consistently improved word-level scores and still enhanced punctuation, although improvements were less pronounced under these severe conditions (Table 3).

For samples with increased speech speed, the performance of both GSR and AI-refined transcripts was generally lower. While 57% of samples without distortions were still judged as subjectively improved, the addition of train noise, crowd noise, or recording distortions led to a substantial decline in word recog-

Table 4. Comparison of evaluation scores for original test phrase with base speed depending on the speaker type and gender.

Speaker		Transcript		AI-improved		
Type	Gender	Words	Punctuation	Words	Punctuation	Is better?
AI	Both	0.875	-2.000	0.750	1.500	0.750
	Female	0.250	-2.000	0.750	1.500	0.750
	Male	1.500	-2.000	0.750	1.500	0.750
Human	Both	-1.083	-2.000	-0.792	1.729	0.396
	Female	-1.375	-2.000	-1.042	1.750	0.375
	Male	-0.792	-2.000	-0.542	1.708	0.417
$\sum$		-0.804	-2.000	-0.571	1.696	0.446

nition. This decrease likely results from GSR failing to capture sufficient context in noisy recordings, leaving the LLM with too little information to meaningfully enhance the transcription. An example GSR transcription of such a challenging sample is shown below:

colors can be quite the bear trap for the unbury Verizon for instance the bear was better still managed to prepare his team in the Friday morning the sentence with Days on the Road e a r animal p a r NAIT the d a r e hunger highlighting the subtle yet significant difference in their meanings

Given the nonsensical nature of such a transcription, the LLM was unable to generate a coherent or meaningful version:

Colors can be quite the bear trap. For instance, the bear was better still managed to prepare his team on Friday morning. The sentence "Days on the Road" is a riddle, highlighting the subtle yet significant difference in their meanings.

Despite this, the AI was able to introduce some coherence into the text. However, the resulting transcription is better characterized as containing hallucinations rather than accurately reflecting the original content. Notably, though, the punctuation was rendered correctly.

Overall, 17.9% of transcripts were judged to be subjectively improved by the AI compared to the original GSR outputs.

Table 4 illustrates the impact of speaker type on transcription results. For AI-generated voices, GSR produced moderately accurate transcriptions (average score = 0.875); however, subsequent AI post-processing occasionally introduced hallucinations, slightly reducing word-level accuracy. In contrast, for human, non-native speakers, AI refinement consistently enhanced transcription quality, likely due to the model's ability to use contextual cues to correct errors – for example, changing "bear his teeth" to the correct "bare his teeth."

This last example underscores the effectiveness of homophone resolution in the proposed approach. While none of the GSR transcriptions correctly rendered "bare his teeth," 43.3% of AI-refined transcriptions did. In the GSR outputs, 131 instances were incorrectly transcribed as "bear his teeth," with 11 occurrences of "wear" and a few others, such as "buy," "prepare," and "spare." The AI-enhanced transcriptions produced 97 correct instances of"bare his teeth," although 39 cases remained as "bear his teeth." Some unintended substitutions, including "sink" or "biting," were also observed, highlighting the ongoing challenge of balancing improvements with the risk of hallucinations.

Finally, generative AI proved highly effective in addressing disfluencies. Among samples containing inserted hesitations and repeated phrases, 74% were successfully corrected, while only 4% showed no measurable improvement.

4 Conclusions

Despite significant advances in automatic speech recognition (ASR), transcription quality remains affected by factors such as background noise, speaker variability, homophones, and speech disfluencies including pauses and repetitions. This study addressed these limitations by exploring the use of generative AI for post-processing ASR-generated transcripts, aiming to improve accuracy and readability, particularly under challenging conditions.

Our methodology employed a three-stage transcription pipeline: (1) audio recording, (2) initial transcription using Google Speech Recognition (GSR), and (3) refinement with a generative AI model guided by a controlled prompt. A diverse dataset of 224 recordings was created, including both human and AI-generated voices, with variations in speech rate and exposure to different types of noise and audio distortions.

The main contributions of this study are as follows:

- Demonstration of a novel application of generative AI to enhance ASR outputs without the need to retrain existing ASR models.
- A comprehensive evaluation across a wide range of controlled distortions, noise conditions, and diverse speaker types.
- Quantitative assessment of improvements using manual Likert-like scale ratings for word accuracy and punctuation, complemented by subjective evaluations of overall transcription quality.

The results indicate that generative AI not only produced subjective improvements in 35% of transcripts but also substantially enhanced punctuation accuracy. Furthermore, the AI demonstrated improved homophone resolution, correctly converting phrases such as "bear his teeth" to "bare his teeth" in 43.3% of cases, and effectively eliminated hesitations and repeated words in 74% of samples. Notably, transcription quality for human, non-native speakers showed greater gains compared to AI-generated speech, highlighting the value of generative models' contextual inference when initial transcriptions are imperfect.

In summary, generative AI post-processing offers a promising, lightweight enhancement to traditional ASR systems, particularly for punctuation restoration and homophone disambiguation in challenging acoustic environments.

Although the findings of this study demonstrate the advantages of using generative AI for transcription enhancement, several challenges remain. A key limitation is the difficulty of applying this post-processing approach to real-time transcription. Because generative AI relies on complete sentences to effectively improve grammar, structure, and punctuation, it introduces latency that is incompatible with live or near-instantaneous speech-to-text systems. Future research should investigate strategies for incremental processing or predictive sentence completion to enable partial, real-time improvements without compromising quality. Additionally, further efforts are needed to optimize prompt design in order to minimize the occurrence of hallucinations.

Acknowledgments. Publication co-financed by Minister of Science under the"Regional Excellence Initiative" Program RID/SP/0046/2024/01.

Disclosure of Interests. The authors have no competing interests to declare that are relevant to the content of this article.

References

1. Broniś, O.: Analiza semantyczna i fonetyczna wybranych fałszywych przyjaciół polsko-włoskich. Wokół homonimii językowej, pp. 175–192 (2017)
2. Busquet, F., Efthymiou, F., Hildebrand, C.: Voice analytics in the wild: validity and predictive accuracy of common audio-recording devices. Behav. Res. Methods **56**(3), 2114–2134 (2024)
3. Chandolikar, N., Joshi, C., Roy, P., Gawas, A., Vishwakarma, M.: Voice recognition: a comprehensive survey. In: 2022 International Mobile and Embedded Technology Conference (MECON), pp. 45–51 (2022). https://doi.org/10.1109/MECON53876.2022.9751903
4. Chao, Y., Bourguet, M.-L.: What speech recognition accuracy is needed for video transcripts to be a useful search interface? In: Karpov, A., Potapova, R., Mporas, I. (eds.) SPECOM 2017. LNCS (LNAI), vol. 10458, pp. 820–828. Springer, Cham (2017). https://doi.org/10.1007/978-3-319-66429-3_82
5. Cohen, P.R., Oviatt, S.L.: The role of voice input for human-machine communication. Proc. Natl. Acad. Sci. **92**(22), 9921–9927 (1995)
6. Conn, N., McTear, M.: Speech technology: a solution for people with disabilities (2000)
7. Edlabadkar, S.R., Joshi, P.P.: Accurate medicine recognition and audio conversion for handwritten prescriptions. In: 2024 International Conference on Emerging Smart Computing and Informatics (ESCI), pp. 01–06. IEEE (2024)
8. El-metwally, F.S., Eldesouky, A.I., Abdel-Hamid, N.B., Elghamrawy, S.M.: Optimized deep neural networks audio tagging framework for virtual business assistant. J. Adv. Inf. Technol. **14**(3) (2023)
9. Fogg, T., Wightman, C.W.: Improving transcription of qualitative research interviews with speech recognition technology (2000)
10. Froyen, D.J., Bonte, M.L., van Atteveldt, N., Blomert, L.: The long road to automation: neurocognitive development of letter-speech sound processing. J. Cogn. Neurosci. **21**(3), 567–580 (2009)
11. Galvez, D., et al.: The people's speech: a large-scale diverse English speech recognition dataset for commercial usage. arXiv preprint: arXiv:2111.09344 (2021)

12. López, G., Quesada, L., Guerrero, L.A.: Alexa vs. Siri vs. Cortana vs. Google assistant: a comparison of speech-based natural user interfaces. In: Advances in Human Factors and Systems Interaction: Proceedings of the AHFE 2017 International Conference on Human Factors and Systems Interaction, 17–21 July 2017, The Westin Bonaventure Hotel, Los Angeles, California, USA 8, pp. 241–250. Springer (2018)
13. O'Connor, J., McDermott, I.: Principles of NLP: What It Is, How It Works. Singing Dragon (2013)
14. Recognition–ASR, A.S.: Automatic speech recognition–ASR (2013)
15. Royale, O.: Homophones: the most confusing words in English (a List with Meanings) (2019). https://www.oxford-royale.com/articles/efl-homophones
16. Shuster, L.I., Durrant, J.D.: Toward a better understanding of the perception of self-produced speech. J. Commun. Disord. **36**(1), 1–11 (2003)
17. Smith, I., Hudson, S.E.: Low disturbance audio for awareness and privacy in media space applications. In: Proceedings of the Third ACM International Conference on Multimedia, pp. 91–97 (1995)
18. Tremblay, C., Champoux, F., Voss, P., Bacon, B.A., Lepore, F., Théoret, H.: Speech and non-speech audio-visual illusions: a developmental study. PLoS ONE **2**(8), e742 (2007)
19. VoiceOvers.com: what is voice over recording?. https://pl.voiceovers.com/blog/what-is-voice-over-recording
20. Walter, S.M., Cofano, M., Ratley, R.J.: Audio processing technology for law enforcement. In: Sensors, C3I, Information, and Training Technologies for Law Enforcement, vol. 3577, pp. 182–187. SPIE (1999)
21. Ziemba, E.W., Maruszewska, E.W., Grabara, D., Renik, K.: Acceptance and use of ChatGPT among accounting and finance higher education students. In: European Conference on Artificial Intelligence, pp. 185–202. Springer (2023)

Extracting Semantic Knowledge Using Generative AI. A Case Study Approach

Ilona Pawełoszek[1(✉)] and Joanna Kołodziejczyk[2]

[1] Czestochowa University of Technology, Częstochowa, Poland
ilona.paweloszek@pcz.pl
[2] National Institute of Telecommunications, Warsaw, Poland
j.kolodziejczyk@il-pib.pl

Abstract. The growing availability of domain-specific industry reports presents an opportunity for enhanced knowledge acquisition and management. This paper examines the application of generative AI in extracting structured semantic knowledge from business texts, utilizing the 2025 Logility Market Report as a case study. By applying a large language model (LLM) and prompt engineering, we automatically generated RDF triples representing key entities and relationships in AI-driven supply chain management. The methodology included both text-based extraction and OCR-based processing of an infographic, demonstrating the potential of multimodal LLMs for transforming visual data into machine-readable structures. The extracted triples were evaluated in terms of semantic coherence, completeness, and interpretability, and compared against a rule-based NLP pipeline implemented in spaCy. The results show that GPT-4 produced richer and more contextually accurate triples, while traditional methods struggled to capture implicit relations and numerical context. Although the graphs lacked formal ontological grounding, they provide a foundation for knowledge management applications and highlight the importance of hybrid approaches that combine LLM-based extraction with ontology alignment. The study also discusses methodological trade-offs between browser-based and API-based access, and outlines limitations such as reproducibility, formatting sensitivity, and scalability. Overall, the findings support the role of generative AI in augmenting semantic knowledge management in business environments.

Keywords: Generative AI · Knowledge Extraction · Semantic Representation

1 Introduction

The widespread adoption of large language models (LLMs) has transformed how organizations interact with unstructured information. Systems like ChatGPT and Claude can summarize, recommend, and answer questions fluently across domains, raising the question: Is structured knowledge still necessary?

Structured semantic knowledge remains crucial not only for reasoning and interoperability, but also because it enables integration into the broader ecosystem of Linked Data.

Published by Springer Nature Switzerland AG 2026
M. Pondel et al. (Eds.): AI4KAM 2025, IFIP AICT 776, pp. 351–363, 2026.
https://doi.org/10.1007/978-3-032-18920-2_33

The concept of Linked Data, described by Bizer, Heath, and Berners-Lee [1] as a set of best practices for publishing and interconnecting structured data to build a global Web of Data, provides a powerful rationale for extracting semantic structures from text. This approach rests on four core principles initially outlined by Tim Berners-Lee: use URIs to identify things, make them dereferenceable, provide useful data in standard formats (e.g. RDF/S), and include links to other URIs to enable discovery and integration.

By aligning extracted triples with Linked Data principles, organizational knowledge can be linked to external datasets and vocabularies, supporting richer analytics, interoperability across systems, and reuse beyond the scope of a single report. However, manually extracting such structured knowledge from textual documents remains a labour-intensive and error-prone process. This raises the question of whether generative large language models can accelerate and partially automate the construction of knowledge graphs consistent with Linked Data principles.

Despite their strengths, LLMs exhibit limitations in transparency, explainability, reasoning, and deterministic querying. Knowledge graphs (KGs), as machine-readable representations of entities and relationships, offer complementary benefits: they enable traceable knowledge, facilitate logical inference, promote semantic integration, and support efficient querying [2–4].

In practice, KGs are already enhancing business operations in supply chain transparency [5] and customer service automation [6]. However, constructing them remains labor-intensive. Recent advances in generative AI offer new opportunities to accelerate and partially automate knowledge graph construction.

This paper examines how generative AI can facilitate the automatic extraction of semantic structures from business reports, aiming to bridge the gap between unstructured content and formal knowledge modeling.

2 Background and Related Work

As large language models gain fluency and generalization capabilities, interest is growing in combining them with structured knowledge representations to improve interpretability, traceability, and domain control in areas where LLMs still struggle due to hallucinations and a lack of transparency. Recent work illustrates this trend: Bian et al. [7] introduced AutoMathKG, a mathematical knowledge graph that integrates curated resources (e.g., ProofWiki, textbooks, arXiv) with LLM-based augmentation to improve reasoning and knowledge completion. Likewise, Park et al. [8] demonstrated that joint KG–LLM models can enhance clinical decision support systems by improving domain accuracy and explainability.

Recent studies demonstrate the benefits of hybrid approaches. Kumar et al. [9] introduce KGFakeNet, a method that fuses knowledge graphs with generative models for detecting fake news. Meroño-Peñuela et al. [10] present KG.GOV, a framework using knowledge graphs for AI-driven data governance, ensuring semantic consistency.

In cybersecurity, Mishra, Sarma, and Saravanan [11] propose PageLLM, which updates a security knowledge graph by combining LLM-based extraction with graph-based ranking, illustrating the value of such pipelines in dynamic, high-risk settings. Ibrahim et al. [12] provide a comprehensive survey of approaches that augment knowledge graphs with large language models, outlining models, evaluation metrics, and challenges, and thereby demonstrating another path to semantic structuring in evolving knowledge ecosystems. Oprea and Bâra [13] apply probabilistic generative models to identify latent topics in e-business transformation, highlighting the complementary roles of generative and semantic techniques in complex domains.

While these studies reveal a growing synergy between generative AI and structured knowledge, few address how such methods can extract semantic relations from management documents, such as industry reports. In the supply chain domain, a recent survey by Samaridi et al. [14] provides a comprehensive overview of supply chain ontologies, highlighting their role in supporting interoperability and decision-making. However, the challenge of automatically enriching such ontologies with knowledge extracted from unstructured texts has not yet been systematically addressed. This paper fills that gap by applying LLM-based extraction to a domain-specific business report, generating a knowledge graph that supports strategic decision-making in supply chain management.

3 Methodology

3.1 Overview

This study proposes a lightweight methodology for extracting semantically structured knowledge from industry reports using a generative large language model. The extracted knowledge is represented in the form of a knowledge graph, composed of entities, relations, and supporting context. The goal is to evaluate the feasibility of LLM-assisted semantic modeling in a managerial context and assess the quality and interpretability of the generated graphs.

The methodology is demonstrated on the Supply Chain Horizons 2025 report published by Logility [15], a 39-page document combining survey data, strategic insights, and technology forecasts on AI adoption in supply chain management. This type of industry report was chosen because it integrates heterogeneous forms of information: narrative text, statistical tables, and visual infographics, which makes it a representative and challenging test case for semantic knowledge extraction.

3.2 Extraction Process

The extraction workflow comprised four stages. First, the report PDF was converted to plain text using PyMuPDF, and non-informative elements such as headers, footers, and disclaimers were removed. The resulting text was segmented into semantically coherent chunks of approximately 500 tokens, which corresponds to roughly 350–400 words in English prose. Token-based segmentation was necessary because large language models process text in units called tokens – subword elements that can represent a whole word, part of a word, or punctuation. A limit of around 500 tokens per chunk was selected as a practical compromise: large enough to capture full thematic sections (e.g., a paragraph describing a technological trend or managerial practice), yet small enough to avoid input size constraints of the model. Segment boundaries were identified manually by detecting topic shifts in the report, such as transitions between survey findings, managerial recommendations, and technology forecasts. This ensured that each prompt presented the model with a coherent and self-contained passage, improving the interpretability of the extracted triples. The chosen threshold was also practical from a cost perspective, as API providers charge per token. While inputs exceeding the context limit are rejected without fees, segmenting the text into smaller chunks ensures predictable costs and successful processing. For demonstration purposes, a single representative text fragment was selected for knowledge extraction. Full-scale application to the entire document was left for future work, given the exploratory nature of this study and the page limits of the current publication.

Next, the selected chunk was submitted to GPT-4 using a zero-shot prompt designed to elicit RDF-like triples of the form (Subject, Predicate, Object), optionally including qualifiers such as context or metric information. This approach builds upon recent insights into the effectiveness of prompt programming for controlling large language models [16]. Extracted triples were parsed into structured CSV and RDF/XML formats, with synonym resolution and entity normalization (e.g., unifying "Generative AI" and "GenAI"). Finally, the resulting knowledge graph was visualized using networkx and Graphviz. Basic graph metrics were computed as part of exploratory validation, although these results are not reported here. In addition to text, visual elements of the report were also considered, which are detailed in the next section.

3.3 Structured Representation from Visual Data

The second part of our experiment applied OCR and semantic parsing to a business infographic included in the report. While visually accessible to humans, the chart lacked a structured, machine-readable format. In this case, the OCR step was performed directly by the large language model: a screenshot of the infographic was pasted into the prompt, and the model transcribed the embedded content into text. This transcription was subsequently parsed into RDF triples and expressed as weighted edges in a knowledge graph, representing country-specific priorities.

While the example graph was reasonably clear, larger applications of this approach may result in dense and less interpretable visualizations, due to overlapping values and network complexity. Nevertheless, the semantic formalization added significant value, as it enabled the extracted information to be queried, reasoned upon, and integrated into broader knowledge systems.

3.4 Prompting Strategies

Different strategies can be applied when designing prompts for large language models, depending on the level of guidance and the desired type of output.

- Zero-shot prompting requires no examples and relies solely on a general instruction. Its advantages are simplicity, reusability of the same prompt across different texts, and minimal preparation effort. However, the output may sometimes be less consistent or contain ambiguous relations.
- Few-shot prompting supplements the instruction with several input–output examples. This usually improves consistency and reduces the risk of malformed triples, but requires careful preparation of examples and tailoring them to the domain, which increases the effort needed for deployment.

- Chain-of-thought prompting encourages the model to generate intermediate reasoning steps before producing the final result. This can improve the capture of implicit or causal relations, but it increases verbosity and complicates post-processing, since reasoning text must be filtered out to obtain valid triples.

In this study, a zero-shot approach was deliberately selected. The goal was to evaluate whether a single, generalizable prompt could reliably extract semantically coherent triples from management texts. Zero-shot prompting offers clear practical benefits: it is fast, reduces preparation time, and the same instruction can be reused across multiple passages without modification. Moreover, its simplicity makes it suitable for managerial contexts, where end users are unlikely to design elaborate prompts but may still benefit from automated knowledge extraction.

3.5 Prompt Design

To guide the language model towards producing structured outputs, a zero-shot prompt was designed that explicitly specified the required output format and the categories of information to be extracted. The instruction asked the model to generate knowledge triples in the form *(Subject, Predicate, Object)*, with the optional inclusion of a [Context] qualifier when additional information, such as metrics or circumstances, was available.

The overall structure of the prompt included four elements:

1. Output format – a clear specification that each output should be a triple.
2. Optional qualifier – [Context] was introduced to capture supporting details (e.g., "by 32%" or "in short-term forecasting"), without forcing them into the core triple structure.
3. Extraction focus – a short list of categories central to managerial analysis:
   - key technologies (e.g., AI, GenAI),
   - use cases (e.g., demand planning),
   - organizational goals (e.g., improve forecast accuracy),
   - outcomes (e.g., 32% improvement).
4. Input text – the case study passage from the Logility report.

The exact prompt is shown below:

Extract knowledge triples from the following case study. Each triple should be in the form:

(Subject, Predicate, Object).

Also include [Context] if available. Focus on identifying:

- *key technologies (e.g., AI, GenAI),*
- *use cases (e.g., demand planning),*
- *organizational goals (e.g., improve forecast accuracy),*
- *and outcomes (e.g., 32% improvement).*

Text:

A major process manufacturer wanted to improve forecast accuracy, spend less time in crisis mode, and make supply planners more efficient. With the help of AI-powered demand planning, they were able to:

- *Improve short-term forecast accuracy by 32%*
- *Reduce working hours to create a plan by 70%*
- *Improve the timing and planning of promotions*

This zero-shot design was chosen to test whether a single, generalizable instruction could reliably produce semantically coherent triples without requiring examples (few-shot learning) or reasoning scaffolds (chain-of-thought reasoning). The [Context] marker was introduced as a lightweight mechanism to capture quantitative or situational details that are often critical in managerial reports but do not naturally fit into a subject–predicate–object structure.

3.6 Evaluation Criteria

The quality of extracted triples was evaluated along three complementary dimensions:

- Semantic coherence – whether the triples captured logically valid and contextually meaningful relations between entities.
- Completeness – whether the main concepts and relationships present in the source passage were represented without major omissions.
- Interpretability – whether the resulting knowledge graph could be readily understood by a business audience and convey the managerial intent of the original text.

To make the evaluation more transparent, the criteria were operationalized as shown in Table 1, which illustrates what constitutes a high versus a low assessment, along with relevant examples. The examples demonstrate how the same passage can produce either highly informative triples or ambiguous, low-quality outputs depending on extraction accuracy.

The evaluation was performed through expert review by the authors, drawing on their domain knowledge in AI and semantic technologies. While this study relied on qualitative, expert-based validation, future work may include more systematic evaluation methods such as inter-annotator agreement, which measures the consistency of human expert judgments [17], and precision/recall metrics, widely used in information extraction tasks [18, 19]. In the broader context of knowledge graph construction, such metrics have also been surveyed by [2].

3.7 Tooling

Experiments were conducted using GPT-4 (ChatGPT interface, March 2024 version) for semantic extraction. Prompts were tested interactively, and the generated triples were subsequently processed in Python 3.10. Text preprocessing employed PyMuPDF, regular expressions, and pandas, while extracted triples were parsed into CSV and RDF/XML formats using csv, rdflib, and json libraries. Knowledge graph construction and visualization used networkx and Graphviz.

Baseline comparison employed spaCy v3.6 (en_core_web_lg) with custom rule-based extractors [20, 21]. The spaCy en_core_web_sm model was selected for its ease of deployment and fast runtime. While it offers limited linguistic depth, it proved sufficient for identifying basic syntactic dependencies. The experiments were performed on a local

Table 1. Operationalization of evaluation criteria for extracted triples.

Criterion	High – strong evaluation	Low – weak evaluation	Example (High)	Example (Low)
Semantic coherence	Relation is logical and semantically valid	Relation is erroneous, random, or inconsistent	*(AI – improves – forecast accuracy)*	*(Accuracy – is – AI)*
Completeness	All key relations from the text are captured	Key relations are missing or only partial	Captured relations on forecast accuracy, planning time reduction, and promotions	Captured only the relation on forecast accuracy
Interpretability	Triple is understandable without additional context	Requires referring back to the source text to make sense	*(AI – improves – supply chain efficiency)*	*(Improves – it – this)*

Windows 11 setup, while Google Colab (Python 3.10 runtime) was used for spaCy-based experiments due to installation issues.

Although the OpenAI API (via the Python SDK) was considered for automating the extraction pipeline, this study relied on interactive testing in the ChatGPT interface. An API-based approach could streamline the workflow by enabling batch processing of document segments, automated integration with parsing scripts, and large-scale replication of experiments. Exploring such automated pipelines is left for future work.

4 Results

4.1 Extraction Outcomes

The proposed zero-shot methodology was applied to a selected passage from the Supply Chain Horizons 2025 industry report. The paragraph described a manufacturing company's use of AI-powered demand planning and reported measurable outcomes, including improved forecast accuracy and reduced planning time.

Using a carefully designed prompt, GPT-4 extracted seven semantic triples representing relationships between actors, technologies, goals, and observed effects. These triples were organized into a directed knowledge graph, which visually depicts the cause-and-effect structure of the managerial intervention (Fig. 1). The graph uses both colors and node shapes to distinguish object types (actors, tools, goals, and outcomes). This redundancy was introduced deliberately: while colors improve visual clarity, shapes ensure accessibility for color-blind readers and preserve interpretability in monochrome prints.

To improve reliability, selected RDF triples were reviewed by the author with domain knowledge in AI and semantic technologies. Minor extraction artifacts, such as slightly ambiguous entities or repeated predicates, were adjusted to ensure semantic consistency in the final knowledge graph.

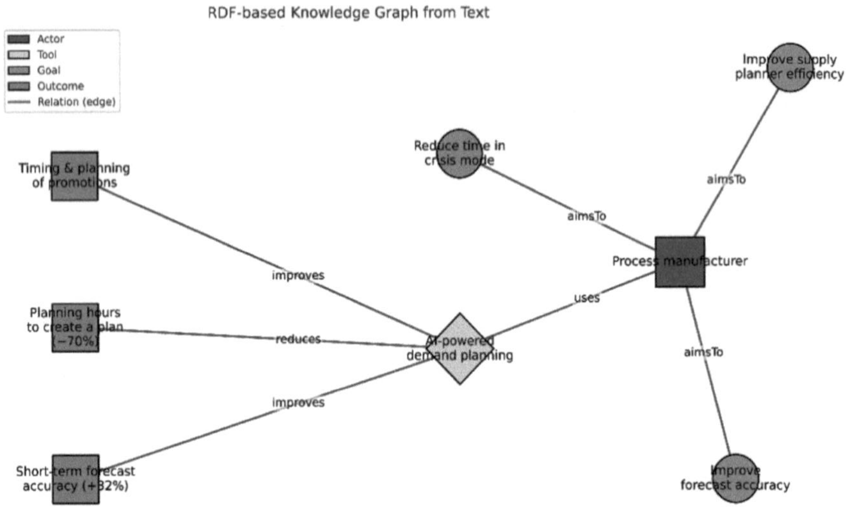

Fig. 1. RDF-based knowledge graph generated from GPT-4 output, representing managerial intent, applied technology, and measurable outcomes.

In addition to the text-based extraction, a second experiment focused on the country-level infographic contained in the report (page 10). OCR with the Python Tesseract library produced several artefacts, such as meaningless character sequences (e.g., eum, aeummmmmmmmm), which likely resulted from the decorative dot-based world map in the background of the chart. In contrast, OCR performed by the LLM on the same screenshot yielded a much cleaner transcription: the model was able to filter out such artefacts and reconstruct coherent categories by leveraging contextual reasoning. On this basis, the underlying percentages were successfully recovered, and a set of RDF triples was generated to represent the priorities of each country. A corresponding knowledge graph was then constructed in Python (Fig. 2), using weighted edges to indicate the strength of emphasis for each use case. The visualization is clear and readable, faithfully reflecting the experimental outcome and demonstrating how visual business data can be transformed into structured semantic knowledge.

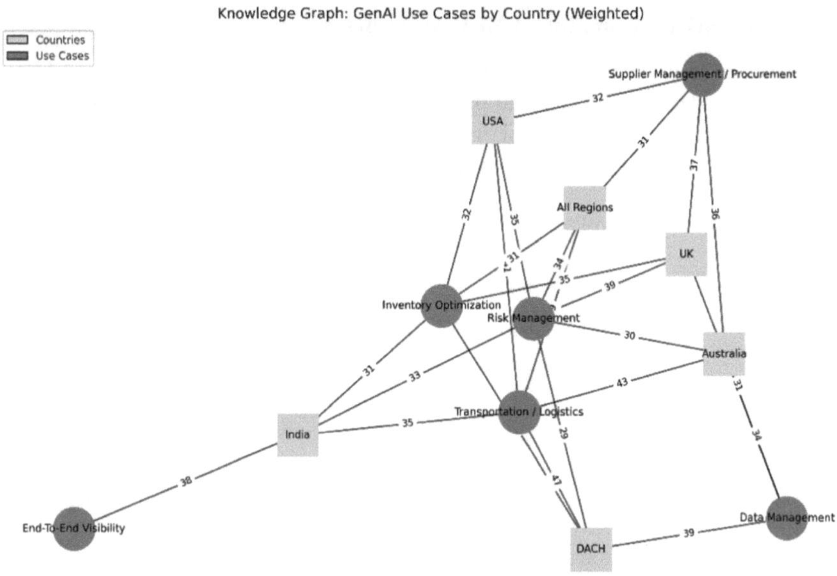

Fig. 2. RDF-based knowledge graph generated from the country-level infographic [15, p. 10]. Percentages were extracted using OCR performed by the LLM, with weighted edges representing the strength of country-specific priorities.

4.2 Quality Assessment

The extracted triples were evaluated across three dimensions: semantic coherence, completeness, and interpretability. All triples were grammatically well-formed and required no corrections. Key entities and outcomes from the original paragraph were captured without significant omissions. The resulting graph was readily understandable by a business audience and effectively traced how AI contributes to performance improvements.

A comparison with rule-based NLP methods (e.g., named entity recognition with spaCy and pattern-based relation extraction) indicated that those approaches failed to capture context-specific relations like *aims to* or *improves*, and often missed implicit connections. To benchmark the effectiveness of our approach, we applied a traditional NLP pipeline using spaCy's en_core_web_lg model combined with simple dependency-based relation templates. While the model was able to detect some named entities (e.g., "forecast accuracy", "promotion"), it failed to capture the nuanced semantic relations present in the text. In particular, context-specific actions such as "improves" and the numerical impact (e.g., "by 32%") were either ignored or misclassified. No complete or interpretable triples could be generated without substantial rule engineering. This highlights the advantage of using generative models for rapid semantic structuring in management contexts.

While spaCy excels at rule-based, fast, and linguistically grounded NLP tasks, it lacks the contextual depth and generative flexibility of LLMs, which are increasingly used for tasks such as semantic knowledge extraction, reasoning, and domain modeling.

4.3 Comparative Evaluation of NLP Approaches

To compare the performance of generative and classical NLP pipelines, we conducted a controlled experiment on a selected paragraph from the Logility report. The input text was processed using (1) a zero-shot prompt in GPT-4 and (2) a rule-based NLP pipeline using spaCy's en_core_web_sm. The results are summarized in Table 2.

Table 2. Comparison of semantic extraction using LLM and classical NLP (spaCy).

Aspect	LLM (GPT-4)	spaCy (en_core_web_sm)
Method	Zero-shot prompt-based semantic parsing	Rule-based named entity recognition and dependency parsing
Input	Full paragraph from industry report	Same paragraph
Named Entities	Process Manufacturer, AI-Powered Demand Planning, forecast accuracy	AI (misclassified as GPE), 32%, 70%
Relation Extraction	7 full semantic triples with subject–predicate–object structure	No valid subject–verb–object triples detected
Example Triple	(AI-Powered Demand Planning, improves, Forecast Accuracy (+32%))	–
Context Understanding	Captures organizational goals and technological outcomes	Does not link entities to actions or context
Interpretability	High – results understandable without post-processing	Low – requires custom rules and domain knowledge
Tooling Requirements	API-based, generative LLM (e.g., ChatGPT, Claude)	Local installation; issues with dependencies on Windows (blis)

As shown, the LLM successfully generated semantically coherent RDF triples, while spaCy failed to extract meaningful relations despite recognizing a few named entities. This contrast underscores the advantage of LLMs in knowledge-rich domains where syntactic structure alone is insufficient.

The semantic triples were exported into both.ttl (RDF Turtle) and.gexf formats for visualization. While the RDF version could be opened in Protégé, the lack of class assignments (e.g., rdf:type) limited its graphical utility. A more legible and publication-ready graph was created using Python and rendered with adjusted node labels for clarity.

5 Discussion

This study demonstrates the potential of generative AI models to extract semantically meaningful knowledge from unstructured business reports. While the extracted RDF triples lack formal ontological typing, they provide a foundational layer for building machine-interpretable knowledge structures. Using large language models (LLMs) enabled the automatic identification of concept–relation–concept structures, which are typically absent from flat textual or visual data.

However, our experiments reveal that extracted triples lack formal constraints such as class hierarchies and logic-bound properties, limiting their direct application in reasoning tasks. This highlights the need for hybrid pipelines where LLM-based extraction is complemented by ontology alignment. While the current study focuses on a single excerpt from an industry report, the proposed method is designed to be scalable and adaptable to full-document extraction scenarios. In future research, the pipeline will be applied to larger corpora to enable cross-document analysis and validation of consistency across extracted graphs.

A comparative experiment using spaCy's en_core_web_sm model showed that the classical rule-based NLP approach failed to extract meaningful triples, detecting only three named entities (with misclassification) and no relational structures. In contrast, GPT-4 produced seven semantically coherent triples without custom rules, illustrating the advantages of generative models in domain-specific knowledge extraction.

Country-specific priorities were extracted and expressed as weighted graph edges using OCR and semantic parsing. The resulting graph was compact but still interpretable, and its semantic formalization makes it possible to perform querying, reasoning, and integration into larger knowledge systems.

Overall, these findings support the growing role of generative AI in AI-driven knowledge management. LLMs can bootstrap domain knowledge extraction, reduce dependency on rigid, manually engineered linguistic rules that often struggle to generalize across domains, and semantically enrich business content. However, limitations such as hallucinations, prompt sensitivity, and inconsistent document formatting remain challenges for future research.

Another methodological consideration concerns the difference between browser-based and API-based access to LLMs. While the present study relied on interactive use of ChatGPT, API-based pipelines offer advantages in terms of scalability and reproducibility. They enable automated batch processing of larger documents, provide explicit control over model parameters (e.g., temperature, token limits), and facilitate easier replication of experiments. At the same time, the use of APIs introduces new trade-offs: a higher technical barrier, since scripting skills are required to orchestrate requests and handle responses, and a usage-based cost model that can become substantial when processing entire reports. In contrast, browser interfaces are easier to adopt in managerial practice and more predictable in cost, but they limit automation and make precise replication more difficult. This distinction highlights an important gap between academic experimentation and practical adoption, and suggests that future research should explore how to balance accessibility with methodological rigor.

6 Conclusion

This study demonstrated the feasibility of using generative AI to extract semantically structured knowledge from business reports. LLMs proved effective in capturing coherent and interpretable triples, although the lack of formal ontological structures limits their direct application in knowledge-based systems. Compared to traditional

NLP pipelines, the generative approach yielded richer and more contextually accurate knowledge graphs.

Nevertheless, several limitations of this study should be acknowledged. First, the experiments were conducted on selected excerpts rather than entire reports, which restricts the scope of the findings. Second, the reliance on interactive prompting in a browser-based interface means that the extraction process is less standardized: small changes in prompt wording or model behavior can lead to different outputs, which reduces reproducibility. Third, the absence of quantitative evaluation metrics means that conclusions about quality remain based on expert judgment rather than statistical validation. Finally, issues such as hallucinations, dependency on specific report formatting, and the costs of large-scale processing remain open challenges for future research.

Future work should also explore the development of automated benchmarking environments that utilize gold-standard ontologies and evaluation metrics, enabling more robust comparisons across methods and domains. At the same time, integrating generative knowledge extraction with domain-specific ontologies is crucial for bridging the gap between informal semantics and formal logic. This may include post-processing pipelines for class assignment, property alignment, and entity linking. Although the extracted triples are not formally grounded in existing ontologies, future versions of the method may incorporate lightweight schemas such as schema.org or domain-specific OWL vocabularies. This would enhance the reasoning capabilities of the graph and improve interoperability with existing semantic web applications.

References

1. Bizer, C., Heath, T., Berners-Lee, T.: Linked data – the story so far. Int. J. Semant. Web Inf. Syst. **5**(3), 1–22 (2009)
2. Ji, S., Pan, S., Cambria, E., Marttinen, P., Yu, P.S.: A survey on knowledge graphs: representation, acquisition, and applications. IEEE Trans. Neural Netw. Learn. Syst. **33**(2), 494–514 (2022). https://doi.org/10.1109/TNNLS.2021.3070843
3. Nickel, M., Murphy, K., Tresp, V., Gabrilovich, E.: A review of relational machine learning for knowledge graphs. Proc. IEEE **104**(1), 11–33 (2016). https://doi.org/10.1109/JPROC.2015.2483592
4. Hogan, A., et al.: Knowledge graphs. ACM Comput. Surv. **54**(4), 1–37 (2021). https://doi.org/10.1145/3447772
5. Zhang, W., Liu, Y., Jiang, L., Shah, N., Fei, X., Cai, H.: The construction of a domain knowledge graph and its application in supply chain risk analysis. In: Lecture Notes on Data Engineering and Communications Technologies, vol. 41, pp. 463–478. Springer, Cham (2020). https://doi.org/10.1007/978-3-030-34986-8_33
6. Landolsi, H., Letaief, K., Taghouti, N., Abdeljaoued-Tej, I.: CAPRAG: a large language model solution for customer service and automatic reporting using vector and graph retrieval-augmented generation. arXiv preprint arXiv:2501.13993 (2025)
7. Bian, R., Geng, Y., Yang, Z., Cheng, B.: AutoMathKG: the automated mathematical knowledge graph based on LLM and vector database. arXiv preprint arXiv:2505.13406 (2025). https://doi.org/10.48550/arXiv.2505.13406
8. Park, C., Lee, H., Lee, S., Jeong, O.: Synergistic joint model of knowledge graph and LLM for enhancing XAI-based clinical decision support systems. Mathematics **13**(6), 949 (2025). https://doi.org/10.3390/math13060949

9. Kumar, A., Kumar, P., Yadav, A., Ahlawat, S., Prasad, Y.: KGFakeNet: a knowledge graph-enhanced model for fake news detection. In: Proceedings of the Generative AI and Knowledge Graph Workshop (GenAIK), Abu Dhabi, pp. 109–122 (2025)
10. Meroño-Peñuela, A., Simperl, E., Kurteva, A., Reklos, I.: KG.GOV: knowledge graphs as the backbone of data governance in AI. J. Web Semant. **85**, Article 100847 (2025). https://doi.org/10.1016/j.websem.2024.100847
11. Mishra, C., Sarma, H., Saravanan, M.: PageLLM: incremental approach for updating a security knowledge graph by using page ranking and large language model. Inf. Process. Manage. **62**(3), Article 104045 (2025). https://doi.org/10.1016/j.ipm.2024.104045
12. Ibrahim, N., Aboulela, S., Ibrahim, A., Kashef, R.: A survey on augmenting knowledge graphs (KGs) with large language models (LLMs): models, evaluation metrics, benchmarks, and challenges. Discov. Artif. Intell. **4**, 76 (2024). https://doi.org/10.1007/s44163-024-00175-8
13. Oprea, S.-V., Bâra, A.: Is artificial intelligence a game-changer in steering e-business into the future? Uncovering latent topics with probabilistic generative models. J. Theor. Appl. Electron. Commer. Res. **20**(1) (2025). https://doi.org/10.3390/jtaer20010016
14. Samaridi, N.E., Karanikolas, N.N., Papoutsidakis, M., Papakitsos, E.C., Papakitsos, C.E.: A survey on supply chain ontologies. Int. J. Prod. Manage. Eng. **11**(1), 89–101 (2023). https://doi.org/10.4995/ijpme.2023.18702
15. Logility: Supply chain horizons 2025: navigating the digital transformation and GenAI journey in supply chain (executive summary). https://www.logility.com/white-papers/sc-horizons-2025-exec-summary/. Accessed 26 Apr 2025
16. Reynolds, L., McDonell, K.: Prompt programming for large language models: beyond the few-shot paradigm. arXiv preprint arXiv:2102.07350 (2021). https://doi.org/10.48550/arXiv.2102.07350
17. Artstein, R., Poesio, M.: Inter-coder agreement for computational linguistics. Comput. Linguist. **34**(4), 555–596 (2008). https://doi.org/10.1162/coli.07-034-R2
18. Chinchor, N.: The statistical significance of the MUC-4 results. In: Proceedings of the 4th Conference on Message Understanding (MUC-4), pp. 30–50 (1992). https://doi.org/10.3115/1072064.1072068
19. Manning, C.D., Raghavan, P., Schütze, H.: Introduction to Information Retrieval. Cambridge University Press (2008)
20. Explosion AI: spaCy models documentation. https://spacy.io/models/en. Accessed 26 Apr 2025
21. Hugging Face: en_core_web_sm model page. https://huggingface.co/spacy/en_core_web_sm. Accessed 26 Apr 2025

Cross-Topic Sentiment Analysis of Wikipedia Articles: A Comparative Study of AI Models

Włodzimierz Lewoniewski(✉), Milena Stróżyna, Izabela Czumałowska, Aleksandra Wojewoda, and Krzysztof Węcel

Poznań University of Economics and Business, al. Niepodległości 10, 61-875 Poznań, Poland
wlodzimierz.lewoniewski@ue.poznan.pl

Abstract. This study compares various methods of verifying adherence of Wikipedia articles to the Neutral Point of View (NPOV) standard. To gauge potential bias, we apply four sentiment analysis models - two lexicon-based (TextBlob, VADER) and two transformer-based (RoBERTa, DistilBERT) - to nearly 7 million articles selected from the English Wikipedia. The articles are pre-processed, categorized by topic and quality, according to proposed methodology. The results showed that sentiment of Wikipedia articles varies by topic and that proper model selection is crucial for accurate NPOV assessment. The paper contributes a practical framework for implementing sentiment analysis on longer texts. We also offer insights into how article quality correlates with sentiment outcomes.

Keywords: sentiment analysis · information quality · Wikipedia · TextBlob · VADER · RoBERTa · DistilBERT · Wikidata

1 Introduction

Assurance of the quality of Wikipedia is essential because of its popularity and wide use. One of its core quality criteria is the Neutral Point of View (NPOV), which requires that information should be presented objectively, without personal bias or subjective opinions, and that all significant viewpoints on a given topic are represented fairly and proportionally. In practice, however, maintaining neutrality is challenging due to Wikipedia's open-editing model, where editors may, often unintentionally, introduce their own bias into article content. Therefore, it is crucial to monitor NPOV compliance in an automated manner. While some studies have attempted to detect bias in Wikipedia articles [6, 12], there is still a lack of research assessing Wikipedia's neutrality through sentiment analysis.

Sentiment analysis enables the categorization of content as positive, neutral, or negative. Today, this technique is commonly used to analyze digital content such as social media posts, reviews, or product feedback [8,9]. It provides valuable insights into consumer opinions, helps monitor public sentiment, identifies reputation risks, and support the evaluation of the credibility of online information [3, 13]. State-of-the-art sentiment analysis models are highly effective in assessing the sentiment [10, 14]. However, these models are typically designed for short, concise texts written in informal language,

Published by Springer Nature Switzerland AG 2026
M. Pondel et al. (Eds.): AI4KAM 2025, IFIP AICT 776, pp. 364–370, 2026.
https://doi.org/10.1007/978-3-032-18920-2_34

such as posts or reviews. In contrast, Wikipedia articles are longer and more structured, which poses a challenge for the direct application of sentiment analysis models that are primarily optimized for shorter forms. Consequently, there is a need for advanced methodologies that can effectively apply sentiment analysis to more complex and extensive types of content.

The goal of this paper is to compare state-of-the-art sentiment analysis methods with regard to their applicability in assessing the sentiment of Wikipedia articles. We aim to determine the extent to which sentiment evaluations provided by different methods are consistent when applied to longer texts like Wikipedia articles. Additionally, we assess how well the NPOV is maintained for Wikipedia articles by examining the prevailing sentiment of articles across different topics and analyzing how article quality influences sentiment outcomes.

This paper's originality comes from employing sentiment analysis to systematically evaluate neutrality of Wikipedia articles, offering a novel approach to verifying if Wikipedia's NPOV standard is met. This study extends our previous research [16] on sentiment analysis that covered only a selected set of Wikipedia articles on companies. In contrast, here we investigated all English Wikipedia articles, added thematic classification and employed a novel method for quality assessment. Additionally, this research focuses on models that previously demonstrated the highest effectiveness and the best performance on longer texts.

2 Methodology

The study was conducted according to the Design Science Research framework [5]. We used the entire English Wikipedia corpus for a systematic comparison of sentiment analysis methods and proposed a novel framework that evaluates both lexicon-based (TextBlob, VADER) and transformer-based models (RoBERTa, DistilBERT). The analyzed articles were further categorized by quality and topic. The sentiment was calculated for each category.

2.1 Models

TextBlob[1] as well as Vader[2] are Python lexicon-based sentiment analysis tools. Provided a text, the former one determines polarity ranging form -1 to 1 and subjectivity within the range 0–1. The latter provides the proportion of three sentiment levels (positive, neutral, negative), and the *compound score* being combined value of all three mentioned before. Because of this difference, we will report compound scores as a variant of the original model, which weights words intensity instead of only classifying them, referring to it as as cVADER.

Task-specific RoBERTa[3] and DistilBERT[4] are BERT-based transformer models, both being fine-tuned for sentiment analysis, provide probability scores for the sentiment labels. RoBERTa specializes in short texts such as tweets and sentiment analysis

[1] https://textblob.readthedocs.io/en/dev/index.html.

[2] https://pypi.org/project/vaderSentiment/.

[3] https://huggingface.co/cardiffnlp/twitter-roberta-base-sentiment.

[4] https://huggingface.co/lxyuan/distilbert-base-multilingual-cased-sentiments-student.

was merely one of the tasks that its training dataset was labeled for [1]. DistilBERT is a distilled model, which can achieve 97% of BERT's original accuracy despite smaller size [2].

2.2 Dataset Preparation and Topic Classification

We used English Wikipedia dumps from March 2025[5]. We extracted only real articles, without redirects, using WikiExtractor library[6]. To determine the topic of a specific Wikipedia article, we used corresponding Wikidata item that can indicate the topic based on properties found in statements [11]. For example, to align an article to the "city" category we select those related Wikidata items that have in "instance-of" (P31) property such values as "city" (Q515), "big city" (Q1549591), "city with powiat rights" (Q925381). The effectiveness of this approach for topic modeling Wikipedia articles has been validated in our recent research [16] when articles about companies were aligned to specific industries. Here we use more general topics like "city", "business", "film"; the list is confined to 18 topics due to keep the paper concise.

The prepared dataset along with sentiment assessment by models is published on HuggingFace[7].

2.3 Quality Measure

In this study, we used a synthetic quality measure dedicated for assessing the quality of Wikipedia articles [11,17]. For every Wikipedia article, we recorded five structural features: length, total number of references, reference density, number of images, and number of sections. Values of features were normalized to the 0–1 range by dividing them by the median value of the respective feature in the language's Featured-Article set (values above the median are capped at 1). Final quality score was calculated according to Eq. 1.

$$QualityScore = \left(\frac{1}{c}\sum_{i=1}^{c}\hat{m}_i\right)(1 - 0.05\,\mathrm{QFT}) \tag{1}$$

where:

- $\hat{m}_i$ – the normalized value of the i-th content-based feature
- c – the total number of article features (here $c = 5$)
- QFT – the count of quality-flaw templates present in the article.

2.4 Implementation

The analysis was conducted using the dedicated Python libraries for each model, namely TextBlob, vaderSentiment, and Transformers. For DistilBERT and RoBERTa, we utlized model versions available on HuggingFace. In the case of TextBlob, which,

[5] https://dumps.wikimedia.org/enwiki/.
[6] https://github.com/attardi/wikiextractor.
[7] https://huggingface.co/datasets/lewoniewski/wikipedia-sentiment.

unlike the other models, provides sentiment values ranging from -1 to 1, we mapped them to classes based on previous studies [7,15], i.e., sentiment scores $(-0.05; 0.05)$ are classified as neutral, scores below -0.05 as negative, and scores above 0.05 as positive.

Due to the fact that some of the selected models have an upper limit on the length of text that can be analyzed, we segmented the content of articles into smaller chunks (referred to as segments later in the article) considering character limit and maintaining contextual integrity. The resulting sentiment scores corresponded to individual text segments rather than entire articles. Similarly to our previous work [16], we used the advanced aggregation method, specified by Eq. 2, to obtain the overall sentiment assessment for whole Wikipedia article.

$$\tilde{p}(j) = \frac{\sum_{i=1}^{N} \ell_i \, p_i(j)}{\sum_{i=1}^{N} \ell_i}, \quad \forall j \in \{positive, neutral, negative\}, \tag{2}$$

where:

- ℓ_i – the length of a text segment i
- $p_i(j)$ – the probability score for one of the sentiment class j in a segment i
- N – the total number of segments in a given Wikipedia article.

The above equation ensures that the overall score for each sentiment class is a weighted average of the segment-level probabilities, where the weights correspond to the segment's length, and that the final values fall within the $[0, 1]$ interval.

Finally, in order to calculate the overall sentiment for each group of Wikipedia articles (selected based on the quality class and topic), additional level of aggregation was performed: having three separate values assigned to each sentiment class for each article, the average value was calculated within the group.

3 Results

The upper section of Fig. 1 presents the share of positive (green), neutral (yellow), and negative (red) sentiment for 18 Wikipedia topics across five sentiment models. The lexicon-based VADER classifies *almost all* articles as neutral regardless of topic, whereas cVADER and DistilBERT reveal a clear topical polarity: technology–oriented pages such as *programming*, *software* and *taxon* exhibit strongly positive sentiment, while *city* and *crypto* articles contain noticeably more negative language.

Aggregated across all topics, the lower section of Fig. 1 shows that article quality influences sentiment polarity differently for each model. For cVADER and DistilBERT, negative content (red) decreases, while positive content (green) increases steadily as we move from stub-class ("0–10") to featured-class ("91–100") pages, suggesting that editorial refinement intended to enhance article quality correlates with more upbeat wording. RoBERTa, in turn, largely shifts negativity into neutrality and VADER remains dominated by neutral classifications. Figure 2 provide a more detailed view, showing the percentage of articles labeled as neutral, negative and positive, respectively, for every topic–quality combination.

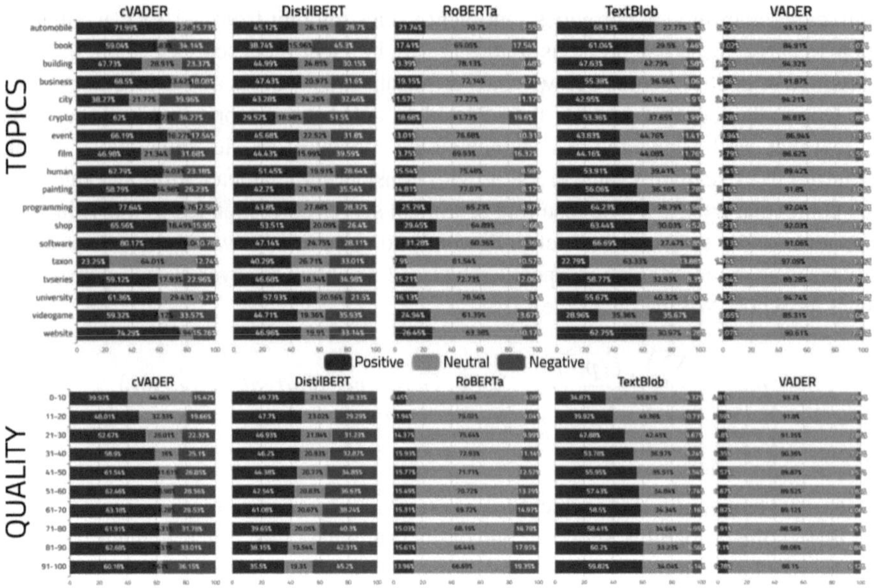

Fig. 1. Sentiment analysis outcomes for Wikipedia articles stratified by topic (upper section) and by quality range (lower section), evaluated using several models. Interactive version at [4].

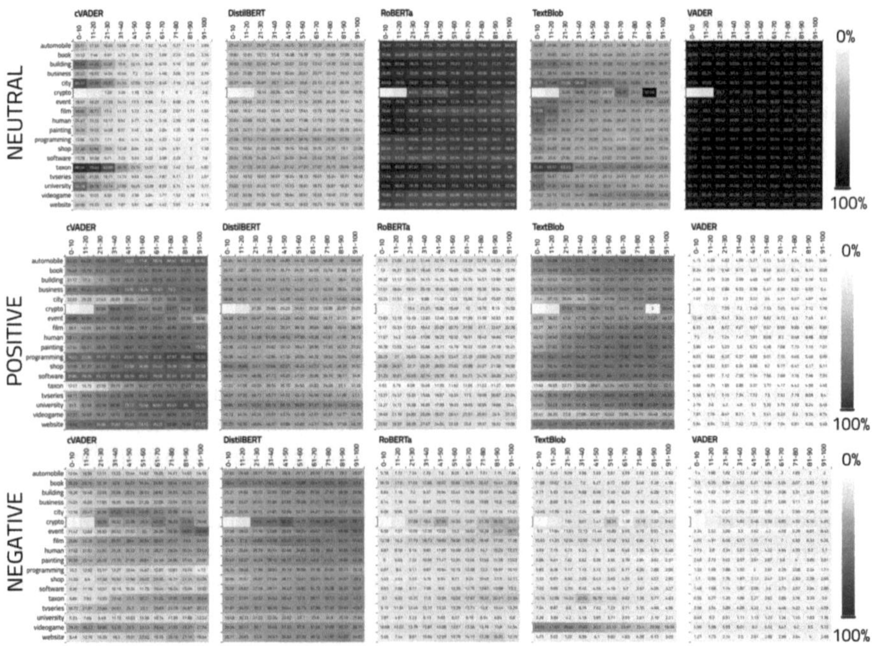

Fig. 2. Share of Wikipedia articles with neutral, positive, negative polarity score across different topics and quality ranges using various models. Interactive version is available at [4].

4 Conclusions and Future Works

In this work, we demonstrated how various AI models can be applied to assess the sentiment of nearly 7 million Wikipedia articles on different topics. We proposed a framework for the assessment and aggregation of sentiment in long texts, providing a scalable, repeatable method for auditing neutrality at scale that can be applied to other forms of long, structured content.

Comparing five AI models, we found that cVADER and DistilBERT indicate substantially more diversification of polarity than RoBERTa, TextBlob or VADER. Results showed, that "Vader" and "cVader" model behave very differently despite the fact that they are derived from the same implementation of the `vaderSentiment` library: the standard VADER labels tokens directly as `neg`, `neu`, `pos` and therefore produces an overwhelmingly neutral picture, whereas our "cVADER" variant maps the compound score to polarity via thresholds, surfacing far richer positive and negative patterns. Transformer models (DistilBERT, RoBERTa) reveal more diverse polarity patterns and fall in between, detecting more polarity than vanilla VADER but less than cVADER.

While the encyclopedia generally avoids overt sentiment, detectable variations in polarity persist, differing by subject area. For example, event and city articles contain the highest share of negative cue words, while technical and biological topics (*programming*, *software*, *taxon*) tilt positive – an ordering that remains stable across the more sensitive detectors (cVADER, DistilBERT). These findings suggest that the NPOV standard, while widely upheld, may not be enforced equally across all content available in Wikipedia.

Our study has several limitations. First, all four models were trained on short, rather informal texts. Fine-tuning on longer and structured content could sharpen their reliability. Second, our token-level aggregation ignores discourse structure – future studies may integrate sentence position, citations or editorial history. Finally, extending the analysis to non-English editions would reveal whether linguistic and cultural factors modulate the sentiment patterns observed here.

Acknowledgements. The research is supported by the project "OpenFact – artificial intelligence tools for verification of veracity of information sources and fake news detection" (INFOSTRATEG-I/0035/2021-00), granted within the INFOSTRATEG I program of the National Center for Research and Development, under the topic: Verifying information sources and detecting fake news.

References

1. Barbieri, F., Camacho-Collados, J., Neves, L., Espinosa-Anke, L.: TweetEval: unified benchmark and comparative evaluation for tweet classification. arXiv preprint arXiv:2010.12421 (2020)
2. Chou, S.: Social media sentiment analysis on multilingual dataset (2024)
3. Danny, A.E., Bashir, S., Muhammad, I.: Enhancing user experience through sentiment analysis for katsina state transport agency: a textblob approach. FUDMA J. Sci. **7**(6), 117–122 (2023). https://doi.org/10.33003/fjs-2023-0706-2057

4. Data.Lewoniewski.info: Supplementary materials for this research (2025). https://data.lewoniewski.info/sentiment
5. Hevner, A.R., March, S.T., Park, J., Ram, S.: Design science in information systems research. MIS Q. 75–105 (2004)
6. Hube, C.: Bias in Wikipedia. In: Proceedings of the 26th International Conference on World Wide Web Companion (2017). https://doi.org/10.1145/3041021.3053375
7. Isnan, M., Elwirehardja, G.N., Pardamean, B.: Sentiment analysis for TikTok review using Vader sentiment and SVM model. Procedia Comput. Sci. **227**, 168–175 (2023)
8. Jahin, M.A., Shovon, M.S.H., Mridha, M.F., Islam, M.R., Watanobe, Y.: A hybrid transformer and attention based recurrent neural network for robust and interpretable sentiment analysis of tweets. Sci. Rep. **14**(1) (2024). https://doi.org/10.1038/s41598-024-76079-5
9. Jojoa, M., Eftekhar, P., Nowrouzi-Kia, B., Garcia-Zapirain, B.: Natural language processing analysis applied to COVID-19 open-text opinions using a DistilBERT model for sentiment categorization. AI Soc. **39**(3), 883–890 (2024)
10. Joshy, A., Sundar, S.: Analyzing the performance of sentiment analysis using BERT, DistilBERT, and RoBERTa. In: 2022 IEEE International Power and Renewable Energy Conference (IPRECON), pp. 1–6 (2022). https://doi.org/10.1109/IPRECON55716.2022.10059542
11. Lewoniewski, W., Węcel, K., Abramowicz, W.: Multilingual ranking of Wikipedia articles with quality and popularity assessment in different topics. Computers **8**(3) (2019). https://doi.org/10.3390/computers8030060
12. Madanagopal, K., Caverlee, J.: Towards ongoing detection of linguistic bias on Wikipedia. In: Companion Proceedings of the Web Conference 2021 (2021). https://doi.org/10.1145/3442442.3452353
13. Rateb, M.N., Alansary, S., Elzouka, M.K., Galal, M.: Sentiment analysis in times of crisis: evaluating SVM, CNN-LSTM, and Psyntimiento. preprint (version 1) (2024). https://doi.org/10.21203/rs.3.rs-3994513/v1
14. Sayeed, M.S., Mohan, V., Muthu, K.S.: BERT: a review of applications in sentiment analysis. HighTech Innov. J. **4**(2), 453–462 (2023)
15. Singgalen, Y.A.: Implementation of perspective, Vader, and TextBlob in toxicity and sentiment analysis of food and tourism. J. Inf. Syst. Res. **5**, 1292–1302 (2024)
16. Stróżyna, M., Lewoniewski, W., Czumałowska, I., Wojewoda, A.: Sentiment analysis of Wikipedia articles about companies: a comparison of different models. (In publication) (2025)
17. Węcel, K., Lewoniewski, W.: Modelling the quality of attributes in Wikipedia infoboxes. In: Abramowicz, W. (ed.) BIS 2015. LNBIP, vol. 228, pp. 308–320. Springer, Cham (2015). https://doi.org/10.1007/978-3-319-26762-3_27

Decision Science, Multi-criteria Methods

Towards Automated Multi-criteria Decision Analysis Using Machine Learning

Aleksandra Bączkiewicz[1(✉)], Jarosław Wątróbski[1,2], Artur Karczmarczyk[3], and Ewa Wanda Ziemba[4]

[1] Institute of Management, University of Szczecin, ul. Cukrowa 8, 71-004 Szczecin, Poland
{aleksandra.baczkiewicz,jaroslaw.watrobski}@usz.edu.pl
[2] National Institute of Telecommunications, ul. Szachowa 1, 04-894 Warsaw, Poland
[3] Department of Computer Science, Westpomeranian University of Technology in Szczecin, ul. Żołnierska 49, 71-210 Szczecin, Poland
artur.karczmarczyk@zut.edu.pl
[4] University of Economics in Katowice, ul. Bogucicka 3, 40-228 Katowice, Poland
ewa.ziemba@ue.katowice.pl

Abstract. As the volume and complexity of data continue to grow, intelligent information systems are increasingly applied to support decision-making. Their expanding use in the domain of multi-criteria decision-making has fostered the development of numerous frameworks that combine diverse methods and tools, including multi-criteria decision analysis (MCDA) techniques and machine learning (ML) models. This study introduces a framework that integrates MCDA with ML models to capture decision-makers' preferences and, based on these, predict country rankings related to the achievement of Sustainable Development Goal 7. The findings demonstrate that the proposed approach provides a practical and effective alternative for multi-criteria evaluation in situations where expert knowledge is unavailable, thereby addressing the limitations of traditional MCDA methods that depend on the direct involvement of decision-makers and stakeholders.

Keywords: Decision support systems · Machine learning · Ranking prediction · Multi-Layer Perceptron Regressor · Multi-criteria decision analysis

1 Introduction

The extensive evolution of intelligent information systems in last years is a response to the increasing amount of input data required for processing. Information overload can lead to wrong interpretations and conclusions when people attempt to process data by themselves without the support of appropriate methods and tools [18]. The development and widespread use of data science is justified by the lack of sufficient human cognitive capacity to process large amounts of

Published by Springer Nature Switzerland AG 2026
M. Pondel et al. (Eds.): AI4KAM 2025, IFIP AICT 776, pp. 373–385, 2026.
https://doi.org/10.1007/978-3-032-18920-2_35

data [2]. Data science brings data mining techniques [30], multi-criteria decision analysis [28], databases, machine learning, and more to information systems [27].

Decision support processes rely on a variety of mathematical transformations, among which multicriteria analysis plays a central role. At the core of most established approaches lies a multicriteria decision analysis (MCDA) method, which serves as the foundation for structuring and evaluating complex decision problems [29]. The integration of multi-criteria decision analysis (MCDA) methods with machine learning (ML) techniques has recently gained increasing attention as a means of enhancing decision-support systems [34]. While MCDA provides a structured framework for evaluating alternatives based on multiple, often conflicting criteria, ML contributes by uncovering hidden patterns in large datasets and improving predictive capabilities. The combination of these approaches enables more robust and data-driven decision-making, particularly in complex environments where expert knowledge is limited or difficult to obtain [9]. Such hybrid models not only enhance the transparency and reliability of the decision-making process but also extend the applicability of MCDA to dynamic domains where continuous learning from historical data is required [24].

Machine learning models may be applied to predict the values of particular indicators based on data, including observations collected for attributes defining these indicators' implementation levels. Predicted values in machine learning arise from algorithms designed to uncover key patterns between predictor variables and outcomes. ML models include techniques among which we can mention artificial neural networks, deep learning models, and architectures such as Convolutional Neural Network (CNN), Recurrent Neural Network (RNN), Deep Belief Networks (DBNs), Generative Adversarial Networks (GANs), Support Vector Machine models (SVM), k-nearest neighbour (KNN), logistic regression (LR), AdaBoost (AB), random forests (RF), genetic algorithms (GA), regression models, like Linear Regression (LR), Bayesian Linear Regression (BLR), and many more [12,25,36].

To enhance usability, machine learning techniques are frequently integrated within intelligent information system frameworks alongside other approaches, including multi-criteria decision-making methods. Guo et al. investigated a hybrid model that incorporates multi-criteria decision support with a Multi-Layer Perceptron Regressor neural network. Their findings suggest that this integration improves predictive accuracy by effectively capturing the relationships between individual attributes and outcomes. The authors note that the proposed method is applicable across diverse fields, such as human resources, medicine, and marketing [14]. In a related study, Nachappa et al. combined two MCDA methods - Analytic Hierarchy Process (AHP) and Analytic Network Process (ANP) - with two ML models, Random Forest (RF) and Support Vector Machine (SVM), to assess flood susceptibility [22].

The aim of this paper is to present a framework for predicting rankings based on previously collected analogous datasets and their expert evaluation. The proposed framework integrates ML techniques and MCDA methods. MCDA methods are appropriate for evaluating datasets requiring consideration of multiple aspects. ML models allow the prediction of the evaluation result for datasets containing values for multiple attributes. MCDA methods have proven their usefulness in a complex evaluation process that requires consideration of many often conflicting factors. However, MCDA methods have limited application in DSSs, because they require the involvement of experts who, using domain knowledge, assign significance values to evaluation criteria [8]. If experts are unavailable, it is possible to use objective weighting methods, which calculate the weights of criteria based on the data contained in the input decision matrix. However, there may be a situation where decision-makers or stakeholders want to evaluate a new dataset, considering expert knowledge despite weights calculated by algorithms based on statistical rules. In such a situation, the introduction of ML models in the presented framework that can predict rankings using expert knowledge contained in historical data fills the gap that occurs in the absence of experts. The usability of the proposed framework is presented on a real-world dataset containing performance values of European countries in relation to implementing the Sustainable Development Goal 7 directives.

The rest of the paper includes a demonstration of the research methodology in Sect. 2, a presentation of research results in Sect. 3, and pointing out conclusions and directions for further work in Sect. 4.

2 Methodology

The purpose of this paper is to show the applicability of the framework based on MCDA and ML procedures for ranking prediction drawing on historical datasets evaluated using domain expert knowledge. The performance of the presented model was demonstrated in the illustrative example of the assessment of European countries with regard to the implementation of Sustainable Development Goal 7 (SDG 7). SDG 7, compiled by the UN, is used to measure the achievement of the Sustainable Development Goals toward reliable, sustainable, and affordable energy systems for all. Within the framework of SDG 7, the UN has identified specific indicators [10].

Among them are eight indicators with the aim of maximization: $\{C_1$ - Primary energy consumption in Tonnes of oil equivalent (TOE) per capita, C_2 - Final energy consumption in TOE per capita, C_3 - Final energy consumption in households per capita in Kilogram of oil equivalent (KGOE), C_4 - Energy productivity in Euro per KGOE, C_5 - Share of renewable energy sources in gross final energy consumption in %, C_6 - Share of renewable energy sources in transport in gross final energy consumption in %, C_7 - Share of renewable energy sources in electricity in gross final energy consumption in %, C_8 - Share of renewable energy sources in heating and cooling in gross final energy consumption in %$\}$. The framework incorporates also two criteria with the aim of minimization: $\{C_9$ - Energy import dependency by products in %, C_{10} - Population unable to keep home adequately warm in %$\}$. These indicators have been adopted in the proposed model as criteria for evaluating European countries against development toward affordable and clean Energy. The evaluation criteria are represented by C_1-C_{10}. These attributes represent features of the input dataset. They are displayed in Table 1 In this study, the target variable represents the evaluation score of each alternative, derived from an MCDA assessment of historical data. Values of input dataset features were acquired from the Eurostat database (accessed on 24 March 2025). Complete data for all evaluation criteria were available for the years 2004–2022. For each year, an assessment of countries with regard to the implementation of SDG 7 was carried out using the Criteria Importance Through Inter-criteria Correlation (CRITIC) [33] weighting method to assign importance values to particular criteria and the Technique for Order of Preference by Similarity to Ideal Solution (TOPSIS) method for evaluation of considered countries.

Table 1. Multi-criteria model evaluation criteria of SDG 7.

C_j	Name	Unit	Objective
C_1	Primary energy consumption	Tonnes of oil equivalent (TOE) per capita	↑
C_2	Final energy consumption	TOE per capita	↑
C_3	Final energy consumption in households per capita	Kilogram of oil equivalent (KGOE)	↑
C_4	Energy productivity	Euro per KGOE	↑
C_5	Share of renewable energy sources in gross final energy consumption	Percentage (%)	↑
C_6	Share of renewable energy sources in transport in gross final energy consumption	Percentage (%)	↑
C_7	Share of renewable energy sources in electricity in gross final energy consumption	Percentage (%)	↑
C_8	Share of renewable energy sources in heating and cooling in gross final energy consumption	Percentage (%)	↑
C_9	Energy import dependency by products	Percentage (%)	↓
C_{10}	Population unable to keep home adequately warm	Percentage (%)	↓

This procedure corresponds to the evaluation of historical datasets by experts. The TOPSIS method was chosen because of its popularity, wide applicability, and algorithm simplicity, which is important when presenting new methodical approaches. TOPSIS was proposed by Hwang and Yoon in 1981 [26]. The TOPSIS algorithm is based on measuring the distances of alternatives from reference solutions, which are Positive and Negative Ideal Solutions. Based on calculated distances, the final score is determined. The highest TOPSIS score value indicates the best-ranked alternative. Basics and assumptions of TOPSIS are provided in detail in [3]. The CRITIC weighting method was proposed by Diakoulaki in 1995 [13]. CRITIC uses the standard deviation to measure the discreteness of alternatives on criteria and adopts the correlation coefficient to reflect conflicts between criteria [33].

The results of the conducted evaluation are score values for each country in each included year, which serve as the target variable. After collecting these data, a dataset with values for 2004–2021 containing 522 observations was extracted.

The fragment of this dataset is visualized in Table 2. This dataset was further split randomly into a training dataset incorporating 80% of the samples and a test dataset including 20% of the samples. A dataset with observations for each considered country collected for 2022 was used as a test dataset for the final step of the research. In order to conduct machine learning procedures, the dataset was preprocessed. The pre-processing procedure includes the normalization of feature values using the Minimum-Maximum normalization method. This procedure is justified by the different units of data for each feature. The piece of the training dataset after pre-processing is shown in Table 3.

Table 2. Fragment of training dataset before preprocessing and target variable (Score).

Symbol	Country	Year	C_1	C_2	C_3	C_4	C_5	C_6	C_7	C_8	C_9	C_{10}	Score
A_1	Belgium	2004	5.03	3.59	882	4.89	1.916	0.642	1.714	2.927	79.774	6.4	0.3926
A_2	Bulgaria	2004	2.36	1.25	270	1.58	9.231	0.971	8.358	14.056	48.34	69.5	0.1059
A_3	Czechia	2004	4.22	2.6	670	2.85	6.773	1.203	3.693	9.918	25.338	9.3	0.3588
A_4	Denmark	2004	3.69	2.84	825	11.15	14.839	0.445	23.754	20.501	-47.12	10.1	0.4986
A_5	Germany	2004	3.92	2.7	785	6.85	6.207	2.136	9.436	7.198	61.158	4.6	0.4279
...	...	...	...	...	...	...	...	...	...	...	...	...	...
A_{29}	Norway	2021	4.86	3.64	864	12.81	74.018	20.66	113.642	32.274	-616.598	1.4	0.6986

Table 3. Fragment of preprocessed dataset and target variable (Score).

Country symbol	Year	C_1	C_2	C_3	C_4	C_5	C_6	C_7	C_8	C_9	C_{10}	Score
A_1	2004	0.3623	0.2920	0.7010	0.3459	0.0309	0.1021	0.0175	0.0369	0.0250	0.9172	0.3926
A_2	2004	0.0683	0.0154	0.0520	0.0000	0.1553	0.1544	0.0853	0.2538	0.0642	0.0000	0.1059
A_3	2004	0.2731	0.1749	0.4761	0.1327	0.1135	0.1913	0.0377	0.1731	0.0928	0.8750	0.3588
A_4	2004	0.2148	0.2033	0.6405	1.0000	0.2506	0.0708	0.2425	0.3794	0.1831	0.8634	0.4986
A_5	2004	0.2401	0.1868	0.5981	0.5507	0.1038	0.3397	0.0963	0.1201	0.0482	0.9433	0.4279
...	...	...	...	...	...	...	...	...	...	...	...	...
A_{29}	2021	0.2346	0.3453	0.5695	0.4638	0.8411	0.6709	1.0000	0.2961	1.0000	0.9824	0.6986

Machine learning procedures include a selection of hyper-parameters for machine learning models, cross-validation, model training, and prediction. The flow of particular procedures forming the proposed framework is presented in Fig. 1. First, domain experts evaluate historical datasets using tools such as multi-criteria decision analysis methods, assigning weights to criteria assessment based on their domain knowledge. The effects of this step are score values for each alternative in each evaluated period that play the role of the target variable. Then, attribute values are normalized in the preprocessing step. The result is a training dataset that can be passed to the machine learning model. The ML model is trained on this dataset. The trained model may be employed to predict scores for other datasets containing data for analogous features for which expert assessment is not available. Predicted scores enable decision-makers or stakeholders to prepare rankings of alternatives that interest them.

The Multi-Layer Perceptron Regressor model was chosen as the base predictive model for this research. MLP Regressor is an artificial neural network model which is a feedforward type [7]. It is composed of layers of nodes, including three layers: an input layer, a hidden layer, and an output layer [19]. Except for the input node, other nodes are neurons that use a nonlinear activation function [17]. MLP uses the backpropagation supervised learning method for training. The construction and principals of MLP Regressor are provided in [21]. Four reference ML models, among which are Linear Regression (LR) [12], Bayesian Ridge Regression (BRR) [25], Kernel Ridge Regression (KRR) [35], and Gradient Boosting Regressor (GBR) [6], were chosen as predictive machine learning models to benchmark with Multi-Layer Perceptron (MLP) Regressor.

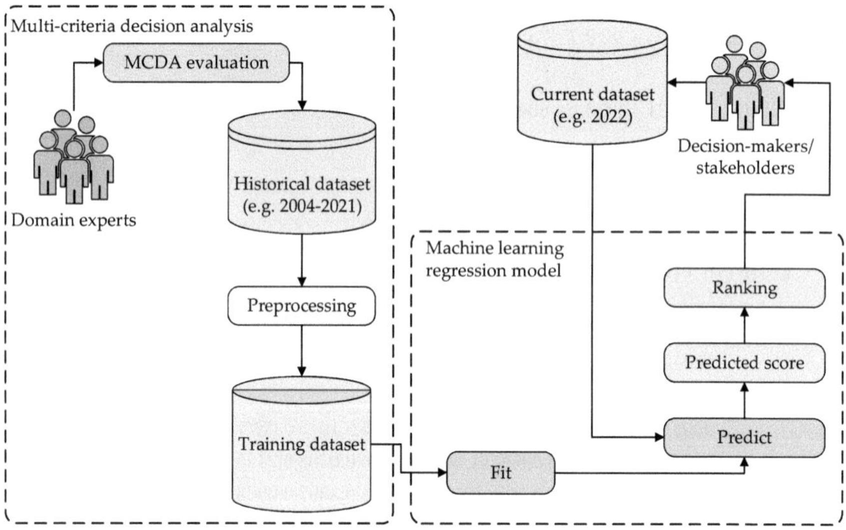

Fig. 1. Framework based on MCDA and ML procedures for ranking prediction.

The most commonly used linear regression model is based on a multivariate statistical technique used to capture the relationship between multiple independent variables and a dependent variable with an explanation and prediction [12]. The Kernel Ridge Regression (KRR) model is based on ridge regression (linear least squares with l2-norm regularization) with the kernel trick [11]. Therefore, it learns a linear function in the space induced by the respective kernel and the data. For non-linear kernels, this corresponds to a non-linear function in the original space. The form of the model trained by KRR is identical to Support Vector Regression (SVR) [15]. However, different loss functions are applied: KRR employs squared error loss while support vector regression relies on epsilon-insensitive loss, both combined with l2 regularization. Unlike SVR, fitting a KRR model can be done in closed-form and is usually faster for medium-sized datasets [35]. Gradient Boosting Regressor (GBR) is an ensemble machine learning algorithm that combines boosting and regression trees. It represents a powerful and popular supervised learning algorithm that is able to capture complex relationships of nonlinear functions effectively. As a result, it is widely used in practical problems requiring consideration of complex relationships between variables [6]. Bayesian regression techniques can be used to include regularization parameters in the estimation procedure: the regularization parameter is not set in a hard sense but tuned to the data at hand. This can be done by introducing uninformative priors over the hyper-parameters of the model. Bayesian Ridge estimates a probabilistic model of the regression problem [25].

The research described in this paper employs models available in the scikit-learn Python library [4,23]. In the next step, hyper-parameters for MLP Regressor using GridSearchCV [1] provided in scikit-learn Python library were selected, k-fold cross-validation [32] and model training and testing were performed. The MLP Regressor results were then compared with reference ML models. The performance of ML models was estimated by comparing the results with factual results achieved through expert multi-criteria evaluation. For this purpose, prediction performance evaluation metrics such as Mean Absolute Error (MAE) [20], Mean Squared Error (MSE) [16], and R-Squared score (R^2) [5] were compared. For this research mentioned metrics implemented in the scikit-learn Python package were employed.

The examination was conducted for a test set containing 105 samples selected randomly and then for a set containing data for 29 samples for the countries under consideration dating back to 2022. For the 2022 data, the rankings generated from the predicted scores were compared in addition to the predicted score values. For comparative analysis of rankings predicted by examined ML models with factual ranking, the Weighted Spearman rank correlation coefficient (r_w) was applied [31].

3 Results

This section gives the results of the following steps achieved for ranking prediction using the proposed MCDA-ML framework. In the first stage, hyper-parameters for the MLP Regressor were selected using the GridSearchCV function. The selected hyper-parameters are shown in Table 4.

Table 4. MLP regressor hyper-parameters selected with GridSearchCV.

Parameters	Values tested in GridSearchCV	Selected hyper-parameters
Hidden layer sizes	(100,), (200,), (500,), (1000,)	(200,)
Activation	'identity', 'logistic', 'tanh', 'relu'	'relu'
Solver	'lbfgs', 'sgd', 'adam'	'lbfgs'
Alpha parameter	0.001, 0.0001, 0.00001	0.00001
Learning rate	'constant', 'invscaling', 'adaptive'	'invscaling'
Maximum iterations number	250, 500, 750, 1000	1000

In the next step, a k-fold cross-validation procedure, where k = 2, was conducted using regression score function R^2 to assess the MLP Regressor model performance. Results obtained for each fold are high and close to 1: {0.9671, 0.9260, 0.9745, 0.9862, 0.9441}. The MLP Regressor's performance was then compared to that of five reference machine learning models. Table 5 includes values of performance evaluation metrics such as MAE, MSE, and R^2 for investigated ML models. The results show that the performance of examined ML models is high and comparable.

Table 5. Values of performance evaluation metrics achieved by examined ML regression models.

Test dataset size, Regression model	MAE	MSE	R^2
105 samples MLP	0.0154	0.0004	0.9785
105 samples LR	0.0165	0.0004	0.9745
105 samples KRR	0.0240	0.0008	0.9494
105 samples GBR	0.0161	0.0004	0.9747
105 samples BRR	0.0165	0.0004	0.9743
29 samples MLP	0.0315	0.0012	0.9410
29 samples LR	0.0343	0.0013	0.9329
29 samples KRR	0.0430	0.0021	0.8919
29 samples GBR	0.0194	0.0006	0.9696
29 samples BRR	0.0346	0.0013	0.9320

In the further stage, the authors focused on the results achieved for the test dataset acquired for the most recent year, which was available, namely 2022. Rankings of considered countries were generated based on scores predicted by each examined ML model and compared with the factual ranking achieved through the expert multi-criteria assessment. The rankings are shown in Table 6. The findings indicate that the top positions across all rankings predicted by the evaluated ML models are consistent. Countries such as Norway, Iceland,

Sweden, Finland, Denmark, Austria, and Luxembourg consistently appear as leaders. This is particularly important, as the upper part of the ranking is typically of greatest relevance to decision-makers and stakeholders. Consequently, ensuring high accuracy in predicting top-ranked alternatives is essential. These observations align with the study by Elavarasan et al., where Iceland, Norway, and Sweden occupied the top three positions with the highest scores among 40 European countries in the context of the proposed SDG 7 composite index [10].

The Weighted Spearman rank correlation coefficient was employed to compare the convergence of predicted rankings with factual ranking, which was considered as reference one and among themselves. The high values of this coefficient that approach 1 prove high convergence of the compared rankings. The obtained correlation values are displayed in Fig. 2.

Table 6. Comparison of factual ranking for 2022 with rankings predicted by different ML regression models.

Country	Factual	MLP	LR	KRR	GBR	BRR	Country	Factual	MLP	LR	KRR	GBR	BRR
Belgium	14	15	14	15	10	14	Luxembourg	7	7	7	7	7	7
Bulgaria	29	29	29	29	29	29	Hungary	16	18	18	18	20	18
Czechia	11	11	13	11	12	13	Malta	21	21	21	21	22	21
Denmark	5	5	5	5	6	5	Netherlands	13	14	11	14	11	11
Germany	15	13	12	13	14	12	Austria	6	6	6	6	5	6
Estonia	8	8	9	8	9	9	Poland	17	16	17	17	17	17
Ireland	10	9	10	10	8	10	Portugal	23	23	23	23	24	23
Greece	27	27	27	28	28	27	Romania	24	24	24	24	23	24
Spain	26	25	25	25	26	25	Slovenia	9	10	8	9	13	8
France	22	22	22	22	18	22	Slovakia	19	20	20	20	19	20
Croatia	18	17	16	16	16	16	Finland	4	4	4	4	4	4
Italy	20	19	19	19	21	19	Sweden	3	3	3	3	3	3
Cyprus	28	28	28	27	27	28	Iceland	2	2	2	2	2	2
Latvia	12	12	15	12	15	15	Norway	1	1	1	1	1	1
Lithuania	25	26	26	26	25	26							

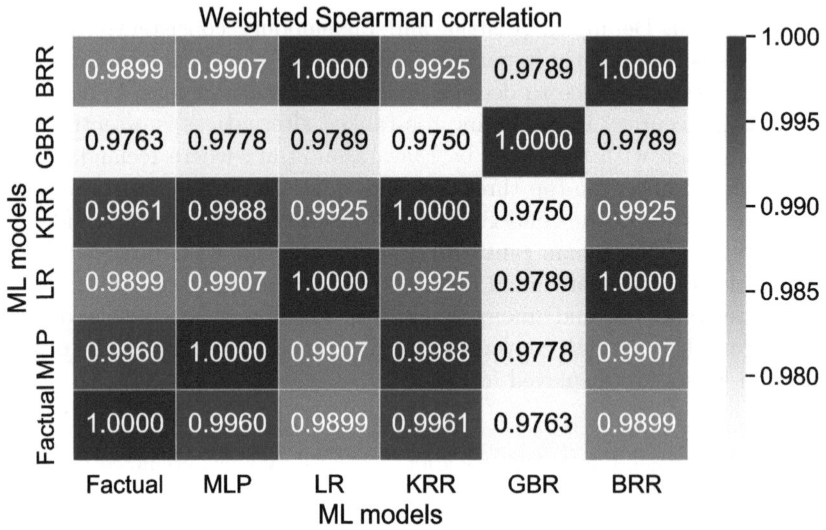

Fig. 2. Correlation of factual ranking in 2022 with rankings predicted by examined ML regression models.

It can be noted that all examined models provided rankings with high consistency with the reference ranking. However, it can be noticed that the highest convergence with factual ranking was demonstrated for rankings predicted by the MLP Regressor and the KRR model. Rankings predicted by the LR and BRR models are also very consistent with the factual ranking. The lowest consistency was revealed for the GBR model. It can be concluded that obtained r_w values presented in Fig. 2 indicate different results rather than performance evaluation metrics values shown in Table 5, where the GBR model demonstrated high convergence compared to the rest of ML models. It is justified by the fact that Table 5 includes results achieved for score values regardless of rankings.

It can be observed that the performance of tested ML models may show differences depending on the score or ranking considered. This justifies the need to compare the performance of ML models for both scores and rankings and to conduct a comparative analysis using multiple models depending on the objectives of the study.

4 Conclusions

The findings indicate that all ML models applied in this study demonstrated strong relevance and performance in predicting rankings derived from previously assessed historical datasets. Incorporating ML models into the proposed framework offers a practical solution to the challenge of limited expert availability when conducting multi-criteria evaluations of new datasets, provided that sufficiently large and analogous datasets, previously assessed by domain experts, are available. The outcomes also emphasize the importance of comparative analysis, as variations among ML models may lead to divergent results. Such analysis enables the validation of consistently strong positions for alternatives of interest to decision-makers, particularly when multiple models identify them as ranking leaders. A limitation of the proposed approach lies in its reliance on historical data, meaning that shifts in decision-makers' preferences are captured with delay, potentially reducing accuracy. Addressing this issue requires regular updates to the training dataset. Future research should therefore focus on extending the training data to enhance the robustness of the model.

Acknowledgments. This research was partially funded by National Science Centre, Poland 2022/45/B/HS4/02960, and Co-financed by the Minister of Science under the "Regional Excellence Initiative" Program RID/SP/0046/2024/01.

Regionalna Inicjatywa Doskonałości

Ministry of Science and Higher Education Republic of Poland

Disclosure of Interests. The authors have no competing interests to declare that are relevant to the content of this article.

References

1. Ahmad, G.N., Fatima, H., Ullah, S., Saidi, A.S., et al.: Efficient medical diagnosis of human heart diseases using machine learning techniques with and without GridSearchCV. IEEE Access **10**, 80151–80173 (2022). https://doi.org/10.1109/ACCESS.2022.3165792
2. Bączkiewicz, A., Wątróbski, J.: Crispyn - a Python library for determining criteria significance with objective weighting methods. SoftwareX **19**, 101166 (2022). https://doi.org/10.1016/j.softx.2022.101166
3. Bera, B., Shit, P.K., Sengupta, N., Saha, S., Bhattacharjee, S.: Susceptibility of deforestation hotspots in Terai-Dooars belt of Himalayan Foothills: a comparative analysis of VIKOR and TOPSIS models. J. King Saud Univ.-Comput. Inf. Sci. **34**(10), 8794–8806 (2022). https://doi.org/10.1016/j.jksuci.2021.10.005
4. Buitinck, L., et al.: API design for machine learning software: experiences from the scikit-learn project. In: ECML PKDD Workshop: Languages for Data Mining and Machine Learning, pp. 108–122 (2013)
5. Cagirici, H.B., Budak, H., Sen, T.Z.: G4Boost: a machine learning-based tool for quadruplex identification and stability prediction. BMC Bioinform. **23**(1), 240 (2022). https://doi.org/10.1186/s12859-022-04782-z
6. Cai, J., Xu, K., Zhu, Y., Hu, F., Li, L.: Prediction and analysis of net ecosystem carbon exchange based on gradient boosting regression and random forest. Appl. Energy **262**, 114566 (2020). https://doi.org/10.1016/j.apenergy.2020.114566
7. Cinar, A.C.: Training feed-forward multi-layer perceptron artificial neural networks with a tree-seed algorithm. Arab. J. Sci. Eng. **45**(12), 10915–10938 (2020). https://doi.org/10.1007/s13369-020-04872-1

8. Cinelli, M., Burgherr, P., Kadziński, M., Słowiński, R.: Proper and improper uses of MCDA methods in energy systems analysis. Decis. Support Syst. **163**, 113848 (2022). https://doi.org/10.1016/j.dss.2022.113848
9. Duan, J., Eben Li, S., Guan, Y., Sun, Q., Cheng, B.: Hierarchical reinforcement learning for self-driving decision-making without reliance on labelled driving data. IET Intel. Transp. Syst. **14**(5), 297–305 (2020). https://doi.org/10.1049/iet-its.2019.0317
10. Elavarasan, R.M., Pugazhendhi, R., Irfan, M., Mihet-Popa, L., Campana, P.E., Khan, I.A.: A novel sustainable development Goal 7 composite index as the paradigm for energy sustainability assessment: a case study from Europe. Appl. Energy **307**, 118173 (2022). https://doi.org/10.1016/j.apenergy.2021.118173
11. Elkhalil, K., Kammoun, A., Zhang, X., Alouini, M.S., Al-Naffouri, T.: Risk convergence of centered kernel ridge regression with large dimensional data. IEEE Trans. Signal Process. **68**, 1574–1588 (2020). https://doi.org/10.1109/TSP.2020.2975939
12. Fiorucci, M., Khoroshiltseva, M., Pontil, M., Traviglia, A., Del Bue, A., James, S.: Machine learning for cultural heritage: a survey. Pattern Recogn. Lett. **133**, 102–108 (2020). https://doi.org/10.1016/j.patrec.2020.02.017
13. Giannakidou, C., Diakoulaki, D., Memos, C.D.: Vulnerability to coastal flooding of industrial urban areas in Greece. Environ. Process. **7**, 749–766 (2020). https://doi.org/10.1007/s40710-020-00442-7
14. Guo, M., Zhang, Q., Liao, X., Chen, F.Y., Zeng, D.D.: A hybrid machine learning framework for analyzing human decision-making through learning preferences. Omega **101**, 102263 (2021). https://doi.org/10.1016/j.omega.2020.102263
15. Hazarika, B.B., Gupta, D.: Affinity based fuzzy kernel ridge regression classifier for binary class imbalance learning. Eng. Appl. Artif. Intell. **117**, 105544 (2023). https://doi.org/10.1016/j.engappai.2022.105544
16. Ho, W.K., Tang, B.S., Wong, S.W.: Predicting property prices with machine learning algorithms. J. Prop. Res. **38**(1), 48–70 (2021). https://doi.org/10.1080/09599916.2020.1832558
17. Hwang, J., Lee, J., Lee, K.S.: A deep learning-based method for grip strength prediction: comparison of multilayer perceptron and polynomial regression approaches. PLoS ONE **16**(2), e0246870 (2021). https://doi.org/10.1371/journal.pone.0246870
18. Injadat, M., Moubayed, A., Nassif, A.B., Shami, A.: Machine learning towards intelligent systems: applications, challenges, and opportunities. Artif. Intell. Rev. **54**(5), 3299–3348 (2021). https://doi.org/10.1007/s10462-020-09948-w
19. Jin, H., Kim, Y.G., Jin, Z., Rushchitc, A.A., Al-Shati, A.S.: Optimization and analysis of bioenergy production using machine learning modeling: multi-layer perceptron, Gaussian processes regression, K-nearest neighbors, and Artificial neural network models. Energy Rep. **8**, 13979–13996 (2022). https://doi.org/10.1016/j.egyr.2022.10.334
20. Kim, K., Lee, H., Kang, M., Lee, G., Jung, K., Kharangate, C.R., Asheghi, M., Goodson, K.E., Lee, H.: A machine learning approach for predicting heat transfer characteristics in micro-pin fin heat sinks. Int. J. Heat Mass Transf. **194**, 123087 (2022). https://doi.org/10.1016/j.ijheatmasstransfer.2022.123087
21. Mozaffaree Pour, N., Oja, T.: Prediction power of logistic regression (LR) and multi-layer perceptron (MLP) models in exploring driving forces of urban expansion to be sustainable in Estonia. Sustainability **14**(1), 160 (2021). https://doi.org/10.3390/su14010160

22. Nachappa, T.G., Piralilou, S.T., Gholamnia, K., Ghorbanzadeh, O., Rahmati, O., Blaschke, T.: Flood susceptibility mapping with machine learning, multi-criteria decision analysis and ensemble using Dempster Shafer Theory. J. Hydrol. **590**, 125275 (2020). https://doi.org/10.1016/j.jhydrol.2020.125275
23. Pedregosa, F., et al.: Scikit-learn: machine learning in Python. J. Mach. Learn. Res. **12**, 2825–2830 (2011)
24. Pham, B.T., et al.: Flood risk assessment using deep learning integrated with multi-criteria decision analysis. Knowl.-Based Syst. **219**, 106899 (2021). https://doi.org/10.1016/j.knosys.2021.106899
25. Saqib, M.: Forecasting COVID-19 outbreak progression using hybrid polynomial-Bayesian ridge regression model. Appl. Intell. **51**(5), 2703–2713 (2021). https://doi.org/10.1007/s10489-020-01942-7
26. Uzun, B., Taiwo, M., Syidanova, A., Uzun Ozsahin, D.: The technique for order of preference by similarity to ideal solution (TOPSIS). Application of multi-criteria decision analysis in environmental and civil engineering, pp. 25–30 (2021). https://doi.org/10.1007/978-3-030-64765-0_4
27. Wątróbski, J., Bączkiewicz, A., Rudawska, I.: A strong sustainability paradigm based analytical hierarchy process (SSP-AHP) method to evaluate sustainable healthcare systems. Ecol. Ind. **154**, 110493 (2023). https://doi.org/10.1016/j.ecolind.2023.110493
28. Wątróbski, J., Bączkiewicz, A., Sałabun, W.: pyrepo-MCDA - reference objects based MCDA software package. SoftwareX **19**, 101107 (2022). https://doi.org/10.1016/j.softx.2022.101107
29. Wątróbski, J., Jankowski, J., Piotrowski, Z.: The selection of multicriteria method based on unstructured decision problem description. In: Hwang, D., Jung, J.J., Nguyen, N.-T. (eds.) ICCCI 2014. LNCS (LNAI), vol. 8733, pp. 454–465. Springer, Cham (2014). https://doi.org/10.1007/978-3-319-11289-3_46
30. Wątróbski, J., Karczmarczyk, A., Bączkiewicz, A.: Using the TOSS method in semi-autonomous passenger car selection. Sustain. Energy Technol. Assess. **58**, 103367 (2023). https://doi.org/10.1016/j.seta.2023.103367
31. Więckowski, J., Kizielewicz, B., Shekhovtsov, A., Sałabun, W.: RANCOM: a novel approach to identifying criteria relevance based on inaccuracy expert judgments. Eng. Appl. Artif. Intell. **122**, 106114 (2023). https://doi.org/10.1016/j.engappai.2023.106114
32. Wong, T.T., Yeh, P.Y.: Reliable accuracy estimates from k-fold cross validation. IEEE Trans. Knowl. Data Eng. **32**(8), 1586–1594 (2019). https://doi.org/10.1109/TKDE.2019.2912815
33. Xu, C., Ke, Y., Li, Y., Chu, H., Wu, Y.: Data-driven configuration optimization of an off-grid wind/PV/hydrogen system based on modified NSGA-II and CRITIC-TOPSIS. Energy Convers. Manage. **215**, 112892 (2020). https://doi.org/10.1016/j.enconman.2020.112892
34. Yilmaz, I., Adem, A., Dağdeviren, M.: A machine learning-integrated multi-criteria decision-making approach based on consensus for selection of energy storage locations. J. Energy Storage **69**, 107941 (2023). https://doi.org/10.1016/j.est.2023.107941
35. Zhang, H., Li, Y., Lu, W., Lin, Q.: On the optimality of misspecified kernel ridge regression. In: International Conference on Machine Learning, pp. 41331–41353. PMLR (2023)
36. Zhu, X., Zhang, P., Xie, M.: A joint long short-term memory and AdaBoost regression approach with application to remaining useful life estimation. Measurement **170**, 108707 (2021). https://doi.org/10.1016/j.measurement.2020.108707

On the Statistical Significance of the Weight Similarity Coefficient in Multi-criteria Decision Analysis

Wojciech Sałabun[1,2](✉), Jean Dezert[3], Andrii Shekhovtsov[1,2], and Jarosław Wątróbski[4]

[1] National Institute of Telecommunications,Warsaw, Poland
{w.salabun,a.shekhovtsov}@il-pib.pl
[2] West Pomeranian University of Technology in Szczecin,Szczecin, Poland
[3] Independent Researcher Member of Int. Society of Information Fusion, l'Hay les Roses, France
jean.dezert@isif.org
[4] University of Szczecin,Szczecin, Poland
jaroslaw.watrobski@usz.edu.pl

Abstract. This paper proposes a method for statistically assessing the similarity between two weight vectors, based on the Weights Similarity Coefficient (WSC). The WSC value measures the degree of similarity between two normalized vectors in the weight space. To determine the significance of this similarity, a non-parametric test is introduced, based on sampling from the Dirichlet distribution, which represents the null hypothesis of no association between the vectors. The p-value is calculated as the proportion of random comparisons in which the WSC exceeds the empirical value. The complete testing procedure, result interpretation, and numerical implementation are presented.

Keywords: Weight Similarity Coefficient · Significance Testing · MCDA

1 Introduction

In Multi-Criteria Decision Analysis (MCDA), recommendation systems, and decision-making models, it is always possible to determine more than one weight vector [6], which generally satisfies the condition $\sum_{i=1}^{n} w_i = 1$, where w_i denotes the i-th weight and n represents the total number of criteria [11]. This requires a comparison of these vectors to evaluate their similarity. Traditional distance measures are often inadequate because they lack normalization, which can hinder their interpretability [1]. Consequently, specialized coefficients that are more straightforward to analyze and interpret are often preferred. One such proposal is the Weight Similarity Coefficient (WSC), specifically developed to compare weight vectors $\mathbf{w}, \mathbf{v} \in [0, 1]^n$ [9].

Published by Springer Nature Switzerland AG 2026
M. Pondel et al. (Eds.): AI4KAM 2025, IFIP AICT 776, pp. 386–392, 2026.
https://doi.org/10.1007/978-3-032-18920-2_36

The WSC coefficient is an asymmetric measure, meaning $\text{WSC}(\mathbf{w}, \mathbf{v}) \neq WSC(\mathbf{v}, \mathbf{w})$. The asymmetry ($WSC(w, v) \neq WSC(v, w)$) results from using the minimum weight ($\min(w)$ or $\min(v)$) in the denominator, which leads to varying sensitivity of the measure depending on the direction of comparison. In practice, asymmetry can be beneficial when one vector serves as a reference point and the other is evaluated in terms of compliance, reflecting hierarchical or directional relationships [4].

For example, in project evaluation, decision maker A with weights $w = [0.5,\ 0.3,\ 0.2]$ may serve as the reference, while decision maker B with weights $v = [0.4,\ 0.5,\ 0.1]$ is compared against it. The value $WSC(w, v) = 0.75$ indicates a moderate level of compliance of B with A, taking into account their differing priorities. In the example above, $WSC(v, w) \approx 0.778$ suggests that the weights of A are slightly more consistent with those of B, which is due to the smaller minimum weight in v (0.1), increasing the flexibility of the measure.

However, in the context of classical similarity measures, symmetry is typically expected, as it enhances clarity, interpretability, and compatibility [10]. To address this, a symmetric variant (WSC_2) has been developed, which satisfies the condition $WSC_2(\mathbf{w}, \mathbf{v}) = WSC_2(\mathbf{v}, \mathbf{w})$. As a result, WSC_2 can be used effectively in applications where an equal and direction-independent comparison of weight vectors is desired [5,9].

The main contribution of the paper is the introduction of a statistical testing procedure that allows one to assess the significance of the WSC in MCDA. Although WSC quantifies the similarity between two weight vectors, its interpretation has so far lacked statistical support. The proposed procedure is based on the Monte Carlo method [8], where the observed WSC value is compared against a distribution of values obtained from random weight vectors generated using the Dirichlet distribution [2]. The test evaluates the following hypotheses: the null hypothesis (H_0) assumes that the observed similarity could have occurred by chance, i.e., that the compared weight vector is randomly drawn and not specifically aligned with the reference vector; the alternative hypothesis (H_1) states that the similarity is significantly higher than would be expected by chance. This allows for the calculation of a p-value, which indicates whether the observed similarity is statistically significant [7]. The proposed approach enhances the interpretability and reliability of WSC-based analyses in the context of multi-criteria decision-making.

The remainder of the paper is structured as follows. Section 2 presents the key preliminaries that serve as the foundation for the subsequent research, along with the proposed statistical test. Section 3 describes the experimental study and provides a brief discussion of the results. Finally, Sect. 4concludes the paper and outlines directions for future research.

2 Methodology

This section presents the methodological foundations of the proposed approach. First, we recall the WSC coefficient and its symmetric variant, which serve as measures of alignment between weight vectors in MCDA. Then, we introduce a statistical procedure based on Monte Carlo simulation [8] to evaluate the significance of the observed similarity values. This provides a robust framework for interpreting weight-based comparisons in decision-making contexts.

2.1 Preliminaries

For two weight vectors $\mathbf{w}, \mathbf{v} \in [0,1]^n$ such that $\sum w_i = \sum v_i = 1$, the Weight Similarity Coefficient (WSC) is defined as:

$$WSC(\mathbf{w}, \mathbf{v}) = 1 - \frac{\sum_{i=1}^{n} |w_i - v_i|}{2(1 - \min(w))} \tag{1}$$

The coefficient takes values in the interval $[0, 1]$, where 1 indicates identical vectors, and lower values reflect decreasing similarity. Due to the presence of $\min(w)$ in the denominator, WSC is an *asymmetric* measure, meaning that $\mathrm{WSC}(\mathbf{w}, \mathbf{v}) \neq \mathrm{WSC}(\mathbf{v}, \mathbf{w})$ in general.

A simple symmetric version of the coefficient, denoted as WSC_2, can be defined as:

$$\mathrm{WSC}_2(\mathbf{w}, \mathbf{v}) = 1 - \frac{\sum_{i=1}^{n} |w_i - v_i|}{2} \tag{2}$$

This version also maps to the interval $[0, 1]$, with 1 corresponding to identical vectors. Unlike WSC, the WSC_2 coefficient is symmetric by construction. A similar effect can be achieved by using $min(w, v)$ in the denominator instead of $min(w)$. This modification also ensures symmetry. It distinguishes the proposed approach from $WSC2$, as it shifts the distribution of the coefficient to the left.

The denominator $2(1-\min(w))$ in WSC depends on the minimum weight in the reference vector w, making the measure more sensitive to the structure of the reference distribution. When $\min(w)$ is small (e.g., close to 0), the denominator increases, which reduces the impact of the absolute differences $\sum |w_i - v_i|$ on the final score, making the measure less strict. Conversely, when $\min(w)$ is relatively large, WSC becomes more sensitive to deviations, reflecting greater attention to small differences under more uniform weight distributions.

In contrast, the denominator in WSC_2 is constant (2), meaning that the measure depends solely on the sum of absolute differences $\sum |w_i - v_i| \in [0, 2]$, making WSC_2 normalized to the interval $[0, 1]$. This makes WSC_2 simpler and more universally applicable, but less flexible in adapting to the characteristics of specific weight.

2.2 The Proposed Approach

To enhance the WSC coefficient in MCDA, we propose a statistical testing procedure that evaluates whether the observed similarity between a reference weight vector $\mathbf{w}$ and a target vector $\mathbf{v}$ is greater than would be expected by chance. Although WSC provides a numerical value that quantifies the similarity between two normalized weight vectors, its raw value alone does not indicate whether the observed alignment is statistically meaningful. To address this, we employ a Monte Carlo simulation-based approach that compares the empirical similarity score to a reference distribution obtained under a null model of randomness. In particular, we generate N random weight vectors $\mathbf{v}^{(k)}$ sampled from the uniform Dirichlet distribution `Dirichlet(1)` (by suing `numpy` library [3]), which ensures that the sampled vectors lie within the same probability simplex as $\mathbf{w}$ and $\mathbf{v}$. Each of these sampled vectors is then compared with the reference vector $\mathbf{w}$ using the WSC formula, yielding a set of similarity scores $\{s^{(1)}, s^{(2)}, \dots, s^{(N)}\}$. The observed similarity $s = \text{WSC}(\mathbf{w}, \mathbf{v})$ is then evaluated against this empirical distribution.

Formally, the test examines the null hypothesis (H_0): the observed similarity could have arisen by chance, assuming $\mathbf{v}$ is randomly aligned with respect to $\mathbf{w}$, against the alternative hypothesis (H_1): the similarity is significantly higher than would be expected under random conditions. The one-sided p-value [7] is computed as the proportion of random scores greater than or equal to the observed score, i.e., $p = \frac{1}{N} \sum_{k=1}^{N} \mathbb{I}\{s^{(k)} \geq s\}$. A low p-value (typically below 0.05) provides evidence against the null hypothesis, suggesting that the alignment between $\mathbf{w}$ and $\mathbf{v}$ is not incidental but statistically significant. The complete procedure is presented in Algorithm 1, which outlines all computational steps required to perform the test in practice.

Algorithm 1. WSC Significance Test

Require: $\mathbf{w}, \mathbf{v} \in [0, 1]^n$; N — number of random samples
1: Compute the empirical similarity score: $s \leftarrow \text{WSC}(\mathbf{w}, \mathbf{v})$
2: **for** $k = 1$ to N **do**
3: Draw a random vector $\mathbf{v}^{(k)} \sim \text{Dirichlet}(\mathbf{1})$
4: Compute its similarity: $s^{(k)} \leftarrow \text{WSC}(\mathbf{w}, \mathbf{v}^{(k)})$
5: **end for**
6: Compute p-value: $p \leftarrow \frac{1}{N} \sum \mathbb{I}\{s^{(k)} \geq s\}$
7: **return** s, p

It is worth noting that the same procedure can also be applied to the symmetric similarity coefficient WSC_2, with a straightforward adaptation consisting of replacing the computed $WSC(\cdot, \cdot)$ values with their $\text{WSC}_2(\cdot, \cdot)$ counterparts. This allows for conducting an analogous significance test in scenarios where a symmetric interpretation of similarity between weight vectors is desired.

3 Exemplary Results of Similarity Testing Using WSC

To demonstrate the practical utility of the proposed WSC-based statistical testing procedure, we present two illustrative examples involving bidirectional similarity analysis. These cases highlight how statistical inference can differentiate between meaningful alignment and coincidental similarity, even when the WSC values appear numerically high.

3.1 Example of One-Sided Significance in Bidirectional WSC Analysis

A histogram-based analysis of the WSC coefficient was conducted to evaluate the statistical significance of similarity between the reference weight vector $\mathbf{w} = [0.5, 0.2, 0.15, 0.1, 0.05]$ and the target vector $\mathbf{v} = [0.3, 0.2, 0.2, 0.15, 0.15]$. The WSC values were computed in both directions, yielding $WSC(\mathbf{w}, \mathbf{v}) = 0.789$ and $WSC(\mathbf{v}, \mathbf{w}) = 0.765$. To assess whether these values indicate significant alignment beyond random expectation, a Monte Carlo simulation was employed, generating 100,000 random weight vectors sampled from the uniform Dirichlet distribution. Each random vector was compared with the fixed vector $\mathbf{w}$ or $\mathbf{v}$, respectively, to produce reference distributions against which the observed WSC values were evaluated.

The resulting histogram (Fig. 1) clearly demonstrates that $WSC(\mathbf{w}, \mathbf{v})$ lies in the upper tail of the corresponding distribution, leading to a low p-value of 0.0340. This result supports the rejection of the null hypothesis, indicating that the similarity between $\mathbf{v}$ and $\mathbf{w}$ is statistically significant when assessed from the perspective of $\mathbf{w}$. In contrast, $WSC(\mathbf{v}, \mathbf{w})$ falls much closer to the center of its distribution, yielding a p-value of 0.3103. This suggests that the observed alignment in this direction may plausibly occur by chance. The statistical significance asymmetry underscores the directional nature of the proposed test and emphasizes the importance of specifying the role of reference and target vectors in similarity analysis.

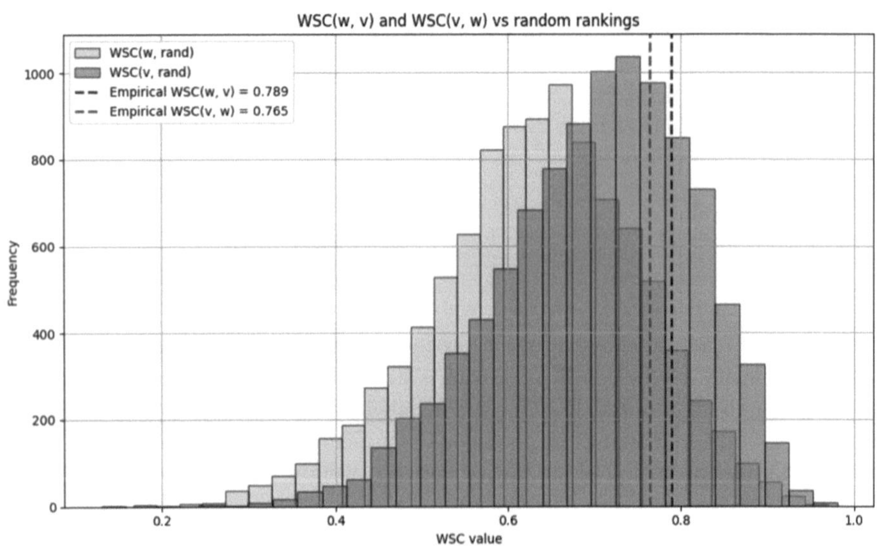

Fig. 1. Empirical distributions of the $WSC(\cdot, \cdot)$ coefficient computed for the vectors $\mathbf{w} = [0.50, 0.20, 0.15, 0.10, 0.05]$ and $\mathbf{v} = [0.30, 0.20, 0.20, 0.15, 0.15]$, based on $n = 100{,}000$ random samples.

3.2 When High WSC Values Do Not Imply Significant Similarity

To further assess the robustness of the WSC-based similarity test, an additional bidirectional analysis was carried out using the same reference vector $\mathbf{w} = [0.5, 0.2, 0.15, 0.1, 0.05]$ and target vector $\mathbf{v} = [0.3, 0.2, 0.2, 0.15, 0.15]$. In this case, the computed values were $WSC(\mathbf{w}, \mathbf{v}) = 0.800$ and $WSC(\mathbf{v}, \mathbf{w}) = 0.812$. As in the previous experiment, the significance of each result was evaluated using a Monte Carlo simulation based on 100,000 random weight vectors drawn from the uniform Dirichlet distribution. Each random vector was compared to the fixed vector in question, producing empirical distributions for both directions of comparison.

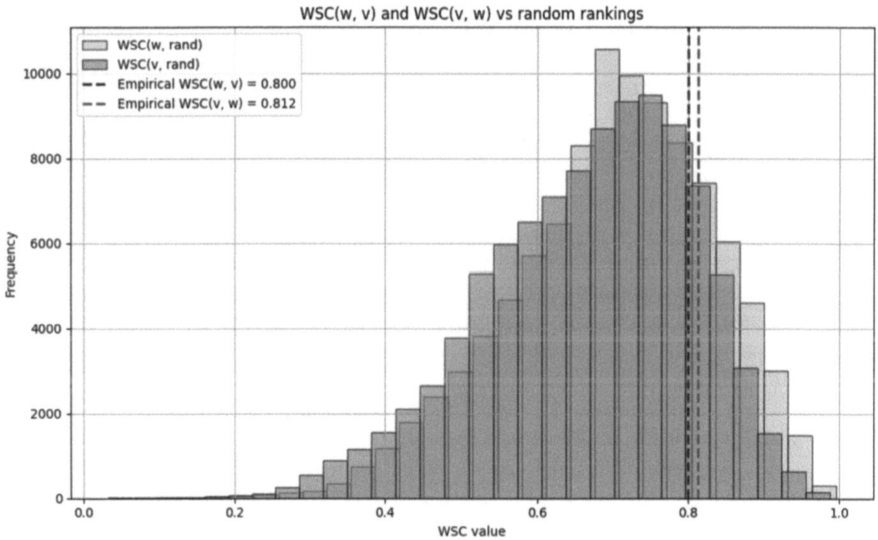

Fig. 2. Empirical distributions of the $WSC(\cdot, \cdot)$ coefficient computed for the vectors $\mathbf{w} = [0.25, 0.25, 0.25, 0.25]$ and $\mathbf{v} = [0.4, 0.2, 0.2, 0.2]$, based on $n = 100{,}000$ random samples.

As illustrated in Fig. 2, both empirical WSC values are situated near the upper region of their corresponding distributions. However, the resulting p-values, 0.2384 for $WSC(\mathbf{w}, \mathbf{v})$ and 0.1412 for $WSC(\mathbf{v}, \mathbf{w})$, do not provide sufficient evidence to reject the null hypothesis in either direction. This outcome demonstrates that relatively high similarity values may not be statistically meaningful unless supported by a clear deviation from the expected distribution under randomness. The result reinforces the importance of statistical inference in validating observed alignment between weight vectors.

4 Conclusion

This study proposed a statistical testing procedure to enhance the interpretability of the WSC in MCDA. By applying Monte Carlo simulations with Dirichlet-distributed random vectors, the method enables significance testing of similarity between weight vectors. The approach is applicable to both asymmetric and symmetric variants of WSC and helps distinguish between meaningful and incidental alignment. Future work will focus on extending the method to other similarity measures and decision-making contexts.

References

1. Choi, S.S., Cha, S.H., Tappert, C.C., et al.: A survey of binary similarity and distance measures. J. Systemics, Cybern. Inf. **8**(1), 43–48 (2010)
2. Devroye, L.: Nonuniform random variate generation. Handb. Oper. Res. Manag. Sci. **13**, 83–121 (2006)
3. Harris, C.R., et al.: Array programming with NumPy. Nature **585**(7825), 357–362 (2020). https://doi.org/10.1038/s41586-020-2649-2
4. He, W., Dutta, B., Liu, Y., Rodríguez, R.M.: Extend Tversky's ratio model to an asymmetric similarity measurement model with three conditional parameters: 3p-ASM model. Int. J. Comput. Intell. Syst. **16**, 113 (2023). https://doi.org/10.1007/s44196-023-00285-8
5. Kizielewicz, B., Sałabun, W.: A robust framework for renewable energy policy evaluation using MCDA and compromise ranking with stochastic weight identification. J. Intell. Manag. Decis. **3**(4), 213–223 (2024)
6. Kizielewicz, B., Tomczyk, T., Gandor, M., Sałabun, W.: Subjective weight determination methods in multi-criteria decision-making: a systematic review. Procedia Comput. Sci. **246**, 5396–5407 (2024). https://doi.org/10.1016/j.procs.2024.09.673
7. Lehmann, E.L., Romano, J.P.: Testing Statistical Hypotheses. Springer Texts in Statistics, 4 edn. Springer, Cham (2022). https://doi.org/10.1007/978-3-030-70578-7
8. Rubinstein, R.Y., Kroese, D.P.: Simulation and the Monte Carlo method. John Wiley & Sons, Hoboken (2016)
9. Shekhovtsov, A.: Evaluating the performance of subjective weighting methods for multi-criteria decision-making using a novel weights similarity coefficient. Procedia Comput. Sci. **225**, 4785–4794 (2023). https://doi.org/10.1016/j.procs.2023.10.478
10. Tao, Y., et al.: A comparative analysis of trajectory similarity measures. GIScience Remote Sens. **58**(5), 643–669 (2021). https://doi.org/10.1080/15481603.2021.1908927
11. Xu, Y.Q., et al.: Weight vector generation in multi-criteria decision-making with basic uncertain information. Mathematics **10**(4), 572 (2022). https://doi.org/10.3390/math10040572

Reducing Pairwise Comparisons in COMET: Novel Approaches for Efficient Decision Model Identification

Andrii Shekhovtsov[1,2](✉)

[1] National Institute of Telecommunications, Warsaw, Poland
[2] West Pomeranian University of Technology in Szczecin, Szczecin, Poland
a.shekhovtsov@il-pib.pl

Abstract. Multi-Criteria Decision Analysis (MCDA) supports decision-making in complex scenarios involving multiple conflicting criteria. The Characteristic Objects METhod (COMET) is a relatively recent MCDA technique that relies on expert-based pairwise comparisons to evaluate alternatives. Despite its growing adoption in MCDA domain the COMET method have some limitations, such as scalability issues due to the rapid increase in the number of required pairwise comparisons as the number of criteria grows. To address this limitation, we propose two extensions to the Triad Support algorithm, which aim to significantly reduce the number of comparisons needed while preserving the model's accuracy. The first enhancement involves reordering Characteristic Objects for easier comparison process, while the second use expert-defined criteria types to predefine characteristic values relations. Under the assumption of transitive and consistent expert preferences, these extensions produce the same decision model as the original COMET method. We also provide comparative experiments that demonstrates how both of proposed methods outperform the original COMET and Triad Support algorithm in terms pairwise comparison reduction. These improvements facilitate the application of COMET to larger and more complex decision problems.

Keywords: COMET · MCDA · Triad Support · Pairwise comparison

1 Introduction

Multi-Criteria Decision Analysis (MCDA) is a broad branch of the operations research domain that focuses on complex decision problems involving multiple conflicting criteria. Such problems are usually difficult to handle without the support of mathematical algorithms and approaches, which makes MCDA methods particularly helpful when addressing challenges in healthcare [1], agriculture [7], sustainability assessments [10], and other domains [2,6].

The Characteristic Objects METhod (COMET) is an MCDA method that relies on fuzzy logic in order to deliver robust and accurate evaluations of alternatives based on expert preferences, which are collected via pairwise comparisons

Published by Springer Nature Switzerland AG 2026
M. Pondel et al. (Eds.): AI4KAM 2025, IFIP AICT 776, pp. 393–399, 2026.
https://doi.org/10.1007/978-3-032-18920-2_37

of the Characteristic Objects (COs) [5]. Although the method is relatively new, it has already been adopted by other researchers and is widely used for various purposes. For example, Habeeb et al. proposed using the COMET method for regional crop yield evaluation [3], while Yulistia et al. proposed using it to avoid rank reversal in the Analytical Hierarchy Process [11].

Despite its advantages, the COMET method also has some limitations, mainly due to the involvement of the pairwise comparison process in model identification. Because of how the COs are generated, the number of pairwise comparisons can grow very quickly depending on the number of criteria. This makes the COMET method difficult to use in more complex decision problems due to the time-consuming comparison process. To address this, various extensions of the COMET method have emerged over time. For example, Kizielewicz et al. proposed using Support Vector Regression to identify the decision model based on available data [6]. Another extension was proposed by Shekhovtsov et al. [9], which utilizes rules of transitivity to automatically resolve some of the pairwise comparisons based on the expert's previous answers. This approach has shown good performance under the assumption that the expert's preferences are in a transitive relation and the answers are consistent.

The main contribution of this study is the proposal of two extensions to the Triad Support algorithm, allowing for an even greater reduction in the number of pairwise comparisons required to identify the COMET model. Both algorithms were compared with the original COMET method and the Triad Support algorithm, demonstrating the superiority of the proposed approaches in various scenarios.

The rest of the paper is structured as follows: in Sect. 2, we briefly describe the methods used in this study in Sect. 3, we propose two extensions of the COMET method and the Triad Support algorithm, which allow for further reduction in the number of pairwise comparisons. Next, in Sect. 4, we demonstrate how these methods perform in different scenarios, and in Sect. 5, we summarize our work and suggest directions for future research.

2 Preliminaries

In this section, we briefly outline methods used in this work, namely Characteristic Objects METhod and Triad Support algorithm which extends it.

2.1 Characteristic Objects METhod

The Characteristic Objects METhod stands out among other MCDA techniques due to its full resistance to the ranking reversal issue [8] and ability to model complex decision-maker's preferences. This approach allows for the creation of an full-domain decision model that consistently delivers accurate and robust results, with no regard to changes of the alternatives' set [5]. The complete algorithm of the COMET method is described in previous works, and a ready-to-use implementation is available in the `pymcdm` library [4].

2.2 Triad Support Algorithm

The Triad Support algorithm is a recently proposed method to reduce the complexity of applying the COMET technique. This algorithm decreases the number of necessary pairwise comparisons and relies on the assumption that when expert assessments are consistent and accurate, the evaluation of Characteristic Objects follows a transitive relation. Using this property, the algorithm avoids unnecessary comparisons and simplifies the identification of the COMET model [9].

3 Proposed Approach

In this section, we explain two interconnected propositions which extend the Triad Support algorithm in order to further reduce required number of the pairwise comparisons.

3.1 Gray Code in COMET

First, we propose changing the order of comparisons of the Characteristic Objects in the Triad Support algorithm. For t COs, the first $t-1$ comparisons are made between neighboring COs in the ordered set. Ordering them in such a way that each pair of consecutive COs differs in only one criterion simplifies these initial comparisons for the expert.

For example, suppose that we have four COs in a two-criteria problem: $[A = [0,0], B = [0,1], C = [1,0], D = [1,1]]$. In this original order, we may be required to compare B and C, which differ on both criteria. However, if we reorder the COs as $[A, B, D, C]$, each pair of consecutive COs differs in only one criterion. This ordering is beneficial, as the initial comparisons become easier for the expert, and the Triad Support algorithm is more likely to find consistent triads with this arrangement of COs.

3.2 Types Definitions in COMET

The next advancement can be applied efficiently only when the characteristic objects are ordered in a Gray Code manner. If, prior to the comparisons, the expert or decision-maker defines the types of criteria or orders the characteristic values for each criterion from best to worst, some comparisons can be made automatically based on this information.

Following the example from the previous section, a comparison between A and B can be made automatically based on whether the second criterion is a profit and should be maximized (in which case $B \succ A$) or a cost and should be minimized (in which case $A \succ B$). Therefore, Gray Code ordering is beneficial, as it facilitates more automatic comparisons. In this example, it is possible to compare not only neighboring COs in the set, but also other objects at later stages of the comparison, such as the pair (A, C).

4 Experiments and Results

In both experiments, to evaluate single characteristic object $CO = \{co_i, i \in 1, 2, \ldots, N\}$, we use a linear combination of functions (1) instead of a human expert to provide simulation results. Depending on the function f_i and weight w_i, criterion C_i will be classified as either a profit ($f_i(co_i) = co_i$) or a cost ($f_i(co_i) = -co_i + 1$) with varying importance w_i. In the experiments, all values fall within the interval $[0, 1]$.

$$f_{exp}(CO) = \sum_{i=1}^{n} w_i f_i(co_i) \tag{1}$$

This allows for the simulation of different scenarios, assessing the number of pairwise comparisons required to identify the model under various preferences.

In the first experiment, we evaluate the number of pairwise comparisons required by a simulated expert to identify the COMET model, depending on the number of criteria n. We assume equal weights in the expert function (1) and three characteristic values $[0.0, 0.5, 1.0]$ for each criterion. Rounded up half of the criteria are treated as profit, and the other are treated as cost.

The results of the experiments are presented in Fig. 1. The number of comparisons in the original COMET method is calculated as $\frac{t(t-1)}{2}$, where $t = 3^n$ is the number of characteristic objects. This value grows rapidly, so it is not presented in the figure. For $n = 5$, the original COMET method requires 29,403 comparisons; for $n = 6$, approximately 265,000 comparisons are needed; and for $n = 8$, the number of comparisons exceeds 21 million. In the $n = 8$ case, we observe that adding the Triad Support algorithm significantly reduces the number of comparisons to 20,758 (0.09% of the original value). The addition of Gray Code ordering further reduces the required number of comparisons to 13,104, and defining the criteria types lowers this to 6,544, demonstrating a 99.97% reduction in queries to the expert.

In contrast, for $n = 2$, identifying the model normally requires 36 pairwise comparisons. However, the addition of the Triad Support algorithm reduces this number to 22. Changing the order of COs to Gray Code reduces the comparisons to 12, and defining the criteria types reduces the number of comparisons to 6, which is 11% of the original value.

In the second experiment, we focused on a smaller number of criteria and different weighting scenarios. To address various scenarios, we decided to identify COMET models with different improvements for all weak orderings for a given number of criteria n and for 10^{n-1} random weight vectors. This ensures a more comprehensive exploration of the scenarios and is adapted to the possible weight combinations depending on number of criteria, allowing for the derivation of generalized insights.

The results of the experiments are presented in Fig. 2. Labels on the X-axis indicate the comparison-reducing techniques used and should be interpreted cumulatively: "+Triads" refers to the original COMET with the Triad Support

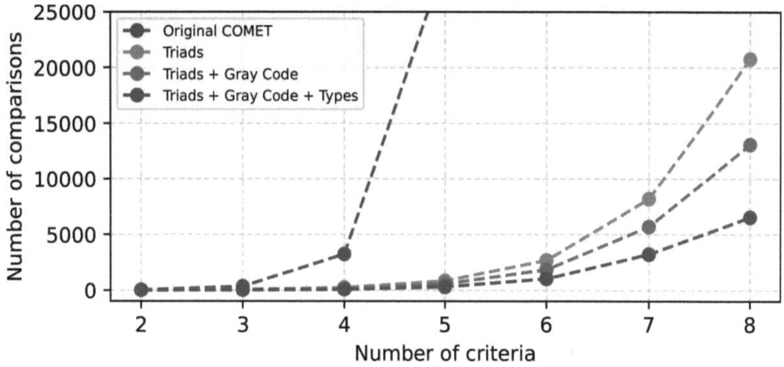

Fig. 1. Dependence of the number of comparisons to identify COMET model depending on number of criteria.

algorithm, "+GC" refers to the original COMET with the Triad Support algorithm and Gray Code ordering, and so on. Dots in the figure represent the median number of comparisons, while the error bars represent the minimum and maximum number of comparisons achieved during the experiment.

It can be seen that for every number of criteria in the experiment, there is a noticeable improvement in the number of comparisons with all the proposed advancements, compared to the original method or the Triad Support algorithm alone. In the case of two criteria, the proposed approach required only 4 or 6 pairwise comparisons, which is 11%–16% of the original value. As the number of criteria increases, the number of comparisons drops significantly: for 3 criteria, the median number of comparisons is 48 (13%); for $n = 4$, it is 276 (8.5%); and for $n = 5$, it is 1,254 (4.3%). The same trend was observed in the first experiment: while the number of comparisons for the original COMET grows rapidly, the addition of the Triad Support algorithm, Gray Code ordering, and criteria type definitions allows for a significant reduction in the number of comparisons, making the method more practical to apply.

However, as shown in Fig. 2, there are some deviations from the median number of comparisons in all cases. This largely depends on the complexity of the expert's preferences. If there are many consistent triads, the number of comparisons will decrease. However, if there is a large number of unknown triads, the number of comparisons will increase. For example, while for $n = 3$ the median number of comparisons with all improvements is 48, it can vary from 20 to 60 depending on the expert's answers. For $n = 4$ with all improvements, the range is from 72 to 358, with a median of 276. For $n = 5$, the number of comparisons can be as low as 242, but in the worst-case scenario, it may require up to 1,666 comparisons.

As shown in the experiments, the level of reduction increases with the number of pairwise comparisons required to identify the model. For example, for $n = 8$, only 0.03% of the original COMET comparisons are needed, although the absolute number of comparisons is still quite large. As the experiments sug-

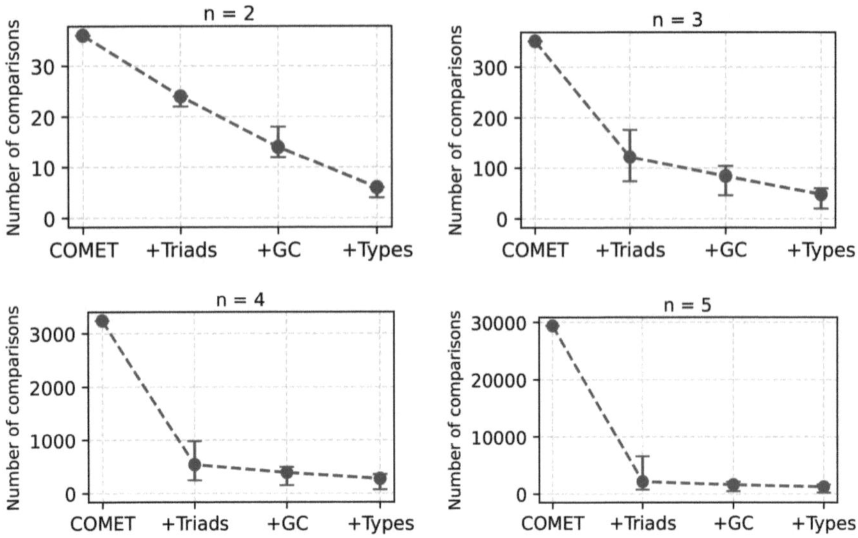

Fig. 2. Results of the second experiment for different n and comparison reducing technic combinations. Dashed line indicates the piecewise declining trend.

gest, the proposed approach is best suited for cases with 2–4 criteria, where the total number of comparisons does not exceed a few hundred.

5 Conclusion

In this paper, we proposed two advancements to the COMET and Triad Support algorithms that enable easier and faster identification of the COMET model. While reordering the Characteristic Objects reduces the number of comparisons when combined with the Triad Support method, further reduction can be achieved by defining criteria types in advance by the expert. Assuming that all comparisons made by the expert are consistent and follow a transitive relation, the results will be the same as those obtained using the original COMET algorithm. This approach allows for faster model identification and, consequently, makes the COMET method more applicable to large decision problems. Additionally, the proposed approach can be combined with the structural COMET method to identify larger models more quickly and efficiently. However, the same limitation applies to the proposed approach as to the Triad Support algorithm: the expert must provide consistent answers in order to achieve the same model as without the proposed extensions.

In future work, we aim to explore the application of this approach in practical decision problems, assessing the quality of the results and the identified models, and comparing them to the original COMET.

References

1. Campolina, A.G., Estevez-Diz, M.D.P., Abe, J.M., de Soárez, P.C.: Multiple criteria decision analysis (MCDA) for evaluating cancer treatments in hospital-based health technology assessment: the paraconsistent value framework. PLoS ONE **17**(5), e0268584 (2022)
2. Dias, L.C., Caldeira, C., Sala, S.: Multiple criteria decision analysis to support the design of safe and sustainable chemicals and materials. Sci. Total Environ. **916**, 169599 (2024)
3. Habeeb, R., Hussain, I., Al-Ansari, N., Sammen, S.S.: A proposed comparative algorithm for regional crop yield assessment: an application of characteristic objects method. Math. Probl. Eng. **2022**(1), 8224953 (2022). https://doi.org/10.1155/2022/8224953
4. Kizielewicz, B., Shekhovtsov, A., Sałabun, W.: pymcdm–the universal library for solving multi-criteria decision-making problems. SoftwareX **22**, 101368 (2023)
5. Kizielewicz, B., Wątróbski, J., Sałabun, W.: Multi-criteria decision support system for the evaluation of UAV intelligent agricultural sensors. Artif. Intell. Rev. **58**(7), 194 (2025)
6. Kizielewicz, B., Więckowski, J., Franczyk, B., Wątróbski, J., Sałabun, W.: Comparative analysis of re-identification methods of multi-criteria decision analysis models. IEEE Access (2024)
7. Psomas, A., Vryzidis, I., Spyridakos, A., Mimikou, M.: MCDA approach for agricultural water management in the context of water-energy-land-food nexus. Oper. Res. Int. J. **21**, 689–723 (2021)
8. Sałabun, W., Piegat, A., Wątróbski, J., Karczmarczyk, A., Jankowski, J.: The COMET method: the first MCDA method completely resistant to rank reversal paradox. Eur. Working Group Ser. **3** (2019)
9. Shekhovtsov, A., Sałabun, W.: The new algorithm for effective reducing the number of pairwise comparisons in the decision support methods. In: Hernes, M., Wątróbski, J. (eds.) ECAI 2023. LNNS, vol. 1079, pp. 243–254. Springer, Cham (2024). https://doi.org/10.1007/978-3-031-66761-9_20
10. Wulf, C., Mesa Estrada, L.S., Haase, M., Tippe, M., Wigger, H., Brand-Daniels, U.: MCDA for sustainability assessment of energy technologies and systems-identifying challenges and opportunities within the Helmholtz Association. Energy Sustain. Soc. (2025)
11. Yulistia, E., Samsuryadi, A.: New AHP improvement using COMET method characteristic to eliminate rank reversal phenomenon. Int. J. Adv. Comput. Sci. Appl. **14**(11) (2023). https://doi.org/10.14569/IJACSA.2023.0141190

Empirical Validation of Simplified Pairwise Comparison in Decision-Making Domain

Anna Shkurina(✉)

University of Szczecin, Szczecin, Poland
anna.shkurina@phd.usz.edu.pl

Abstract. Pairwise comparison is a common technique in multi-criteria decision analysis (MCDA), with simplified methods like the RANking COMparison (RANCOM) aiming to reduce cognitive load by using only directional or equality judgments. This study tested the perceptual validity of such comparisons through an experiment with 81 participants evaluating which of two grids had more black squares, or if they were equal, at varying difficulty levels. Accuracy exceeded 97% for differences above 10% but dropped to 54.7% for differences below 5%. A logistic psychometric model confirmed this pattern and allowed estimation of Just Noticeable Differences (JND) at various confidence levels. The findings support simplified pairwise methods under moderate and large differences, while highlighting reduced reliability in nearly equal cases.

Keywords: Pairwise comparision · validation · RANCOM · MCDA

1 Introduction

Decision making remains a central challenge in both scientific research and practical applications, particularly in areas such as management, engineering, and others. Multi-Criteria Decision Analysis (MCDA) methods provide structured frameworks to tackle such problems, allowing systematic evaluation of alternatives based on multiple, often conflicting criteria [6]. Among these methods, pairwise comparison stands out as a widely used and intuitive mechanism. It allows decision-makers to express preferences between two elements at a time, facilitating the gradual construction of consistent judgment structures [2].

Prominent examples of MCDA methods that rely on pairwise comparisons include the Analytic Hierarchy Process (AHP) and the recently introduced RANking COMparison (RANCOM) method [5]. In AHP, comparisons are made using the Saaty scale, which requires decision makers to evaluate the strength of preference between multiple levels of intensity. Although this scale offers a rich expression of preferences, it also imposes a considerable cognitive burden. Distinguishing between subtle differences on a 9-point scale can be difficult, especially when the elements being compared are similar, potentially leading to inconsistencies and reduced reliability of the judgments [1].

Published by Springer Nature Switzerland AG 2026
M. Pondel et al. (Eds.): AI4KAM 2025, IFIP AICT 776, pp. 400–406, 2026.
https://doi.org/10.1007/978-3-032-18920-2_38

In contrast, the RANCOM method significantly simplifies the comparison process. Instead of assessing the magnitude of preference, the decision maker is only required to indicate the direction of preference, i.e., which alternative is better, or declare a tie. This binary (or ternary) comparison model aims to reduce the cognitive load and increase the robustness of the elicited preferences. The underlying assumption of RANCOM is that such simplified comparisons are not only easier but also more accurate, making the method accessible and effective [7].

However, this assumption has not been thoroughly validated. It remains an open question whether individuals can consistently make accurate pairwise comparisons, even when the task is simplified. Human judgment is influenced by numerous factors, including perceptual limits, fatigue, cognitive biases, and the context in which comparisons are made [4]. Therefore, the central motivation of this study is to empirically investigate the validity of simplified pairwise comparison models, such as that employed in RANCOM.

To this end, we designed an experiment to isolate and test the perceptual component of pairwise comparison: the ability to detect and assess differences in a single well-defined attribute. The participants were presented with 20×20 grids (each containing 400 cells), varying only in the number of black squares. In each trial, two grids were displayed side by side, and the participant's task was to indicate which one contained more black squares or whether the two appeared approximately equal. This setup mirrors the decision context of RANCOM by eliminating domain knowledge and focusing purely on perceptual discrimination.

The differences between the grids were systematically varied in four predefined levels, ranging from very small (up to 5%) to large (more than 20%). Each of the 81 participants completed 40 comparisons. The resulting data set enables a detailed analysis of how perceptual accuracy depends on the magnitude of the difference, thus shedding light on the limits and potential of simplified decision-making approaches based on pairwise judgments.

The rest of the paper is as follows. Section 2 describes the experimental design, including the structure of the task, the controlled variation in difficulty levels, and the data collection procedure. Section 3 presents the empirical findings, focusing on the precision of the participants in different comparison categories. In Sect. 4, we interpret the results in the context of simplified pairwise comparison methods, such as RANCOM, and discuss cognitive factors that may influence judgment quality. Finally, Sect. 5 summarizes the main contributions of the study and outlines potential directions for future research.

2 Methodology

To empirically assess the reliability of simplified pairwise comparisons, we designed a visual experiment centered on perceptual discrimination under controlled conditions. The experimental task simulates a decision-making scenario in which participants compare two alternatives, represented as grids populated with black squares, and select the one that contains more black elements or indicates that the two appear equal. The procedure is presented as the Algorithm 1.

Algorithm 1. Black Square Comparison Experiment

```
Parameters: GRID_SIZE ← 20, NUM_TRIALS ← 40, SQUARE_SIZE ← 15
Define: DIFF_CATEGORIES ← {under_5, 5_10, 10_20, above_20}
category_sequence ← evenly distributed and shuffled list of categories
results ← empty list
current_trial ← 0
while current_trial < NUM_TRIALS do
    current_category ← category_sequence[current_trial]
    left_count, right_count ← GenerateCounts(current_category)
    DrawGrid(canvas_left, left_count)
    DrawGrid(canvas_right, right_count)
    response ← user input (left, right, equal)
    Append(results, trial data with response and counts)
    current_trial ← current_trial + 1
end while
SaveToCSV(results)
```

The core concept of this experiment is to model a simplified pairwise comparison task in a neutral domain-independent setting. Each grid is a matrix 20×20 (400 cells in total) with a variable number of randomly distributed black squares. This format ensures perceptual engagement while remaining sufficiently abstract to isolate visual judgment processes from contextual or semantic influence (Fig. 1).

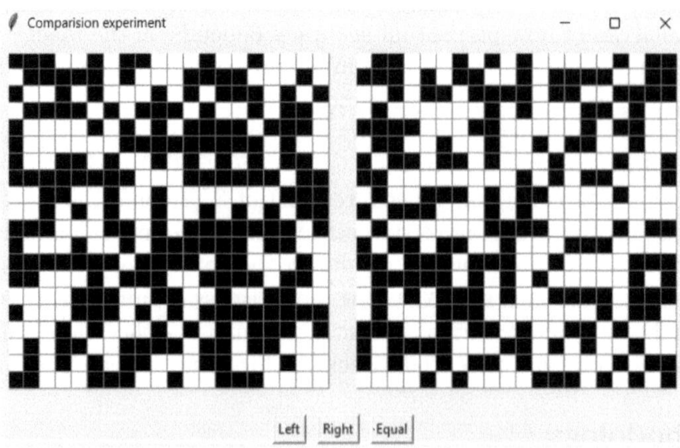

Fig. 1. Sample trial showing a screenshot from the program with one query to the user. The left grid contains 215 black squares and the right grid contains 183.

The key independent variable is the difference in the number of black squares between the two grids, classified into four levels: `under 5%`, `5-10%`, `10-20%`, and `above 20%`. These intervals were selected to capture a spectrum of perceptual difficulty, allowing analysis of how accuracy varies with stimulus disparity and allowing estimation of thresholds such as the Just Noticeable Difference (JND).

Each participant performs 40 tests, with categories evenly represented and randomly ordered to minimize bias and avoid order effects. During each trial, participants choose one of three response options: `left`, `right`, or `equal`, mirroring the decision structure of methods like RANCOM. This design allows for the investigation of both directional judgments and the strategic use of neutrality when perceptual differences are ambiguous.

The data collected provide a basis for modeling psychometric response functions and assessing perceptual accuracy across difference categories. By abstracting the comparison process from real-world semantics, the experiment yields empirical insight into the cognitive plausibility of simplified pairwise methods and the robustness of human judgment when faced with reduced comparison complexity.

3 Results

The aggregate results of all participants indicate a high general level of precision in simplified pairwise comparisons, with a mean success rate of 85.06%. The participants declared a draw in 10.46% of all trials. Performance varied systematically with the magnitude of the difference between the compared grids. When the difference exceeded 10%, the precision was very high: 97. 9% for the range of 10 to 20% and 99. 0% for differences greater than 20%. In the category 5–10%, the precision remained high at 90.1%, although slightly lower, suggesting a transition zone where comparisons become more challenging.

The most substantial drop occurred in the lowest range of differences (under 5%), where accuracy declined to 54.7%, only slightly above the chance level. This category also saw the highest number of draw declarations, accounting for 260 out of 810 responses. Errors in this range were more frequent and less balanced between incorrect "left" and "right" selections. In contrast, for larger differences, incorrect responses were rarer and more evenly distributed. These patterns confirm that, while simplified pairwise comparisons are robust and reliable when perceptual differences are pronounced, their accuracy diminishes substantially near the threshold of discriminability.

The overall psychometric trend shows a steep and monotonic increase in precision with increasing stimulus disparity. This confirms that participants were highly sensitive to visual asymmetry when differences were moderate or large, but struggled to make consistent judgments in near-equal cases, even under controlled conditions.

4 Disscussion

To further investigate the structure of participant responses, we analyzed the confusion matrix summarizing how declared responses matched the ground truth. Table 1 presents the frequency of each response category relative to the actual correct answer.

Table 1. Confusion matrix: correct answers vs. user responses

Correct Answer	Left	Right	Equal
Left	1415	61	125
Right	69	1331	192
Equal	15	10	22

The matrix shows a strong overall alignment between the correct and declared responses, with the highest frequencies concentrated along the diagonal. This confirms that the participants accurately identified the side with more black squares when the difference was sufficiently large. However, a notable number of off-diagonal entries, particularly the 125 cases where a draw was declared instead of selecting the correct side, suggest that participants were often uncertain or overly conservative in borderline conditions.

Interestingly, trials in which the correct answer was `equal` were both infrequent and frequently misclassified, indicating that perceptual equality is difficult to judge or rarely occurs in practice. These findings reinforce the need to account for uncertainty and bias in simplified pairwise comparison methods.

To more precisely quantify perceptual sensitivity, we fitted a logistic psychometric function to the observed accuracy as a function of stimulus difference, defined as the signed percentage difference in the number of black squares between the left and right grids. The model takes the form $P(\text{correct}) = \frac{1}{1+e^{-(x-\alpha)/\beta}}$, where x is the difference in stimulus, α is the point of subjective equality (50% accuracy), and β determines the slope of the curve [8]. Based on this model, we estimate the (JND) for a range of target accuracy levels using the inverse of the logistic function [3]:

$$\text{JND}_p = \alpha + \beta \cdot \ln\left(\frac{p}{1-p}\right),$$

where p denotes the desired probability of a correct response.

To explore how JND varies with the selected performance threshold, we calculated values for accuracy levels ranging from 0.50 to 0.95 in increments of 0.05. As shown in Fig. 2, the estimated JND increased monotonically with the accuracy threshold. For example, $\text{JND}_{0.75}$ was estimated at 4.78%, indicating that a 4.78% difference between the grids was required to achieve 75% precision. In contrast, $\text{JND}_{0.90}$ was estimated at 7. 53%, demonstrating that a nearly a 8% difference was needed for participants to respond correctly 90% of the time. These results emphasize that, while small differences may be sufficient for moderate confidence, larger disparities are necessary to achieve high decision certainty. The findings highlight that the JND is not fixed but depends on the chosen criterion, underscoring the importance of explicitly defining performance thresholds in perceptual judgment tasks.

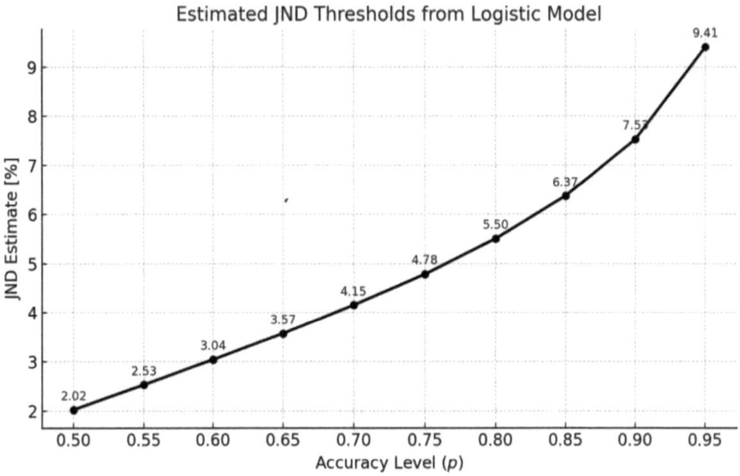

Fig. 2. Estimated JND thresholds at varying levels of response accuracy. The curve is based on the inverse of the fitted logistic psychometric function.

5 Conclusion

The results of this study highlight important considerations for the design and interpretation of simplified pairwise comparison methods. In particular, it should be emphasized that the concept of a "draw" or equality judgment may not correspond to a fixed threshold, but rather varies depending on both the perceptual capabilities of the individual and the level of decision accuracy assumed in a given context.

The psychometric modeling conducted in this experiment illustrates this point clearly: in our specific case, a difference of approximately 4.8% was required to reach 75% accuracy, while nearly 7.5% was needed to achieve 90% certainty. These values are not generalizable across all settings, but they demonstrate that the region in which a participant tends to declare equality is not static, it shifts based on how much precision is expected. This observation reinforces the idea that the treatment of perceptual equality is a non-trivial and context-sensitive element in decision-making frameworks.

Therefore, in applications such as MCDA that utilize ternary pairwise judgments, it is essential to account for this perceptual variability. Systems should avoid hard-coded equality thresholds and instead consider incorporating context aware or user-adaptive mechanisms to model uncertainty. The interpretation of "equal" should not be treated as absolute, but as dependent on both the structure of the stimulus and the confidence level that the system or user is designed to tolerate.

References

1. Franek, J., Kresta, A.: Judgment scales and consistency measure in AHP. Procedia Econ. Financ. **12**, 164–173 (2014). https://doi.org/10.1016/S2212-5671(14)00332-3
2. Kou, G., Ergu, D., Chen, Y., Lin, C.: Pairwise comparison matrix in multiple criteria decision making. Technol. Econ. Dev. Econ. **22**(5), 738–765 (2016). https://doi.org/10.3846/20294913.2016.1210694
3. Luce, R.D., Edwards, W.: The derivation of subjective scales from just noticeable differences. Psychol. Rev. **65**(4), 222–237 (1958). https://doi.org/10.1037/h0039821
4. Pronin, E.: Perception and misperception of bias in human judgment. Trends Cogn. Sci. **11**(1), 37–43 (2007). https://doi.org/10.1016/j.tics.2006.11.001
5. Więckowski, J., Kizielewicz, B., Shekhovtsov, A., Sałabun, W.: RANCOM: a novel approach to identifying criteria relevance based on inaccuracy expert judgments. Eng. Appl. Artif. Intell. **122**, 106114 (2023). https://doi.org/10.1016/j.engappai.2023.106114
6. Więckowski, J., et al.: Recent advances in multi-criteria decision analysis: A comprehensive review of applications and trends. Int. J. Knowl.-Based Intell. Eng. Syst. **27**(4), 367–393 (2023). https://doi.org/10.3233/KES-230487
7. Więckowski, J., Wątróbski, J., Shkurina, A., Sałabun, W.: Adaptive multi-criteria decision making for electric vehicles: a hybrid approach based on RANCOM and ESP-SPOTIS. Artif. Intell. Rev. **57**(10), 270 (2024). https://doi.org/10.1007/s10462-024-10901-4
8. Wu, J., Shi, G., Lin, W.: Survey of visual just noticeable difference estimation. Front. Comp. Sci. **13**(1), 4–15 (2019). https://doi.org/10.1007/s11704-016-6213-z

Sensitivity to the Damping Factor in Local Markov Consensus Methods

Joanna Kołodziejczyk[1,2(✉)] and Andrii Shekhovtsov[1,2]

[1] National Institute of Telecommunications, ul. Szachowa 1, 04-894 Warsaw, Poland
{j.kolodziejczyk,a.shekhovtsov}@il-pib.pl

[2] Research Team on Intelligent Decision Support Systems, Department of Artificial Intelligence and Applied Mathematics, Faculty of Computer Science and Information Technology, West Pomeranian University of Technology in Szczecin, Szczecin, Poland

Abstract. This paper examines the sensitivity of the Local Markovian Consensus method with weak ordinal dominance (LMC-WOD) to changes in the damping factor d. The factor controls the balance between local preferences dominance and global uniform influence. Experiments are conducted on synthetic datasets generated via the Mallows model, covering various levels of ranking noise, dimensionality, and data volume. Aggregation outcomes are evaluated using Kendall tau distance, pairwise agreement, and the WS coefficient, each capturing distinct aspects of ranking consistency. The results show that LMC-WOD is robust to changes in d across all settings, consistently producing high-quality aggregated rankings in non-conflicting environments. Moreover, the method maintains stable consensus rankings even in the presence of preference conflicts or a large number of alternatives.

Keywords: Ranking aggregation · Markov chain methods · Sensitivity

1 Introduction

Parameter-dependent methods require careful configuration, as the output often depends on internal variables. This issue is addressed through parameter tuning, which identifies an optimal set of values for a specific task. Alternatively, it may be approached through sensitivity analysis, which examines how variations in parameter values influence the method's output [1,3,4].

In ranking aggregation methods based on the Markov model, the positional information of alternatives across the entire set of rankings is represented as probability. The first such approach, which is Markov Chain (MC) [2], models each alternative as a state. A ranking consists of a set of states and a transition probability matrix. The main idea was to obtain a stationary distribution to define the aggregated consensus.

The other Markov-based ranking aggregation applied to web search engines is PageRank [9,12], which creates the final aggregated ranking based on the recursive references between nodes that are ranked based on the number of links.

Published by Springer Nature Switzerland AG 2026
M. Pondel et al. (Eds.): AI4KAM 2025, IFIP AICT 776, pp. 407–417, 2026.
https://doi.org/10.1007/978-3-032-18920-2_39

This method applied an inner parameter, the damping factor d, which controls the influence of output links versus random surfer uniform choice.

The Local Markovian Consensus (LMC) [7] method and its extension, LMC-WOD [6], also utilize the Markov model to search for the aggregated consensus. Both LMC and LMC-WOD utilize the damping factor d to regulate the extent to which the aggregation favors subjective/local preference domination over global/uniform alternative influence. However, the impact of d on the stability and accuracy of LMC-WOD has not yet been systematically evaluated.

This paper fills this gap through a sensitivity analysis of LMC-WOD over varying d values. Simulated empirical experiments are conducted over a predefined range of damping values. The robustness of aggregated rankings across different noise levels, ranking lengths, and sample sizes was assessed using synthetic datasets generated via the Mallows model. The resulting aggregated rankings are evaluated using three metrics: Kendall tau distance, pairwise agreement, and WS Coefficient.

The main results provide insights into how the LMC-WOD method responds to preference inconsistency, the ranking size for varying d factor, and offer practical rules for the initial setup.

2 Preliminaries

The Local Markovian Consensus (LMC) method [7] aggregates a set of individual rankings into a consensus ranking using Markovian transitions derived from local pairwise preferences. It was inspired by Markov Chain-based ranking (MC) [2] and the PageRank algorithm [9], but differs in how it constructs the transition matrix, focusing on direct dominance relations rather than global preference/link structures.

The LMC-WOD extension [6] generalizes the LMC method to handle both strict and weak ordinal rankings, allowing alternatives to share the same position (ties), which makes the model applicable to a broader range of real-world rank aggregating datasets. The method assumes complete input rankings. To satisfy the ergodic assumption of the transition matrix, the method employs an additional uniform matrix to model the global alternative uniformity. The damping factor d balances the dominance of the alternatives with global uniformity, representing a non-preference scenarios.

To provide a foundation for the sensitivity analysis presented later in this work, the following section introduces the theoretical background of the LMC-WOD framework. It formalizes the aggregation task and defines the probabilistic transition graph over the ranking space, outlining the mathematical constructs that underpin the method's iterative consensus formation process.

Let $A = \{a_1, a_2, \dots, a_m\}$ be the set of alternatives. A set of n complete rankings is represented by a matrix $\mathbf{R} \in \mathbb{N}^{n \times m}$, where each row $\mathbf{r}_k$ assigns a rank $r_k(a_i) \in \{1, \dots, z\}$ to each alternative a_i:

$$\mathbf{R} = \begin{array}{c} \\ \mathbf{r_1} \\ \mathbf{r_2} \\ \dots \\ \mathbf{r_n} \end{array} \!\! \begin{array}{c} \begin{array}{cccc} a_1 & a_2 & \dots & a_m \end{array} \\ \begin{bmatrix} r_1(a_1) & r_1(a_2) & \dots & r_1(a_m) \\ r_2(a_1) & r_2(a_2) & \dots & r_2(a_m) \\ \dots & \dots & \dots & \dots \\ r_n(a_1) & r_n(a_2) & \dots & r_n(a_m) \end{bmatrix} \end{array} \quad (1)$$

The LMC-WOD goal is to produce a consensus ranking vector $\mathbf{r}*$ which representd the order of m alternatives:

$$\mathbf{r}^* = [r(a_1), r(a_2), ..., r(a_m)]; \quad where \quad r(\cdot) \in \{1, \dots, z\} \wedge a_i \in A.$$

Each $r(a_i)$ is an integer indicating the rank of alternative a_i, where a rank of 1 denotes the highest preference, and $z = \max\{\mathbf{r}\}$ is the lowest (worst) rank. The length of the vector is $|\mathbf{r}^*| = |A| = m$.

It is assumed that each ranking $r(\cdot)$ expresses ordinal preferences over the alternatives. If $r(a_i) < r(a_j)$, then the alternative a_i is strictly preferred to a_j or equivalently, strictly ordinally dominates a_j. If $r(a_i) \leq r(a_j)$, then a_i is at least as preferred as a_j or weakly ordinally dominates a_j.

The LMC-WOD method constructs a weighted directed graph over the set of alternatives $A = \{a_1, a_2, \dots, a_m\}$, where each node corresponds to an alternative and each directed edge encodes the frequency of direct (local) alternative connection or dominance. The graph is used to define a Markov chain whose transition matrix captures local consensus information aggregated from the input rankings.

Let $\mathbf{R} \in \mathbb{N}^{n \times m}$ be the input ranking matrix (Eq. 1). For each ordered pair of alternatives (a_i, a_j), we define a dominance frequency $g(a_i, a_j)$, representing the frequency with which a_i is one rank more preferred as a_j, which means (a_i directly dominates a_j) across the set of rankings:

$$g(a_i, a_j) = \begin{cases} \sum_{k=1}^{n} \mathbf{1} \text{ if } r_k(a_i) - r_k(a_j) = 1 \\ \sum_{k=1}^{n} \mathbf{1} \text{ if } r_k(a_i) = z \wedge i = j \\ 0 \quad \text{otherwise} \end{cases}, \quad (2)$$

where z is the worst rank (in strict ordinal rankings $z = m$). The first case counts direct sequential dominance; the second ensures that the lowest-ranked items maintain self-loops for stochastic completeness.

This results in a dominance matrix $\mathbf{G} \in \mathbb{N}^{m \times m}$, with each entry $g_{ij} = g(a_i, a_j)$. A stochastic transition matrix $\mathbf{P} \in [0,1]^{m \times m}$ is developed by normalizing the rows of $\mathbf{G}$ and using a damping factor $d \in (0,1)$ to model non-dominant influence:

$$\mathbf{P} = d \cdot \mathbf{G}_{\text{norm}}^{\top} + (1 - d) \cdot \frac{1}{m}\mathbf{1}, \quad (3)$$

where:

- $\mathbf{G}_{\text{norm}}$ is the row-normalized version of $\mathbf{G}$, i.e., each row i is divided by $\sum_{j=1}^{m} g_{ij}$,
- $\frac{1}{m}\mathbf{1}$ is a matrix of ones representing a uniform connection to any alternative.

- The transpose $\mathbf{G}_{\mathrm{norm}}^{\top}$ ensures compatibility with left eigenvector interpretation.

The transition matrix $\mathbf{P}$ defines a discrete-time Markov chain which stationary distribution $\boldsymbol{\pi}$ which stationary distribution $\boldsymbol{\pi}$, satisfying $\boldsymbol{\pi}\mathbf{P} = \boldsymbol{\pi}$ and $\sum_{i=1}^{m} \pi_i = 1$, encodes the aggregated consensus. Alternatives are ranked by sorting $\boldsymbol{\pi}$ in descending order producing the final consensus ranking $\mathbf{r}^*$. The damping factor d regulates the influence between local dominance represented by $\mathbf{G}_{\mathrm{norm}}$ and non-dominant scenarios represented by $(\frac{1}{m} \cdot \mathbf{1})$. The latter ensures ergodicity and stability.

3 Experiments on Sensitivity to Damping Factor

The damping factor d in the LMC and LMC-WOD methods plays a central role in regulating the balance between local dominance in preference propagation and uniform preference equality. Specifically, it determines the degree to which the final ranking reflects direct pairwise preferences versus 'non-preference' scenarios. The sensitivity analysis examines the influence of local preference changes relative to the uniform distribution.

In LMC-WOD the parameter $d \in (0, 1)$ plays a role analogous to the damping factor in the PageRank algorithm: the larger the value of d, the more the aggregated ranking relies on transitions derived from local dominance; the smaller the value of d, the greater the contribution of the uniform preferences.

The following experiments investigate the sensitivity of LMC-WOD outcomes to different values of d, particularly in contexts with varying levels of ranking heterogeneity.

3.1 Similarity Measures

To assess the consistency between individual rankings and an aggregated consensus ranking, three complementary evaluation measures are used. Each captures a different perspective on similarity and disagreement in ordinal ranking data:

1. Kendall Tau Distance (Kτ):
   This measure quantifies the number of pairwise disagreements between two rankings. It is symmetric, widely used, and provides a general indicator of deviation from the reference ranking (in this case, the aggregated ranking $\mathbf{r}^*$). Due to its theoretical robustness Kendall tau serves as a standard baseline in rank aggregation studies [10].
2. Pairwise Agreement Coefficient (PA): This measure calculates the proportion of item pairs that are ordered identically in both rankings. It complements Kendall tau by measuring agreement instead of disagreement, offering a more intuitive scale (with values closer to 1 indicating higher consistency), making it especially useful for visualizing the internal coherence of rankings [5].

3. WS Coefficient (WS): This metric is related to the difference between two rankings on particular positions, which gives greater importance to correct ordering among top-ranked items, which are often more critical in practical applications (e.g., recommendation, decision support). WS is beneficial when the relative importance of higher positions outweighs that of lower-ranked alternatives, reflecting position-sensitive relevance in the evaluation [11,13].

Together, these measures offer a comprehensive evaluation framework. Kendall Tau (Kτ) captures overall structural deviations, pairwise agreement (PA) provides an intuitive measure of consistency, and WS Coefficient highlights agreement where it matters most.

3.2 Experimental Setup

The experiments were conducted on synthetically generated datasets, produced within a controlled environment using the Mallows model as introduced in [8]. The Mallows model is parametric framework for ranking generation, in which permutations are distributed around a central (reference) ranking according to a dispersion parameter $\phi \geq 0$. A smaller value of ϕ indicates stronger agreement with the central ranking, while $\phi = 0$ corresponds to identical rankings across all samples. This model is commonly employed in preference elicitation tasks, as it reflects a plausible process in which individual rankings are centered around a consensus or "ground truth" order. To simulate the dispersion behavior of the Mallows model for practical experimentation, we adopt the Gumbel noise approximation.

Interpretation of dispersion parameter ϕ in Mallow model:

- $\phi = 0$: All sampled rankings are identical to the central ranking (maximum consistency).
- $\phi \to \infty$: Rankings are uniformly random (maximum inconsistency).
- $\phi \in (0, 1)$: Rankings are close to the central permutation, with low variability.
- $\phi > 1$: Rankings exhibit higher dispersion and entropy.

The experiments were designed to assess the sensitivity of aggregated rankings consistency to variations in problem size, data volume, and ranking noise.

Three levels of dimensionality were considered by varying the number of alternatives: 5, 10, and 50. These correspond to typical small, medium, and large-scale ranking tasks. The case with five alternatives simulates constrained environments such as expert panels or focused decision-making contexts. The medium-scale case with ten alternatives reflects the typical size of user preference datasets or comparative studies in behavioral sciences. Finally, the inclusion of fifty alternatives captures conditions representative of large-scale, real-world applications such as product recommendation systems, information retrieval tasks, or multi-agent consensus formation.

Additionally, the number of individual rankings was set to 10, 100, and 500, reflecting scenarios with low, moderate, and high data availability. The setting with ten rankings reflects limited-data contexts, often encountered in domains with costly or expert-generated inputs. One hundred rankings represent a realistic volume for mid-sized survey-based studies, while five hundred rankings simulate high-volume crowd sourced or system-logged preference data.

To control the degree of consistency among rankings, the Mallows model was used with three dispersion parameters: $\phi = 0.1$, $\phi = 1$, $phi = 3$. These values represent high, moderate, and low levels of agreement with the central ranking, respectively.

This factorial design enables systematic evaluation of how the number of alternatives, the volume of ranking data, and the degree of consensus affect the performance of ranking-based methods.

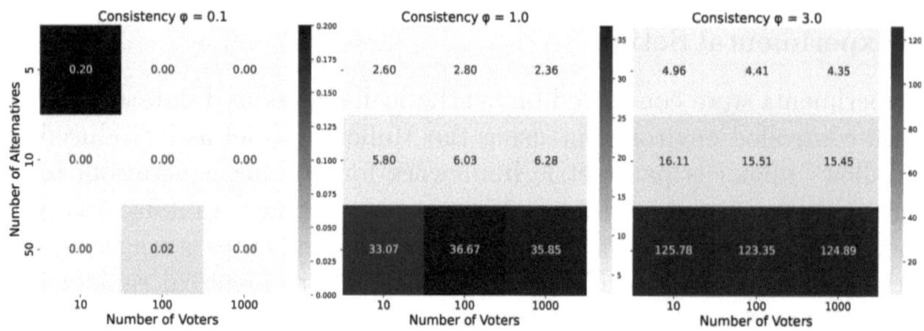

Fig. 1. Average Kendall tau distances for 27 synthetic datasets. Each heatmap corresponds to a different level of dispersion: high agreement ($\phi = 0.1$, blue), moderate ($\phi = 1.0$, green), and low agreement ($\phi = 3.0$, orange). (Color figure online)

In Fig. 1, three heatmaps illustrate the consistency of synthetic datasets generated under varying levels of preference dispersion, governed by the Mallows model parameter ϕ, and evaluated using the average Kendall tau distance. The heatmap corresponding to $\phi = 0.1$ (shown in blue) reflects near-perfect agreement among voters, as expected in a low-conflict setting with strong central consensus. In contrast, the heatmaps for $\phi = 1$ and $\phi = 3$, which represent moderate and high levels of preference dispersion, respectively, exhibit different patterns. Notably, the average Kendall tau values remain nearly constant across different numbers of voters, indicating that the degree of pairwise disagreement is primarily driven by preference heterogeneity rather than dataset size. However, these values increase with the number of alternatives, suggesting that variability in individual rankings becomes more pronounced as the size of the choice set grows.

3.3 Results

This section evaluates the sensitivity of the LMC-WOD method to the damping factor d using synthetic datasets generated under varying levels of preference dispersion ϕ, ranking length (alternatives), and ranking set size (voters). The analysis assesses the average similarity between the aggregated ranking produced by LMC-WOD and the corresponding input rankings by K$\mathcal{T}$, PA, and the WS Coefficient.

3.4 Influence of the Damping Factor D on Aggregation Consistency

Figure 2 illustrates how the similarity metrics, computed as the average similarity between the output ranking produced by the LMC-WOD and individual rankings in the corresponding dataset, vary with the damping factor d, depending on the number of alternatives, input rankings, and the dispersion level. The observed trends reveal how each measure responds to changes in d, offering insight into the robustness of the LMC-WOD aggregated ranking.

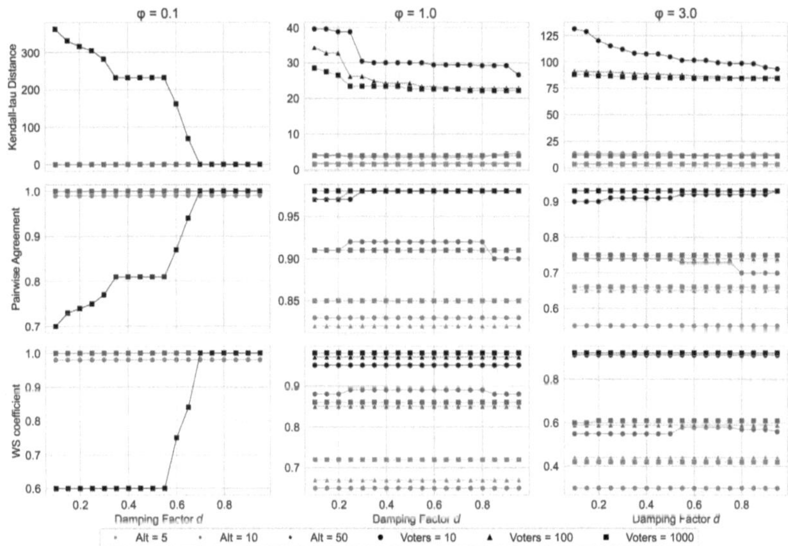

Fig. 2. Mean values of Kendall tau (K$\mathcal{T}$), pairwise agreement (PA), and WS coefficient between LMC-WOD aggregated rankings and input rankings, across varying damping factors d, levels of dispersion ϕ, and problem dimensions.

The results demonstrate that the influence of d varies substantially depending on the disagreement level in the data:

- For highly conflicting scenarios ($\phi = 3$), all metrics remain nearly constant across the full range of d, forming horizontal curves. This suggests that, in heterogeneous environments, the balance between local preference propagation and global smoothing has a limited effect on the resulting consensus ranking.
- In contrast, for moderate dispersion ($\phi = 1.0$), metric values show slightly greater sensitivity to d in their lower values ($d < 0.3$), especially when the number of alternatives is large (Alt=50). Higher d values tend to improve $\mathbf{r}^*$ alignment with the input set of preferences (K$\mathcal{T}$ and PA).

- The highly consistent settings ($\phi = 0.1$), with perfect agreement, show that the damping factor has a greater influence in large datasets (e.g., 1000 voters, 50 alternatives) (dark green curve with squares). For large datasets, the recommendation is to set d with a value greater than 0.7 to maintain a perfect $\mathbf{r}^*$ match. For smaller d, the consensus ranking is increasingly affected by the uniform (non-preference) component and does not express the local dominance.
- Surprisingly, in conflicting settings with many alternatives (e.g., 50 items, $\phi = 3$), the average similarity improves. This result may seem counterintuitive, but the combinatorial structure of longer rankings can explain it. Although consensus is challenging to achieve globally, the aggregation process spreads disagreement more uniformly, resulting in a higher average pairwise alignment across many positions.
- The WS coefficient, which emphasizes top-ranked consistency in long rankings, confirms that LMC-WOD maintains stable performance even under significant disagreement and variation in d, reflecting robustness to local rank changes.

3.5 Evaluation of Aggregation Consistency

Table 1 provides summary statistics for all experimental configurations, reporting mean, standard deviation, and extreme values of $\mathrm{K}\mathcal{T}$, PA, and WS overall values of d.

The empirical findings can be summarized as follows:

- Most standard deviations in the table, except $K\mathcal{T}$ for 50 alternatives, are close to or equal to zero, indicating that the aggregated ranking produced by the LMC-WOD is generally robust to the choice of damping factor d.
- Higher agreement among input rankings leads to better alignment of the aggregated ranking with the overall preference structure, as reflected in the bolded values (Table 1).
- The number of voters has a limited effect: increasing the number of input rankings tends to improve stability, but it does not significantly influence the average similarity values.
- Instead, the data structure, namely the level of dispersion and the number of alternatives, plays a more decisive role in shaping the aggregation outcome.
- As expected, high value of $\phi = 3.0$ results in greater disagreement between the LMC-WOD output and the source rankings. This reflects the increased difficulty of reaching a meaningful consensus under more conflicting preferences.
- High values of the pairwise agreement indicators, such as PA and WS reflect a strong agreement among the input rankings and the consensus solution. When these measures approach unity, the rankings exhibit a nearly identical ordering of alternatives, indicating that the aggregation process preserves local preference structures consistently across the dataset. This phenomenon illustrates the consensus stabilization effect inherent in the LMC-WOD method.

Table 1. ummary statistics of LMC-WOD performance for three evaluation measures ($K\tau$, PA, and WS): mean, standard deviation, minimum, and maximum values computed over all d for each experimental configuration.

Alt	**Voters**	ϕ	**K•**				**PA**				**• •**			
			Mean	STD	Min	Max	Mean	STD	Min	Max	Mean	STD	Min	Max
5	10	0.10	0.10	0.00	0.10	0.10	0.99	0.00	0.99	0.99	0.98	0.00	0.98	0.98
5	10	1.00	1.70	0.00	1.70	1.70	0.83	0.00	0.83	0.83	0.65	0.00	0.65	0.65
5	10	3.00	3.70	0.00	3.70	3.70	0.55	0.00	0.55	0.55	0.30	0.00	0.30	0.30
5	100	0.10	**0.00**	**0.00**	**0.00**	**0.00**	**1.00**	**0.00**	**1.00**	**1.00**	**1.00**	**0.00**	**1.00**	**1.00**
5	100	1.00	1.84	0.00	1.84	1.84	0.82	0.00	0.82	0.82	0.67	0.00	0.67	0.67
5	100	3.00	3.45	0.00	3.45	3.45	0.65	0.00	0.65	0.65	0.44	0.00	0.44	0.44
5	1000	0.10	**0.00**	**0.00**	**0.00**	**0.00**	**1.00**	**0.00**	**1.00**	**1.00**	**1.00**	**0.00**	**1.00**	**1.00**
5	1000	1.00	1.51	0.00	1.51	1.51	0.85	0.00	0.85	0.85	0.72	0.00	0.72	0.72
5	1000	3.00	3.42	0.00	3.42	3.42	0.66	0.00	0.66	0.66	0.42	0.00	0.42	0.42
10	10	0.10	**0.00**	**0.00**	**0.00**	**0.00**	**1.00**	**0.00**	**1.00**	**1.00**	**1.00**	**0.00**	**1.00**	**1.00**
10	10	1.00	3.74	0.37	3.50	4.50	0.91	0.01	0.90	0.92	0.89	0.00	0.88	0.89
10	10	3.00	12.67	0.71	11.70	13.30	0.73	0.02	0.70	0.74	0.56	0.01	0.55	0.58
10	100	0.10	**0.00**	**0.00**	**0.00**	**0.00**	**1.00**	**0.00**	**1.00**	**1.00**	**1.00**	**0.00**	**1.00**	**1.00**
10	100	1.00	3.90	0.00	3.90	3.90	0.91	0.00	0.91	0.91	0.85	0.00	0.85	0.85
10	100	3.00	11.98	0.00	11.98	11.98	0.74	0.00	0.74	0.74	0.59	0.00	0.59	0.59
10	1000	0.10	**0.00**	**0.00**	**0.00**	**0.00**	**1.00**	**0.00**	**1.00**	**1.00**	**1.00**	**0.00**	**1.00**	**1.00**
10	1000	1.00	3.95	0.00	3.95	3.95	0.91	0.00	0.91	0.91	0.86	0.00	0.86	0.86
10	1000	3.00	11.28	0.06	11.26	11.46	0.75	0.00	0.75	0.75	0.61	0.00	0.60	0.61
50	10	0.10	166.11	136.87	0.00	363.00	0.86	0.11	0.70	1.00	0.76	0.19	0.60	1.00
50	10	1.00	31.61	4.25	26.60	39.60	0.98	0.00	0.97	0.98	0.95	0.00	0.95	0.95
50	10	3.00	106.73	10.97	93.40	131.40	0.91	0.01	0.90	0.93	0.91	0.00	0.91	0.91
50	100	0.10	166.12	136.87	0.01	362.99	0.86	0.11	0.70	1.00	0.76	0.19	0.60	1.00
50	100	1.00	25.36	3.81	22.85	34.35	0.98	0.00	0.97	0.98	0.97	0.00	0.97	0.97
50	100	3.00	87.64	2.22	84.77	90.91	0.93	0.00	0.93	0.93	0.92	0.00	0.92	0.92
50	1000	0.10	166.11	136.87	0.00	363.00	0.86	0.11	0.70	1.00	0.76	0.19	0.60	1.00
50	1000	1.00	23.55	1.91	22.20	28.53	0.98	0.00	0.98	0.98	0.98	0.00	0.98	0.98
50	1000	3.00	85.37	1.19	84.40	88.02	0.93	0.00	0.93	0.93	0.92	0.00	0.92	0.92

When the number of alternatives ($Alt = 50$) is large and the ranking set contains numerous conflicts, local dominance differences tend to average out through the Markovian transition process. As a result, the damping parameter d becomes less influential, since both the local dominance component and the global uniformity component converge toward a similar stationary distribution.

4 Conclusions

The results demonstrate that the LMC-WOD method offers robust and consistent performance across a wide spectrum of experimental settings. The aggregated rankings remain closely aligned with individual preference structures under low-dispersion, consensus-oriented conditions, particularly when the number of alternatives is small ($Alt \leq 10$). Even in high-conflict environments or when the number of alternatives increases substantially, the method exhibits remarkable stability, showing minimal sensitivity to variations in the damping factor d. This behavior reflects an inherent equilibrium between local dominance and global uniformity, allowing the Markovian process to converge toward a stable stationary distribution despite fluctuations in the underlying preference space.

The robustness of LMC-WOD underscores its potential for applications involving heterogeneous rankings, with a large number of alternatives, or homogeneous, with a small to medium number of alternatives. Consequently, the algorithm is well suited for complex decision-making contexts such as expert consensus modeling, multi-criteria evaluation, recommender systems, and social choice analysis.

Future research may further advance the understanding and applicability of LMC-WOD by analyzing its convergence properties in the presence of ties, validating its performance on heterogeneous real-world datasets, and benchmarking it against alternative probabilistic or graph-based aggregation techniques. Moreover, exploring the influence of diverse similarity and distance measures, extending the method to incomplete or partially ordered data, and investigating its theoretical limits and integration with learning-based preference models represent promising directions for continued development.

References

1. Borgonovo, E., Plischke, E.: Sensitivity analysis: a review of recent advances. Eur. J. Oper. Res. **248**, 869–887 (2016). https://doi.org/10.1016/j.ejor.2015.06.032
2. Dwork, C., Kumar, R., Naor, M., Sivakumar, D.: Rank aggregation methods for the web. In: Proceedings of the 10th International Conference on World Wide Web (WWW), pp. 613–622. ACM (2001). https://doi.org/10.1145/371920.372165
3. He, H., Cai, J., Zhang, J., Tao, D., Zhuang, B.: Sensitivity-aware visual parameter-efficient fine-tuning. In: 2023 IEEE/CVF International Conference on Computer Vision (ICCV), pp. 11791–11801 (2023). https://doi.org/10.1109/ICCV51070.2023.01086
4. Huang, C., Li, Y., Yao, X.: A survey of automatic parameter tuning methods for metaheuristics. IEEE Trans. Evol. Comput. **24**(2), 201–216 (2020). https://doi.org/10.1109/TEVC.2019.2921598
5. Kemeny, J.G., Snell, J.L.: Mathematical Models in the Social Sciences. Ginn and Company (1962)
6. Kołodziejczyk, J.: Local Markovian consensus for ranking aggregation: a novel approach to weak ordinal dominance. In: Paszynski, M., Barnard, A.S., Zhang, Y.J. (eds.) Computational Science – ICCS 2025 Workshops. ICCS 2025. LNCS, vol. 15910, pp. 366–380. Springer, Cham (2025). https://doi.org/10.1007/978-3-031-97567-7_29
7. Kołodziejczyk, J., Szyjewski, Z., Sałabun, W.: Markovian consensus based on local alternatives' relation. In: Hernes, M., Wątróbski, J., Rot, A. (eds) Emerging Challenges in Intelligent Management Information Systems. ECAI 2024. LNNS, vol. 1218, pp. 224–236. Springer, Cham (2024). https://doi.org/10.1007/978-3-031-78468-2_17

8. Mallows, C.L.: Non-null ranking models. I. Biometrika **44**(1–2), 114–130 (1957). https://doi.org/10.1093/biomet/44.1-2.114
9. Page, L., Brin, S., Motwani, R., Winograd, T.: The pagerank citation ranking: bringing order to the web. Technical report 1999-66, Stanford InfoLab (1999). http://ilpubs.stanford.edu:8090/422/, previous number = SIDL-WP-1999-0120
10. Puka, L.: Kendall's Tau, pp. 713–715. Springer, Berlin, Heidelberg (2011). https://doi.org/10.1007/978-3-642-04898-2_324
11. Sałabun, W., Urbaniak, K.: A new coefficient of rankings similarity in decision-making problems. In: Krzhizhanovskaya, V.V., et al. (eds.) Computational Science – ICCS 2020. ICCS 2020. LNCS, vol. 12138, pp. 632–645. Springer, Cham (2020). https://doi.org/10.1007/978-3-030-50417-5_47
12. Wills, R.S., Ipsen, I.C.F.: Ordinal ranking for Google's PageRank. SIAM J. Matrix Anal. Appl. **30**(4), 1677–1696 (2009). https://doi.org/10.1137/070698129
13. Wątróbski, J., Bączkiewicz, A., Sałabun, W.: pyrepo-mcda — reference objects based MCDA software package. SoftwareX **19**, 101107 (2022). https://doi.org/10.1016/j.softx.2022.101107

From Data to Preferences: Hybrid Swarm Intelligence and MCDA for Criteria Weight Identification

Jarosław Wątróbski[1,2](✉), Aleksandra Bączkiewicz[1], Artur Karczmarczyk[3], and Wojciech Drożdż[1]

[1] Institute of Management, University of Szczecin, ul. Cukrowa 8, 71-004 Szczecin, Poland
{jaroslaw.watrobski,aleksandra.baczkiewicz,wojciech.drozdz}@usz.edu.pl
[2] National Institute of Telecommunications, ul. Szachowa 1, 04-894 Warsaw, Poland
[3] Department of Computer Science, Westpomeranian University of Technology in Szczecin, ul. Żołnierska 49, 71-210 Szczecin, Poland
artur.karczmarczyk@zut.edu.pl

Abstract. This study investigates a hybrid methodology that integrates stochastic optimization algorithms with multi-criteria decision analysis (MCDA) methods to identify decision-makers' preferences for criteria weights. Specifically, Particle Swarm Optimization (PSO) and Cat Swarm Optimization (CSO) were combined with the TOPSIS (Technique for Order Preference by Similarity to Ideal Solution) method to approximate criteria weights when expert-derived reference values are unavailable. A two-dimensional decision matrix comprising 30 alternatives evaluated across nine criteria was generated, with reference weights computed using the CRITIC objective weighting method. The stochastic algorithms were then employed to iteratively determine weights that minimize the Euclidean distance between the generated TOPSIS rankings and the reference ranking.

The results demonstrate that both PSO and CSO effectively identify decision-makers' preferences, with CSO achieving superior accuracy. Convergence analyses and fitness function evaluations revealed that CSO maintains consistent progress throughout iterations, ultimately producing weights and rankings more closely aligned with reference values. Both algorithms successfully identified the top-ranked alternative, confirming their practical utility.

This approach provides a significant advantage over traditional MCDA methods by capturing non-linear preferences and enabling automated preference identification without direct expert input. The study highlights the potential of stochastic optimization for enhancing MCDA frameworks.

Keywords: Multi-criteria decision analysis · Preference identification · Expert knowledge · Decision support systems

M. Pondel et al. (Eds.): AI4KAM 2025, IFIP AICT 776, pp. 418–427, 2026.
https://doi.org/10.1007/978-3-032-18920-2_40

1 Introduction

Multi-criteria decision analysis (MCDA) techniques are widely applied across disciplines to address problems characterized by multiple, and often competing, evaluation criteria. The majority of MCDA approaches require the problem to be formalized as a decision matrix, where each alternative is described according to the predefined assessment criteria [5,14]. In parallel, criteria weights must be specified to reflect their relative importance from the perspective of decision makers [11]. Nevertheless, reliance on expert knowledge can limit the practicality of such models, especially in contexts where expert input is scarce or inconsistent [6]. These challenges emphasize the importance of refining and extending MCDA methods so that they can reconstruct decision models while also incorporating stakeholders' weighting preferences [15].

This study introduces an approach that combines stochastic, swarm intelligence–based algorithms – namely Particle Swarm Optimization (PSO) [8] and Cat Swarm Optimization (CSO) [2] – with the multi-criteria decision-making technique TOPSIS (Technique for Order Preference by Similarity to Ideal Solution) [13]. The integration is particularly advantageous in cases where a dataset comparable to the one under consideration is available, together with outcomes from earlier expert-based evaluations using a chosen MCDA method. A key challenge, however, lies in the fact that the exact relative importance weight values attributed to the criteria by experts in prior assessments are typically unknown [4,9]. To address this, stochastic optimization is applied to infer decision-makers' weighting preferences, thereby improving the evaluation process by embedding the reconstructed expert priorities.

2 Methodology

Stochastic methods draw upon sophisticated population-based metaheuristic algorithms, which provide an effective means of approximating solutions to mathematical optimization problems and are capable of efficiently locating near-global optima [10]. Within this domain, population-based metaheuristics (PMAs) represent the most prominent class of optimization techniques [12]. These algorithms encompass a wide spectrum of inspirations, including evolutionary mechanisms, collective swarm behavior, physical processes, human activities, and biological phenomena [16]. Owing to their versatility, PMAs have found applications across numerous disciplines, such as electrical and electronic engineering, control and automation, communications, operations research, mechanical and chemical engineering, energy systems, and bioengineering [1,3].

The objective of this study was to evaluate the capability of two stochastic optimization methods – Particle Swarm Optimization (PSO) and Cat Swarm Optimization (CSO) – in deriving decision-makers' preferences for criteria weighting. For this purpose, a two-dimensional decision matrix was generated, consisting of 30 alternatives assessed across 9 criteria, with values randomly assigned within the range of 1 to 100. To establish a benchmark, the

CRITIC method [18] (Criteria Importance Through Inter-criteria Correlation) was applied to compute reference weights. Using these weights, the matrix was subsequently analyzed with the TOPSIS approach, yielding both reference scores and rankings.

Subsequently, both PSO and CSO were executed for 100 iterations to identify the set of criteria weights that would produce a TOPSIS ranking most closely aligned with the reference ranking. The degree of similarity between the generated and reference rankings was assessed using Euclidean distance, which served as the fitness function. The optimization objective was to minimize this distance, thereby ensuring the closest possible correspondence between the two rankings.

In the subsequent phase, the outcomes were validated by comparing the criteria weights and rankings produced through the PSO and CSO algorithms with the established reference weights and ranking. The degree of convergence toward the reference values was examined using correlation measures: the Pearson correlation coefficient was applied to evaluate the weights, while the Weighted Spearman correlation coefficient was used to assess the rankings. Below we provide a concise overview of the methods adopted in this study, namely the TOPSIS approach together with the stochastic algorithms PSO and CSO.

The TOPSIS Method. The Technique for Order Preference by Similarity to an Ideal Solution (TOPSIS) relies on the concept of reference points, namely the Positive Ideal Solution and the Negative Ideal Solution. Alternatives are evaluated according to their relative proximity to these two landmarks. A detailed description of the method and its algorithmic procedure can be found in [13,17].

Particle Swarm Optimization (PSO). Proposed by Kennedy and Eberhart in 1995 [7], PSO is a population-based optimization technique characterized by its straightforward implementation and relatively few control parameters. The algorithm draws inspiration from the collective behavior of bird flocks and fish schools, where individuals follow the member nearest to a food source. This natural phenomenon is modeled algorithmically by treating each particle in the swarm as a potential solution, collectively exploring the search space to identify the optimal outcome. In PSO, particles update their positions based on both individual experience and social interaction, governed by three key parameters: inertia weight, cognitive learning rate, and social learning rate. A comprehensive explanation of PSO, including its algorithmic formulation, is provided in [8].

Cat Swarm Optimization (CSO). Cat Swarm Optimization, introduced by Shu-Chuan, Pei-Wei, and Jeng-Shyang in 2006 [2], is a swarm intelligence technique inspired by the natural behavior of cats. In this algorithm, each cat is defined by its position within an M-dimensional search space, a velocity vector, and a fitness value that reflects its performance. The search process alternates between two behavioral submodels: the seeking mode, which imitates a cat at rest while exploring possible moves, and the tracing mode, which simulates active pursuit of a target. The optimization outcome is determined by the best position achieved among all cats after a series of iterations. A full description of the algorithm's assumptions, mathematical formulations, and pseudocode is provided in [2].

3 Results

Figure 1 presents a comparison between the reference criteria weights and those obtained using the PSO and CSO stochastic algorithms. The PSO-derived weights showed the closest alignment with the reference values for criteria C_1, C_4, and C_5. In contrast, the CSO-derived weights exhibited stronger agreement with the reference for criteria C_2, C_3, C_6, C_7, C_8, and C_9. Overall, these results suggest that CSO more accurately captures the actual criteria weights compared to PSO. This observation is further supported by Pearson's correlation coefficient, which equals 0.9339 for CSO versus the reference weights, while PSO achieves a notably lower correlation of 0.6954.

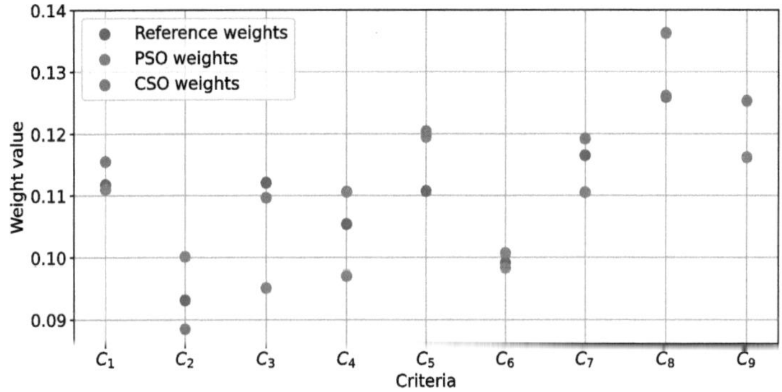

Fig. 1. Criteria weights identified with PSO and CSO.

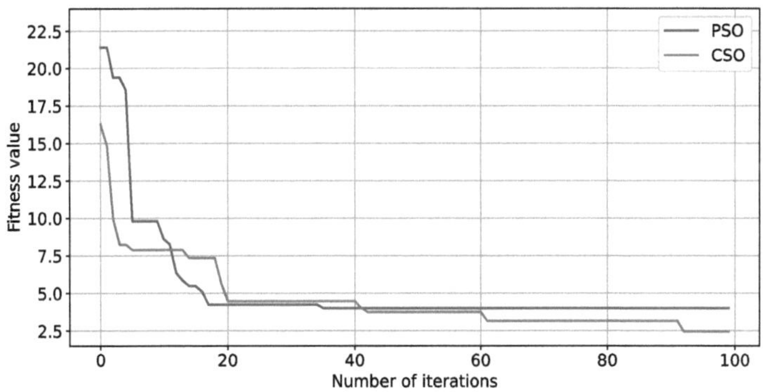

Fig. 2. Matching the fitness function for PSO and CSO.

Figure 2 depicts the evolution of the fitness function, defined as the Euclidean distance between the reference rankings and the rankings obtained using criteria weights derived from the PSO and CSO algorithms, across successive iterations. For the CSO algorithm, a notable reduction in the fitness function was observed as early as the 5th iteration, whereas the PSO algorithm exhibited a comparable decrease only by the 11th iteration. In both cases, the fitness value continued to decline in subsequent iterations. The PSO algorithm reached a stable fitness value by the 35th iteration, while CSO continued to improve, achieving its minimum at the 92nd iteration. After completing all 100 iterations, the CSO algorithm produced a lower fitness value than PSO, indicating a closer alignment of the derived weights with the reference rankings.

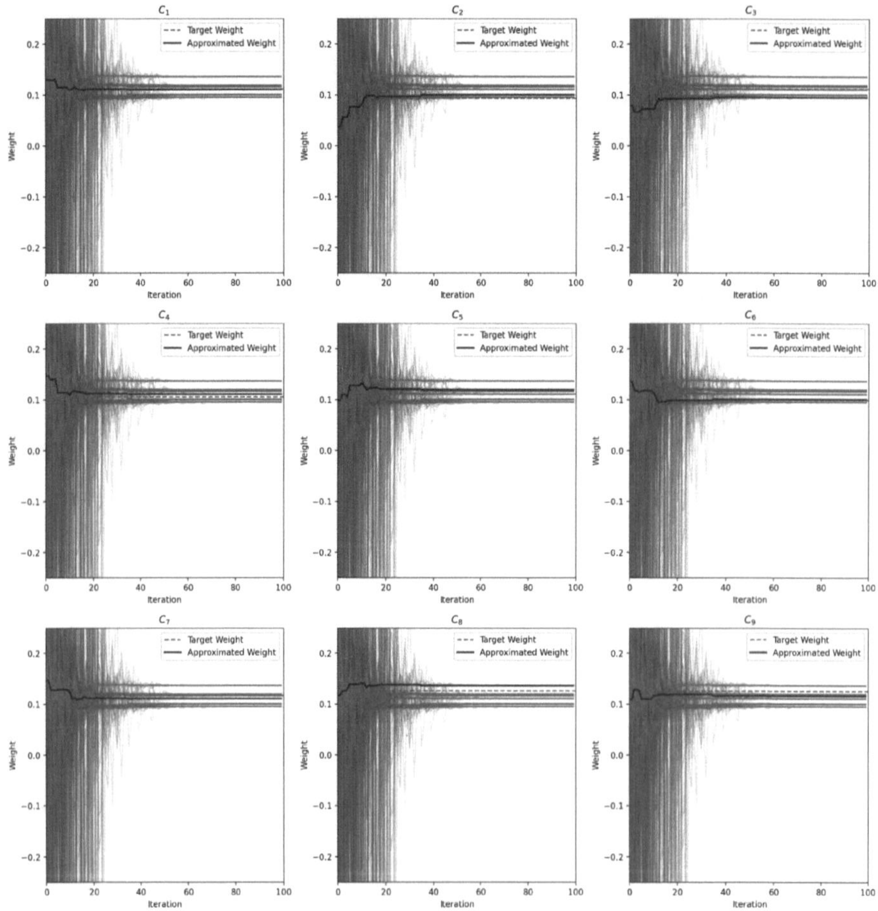

Fig. 3. Weights approximation throughout all 100 iterations with the PSO approach.

Figure 3 illustrates the trajectories of all 30 particles in the PSO algorithm across 100 iterations as they approximated the criteria weights. In the initial iterations, several particles exhibited extreme deviations from the target, with some values exceeding 50 or dropping below –50, far beyond the expected weight range of 0 to 1. For better comparison with the CSO results, the plots were cropped to the range of –0.25 to 0.25. In the figure, the dashed red line indicates the target weight for each criterion, the solid blue line represents the best approximation at each iteration, and the semi-transparent lines in the background correspond to all 30 particles. The solid blue lines generally confirm substantial convergence toward the target weights by approximately the 11th iteration. After reaching the target, the weights remain stable with no further changes observed throughout the remaining iterations.

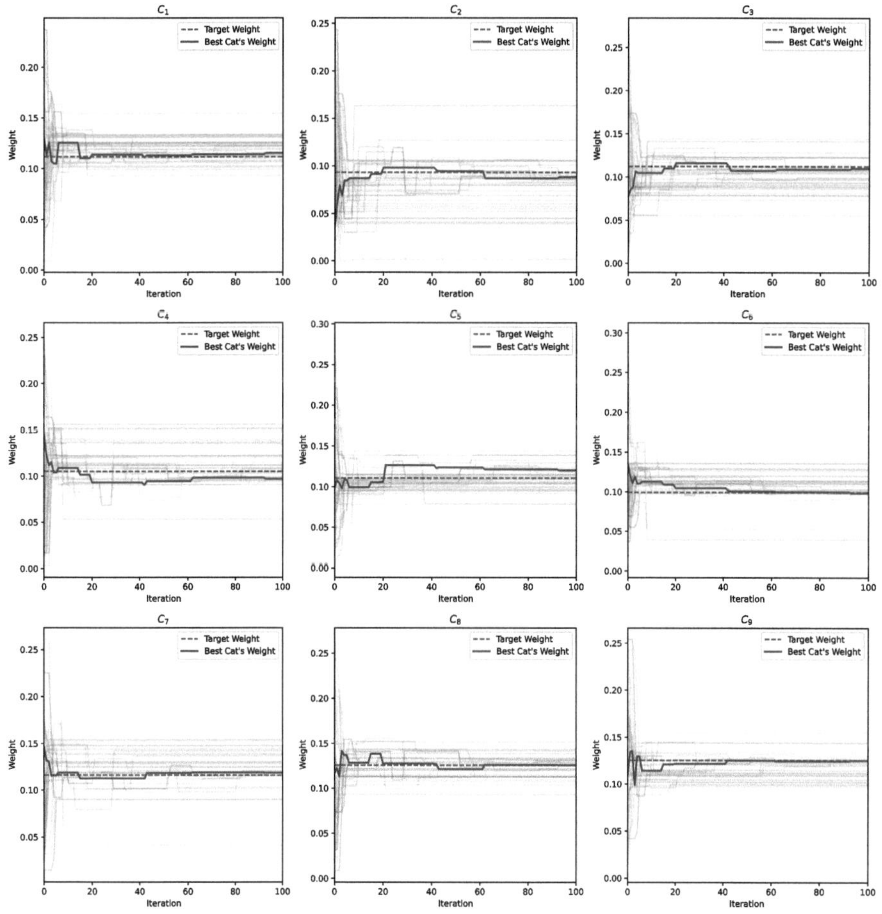

Fig. 4. Weights approximation throughout all 100 iterations with the CSO approach.

Figure 4 illustrates the trajectories of all 30 cats in the CSO algorithm as they approached the target criteria weights. As in the PSO plots, the dashed red line indicates the target weight, the solid blue line shows the best approximation at each iteration, and the semi-transparent lines in the background represent all 30 cats. Compared to PSO, the cats exhibited more target-oriented behavior, with deviations rarely exceeding 0.25 within the 0–1 weight range. This resulted in clearer visualizations that allowed tracking of individual cats' paths. Notably, adjustments to the weights continued as late as the 90th iteration, consistent with the gradual reduction in the fitness function observed in Fig. 2. Unlike PSO, even after achieving a near-optimal fitness, some weights continued to fluctuate. For example, criterion C_8 initially stabilized close to the target (0.1278 vs. 0.1259 target value) between iterations 20 and 40, decreased to 0.1216 during iterations 40–60, and subsequently oscillated around 0.1261 (0.002 difference from the target value). This indicates that while CSO converged more slowly than PSO, it ultimately achieved closer alignment with the true target weights.

The criteria weights obtained from the PSO and CSO algorithms were then applied to compute TOPSIS rankings, which were subsequently compared with the reference ranking. A visual comparison of these rankings is presented in Fig. 5.

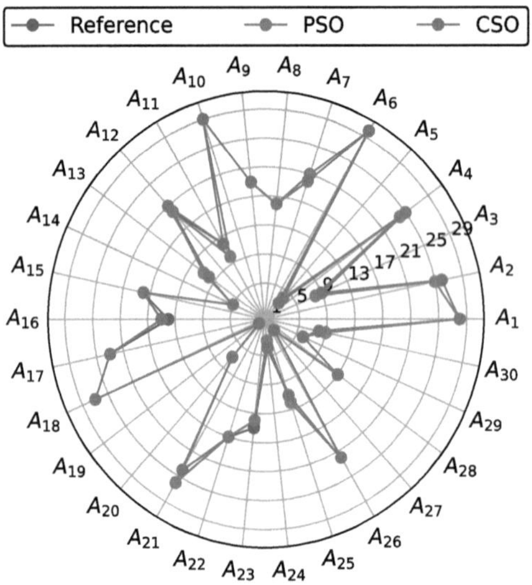

Fig. 5. Rankings generated with criteria weights identified with PSO and CSO.

The TOPSIS rankings generated using the weights derived from the CSO algorithm exhibited a closer alignment with the reference ranking compared to those obtained from the PSO algorithm. This stronger correspondence is reflected in the Weighted Spearman correlation coefficient (r_w), which equals 0.9986 for CSO. By comparison, the r_w value for the PSO-based rankings is slightly lower, at 0.9965, indicating a marginally weaker agreement with the reference ranking.

The discrepancies between the rankings produced by the CSO and PSO algorithms relative to the reference ranking are minimal, demonstrating that both stochastic approaches are effective in capturing decision-makers' preferences for criteria weights. Among the two, the CSO algorithm proved slightly more precise. Importantly, both methods successfully identified the top-ranked alternative, A_{19}.

4 Conclusions

This study explored the effectiveness of a hybrid approach that integrates stochastic algorithms with multi-criteria decision analysis methods for identifying decision-makers' preferences for criteria weights. The results demonstrate that both Cat Swarm Optimization and Particle Swarm Optimization are capable of accurately approximating criteria weights, thereby supporting automated preference identification and enhancing traditional MCDA models with insights derived from previously collected data. By capturing non-linear preferences, the proposed approach overcomes a key limitation of conventional MCDA methods, which are often constrained to linear relationships. Furthermore, the approach allows the re-identification of decision-makers' preferences without relying on direct expert input, making it particularly useful in decision support systems where expert availability is limited. Among the algorithms studied, CSO provided a more precise alignment with reference weights and rankings than PSO, as evidenced by higher correlation coefficients and closer convergence in particle trajectories and fitness function optimization. Both algorithms successfully identified the top-ranked alternative, confirming their effectiveness in practical ranking tasks.

While the study presents a promising initial approach for preference identification using stochastic optimization, several areas for future research were identified. Subsequent work should investigate the influence of algorithmic parameters on performance, as variations in these values can significantly affect outcomes. Expanding the comparative analysis to include additional stochastic optimization methods would provide broader insights into the suitability of different approaches. Further simulations are also needed to evaluate the repeatability and robustness of the results. Moreover, addressing uncertain or imprecise data through the incorporation of fuzzy sets could enhance the applicability of the method to more complex and ambiguous decision-making scenarios. Finally, testing the proposed approach on real-world decision problems will be essential for a comprehensive assessment of its practical utility and scalability.

In summary, this research establishes a foundation for leveraging stochastic optimization in MCDA to identify decision-makers' criteria preferences, highlights the superior performance of CSO in this context, and opens pathways for future methodological and applied developments.

Acknowledgments. This research was partially funded by National Science Centre, Poland 2022/45/B/HS4/02960, and Co-financed by the Minister of Science under the "Regional Excellence Initiative" Program RID/SP/0046/2024/01.

Regionalna Inicjatywa Doskonałości

Ministry of Science and Higher Education Republic of Poland

Disclosure of Interests. The authors have no competing interests to declare that are relevant to the content of this article.

References

1. Agushaka, J.O., Ezugwu, A.E.: Initialisation approaches for population-based metaheuristic algorithms: a comprehensive review. Appl. Sci. **12**(2), 896 (2022). https://doi.org/10.3390/app12020896
2. Ahmed, A.M., Rashid, T.A., Saeed, S.A.M.: Cat swarm optimization algorithm: a survey and performance evaluation. Comput. Intell. Neurosci. **2020**(1), 4854895 (2020). https://doi.org/10.1155/2020/4854895
3. Amirghafouri, F., Neghabi, A.A., Shakeri, H., Sola, Y.E.: Nature-inspired metaheuristic algorithms for resource allocation in the internet of things. Int. J. Commun. Syst. **38**(5), e6141 (2025). https://doi.org/10.1002/dac.6141
4. Bączkiewicz, A., Wątróbski, J.: Crispyn - a python library for determining criteria significance with objective weighting methods. SoftwareX **19**, 101166 (2022)
5. Bączkiewicz, A., Wątróbski, J., Karczmarczyk, A.: A novel multi-criteria temporal decision support method-sustainability evaluation case study. In: Franco, L., de Mulatier, C., Paszynski, M., Krzhizhanovskaya, V.V., Dongarra, J.J., Sloot, P.M.A. (eds.) Computational Science – ICCS 2024. ICCS 2024. LNCS, vol. 14833, pp. 189–203. Springer, Cham (2024). https://doi.org/10.1007/978-3-031-63751-3_13
6. Bohanec, M.: DEX (Decision EXpert): a qualitative hierarchical multi-criteria method. In: Kulkarni, A.J. (eds.) Multiple Criteria Decision Making. Studies in Systems, Decision and Control, LNCS, vol. 407, pp. 39–78. Springer, Singapore (2022). https://doi.org/10.1007/978-981-16-7414-3_3
7. Freitas, D., Lopes, L.G., Morgado-Dias, F.: Particle swarm optimisation: a historical review up to the current developments. Entropy **22**(3), 362 (2020). https://doi.org/10.3390/e22030362
8. Houssein, E.H., Gad, A.G., Hussain, K., Suganthan, P.N.: Major advances in particle swarm optimization: theory, analysis, and application. Swarm Evol. Comput. **63**, 100868 (2021). https://doi.org/10.1016/j.swevo.2021.100868
9. Karczmarczyk, A., Wątróbski, J., Bączkiewicz, A., Mróz-Malik, O., Drożdż, W.: New robust multi-criteria decision-making method for wind farm location problems. Appl. Energy **398**, 126401 (2025). https://doi.org/10.1016/j.apenergy.2025.126401, https://www.sciencedirect.com/science/article/pii/S0306261925011316
10. Kim, Y.J., Kim, B.S.: Population-based meta-heuristic algorithms for integrated batch manufacturing and delivery scheduling problem. Mathematics **10**(21), 4127 (2022). https://doi.org/10.3390/math10214127
11. Kumar, P., Sharma, R., Bhaumik, S.: MCDA techniques used in optimization of weights and ratings of DRASTIC model for groundwater vulnerability assessment. Data Sci. Manag. **5**(1), 28–41 (2022). https://doi.org/10.1016/j.dsm.2022.03.004

12. Navarro, M.A., Oliva, D., Ramos-Michel, A., Morales-Castaneda, B., Zaldívar, D., Luque-Chang, A.: A review of the use of quasi-random number generators to initialize the population in meta-heuristic algorithms. Arch. Comput. Methods Eng. **29**(7), 5149–5184 (2022). https://doi.org/10.1007/s11831-022-09759-y
13. Sahabuddin, M., Khan, I.: Multi-criteria decision analysis methods for energy sector's sustainability assessment: robustness analysis through criteria weight change. Sustain. Energy Technol. Assess. **47**, 101380 (2021). https://doi.org/10.1016/j.seta.2021.101380
14. Shao, M., Han, Z., Sun, J., Xiao, C., Zhang, S., Zhao, Y.: A review of multi-criteria decision making applications for renewable energy site selection. Renew. Energy **157**, 377–403 (2020). https://doi.org/10.1016/j.renene.2020.04.137
15. Vieira, A.C., Oliveira, M.D., e Costa, C.A.B.: Enhancing knowledge construction processes within multicriteria decision analysis: the collaborative value modelling framework. Omega **94**, 102047 (2020). https://doi.org/10.1016/j.omega.2019.03.005
16. Wątróbski, J., Bączkiewicz, A., Rudawska, I.: A strong sustainability paradigm based analytical hierarchy process (SSP-AHP) method to evaluate sustainable healthcare systems. Ecol. Ind. **154**, 110493 (2023)
17. Wątróbski, J., Bączkiewicz, A., Sałabun, W.: pyrepo-mcda - reference objects based MCDA software package. SoftwareX **19**, 101107 (2022)
18. Wątróbski, J., Karczmarczyk, A., Bączkiewicz, A.: Using the TOSS method in semi-autonomous passenger car selection. Sustain. Energy Technol. Assess. **58**, 103367 (2023)

Pairwise Outperformance Aggregations in Multiple-Criteria Analysis of Alternatives

Janusz Granat[1,2](✉) and Marek Makowski[3,4]

[1] National Institute of Telecommunications, Warsaw, Poland
[2] Warsaw University of Technology, Warsaw, Poland
J.Granat@il-pib.pl
[3] International Institute for Applied Systems Analysis, Laxenburg, Austria
marek@iiasa.ac.at
[4] Systems Research Institute of the Polish Academy of Sciences, Warsaw, Poland

Abstract. Numerous methods have been developed for multiple criteria analysis of a given set of discrete alternatives. Only a small fraction of them are suitable for users with limited modeling skills and for problems with a large number of alternatives and criteria, optionally organized into hierarchical structures. This paper presents new methods effectively supporting analysis of such problems, as well as using a very simple, but yet effective, way of interactive specification of preferences suitable also for a large number of diversified users without experience in analytical methods. The approach is based on the equitable aggregations concept and the developed pairwise outperformance aggregations considering inter-alternative relations. The methods have been implemented for, and applied to analyzing real-life problems, in particular multiple-criteria analysis of future energy technologies, which involves large numbers of both alternatives and criteria.

Keywords: multiple criteria analysis · equitable aggregations · outperformance measures

1 Introduction

Multiple-criteria analysis is a well-established area of applied science, which has been developed in response to problems that could not be met by single-criterion optimization methods. A sample of diverse approaches and the corresponding tools can be found e.g., in [2,3,5,16], and references therein.

Representation of the user preferences is a key characteristics of every method of multiple-criteria analysis. Such representations consist of two main parts. First, a specification of the user preferences, e.g., by defining relative importance of criteria, or derived from comparisons of alternatives. Second, an aggregation of the user specified preferences into an outperformance measure used for selecting the alternative fitting best the specified preferences. Various representations of preferences aim at meeting needs of diverse audiences/users and problem characteristics.

Published by Springer Nature Switzerland AG 2026
M. Pondel et al. (Eds.): AI4KAM 2025, IFIP AICT 776, pp. 428–435, 2026.
https://doi.org/10.1007/978-3-032-18920-2_41

2 Preliminaries

2.1 Basic Concepts and Notations

The paper deals with the problem of multiple-criteria analysis of a given set of alternatives (objects) o_j, $j \in J = \{1, 2, \ldots, m\}$ evaluated in terms of criteria c_i, $i \in I = \{1, 2, \ldots, n\}$. Indices i and j, as well as the corresponding sets I and J are consistently used for criteria and alternatives, respectively. Additional indices are defined locally, whenever needed.

The set of alternatives are referred to as $Q = \{o_j : j \in J\}$. Each alternative o_j is characterized by the corresponding numerical attributes (aka outcomes, criteria values) denoted by $q_{ij} \in R$ for each pair $(i \in I, j \in J)$.

Vectors are generally denoted through using the ***bold*** font. In particular, vector of attributes of alternative o_j is defined as $\boldsymbol{q}^j = [q_{1j}, q_{2j}, \ldots, q_{nj}]^T$, and vector of attributes of criterion c_i is defined as $\boldsymbol{q}^i = [q_{i1}, q_{i2}, \ldots, q_{im}]$. The matrix Q is composed of all values of attributes, columns and rows of Q are composed of the values of attributes of the corresponding alternative or criterion, respectively.

In order to measure the outcomes (attribute values) of all alternatives in a uniform scale, as well as to uniformly handle maximized and minimized criteria, we apply the below defined linear mapping $map_i(\cdot)$, which is based on the concept of criteria achievement function described in detail e.g., in [4]:

$$map_i(x) = \begin{cases} \frac{x - xmin_i}{xmax_i - xmin_i}, & \text{for maximized criterion;} \\ \frac{xmax_i - x}{xmax_i - xmin_i}, & \text{for minimized criterion.} \end{cases} \quad (1)$$

where $xmin_i$ and $xmax_i$ are the minimum and maximum values of i-th attribute amongst all alternatives, i.e., elements of $\boldsymbol{q}^i$ vector. In other words, linear mappings $map_i(\cdot) : R \to R$ enable measure of achievements of all outcomes in a uniform scale [0, 1], where values of 0 and 1 correspond to the worst and best (for each criterion) outcome performance, respectively.

Individual (in terms of criteria) achievements of each alternative are denoted by a_{ij} and defined by:

$$a_{ij} = map_i(q_{ij}), \quad i \in I, j \in J, \quad (2)$$

and the achievement vector of alternative j is denoted by $\boldsymbol{a}^j = [a_{1j}, a_{2j}, \ldots, a_{nj}]^T$; similarly, the achievement vector of criterion i is denoted by $\boldsymbol{a}^i = [a_{i1}, a_{i2}, \ldots, a_{im}]$.

3 Pairwise Outperformance Measure

3.1 POA Concept

In order to explain the Pairwise Outperformance Aggregation (POA) concept let us consider two pairs of alternatives defined in Table 1, namely (o_6, o_8) and (o_3, o_7). Let us denote the differences of achievements in criterion c_i by $\Delta_{ijl} = a_{ij} - a_{il}$, and recall the simple calculations shown in Table 1: $\Delta_{168} =$

Table 1. Values of achievements a_1 and a_2 for nine alternatives, as well as differences of achievements for the two selected pairs of alternatives: $(o_6,\ o_8)$ and $(o_3,\ o_7)$.

	b_1	b_2	o_1	o_2	o_3	o_4	o_5	o_6	o_7	o_8	o_9	o_6 - o_8	o_3 - o_7
c_1	0.0	0.1	0.50	0.40	0.20	0.10	0.60	0.80	0.40	1.00	0.00	-0.20	-0.20
c_2	1.0	0.9	0.50	0.40	0.75	0.90	0.30	0.05	0.70	0.00	1.00	0.05	0.05

$\Delta_{137} = -0.2$, $\Delta_{268} = \Delta_{237} = 0.05$. In other words, the first and the second alternative in each pair has better achievement c_2 and c_1, respectively.

Further, let us point out that although the differences of the corresponding achievements are the same for each criterion, the considered achievements differ substantially. Namely, the c_1 achievements $a_{16} = 0.8, a_{18} = 1$, and $a_{13} = 0.2, a_{17} = 0.4$ are high and low, for the first and the second pair, respectively. The c_2 achievements $a_{26} = 0.05, a_{28} = 0$, and $a_{23} = 0.75, a_{27} = 0.7$ are very poor and good, for the first and the second pair, respectively. Any linear aggregation of achievements provides the same comparison of both pairs of alternatives, i.e., either $((o_6 \succ o_8)$ and $(o_3 \succ o_7))$ or $((o_6 \prec o_8)$ and $(o_3 \prec o_7))$ or $((o_6 \sim o_8)$ and $(o_3 \sim o_7))$.

In other words, linear aggregations treat equally improvements and worsening of achievements by the same value, irrespective the corresponding criteria performance. However, improvement of a weak performance is typically preferred over worsening a better performance by the same value. This is formalized in the property of decreasing the marginal achievement utility. In order to conform to this property it is necessary to apply a non-linear aggregation of achievements.

We now explain step-by-step how such principles are implemented. Let us start with recalling the frequently used elements of the notations (defined in detail in Sect. 2.1):

- Specific criteria and alternatives are denoted by c_i, and o_j, respectively; where $i \in I$, $j \in P$, and $P \subset J$ denotes set of Pareto-efficient alternatives. Other indices are locally defined, whenever necessary.
- To simplify the notations, alternatives not specifically referred to are denoted by vectors of their achievements, e.g., $\boldsymbol{x}$, where x_i stands for achievement of criterion c_i.
- Values of achievements are denoted by $a_{ij} \in [0,\ 1]$ and by $x_i \in [0,\ 1]$ (for alternative o^j and alternative represented by $\boldsymbol{x}$, respectively), where values of 0 and 1 stand for the worst and best achievement of i-th criterion.
- The preference of alternative $\boldsymbol{x}$ over $\boldsymbol{y}$ is denoted by $\boldsymbol{x} \succ \boldsymbol{y}$ (equivalently by $\boldsymbol{y} \prec \boldsymbol{x}$). The indifference of alternatives is denoted by $\boldsymbol{x} \sim \boldsymbol{y}$.
- Criteria importance weights are denoted by w_i.

3.2 Pairwise Outperformance Measure

The developed methods are based on pairwise comparisons of alternatives. This process is composed of two steps:

- measuring the outperformance for each criterion separately, and
- aggregating such individual criteria measures into an outperformance measure.

We now discuss the first step. For the already explained reasons, we compare only Pareto alternatives. Moreover, since we consider now only one criterion, we disregard here the weights representing criteria importance.

Let us consider a pair of Pareto alternatives ($\boldsymbol{x} \in P$, $\boldsymbol{y} \in P$), where elements of each vector are defined by the corresponding achievements. The Pareto alternatives are rationally indifferent because one of them will perform better in at least one criterion while the other alternative will be better in terms of another criterion, i.e.:

$$\exists\, i,\ k\ \in I\ :\ y_i < x_i \text{ and } x_k < y_k. \tag{3}$$

Therefore, in order to prefer one of the Pareto alternatives pair, one needs to measure for each criterion the pairwise outperformance, based on the difference of values of the corresponding achievements.

To illustrate this issue let us consider performance of i-th criterion in the example defined by (3). Considering the above together with the principle of decreasing marginal value, it is justified to define:

$$\alpha = \beta(x_i) + \beta(y_i), \tag{4}$$

where function $\beta(\cdot)$ represents the marginal value of the corresponding achievement, i.e., it is equal to the derivative of the value function. Such principles can be represented by strictly increasing concave value functions of achievements, which – thanks to their properties – have the decreasing marginal value property.

Let us note that the role of the function $\alpha(\cdot)$ is to scale the impact of the achievements according to their performance in i-th criterion. Instead of attempting to define a value function and its marginal value, we directly use the latter, i.e., its derivative. Let us note that the derivative of a strictly concave strictly increasing function is a strictly convex strictly decreasing positive function, which also represents the marginal value of the achievements. We denote such a function by $\beta(\cdot)$.

In order to equitably treat all criteria and all alternatives, we apply the same scaling function for all achievements. The scaling factor is equal to the average of the marginal values of the corresponding achievements of the two compared alternatives. For simplifying the presentation, we skip the division by two as it does not change the resulting pairwise comparison.

Therefore, the pairwise outperformance measure in respect of i-criterion can be reformulated to

$$cmp_i(x_i, y_i) = (\beta(x_i) + \beta(y_i)) \cdot (x_i - y_i), \qquad i \in I. \tag{5}$$

One may also note the following properties of (5):

- For a large absolute value of $(x_i - y_i)$ one element of $(\beta(x_i) + \beta(y_i))$ is also large, and thus the value of $cmp_i(\cdot)$ is large.

- For a small absolute value of $(x_i - y_i)$ the $cmp_i(\cdot)$ value is either large or small, for weak or strong achievements, respectively.
- $cmp_i(x_i, y_i) = -cmp_i(y_i, x_i)$; moreover, the sign of $cmp_i(\cdot)$ is determined by the difference of achievements of the alternatives, i.e., a positive value shows that alternative $\boldsymbol{x}$ performs on i-the criterion better than $\boldsymbol{y}$.

4 Pairwise Outperformance Aggregation (POA)

4.1 POA Background

The pairwise outperformance measure was defined in Section 3 with the assumption of equally important criteria, i.e., taking into account the user preferences regarding the relative criteria importance.

We now aggregate the components defined for each criterion by (5) into the Pairwise Outperformance Aggregation (POA) measure of comparison of alternatives $(\boldsymbol{x}, \boldsymbol{y})$:

$$POA(\boldsymbol{x}, \boldsymbol{y}) = \sum_{i \in I} w_i \cdot cmp_i(\boldsymbol{x}, \boldsymbol{y}). \tag{6}$$

The aggregated outperformance measure (6) enables building the corresponding valued preference relation:

$$\begin{array}{l} POA(\boldsymbol{x}, \boldsymbol{y}) > 0 \quad \Rightarrow \quad \boldsymbol{x} \succ \boldsymbol{y} \quad \text{and} \quad POA(\boldsymbol{x}, \boldsymbol{y}) < 0 \quad \Rightarrow \quad \boldsymbol{y} \succ \boldsymbol{x} \\ \text{and} \quad POA(\boldsymbol{x}, \boldsymbol{y}) = 0 \quad \Rightarrow \quad \boldsymbol{x} \sim \boldsymbol{y}. \end{array} \tag{7}$$

We refer to the preference model (7) as the outperformance relation.

Respectively, the following relation defines the weekly outperformance:

$$POA(\boldsymbol{x}, \boldsymbol{y}) \geq 0 \quad \Rightarrow \quad \boldsymbol{x} \succeq \boldsymbol{y}. \tag{8}$$

Such weak outperformance relation is quite different from the commonly used outranking relations. However, it is similar in the sense of a binary relation defined on $Q \times Q$ such that $\boldsymbol{x} \succeq \boldsymbol{y}$, if there are enough arguments to decide that $\boldsymbol{x}$ is at least as good as $\boldsymbol{y}$, while there is no essential reason to refuse that statement. In order to account for both the differences and similarity, we use the term outperformance instead of the established outranking term.

Let us consider a comparison of the i-th criterion performance in a pair of alternatives $\{o_j,\ o_l\}$, and introduce the comparison measure dc'_{jli} defined by:

$$dc'_{jli} = \beta(a_{ij})(a_{ij} - a_{il}), \tag{9}$$

where $\beta(\cdot)$ is a convex, decreasing, and positive internal scaling function. The role of $\beta(\cdot)$ is to amplify the impact of the difference of the achievements on the value of dc'_{jli} differently depending on the a_{ij} performance. Due to the $\beta(\cdot)$ properties, the amplification of improvements of weak achievements is stronger than that of the strong achievements.

The difference of the measure (9) applied to both alternatives defines the outperformance aggregation component dc_{jli} representing the i-criterion:

$$dc_{jli} = dc'_{jli} - dc'_{lji} = (\beta(a_{ij}) + \beta(a_{il}))(a_{ij} - a_{il}). \tag{10}$$

One may also note the following properties of (10):

- For a large absolute value of $(a_{ij} - a_{il})$ one element of $(\beta(a_{ij}) + \beta(a_{il}))$ is also large, and thus the value of dc_{jli} is large.
- For a small absolute value of $(a_{ij} - a_{il})$ the dc_{jli} value is large or small, for weak or strong achievements, respectively.
- $dc_{jli} =$ - dc_{lji}; moreover, $dc_{jli} > 0$ means that the performance of a_{ij} is better than that of a_{il}.

We now aggregate the components defined for each criterion by (10) into the Pairwise Outperformance Aggregation (POA) measure of comparison of alternatives (o_j, o_l). The measure is denoted equivalently by $POA(o_j, o_l)$ and d_{jl}, and it is defined by:

$$POA(o_j, o_l) = d_{jl} = \sum_{i \in I} w_i dc_{jli} = \sum_{i \in I} w_i(\beta(a_{ij}) + \beta(a_{il}))(a_{ij} - a_{il}). \tag{11}$$

The aggregated outperformance measure (11) enables building the corresponding valued preference relation:

$$POA(o_j, o_l) > 0 \quad \Rightarrow \quad o_j \succ o_l \quad \text{and} \quad POA(o_j, o_l) = 0 \quad \Rightarrow \quad o_j \sim o_l. \tag{12}$$

The aggregated outperformance measure (11) enables building the corresponding valued preference relation. We will refer to the preference model (12) as the outperformance relation. Respectively, $o_j \succeq o_l$ (i.e., alternative o_j weakly outperforms alternative o_l) if $d_{jl} \geq 0$. The outperformance relation can be lexicographically enhanced, if the corresponding differences scaled by $\beta(\cdot)$ lead to the equivalent result as obtained by applying the original differences:

$$\begin{array}{lcll} o_j \succ_e o_l & \Leftrightarrow & d_{jl} > 0 & \text{or} \quad \left(d_{jl} = 0 \quad \text{and} \quad \sum_{i \in I} w_i(a_{ij} - a_{il}) > 0 \right), \\ o_j \sim_e o_l & \Leftrightarrow & d_{jl} = 0 & \text{and} \quad \sum_{i \in I} w_i(a_{ij} - a_{il}) = 0. \end{array} \tag{13}$$

Note that while the enhancement narrows the indifference relation, it does not affect the weak outperformance relation.

5 Conclusions

The paper presents outperformance aggregations of alternatives that take into account the inter-alternative relations. Such aggregations use the pairwise alternative comparisons based on evaluation of the differences of the corresponding criteria values. The approach has been used in numerous applications, including those with large numbers of alternative as well as of criteria.

Acknowledgments. Włodzimierz Ogryczak of the Warsaw University of Technology developed many ideas adapted to this paper.

Disclosure of Interests. The authors have no competing interests.

References

1. Bouyssou, D.: Ranking methods based on valued preference relations: a characterization of the net flow method. Eur. J. Oper. Res. **60**, 61–67 (1992)
2. Bouyssou, D., Marchant, T., Pirlot, M., Tsoukias, A., Vincke, P.: Evaluation and Decision Models with Multiple Criteria: Stepping Stones for the Analyst. Springer, New York (2006). https://doi.org/10.1007/0-387-31099-1
3. Cinelli, M., Kadzinski, M., Gonzalez, M., and Slowinski, R.: How to support the application of multiple criteria decision analysis? Let us start with a comprehensive taxonomy. Omega **96** (2020)
4. Granat, J., Makowski, M.: Interactive specification and analysis of aspiration-based preferences. Eur. J. Oper. Res. **122**(2), 469–485 (2000)
5. Greco, S., Ehrgott, M., Figueira, J.: Multiple Criteria Decision Analysis: State of the Art Surveys. Springer, New York (2016). https://doi.org/10.1007/b100605
6. Makowski, M.: Management of attainable tradeoffs between conflicting goals. J. Comput. **4**(10), 1033–1042 (2009). ISSN 1796-203X
7. Makowski, M., Granat, J., Ogryczak, W.: Overview of methods implemented in MCA: multiple criteria analysis of discrete alternatives with a simple preference specification. Interim Report IR-09-24, International Institute for Applied Systems Analysis, Laxenburg, Austria, 2009
8. Makowski, M., Granat, J., Ren, H.: User guide to MCA: multi-criteria analysis of discrete alternatives with a simple preference specification. Interim Report IR-09-22, International Institute for Applied Systems Analysis, Laxenburg, Austria, 2009
9. Ogryczak, W.: WOWA enhancement of the preference modeling in the reference point method. In: Torra, V., Narukawa, Y. (eds.) Modeling Decisions for Artificial Intelligence. MDAI 2008. LNCS, vol. 5285, pp. 38–49. Springer, Berlin, Heidelberg (2008). https://doi.org/10.1007/978-3-540-88269-5_5
10. Ogryczak, W.: On principles of fair resource allocation for importance weighted agents. In: International Workshop on Social Informatics (Warsaw, 2009), pp. 57–62 (2009)
11. Ogryczak, W., Opolska-Rutkowska, M.: SSD consistent criteria and coherent risk measures. In: Ceragioli, F., Dontchev, A., Futura, H., Marti, K., Pandolfi, L. (eds.) System Modeling and Optimization. CSMO 2005. IFIP International Federation for Information Processing, LNCS, vol. 199, pp. 227–237. Springer, Boston, MA (2006). https://doi.org/10.1007/0-387-33006-2_21
12. Ogryczak, W., Śliwiński, T. : On optimization of the importance weighted OWA aggregation of multiple criteria. In: Gervasi, O., Gavrilova, M.L. (eds.) Computational Science and Its Applications – ICCSA 2007. ICCSA 2007. LNCS, vol. 4705, pp. 804–817. Springer, Berlin, Heidelberg (2007). https://doi.org/10.1007/978-3-540-74472-6_66
13. Ozturk, M., Tsoukias, A., Vincke, P.: Preference modelling. Technical report 2003-34, DIMACS, Rutgers University, 2003

14. Rawls, J.: A Theory of Justice. Oxford University Press, London, Oxford, New York (1973). ISBN 0-19-881301-5
15. Wierzbicki, A., Granat, J., Makowski, M.: Discrete decision problems with large number of criteria. Interim Report IR-07-25, International Institute for Applied Systems Analysis, Laxenburg, Austria, 2007
16. Wierzbicki, A., Makowski, M., Wessels, J. (Eds.): Model-Based Decision Support Methodology with Environmental Applications. Series: Mathematical Modeling and Applications. Kluwer Academic Publishers, Dordrecht, 2000. ISBN 0-7923-6327-2

Multi-criteria Assessment of the Capabilities of Latest Large Language Models in Tax Law Analysis and Validation of the 'LLM as a Judge' Concept

Tomasz Strąk[2,3], Mateusz Piwowarski[1,3(✉)], Michał Tuszyński[1,3], and Kesra Nermend[1,3]

[1] Institute of Management, University of Szczecin, Cukrowa 8, 71-004 Szczecin, Poland
mateusz.piwowarski@usz.edu.pl

[2] Institute of Economics and Finance, University of Szczecin, Mickiewicza 64/66, 71-101 Szczecin, Poland

[3] Enelpe Sp. z o.o., Tomasza Zana 43/2.1, Lublin, Poland

Abstract. The development of generative large language models (LLMs) opens up new possibilities in professional domains, including law. Tax law, in particular, poses unique challenges due to its dynamism, complexity, and susceptibility to errors ("hallucinations"). There is a need for a systematic evaluation of LLM capabilities in this context, especially for Polish tax law, and validation of tools supporting result verification. The aim of this study is a multi-faceted evaluation of the latest LLM in the analysis of Polish private tax ruling and assessment of the potential of the "LLM as a Judge" concept as a quality control tool. A two-stage experimental study was conducted. In Stage 1, four experts evaluated the quality of analyses for 100 tax interpretations performed by four LLM (Gemini 2.5 Pro/Flash, GPT-4o, DeepSeek R1) using a proprietary Legal Quality Index (LQI). In Stage 2, the model quality rankings established by experts were compared with the rankings created by the "LLM as a Judge" (based on the OpenAI o3 model), measuring concordance using Spearman's rank correlation coefficient, Rank-Biased Overlap (RBO), and Cohen's κ coefficient. The impact of the "critique + rerank" procedure on the quality of analyses was also examined. Stage 1 results showed varying model performance, with Gemini achieving the highest LQI. Stage 2 results showed very high concordance between expert evaluations and the "LLM as a Judge" (Spearman 0.861, RBO > 0.9 for low p, $\kappa = 0.88$). This confirms the reliability of automated evaluation, especially in identifying models from the top ranks. The "critique + rerank" procedure significantly improved the quality of the analyses. The obtained results suggest that LLM-based tools can serve as reliable support for quality control of tax law analyses, leading to process optimization and potential cost savings in law firms. Further research should focus on agent-based approaches, utilizing domain-specific knowledge bases, and analyzing the impact of AI on task completion time.

Keywords: Large Language Models · Tax Law · Private Tax Rulings · LLM-as-a-Judge · Legal Quality Index · LegalTech Automation

Published by Springer Nature Switzerland AG 2026
M. Pondel et al. (Eds.): AI4KAM 2025, IFIP AICT 776, pp. 436–448, 2026.
https://doi.org/10.1007/978-3-032-18920-2_42

1 Introduction

Over the past few months, generative large language models (LLMs) have evolved into sophisticated systems showing potential across domains, including law. However, their deployment in legal practice, especially complex areas like tax law, presents unique challenges. Tax law's dynamism, detailed regulations, and connection to numerical data make it particularly sensitive to AI errors or "hallucinations." [1, 2] Preliminary studies suggest LLMs can handle some tax cases but highlight significant risks of generating incorrect or non-existent legal information. The concept of using LLMs to verify or evaluate content ("LLM as a Judge") shows promise in improving reliability [3].

Despite these advancements, there is a lack of systematic, comprehensive research evaluating LLM capabilities in tax law analysis, particularly for real Polish individual tax rulings, using the latest models, and applying multidimensional quality criteria. Existing evaluation methods, often relying on automatic NLP metrics or human judges, face challenges in capturing the nuances of legal correctness and are resource-intensive. Accurate and reliable evaluation methods and validation of AI verification tools are urgently needed.

From a law firm management perspective, implementing LLMs offers potential benefits like optimizing resource allocation and increasing throughput, but also involves significant challenges, including risk management, data security, and the need for new digital competencies. [4, 5] Therefore, the ability to reliably assess LLM performance and utilize verification support is crucial for safe and effective integration.

The aim of this article is to fill the identified research gap by conducting a multi-aspect evaluation of the capabilities of the latest LLM in tax law analysis and assessing the potential of tools supporting their verification. In particular, the study addresses three key research questions:

C1. What accuracy and completeness do four leading LLM (Gemini 2.5 Pro Preview 03–25, Gemini 2.5 Flash Preview 04 – 17, GPT-4o, DeepSeek R1) achieve across a set of 11 predefined analytical tasks on a representative corpus of 100 Polish private tax rulings, measured using a proprietary Legal Quality Index?

C2. Does adding an LLM-based verification module ("LLM Judge," utilizing the ChatGPT model technology with a "critique + rerank" procedure) increase the Legal Quality Index value compared to results obtained directly from raw models?

C3. To investigate the concordance between the quality assessment of legal analyses performed by an experienced tax advisor expert and the assessment performed by the "LLM Judge" module (measured using Cohen's κ coefficient, Spearman's correlation, and the RBO metric)?

2 Theoretical Background

This section provides a theoretical overview of key concepts and existing research relevant to the application and evaluation of large language models (LLMs) in legal analysis, with a particular focus on tax law. It discusses the capabilities of LLMs in the legal domain, highlights the unique challenges posed by tax law texts, reviews previous work on evaluating LLM performance in legal tasks, and introduces the concept of LLM-based evaluation methods.

2.1 Large Language Models in the Legal Domain

Large language models (LLMs), built on transformer architectures, have shown remarkable capabilities in various natural language processing tasks and are increasingly applied in professional domains, including law [6, 7]. While legal tasks, characterized by extensive documentation and complex reasoning, present opportunities, their deployment in law requires careful consideration due to the domain's specific demands for high accuracy, reliability, and explainability. Critically, errors known as "hallucinations" [8, 9], where models generate factually inconsistent or non-existent information – are particularly problematic in legal contexts, potentially leading to incorrect advice or severe consequences.

2.2 Specific Challenges of Tax Law for LLMs

Tax law is a particularly demanding legal sub-domain due to its complexity, stemming from its high dynamism and frequent legislative changes, detailed and specific regulations requiring precise interpretation, connection to numerical data, and reliance on administrative interpretations and case law (like PTRs and court rulings). Preliminary studies on LLM performance in tax analysis highlight these challenges; while LLMs can handle some hypothetical cases, they show a significant tendency to hallucinate non-existent provisions and do not yet achieve the precision and reliability required by professional tax advisors. However, techniques like few-shot prompting and RAG show promise in significantly improving their performance, suggesting potential for supportive applications in legal practice.

2.3 Evaluating LLM Performance in Legal Tasks

Research in this area focuses on human evaluation by legal experts as the gold standard for assessing LLM outputs in legal contexts [10]. Human evaluations verify factual correctness, legal reasoning, and overall quality from a practical perspective. However, human evaluation is resource-intensive and requires careful design to ensure consistency and minimize subjectivity. Developing robust, domain-specific criteria for nuanced legal analysis remains an ongoing challenge [11]. While benchmarks like LegalBench [12] exist, the specificity and dynamism of tax law often necessitate more tailored approaches.

2.4 LLM as a Judge and Automated Verification

The limitations of current large language models (LLMs – particularly their reliability issues and propensity to hallucinate – together with the challenges of human evaluation have spurred the search for alternative verification methods. A promising avenue is to employ LLMs themselves to assess or refine the outputs of other LLMs, a paradigm known as "LLM as a Judge" or AI-assisted evaluation [13], using LLMs to critique, score, or re-rank generated answers based on predefined criteria or source materials. Techniques such as "critique and rerank" are used [13, 14]. However, the reliability and validity of LLM-based evaluation in fields like tax law – where accuracy is paramount – must be rigorously validated against human-expert judgment to establish the credibility and practical utility of such automated tools.

Building on this theoretical foundation, our study sets out to (i) conduct an empirical assessment of the latest LLMs in tax-law analysis, (ii) apply domain-specific evaluation frameworks, and (iii) validate the potential of the LLM-as-a-Judge approach against human-expert appraisal, thereby addressing the gaps identified in current research.

3 Methodology

The study adopted an experimental methodology that was implemented in two main stages and that corresponded to the stated research objectives (C1–C3). The research design was crafted to assess the quality of tax law analyses generated by the latest LLM.

Stage 1: Quality assessment of LLM-generated tax interpretations (objective C1)

1. **LLMs:** For the study, four leading models with advanced reasoning and language processing capabilities were selected: Gemini 2.5 Pro (version from 25.03.2025), Gemini 2.5 Flash (version from 17.04.2025), GPT-4o, DeepSeek R1. The models were accessed via their official APIs, using default parameters. Prompts (instructions for the models) were uniform across all models and were carefully designed to clearly define each analytical task.
2. **Research Material:** We analysed a dataset of 100 private tax rulings (PTRs) issued by the National Revenue Information from 2022–2024 (Table 1) was used. PTRs have a special place in these sources of information. According to the IMF a PTR consists of advice that a taxpayer may seek from the tax authority in relation to the application of the tax law to their particular arrangement. PTRs are a valuable instrument for limiting or even removing tax risks of significant economic transactions. They expose primary reasoning of tax authorities and often form a basis of later interpretations. The selection was deliberate, focusing on interpretations concerning particularly complex, contentious, or frequently challenged tax law issues. A selection criterion was the confirmation that the interpretation was subject to a complaint before the Voivodship Administrative Courts (VAC), and where possible, also to a cassation complaint before the Supreme Administrative Court (SAC). Diversity of interpretations in terms of tax type and thematic area was ensured.

Table 1. Structure of research material by taxes.

Tax	Number of observations	%
Excise tax	10	10%
CIT	30	30%
PCC	10	10%
PIT	20	20%
VAT	30	30%
Suma	100	100%

Source: Own study

3. **Analytical Tasks:** For each of the 100 interpretations, each of the four LLM was tasked with performing eleven (11) predefined analytical tasks, reflecting key aspects of the legal document analysis:
   a) identification of all cited court rulings and individual interpretations in the taxpayer's position;
   b) identification of all cited court rulings and individual interpretations in the position of the National Revenue Information (NRI);
   c) compilation of a list of all legal provisions cited in the interpretation, including the full name of the legal act (statute/regulation) and editorial unit;
   d) identification and description of the main legal problem / issue to which the interpretation pertains;
   e) indication of key legal provisions fundamental for resolving the analyzed problem;
   f) listing a maximum of ten (10) most important terms / phrases from the interpretation that are statutory terms or are crucial for the analyzed problem and require precise interpretation in tax law;
   g) stating significant differences in argumentation and legal position presented by the taxpayer in the submitted request and by the tax authority (NRI);
   h) stating significant differences in the legal position of the tax authority (NRI) and the VAC, if the case was subject to court proceedings;
   i) formulating the position of the tax authority in the form of concise, synthetic legal statements (similar to court ruling theses);
   j) formulating the position of the VAC in the form of concise, synthetic legal statements, if the case was subject to court proceedings;
   k) identification and description of the methods (e.g., linguistic, systemic, teleological) and legal interpretation arguments applied by the tax authority in the justification of the interpretation.
4. **Evaluation of Analysis Results:** The quality of results generated by LLM for each of the 11 tasks (for each of the 100 interpretations) was evaluated by four experienced specialists in tax law (tax advisors or legal counsels with a minimum of five years of practice). The specialists were trained in the application of uniform evaluation criteria and the research protocol. The evaluation process was conducted "blind," meaning the evaluators did not know which model generated a given analysis. Each of the 100 interpretations (and the corresponding $4\text{x}11 = 44$ analyses generated by LLMs) was evaluated by at least one specialist, with work divided (each evaluated 25 full cases). In case of doubts or significant discrepancies among evaluators, a consensus procedure was applied.
5. **Evaluation Criteria (Mi) and Legal Quality Index (LQI):** For each of the 11 analytical tasks (M1-M11) and for each LLM, a specialist assigned an evaluation value according to precisely defined, domain-specific criteria:
   a) M1 - M3 (Extraction of Rulings/Provisions): Evaluation by proportion metric: (number of correctly identified elements) / (total number of actually present elements in the interpretation). Value in the range;
   b) M4, M5, M6, M7, M8, M9, M10 (Identification/Description/Synthesis): Evaluation on a 3-point scale: 1 - Correct and precise / Essence captured; 0.5 - Partially correct / Not fully precise / Essence partially captured; 0 - Incorrect / No response;

c) M11 (Interpretation Method): Evaluation on a 5-point scale, reflecting the complexity of the task: 1 - Correct and exhaustive; 0.75 - Correct, but not fully exhaustive; 0.5 - Partially correct; 0.25 - Minimal correctness; 0 - Incorrect.

Based on the Mi evaluations, a Proprietary Legal Quality Index (LQI) was calculated for each analysis generated by the LLM:

$$LQI = 0{,}08 \times (M_1 + M_2 + M_3 + M_6 + M_7 + M_8 + M_9 + M_{10}) + 0{,}12 \times (M_4 + M_5 + M_{11})$$

The weights (0.08 and 0.12) were determined based on expert evaluation of the relative difficulty and substantive importance of individual analytical tasks in practice, with M3, M4, and M10 considered more cognitively demanding and crucial for a correct legal assessment. The maximum LQI value is 1.

Stage 2: Evaluation and Verification of Analyses by LLM as a Judge (Realization of C2 & C3)

1. **LLM as a Judge Configuration:** The LLM (ChatGPT o3) was utilized. This model served as the "LLM Judge".
2. **Evaluation Process by LLM Judge:** The LLM Judge was programmed to evaluate each of the analyses generated by the models in Stage 1, based on the same analytical tasks (1–11) and attempting to apply a similar evaluation logic to the Mi/LQI indicators. The Judge evaluated the results of each of the four models for a given interpretation and assigned an LQI value based on this.
3. **Concordance Assessment (Realization of C3):** To assess the reliability of the LLM Judge, the LQI scores assigned to the analyses by human experts (Stage 1) and by the LLM Judge (Stage 2) were compared. The concordance of LQI scores was measured using Spearman's rank correlation coefficient. Additionally, the concordance of the quality rankings of LLM (from best to worst), established independently by experts (based on their LQI scores) and by the LLM Judge (based on its LQI scores), was evaluated using standard ranking concordance metrics: Rank-Biased Overlap (RBO), supplemented by Cohen's κ coefficient.
4. **Analysis Improvement Procedure:** For each case, the Judge analyzed the results of all four models and identified the "best" analysis according to its own LQI evaluation. Then, using the critique + rerank procedure, the Judge proposed modifications or improvements for this best analysis, aiming to enhance its precision, completeness, and consistency with the source interpretation. The outcome was an improved version of the analysis for each case.

4 Results

In the implementation of Stage 1, experts evaluated 4400 results of analyses performed by LLM (44 analyses per interpretation). The results of the expert assessment of the quality of legal analyses performed by the four LLMs were presented in Table 2.

Table 2. Results of expert assessment of the quality of legal analyses performed by four LLMs.

Criterion	ChatGPT 4o	DeepSeek R1	Gemini 2.5 Pro	Gemini 2.5.Flash	Average
1. Rulings cited by the taxpayer	78,50%	59,50%	99,00%	100,00%	84,25%
2. Rulings cited by NRI	75,50%	56,10%	99,00%	100,00%	82,65%
3. All provisions cited	71,55%	52,79%	92,89%	88,89%	76,53%
4. Legal issue/question addressed by the interpretation	95,00%	90,00%	96,00%	98,00%	94,75%
5. Key legal provisions	89,00%	83,00%	92,00%	92,00%	89,00%
6. 10 statutory terms	85,00%	71,00%	90,00%	90,00%	84,00%
7. Difference between tax authority and taxpayer	91,50%	83,00%	100,00%	100,00%	93,63%
8. Difference between tax authority and VAC	78,00%	57,00%	95,00%	95,00%	81,25%
9. Tax authority's holdings/theses	80,00%	73,00%	94,00%	92,00%	84,75%
10. VAC's holdings/theses	70,00%	53,00%	96,00%	95,00%	78,50%
11. Method of legal interpretation	72,50%	64,50%	84,00%	85,00%	76,50%
Legal Quality Index	81,18%	68,93%	93,91%	93,87%	84,47%

Source: Own study

The best evaluation results were obtained for the Gemini Flash / Gemini Pro models. Both variants of the Google model almost provided full lists of editorial units and accurately reproduced the procedural dispute (interpretation denial → WSA → NSA).

The weakest results were obtained for the DeepSeek R1 model. This model did not perform comprehensive analysis. However, the analyses performed by this model are useful when a brief outline is needed, but requires supplementation in terms of methodology and case law.

ChatGPT 4o was a model for which significantly worse results were obtained compared to Gemini models for both simple and complex analyses.

Among the types of analyses performed, the Gemini Flash model completed three of them flawlessly (analyses 1, 2, and 7).

In the case of analysis 3, LLMs indicated the most important provision, but omitted some of the provisions used in PTRs. However, this is the type of analysis for which correct results could be obtained with minor modifications to the research schemes.

Generally, the quality of results obtained by Gemini models indicates their high practical utility in performing both simple and complex tax law analyses.

In Stage 2 of the study, the OpenAI o3 model (OpenAI reasoning model – generation 3) was used as a Judge. The results of the quality assessment of legal analyses performed by the four LLMs, as conducted by OpenAI o3, are presented in Table 3.

Table 3. Results of the quality assessment of legal analyses performed by four LLMs, as conducted by OpenAI o3.

Criterion	ChatGPT 4o	DeepSeek R1	Gemini 2.5 Pro	Gemini 2.5.Flash	Average
1. Rulings cited by the taxpayer	72,50%	64,50%	89,00%	90,00%	79,00%
2. Rulings cited by NRI	70,00%	53,00%	99,00%	95,00%	79,25%
3. All provisions cited	80,00%	73,00%	100,00%	100,00%	88,25%
4. Legal issue/question addressed by the interpretation	78,00%	57,00%	99,00%	100,00%	83,50%
5. Key legal provisions	91,50%	83,00%	100,00%	100,00%	93,63%
6. 10 statutory terms	85,00%	71,00%	90,00%	90,00%	84,00%
7. Difference between tax authority and taxpayer	89,00%	83,00%	100,00%	100,00%	93,00%
8. Difference between tax authority and VAC	100,00%	90,00%	100,00%	100,00%	97,50%

(*continued*)

(continued)

Criterion	ChatGPT 4o	DeepSeek R1	Gemini 2.5 Pro	Gemini 2.5.Flash	Average
9. Tax authority's holdings/theses	71,55%	52,79%	92,89%	88,89%	76,53%
10. VAC's holdings/theses	75,50%	56,10%	99,00%	100,00%	82,65%
11. Method of legal interpretation	78,50%	59,50%	99,00%	100,00%	84,25%
Legal Quality Index	81,78%	68,93%	96,99%	96,71%	86,10%

Source: Own study

In Table 4, the LQI value obtained from expert evaluation and OpenAI o3 was compred.

Table 4. Comparison of LQI value obtained from expert evaluation and OpenAI o3.

Judge	ChatGPT 4o	DeepSeek R1	Gemini 2.5 Pro	Gemini 2.5.Flash
Eksert	0,8118	0,6893	0,9391	0,9387
OpenAI o3	0,8178	0,6893	0,9699	0,9671
Difference	-0,01	0,00	-0,03	-0,03
Difference in %	-0,74%	0,00%	-3,28%	-3,03%

Source: Own study

The differences in the evaluation made by the expert and by OpenAI o3 were very small. The ranking of LLM quality was the same in both cases: the best model was Gemini 2.5 Flash, followed by Gemini 2.5 Pro (slightly worse evaluation), ChatGPT 4o, and finally DeepSeek R1.

In the following step, the concordance of the obtained results was evaluated using: Spearman's rank correlation coefficient, Rank-Biased Overlap (RBO), Cohen's κ coefficient.

The value of Spearman's rank correlation coefficient for evaluations made by the experts and by OpenAI o3 was presented in Table 5.

Table 5. Value of Spearman's rank correlation coefficient.

Model	Wartość wskaźnika
ChatGPT 4o	0,722

(continued)

(*continued*)

Model	Wartość wskaźnika
DeepSeek R1	0,791
Gemini 2.5 Pro Preview	0,943
Gemini 2.5.Flash Preview	0,939
Razem	0,861

Source: Own study

The values of Spearman's rank correlation coefficient were positive and were close to + 1. This indicated a strong, positive monotonic relationship between the evaluated scores.

Rank-Biased Overlap (RBO) was a statistical measure used to assess the similarity or concordance between two rankings (ordered lists). It was a measure frequently used in fields such as information retrieval, recommender systems, and expert ranking analysis.

The RBO coefficient took values from 0 to 1, where: 1 meant ideal agreement between rankings (elements appeared in the same order) and 0 meant no agreement (agreement at a random level).

RBO had a parameter p (where $0 \leq p < 1$) that controlled how quickly the weight of agreement decreased down the ranking. The closer p was to 1 (e.g., 0.98, 0.99), the slower the weight decreased, meaning that agreement on lower positions still had a significant impact on the result. The closer p was to 0 (e.g., 0.8, 0.9), the faster the weight decreased, meaning that the RBO result was almost entirely determined by agreement on the very top positions of the ranking.

The following RBO index results were obtained:

RBO ($p = 0.95$): ≈ 0.9098;

RBO ($p = 0.80$): ≈ 0.9600;

RBO ($p = 0.50$): ≈ 0.9961;

RBO ($p = 0.24$): ≈ 0.9990.

Obtaining very high RBO values (above 0.9, and even above 0.99) for low p values (like 0.50 or 0.24) strongly confirmed the assumption of very high concordance in evaluation at the key, highest positions (top 1 and top 2), as well as high concordance at positions 3 and 4.

Higher RBO values for lower p (0.50, 0.24) compared to higher p (0.95) in this specific scenario (where disagreement only appeared at later positions) were direct evidence that discrepancies in rankings occurred mainly at lower positions, while the top positions were evaluated very consistently.

The final stage of assessing the concordance of the obtained results was the determination of Cohen's κ. For the analyzed example, Cohen's κ coefficient was 0.88 which meant that a very high / almost perfect concordance of evaluations was obtained.

Thus, a result of $\kappa = 0.88$ indicated a very high concordance between the ranking of LLM prepared by the model (or resulting from its operation) and the ranking prepared by the expert, taking into account the concordance in assigning models to specific rank categories (1st, 2nd, 3rd place, etc.), corrected for chance agreement.

Summarizing the obtained results of the concordance assessment using three different methods, the following conclusions could be drawn:

1) a high value of Spearman's rank correlation coefficient and also Cohen's κ indicated that both LLMs and experts largely agreed on which models deserved specific positions in the quality ranking (e.g., they agreed on which model was best, which was second, which was third, etc.);
2) this agreement was significantly higher than would be expected if rank assignment had been completely random;
3) high RBO values, especially for low p, indicated very strong concordance in the order of elements at the top of the ranking and strong concordance at subsequent positions.

At the conclusion of Stage II, the OpenAI o3 model, based on its evaluation of the results of the analyses of the four LLMs, proposed a new version of each of the 11 analyses based on the analysis performed by the highest-rated model for a given PTR. These generated analyses were then resubmitted for expert evaluation. The evaluation results were presented in Table 6.

Table 6. Results of expert assessment of the quality of legal analyses performed by the improved LLM.

Criterion	Gemini 2.5.Flash
1. Rulings cited by the taxpayer	100,00%
2. Rulings cited by NRI	100,00%
3. All provisions cited	90,00%
4. Legal issue/question addressed by the interpretation	100,00%
5. Key legal provisions	100,00%
6. 10 statutory terms	90,00%
7. Difference between tax authority and taxpayer	100,00%
8. Difference between tax authority and VAC	100,00%
9. Tax authority's holdings/theses	100,00%
10. VAC's holdings/theses	100,00%
11. Method of legal interpretation	90,00%
Legal Quality Index	**97,20%**

Source: Own study

The use of OpenAI o3 as a Judge allowed for the improvement of the quality of legal analysis results, measured by LQI by 3.33% (from 93.87% to 97.20%), and for obtaining a 100% evaluation score for 8 out of 11 conducted analyses.

The quality of the conducted analyses indicated that the proposed research approach allowed for the development of an operational scheme enabling full automation of the proposed types of analyses, which potentially meant significant cost savings in law firm operations.

5 Discussion and Conclusion

This study aimed to comprehensively evaluate the capabilities of the latest LLM in tax law analysis and assess the potential of the "LLM as a Judge" concept as a tool supporting quality control, addressing a research gap in systematic evaluation on real Polish private tax rulings with multidimensional criteria.

In Stage 1, four experts evaluated the quality of analyses by four leading LLMs (Gemini 2.5 Pro/Flash, GPT-4o, DeepSeek R1) on 100 Polish PTRs using LQI. Expert results showed varying performance, with Gemini models achieving the highest LQI and demonstrating high potential, especially in key element extraction.

Stage 2 validated "LLM as a Judge". High concordance (Spearman 0.861, RBO > 0.9 for low p, $\kappa = 0.88$) was found between expert and "LLM Judge" LQI scores and rankings, confirming the reliability of automated evaluation, particularly for identifying the best and second-best models. The "critique + rerank" procedure also improved analysis quality, suggesting LLM tools can enhance content.

The high concordance indicates LLM-based tools can serve as reliable quality control support for tax law analyses, optimizing resource allocation and potentially saving costs in law firms by assisting or automating initial evaluations and verifications.

Despite these results, further research is needed. Future work should explore agent-based approaches, utilizing domain-specific legal knowledge bases to reduce "hallucinations," and analyzing the real impact on task completion time in legal practice. Expanding the research scope to broader tasks and LLM types is also important.

In summary, this study provides initial positive evidence for the potential of the latest LLMs in Polish tax law analysis. Crucially, it validates "LLM as a Judge" as a promising, reliable automated quality control tool essential for the safe and effective implementation of AI-based LegalTech in legal practice.

Acknowledgments. This study was co-financed by the Minister of Science under the "Regional Excellence Initiative".

Regionalna Inicjatywa Doskonałości

Ministry of Science and Higher Education Republic of Poland

Disclosure of Interests. The authors have no competing interests to declare that are relevant to the content of this article.

References

1. Choi, J.H., Ahn, J., Nay, J., Bommarito, M.J.II: Large language models as tax attorneys. Philos. Trans. Roy. Soc. A **381**(2270) (2023). https://doi.org/10.1098/rsta.2023.0154
2. Zhou, R., Xu, A., Liao, J., Chen, S., Wang, B.: Rule Arena: a benchmark for rule-guided reasoning. arXiv Preprint: arXiv:2412.08972 (2024)
3. Daniel, K., Bommarito, M.J. II, Hartung, D., Katz, D.M.: LegalBench: a collaboratively built benchmark for legal reasoning. arXiv Preprint: arXiv:2308.16216 (2023)
4. Susskind, R.: Tomorrow's Lawyers: An Introduction to Your Future, 2nd edn. Oxford University Press, Oxford (2013)

5. PwC. Law firms' survey 2024. PricewaterhouseCoopers (2024)
6. Katz, D.M., Hartung, D., Gerlach, L., Jana, A., Bommarito, M.J.II: Natural language processing in the legal domain. arXiv Preprint: arXiv:2302.12039 (2023)
7. Medvedeva, M., McBride, P.: Legal judgment prediction: if you are going to do it, do it right. In: Proceedings of the Natural Legal Language Processing Workshop, pp. 73–84 (2023). Association for Computational Linguistics. https://doi.org/10.18653/v1/2023.nllp-1.9
8. Magesh, V., Dahl, M., Suzgun, M., Ho, D.E.: Hallucination-free? Assessing the reliability of leading AI legal research tools. arXiv Preprint: arXiv:2405.20362 (2024)
9. Dahl, M., Magesh, V., Suzgun, M., Ho, D.E.: Large legal fictions: profiling legal hallucinations in large language models. J. Legal Anal. **16**(1), 64–93 (2024). https://doi.org/10.1093/jla/laae003
10. Nay, J., Ahn, J., Choi, J.H., Bommarito, M.J.II: Large language models as tax attorneys: a case study in legal capabilities emergence. Philos. Trans. Roy. Soc. A **382**(2270) (2024). https://doi.org/10.1098/rsta.2024.0142
11. Pipitone, N., Houir Alami, G.: LegalBench-RAG: a benchmark for retrieval-augmented generation in the legal domain. arXiv Preprint: arXiv:2408.10343 (2024)
12. Guha, N., Bommarito, M.J.II, Katz, D.M., Hartung, D.: LegalBench: a collaboratively built benchmark for measuring legal reasoning in large language models. arXiv Preprint: arXiv:2308.11462 (2023)
13. Akarajaradwong, P., Pothavorn, P., Chaksangchaichot, C., Tasawong, P., Nopparatbundit, T., Nutanong, S.: NitiBench: a comprehensive study of LLM framework capabilities for Thai legal question answering. arXiv Preprint: arXiv:2502.10868 (2025)
14. Kannappan, G.: LLM-as-a-judge: crafting the model and navigating its challenges. Medium (2024). https://medium.com/@ganeshkannappan/llm-as-a-judge-crafting-the-model-and-navigating-its-challenges-6dbc21029764

Multidimensional Intuitionistic Fuzzy-Based Framework for Modeling Artificial Intelligence Acceptance in Services

Bartłomiej Jefmański[1], Elwira Gross-Gołacka[2](✉), Marta Kusterka-Jefmańska[3], and Iwona Chomiak-Orsa[3]

[1] Wroclaw University of Economics and Business, Wrocław, Poland
bartlomiej.jefmanski@ue.wroc.pl
[2] University of Warsaw, Warszawa, Poland
egross@wz.uw.edu.pl
[3] Wroclaw University of Economics and Business, Wrocław, Poland

Abstract. The paper presents a multidimensional approach to modeling artificial intelligence (AI) acceptance in services based on intuitionistic fuzzy sets, integrating the Artificially Intelligent Device Use Acceptance model (AIDUA) with the Intuitionistic Fuzzy Synthetic Measure (IFSM). The proposed methodology consists of three main stages: transformation of responses into intuitionistic fuzzy sets, aggregation of evaluations within attitudinal dimensions, and calculation of a synthetic AI acceptance index. The method supports more flexible and accurate modeling of user attitudes toward AI technologies in the context of service acquisition. It can be applied in academic research and the practical management of AI implementations.

Keywords: AI acceptance in services · artificial intelligence · Gen Z · AIDUA · Intuitionistic Fuzzy Synthetic Measure · intuitionistic fuzzy sets · fuzzy measurement

1 Introduction

AI is increasingly important in service delivery, with applications including chatbots, self-service kiosks, recommendation systems, and humanoid robots supporting customer service. The growing presence of AI technologies across various sectors, such as retail, tourism, healthcare, and public administration, requires a deeper understanding of user attitudes toward their use. The level of user acceptance is one of the key factors influencing the successful implementation and continued use of these technologies. This need to assess acceptance levels has prompted researchers and practitioners to seek reliable tools for evaluating user attitudes toward AI-based solutions (Xiao and Benbasat, 2007; Wirtz et al., 2018; Belanche et al., 2020).

AI acceptance is shaped by cognitive evaluations and emotional responses, as highlighted in the AIDUA framework, which emphasizes the sequential role of attitudes and

M. Pondel et al. (Eds.): AI4KAM 2025, IFIP AICT 776, pp. 449–460, 2026.
https://doi.org/10.1007/978-3-032-18920-2_43

emotions in shaping willingness to adopt AI (Gursoy et al., 2019). Studies show that factors such as trust, perceived usefulness, and ease of use significantly influence acceptance across sectors, including banking, e-commerce, and education (Belanche et al., 2020; Turnadžić et al., 2024; Ionescu & Ardelean, 2024; Runge et al., 2025). Moreover, demographic variables like age and digital familiarity play critical roles, with younger generations like Generation Z showing higher openness to AI-driven interactions (Gerlich, 2023; Abed, 2024; Gross-Gołacka et al., 2025). These multidimensional factors underscore the need for robust, context-sensitive frameworks, such as the proposed multidimensional intuitionistic fuzzy-based framework, to model and predict AI acceptance in service settings comprehensively.

In research on technology acceptance, classical models such as the Technology Acceptance Model (TAM) (Davis et al., 1989) and the Unified Theory of Acceptance and Use of Technology (UTAUT) (Venkatesh et al., 2012) have played a dominant role. While these approaches have been valuable in explaining user behavior toward functional technologies, they are insufficient for analyzing systems based on artificial intelligence. As Gursoy et al. (2019) noted, traditional models fail to account for the multi-dimensional nature of interactions with intelligent devices, which increasingly serve social, affective, and relational functions. AI supports users in task completion, simulates human behaviors, and evokes emotional responses that constructs such as "ease of use" or "perceived usefulness" cannot fully capture.

Gursoy et al. (2019) proposed the AIDUA model, grounded in Cognitive Appraisal Theory, to address these limitations. The model assumes a three-stage process: primary appraisal, secondary appraisal, and emotional response, leading to a decision to accept or reject AI. Thus, AIDUA encompasses cognitive factors (e.g., performance expectancy), emotional dimensions (e.g., fear, satisfaction), and harmful components such as objection to technology (see Appendix).

Given the multi-dimensional nature of user attitudes captured in the AIDUA model and the presence of uncertainty and indecision in respondents' evaluations, this article proposes a hybrid methodology that integrates the AIDUA model with IFSM. This approach enables a synthetic assessment of AI acceptance levels by aggregating fuzzy evaluations across attitude dimensions while accounting for incomplete or imprecise data. It allows for more flexible and accurate modeling of complex user attitudes toward AI technologies in service contexts. An intuitionistic fuzzy conversion scale was developed and applied to transform ordinal measurement scale responses into intuitionistic fuzzy sets.

2 Methodology

The proposed methodology is based on the AIDUA model introduced by Gursoy et al. (2019), as well as the IFSM as characterized in the studies by Jefmański (2020) and Jefmański, Roszkowska, and Kusterka-Jefmańska (2021). The AIDUA model explains the three-stage process of people's acceptance or objection to using AI devices in services and includes a set of eight dimensions (social influence, hedonic motivation, anthropomorphism, performance expectancy, effort expectancy, emotion, willingness to accept the use of AI devices, and objection to the use of AI devices). In the first stage, consumers assess the appropriateness of AI based on social influence (Jeon et al., 2028), hedonic motivation (Fryer et al., 2017), and anthropomorphism (Alsaad, 2023), forming initial attitudes. In the second stage, they conduct a more deliberate, context-specific evaluation focused on performance (Bhuiyan et al., 2024) and effort (Lu et al., 2019) expectancy, which generates emotions (West et al., 2018). In the final stage, these emotions determine people's willingness (Gursoy 2019) or objection to accept AI in service interactions. The AIDUA includes a set of eight dimensions related to the acceptance of AI device usage in service delivery: social influence, hedonic motivation, anthropomorphism, performance expectancy, effort expectancy, emotion, willingness to accept the use of AI devices, and objection to the use of AI devices. These dimensions are measured using 34 statements. Items used to assess emotions were measured on a 5-point ordinal scale (e.g., 1 – unsatisfied, 5 – satisfied). Items to assess the remaining constructs were measured on a 5-point Likert scale (1 – strongly disagree, 5 – strongly agree).

The IFSM is a method for measuring complex phenomena in situations where the evaluations of criteria/variables are expressed as intuitionistic fuzzy sets (IFS). IFS, proposed by Atanassov (1986), represents an extension of classical fuzzy set theory. This extension takes into account not only the degree of membership μ, but also the degree of non-membership ν and the degree of uncertainty π. Such an approach allows for a better modeling of uncertainty and ambiguity in evaluations, which is particularly useful in social research, opinion analysis, decision-making processes, and expert systems.

The proposed method for evaluating attitudes toward AI consists of three main stages:

1. Transformation of individual responses into intuitionistic fuzzy sets using the intuitionistic fuzzy conversion scale,
2. Averaging respondents' opinions at the level of individual dimensions,
3. Assessment of respondents' attitudes toward artificial intelligence using the IFSM.

The general framework of the proposed research procedure is presented in Fig. 1, followed by a detailed description of its individual stages and steps.

Stage 1. Transformation of individual responses into intuitionistic fuzzy sets using the intuitionistic fuzzy conversion scale

Let R_i $i = 1, 2, \ldots, n$, denote the i-th respondent, and let A_{jl} where $j = 1, 2, \ldots, m$ and $l = 1, 2, \ldots, k_j$ represent the l-th item of the j-th attitudinal dimension. Each item is assessed using a five-point ordinal scale. Responses are transformed into intuitionistic fuzzy sets $IFS_{ijl} = (\mu_{ijl}, \nu_{ijl}, \pi_{ijl})$ using the intuitionistic fuzzy conversion scale presented in Table 1.

It should be noted that missing responses are also encoded as intuitionistic fuzzy sets, which allows them to be included in further analyses. For reverse-coded questions/items, the intuitionistic fuzzy conversion scale presented in Table 2 should be applied.

Stage 2. Averaging respondents' opinions at the level of individual dimensions

For each respondent i and dimension D_j the average intuitionistic fuzzy evaluation is computed across all k_j items:

$$\mu_{ij} = \frac{1}{k_j} \sum_{l=1}^{k_j} \mu_{ijl}, \tag{1}$$

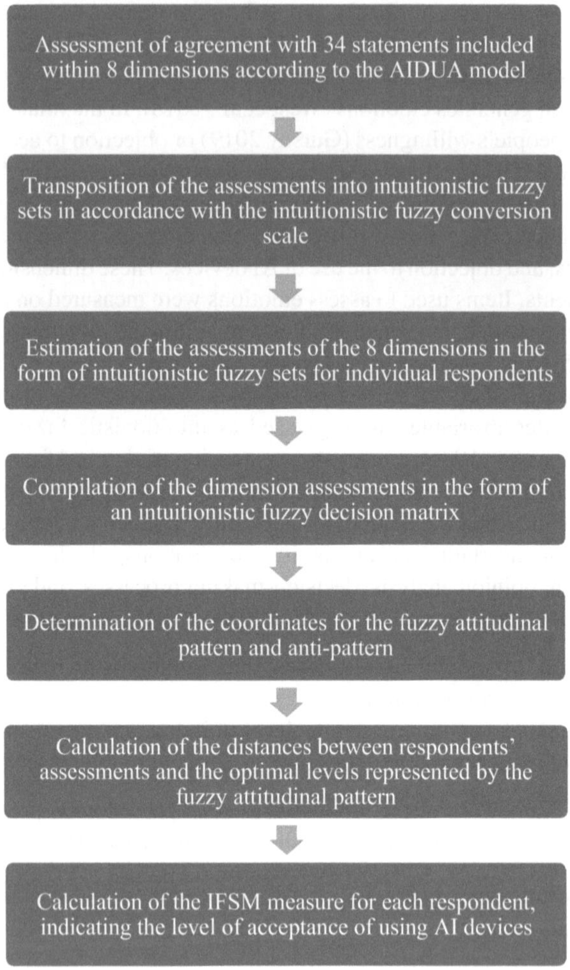

Fig. 1. Flow chart of the research procedure

$$\nu_{ij} = \frac{1}{k_j}\sum_{l=1}^{k_j} \mu_{ijl}, \tag{2}$$

$$\pi_{ij} = \frac{1}{k_j}\sum_{l=1}^{k_j} \mu_{ijl}. \tag{3}$$

This results in m evaluations per respondent, each in the form $IFS_{ij} = \left(\mu_{ij}, \nu_{ij}, \pi_{ij}\right)$.

Stage 3. Assessment of respondents' attitudes toward artificial intelligence using the $IFSM$, divided into four steps

Step 1. Determination of the decision matrix

Table 1. Intuitionistic fuzzy conversion scale

Response	μ	ν	π
1 – Strongly disagree	0.05	0.95	0.00
2 – Disagree	0.25	0.65	0.10
3 – Neither agree nor disagree	0.40	0.40	0.20
4 – Agree	0.65	0.25	0.10
5 – Strongly agree	0.95	0.05	0.00
Non-response	0.00	0.00	1.00

Table 2. Intuitionistic fuzzy conversion scale for reverse-coded questions

Response	μ	ν	π
1 – Strongly disagree	0.95	0.05	0.00
2 – Disagree	0.65	0.25	0.10
3 – Neither agree nor disagree	0.40	0.40	0.20
4 – Agree	0.25	0.65	0.10
5 – Strongly agree	0.05	0.95	0.00
Non-response	0.00	0.00	1.00

The results of measuring m attitudinal dimensions for each of n respondents, represented as $IFS_{ij} = \left(\mu_{ij}, \nu_{ij}, \pi_{ij}\right)$, should be recorded in the form of the following decision matrix:

$$\boldsymbol{D} = \begin{bmatrix} (\mu_{11}, \nu_{11}) & (\mu_{12}, \nu_{12}) & \ldots & (\mu_{1m}, \nu_{1m}) \\ (\mu_{21}, \nu_{21}) & (\mu_{22}, \nu_{22}) & \ldots & (\mu_{2m}, \nu_{2m}) \\ \ldots & \ldots & \ldots & \ldots \\ (\mu_{n1}, \nu_{n1}) & (\mu_{n1}, \nu_{n1}) & \ldots & (\mu_{nm}, \nu_{nm}) \end{bmatrix} \tag{4}$$

Step 2a. Determination of the intuitionistic fuzzy attitudinal pattern.

The coordinates of the intuitionistic fuzzy attitudinal pattern, denoted as P^{+}, are determined as follows:

$$P^{+} = [(1, 0), \ldots, (1, 0)] \tag{5}$$

Step 2b. Determination of the intuitionistic fuzzy attitudinal anti-pattern.

The coordinates of the intuitionistic fuzzy attitudinal anti-pattern, denoted as P^{-}, are determined as follows:

$$P^{-} = [(0, 1), \ldots, (0, 1)] \tag{6}$$

Step 3. Determination of the distance between the respondent's intuitionistic fuzzy evaluation of the attitudinal dimension and the intuitionistic fuzzy attitudinal pattern

The distance between the intuitionistic fuzzy evaluation of the attitudinal dimension by the $i - th$ respondent and the intuitionistic fuzzy attitudinal pattern P^{+} is determined according to the following formula:

$$d\left(P^{+}, P_i\right) = \sqrt{\frac{1}{2m}\sum_{j=1}^{m}\left[\left(\mu_{Pj^{+}} - \mu_{P_{ij}}\right)^2 + \left(\nu_{Pj^{+}} - \nu_{P_{ij}}\right)^2 + \left(\pi_{Pj^{+}} - \pi_{P_{ij}}\right)^2\right]} \quad (7)$$

Step 4. Calculation of the IFSM

The $IFSM$ value for the i-th respondent is calculated according to the formula:

$$IFSM_i = 1 - \frac{d\left(P^{+}, P_i\right)}{d\left(P^{+}, P^{-}\right)}, \quad (8)$$

where $d\left(P^{+}, P^{-}\right)$ is a distance between intuitionistic fuzzy attitudinal pattern and intuitionistic fuzzy attitudinal anti-pattern according to the formula:

$$d\left(P^{+}, P^{-}\right) = \sqrt{\frac{1}{2m}\sum_{j=1}^{m}\left[(\mu_{P^{+}} - \mu_{P^{-}})^2 + (\nu_{P^{+}} - \nu_{P^{-}})^2 + (\pi_{P^{+}} - \pi_{P^{-}})^2\right]} \quad (9)$$

The $IFSM$ values are normalized within the [0–1] range, with higher values indicating a higher level of acceptance of using AI devices in service provision among respondents.

3 Results

The proposed methodology was applied in a study conducted among Generation Z representatives. Generation Z refers to individuals born from 1994 onward, known for growing up fully immersed in digital technology (Al-Sharafi et al., 2023). This group was chosen for the study because they strongly influence the adoption of new technologies and define emerging consumer trends (Twenge, 2017; Ruiz-Viñals, 2024). The survey was conducted using the CAWI method (Computer-Assisted Web Interviewing) between May 1 and May 31, 2023. The research sample consisted of 548 respondents and was selected purposively, including students participating in courses taught by the authors of this paper. A prerequisite for participation in the study was a declaration of having used AI-based devices to acquire services.

According to Stage 2 of the proposed procedure, the evaluations expressed in the form of intuitionistic fuzzy sets were averaged for each dimension. The average evaluations for the first dimension and the first 10 respondents are presented in Table 3.

A fragment of the decision matrix (4), the determination of which constitutes the first step in calculating the IFSM measure for each respondent, is as follows:

$$\boldsymbol{D} = \begin{bmatrix} (0.325, 0.558) & (0.390, 0.470) & \ldots & (0.338, 0.563) \\ (0.567, 0.300) & (0.650, 0.250) & \ldots & (0.950, 0.050) \\ \ldots & & \ldots & \ldots \\ (0.200, 0.717) & (0.600, 0.280) & \ldots & (0.050, 0.950) \end{bmatrix} \quad (10)$$

Table 3. Average evaluations of the "Social influence" dimension

Respondent	μ	ν	π
1	0.325	0.558	0.117
2	0.050	0.950	0.000
3	0.567	0.300	0.133
4	0.442	0.375	0.183
5	0.950	0.050	0.000
6	0.083	0.900	0.017
7	0.333	0.550	0.117
8	0.050	0.950	0.000
9	0.083	0.900	0.017
10	0.583	0.283	0.133

Subsequently, the distances between each respondent's evaluations and the intuitionistic fuzzy attitudinal pattern were calculated, which then allowed the IFSM values to be determined. The distances and IFSM values for the first 10 respondents are presented in Table 4.

Table 4. Distances and IFSM values

Respondent	$d(P^+, P_i)$	IFSM
1	0.593	0.407
2	0.823	0.177
3	0.477	0.523
4	0.609	0.391
5	0.220	0.780
6	0.945	0.055
7	0.598	0.402
8	0.818	0.182
9	0.854	0.146
10	0.632	0.368

The average level of acceptance of AI device use in service delivery in the research sample was 0.441. The IFSM values for respondents with the lowest and highest acceptance levels were 0.055 and 0.809, respectively. Table 5 presents the measurement results divided into five acceptance level classes.

A high level of AI acceptance was reported by only 5.292% of respondents, and a very high level by just 0.182%. The distribution of results indicates limited trust and moderate

Table 5. Distribution of respondents by acceptance level of AI device use

Level of acceptance	[%]
very low [0–0.2)	3.467
low [0.2–0.4)	27.372
medium [0.4–0.6)	63.686
high [0.6–0.8)	5.292
very high [0.8–1]	0.182

acceptance of AI technologies among the respondents, with a clear concentration of attitudes around average values. A more detailed perspective on the results is provided by the summary of average evaluations of individual dimensions, including uncertainty in their measurement (see Table 6).

Table 6. Average intuitionistic fuzzy values for AIDUA model dimensions related to the use of AI devices

Dimension	μ	ν	π
Social influence	0.407	0.486	0.107
Hedonic motivation	0.510	0.381	0.109
Anthropomorphism	0.255	0.666	0.078
Performance expectancy	0.479	0.404	0.117
Effort expectancy	0.528	0.358	0.113
Emotion	0.456	0.413	0.131
Willingness to accept the use of AI Devices	0.533	0.357	0.110
Objection to the use of AI devices	0.316	0.580	0.104

The analysis of average intuitionistic fuzzy values for the individual AIDUA model dimensions reveals varied respondent attitudes toward using AI devices in service contexts. The highest degree of membership ($\mu = 0.533$) was recorded for the dimension "willingness to accept the use of AI devices", indicating the most significant level of declared openness toward AI adoption. A similarly high value was observed for "effort expectancy" ($\mu = 0.528$), suggesting that users do not perceive AI technologies as difficult to use.

In contrast, the lowest level of acceptance was associated with the "anthropomorphism dimension" ($\mu = 0.255$, $\nu = 0.666$), indicating an evident reluctance among respondents to attribute human-like qualities to AI devices. The dimension "objection to the use of AI devices" also received a relatively low membership value ($\mu = 0.316$), although with a lower degree of non-membership than "anthropomorphism".

The uncertainty values (π) were relatively consistent across all dimensions, not exceeding 0.131. The highest level of uncertainty was observed in the "emotion dimension" ($\pi = 0.131$), which may reflect the difficulty respondents experience in clearly identifying their emotional reactions to AI.

4 Conclusions

The article presents a multidimensional approach to assessing the acceptance of artificial intelligence-based devices in service contexts, integrating the AIDUA model with the Intuitionistic Fuzzy Synthetic Measure. The proposed framework enables more flexible and precise modeling of user attitudes toward AI by simultaneously accounting for cognitive and emotional components and the uncertainty present in user evaluations.

The analysis results indicate a moderate acceptance of AI use in service delivery among the surveyed respondents, with a clear dominance of attitudes of medium intensity. The highest scores were observed for willingness to accept AI and effort expectancy. In contrast, the lowest acceptance levels were associated with anthropomorphism and objection to AI use, which may reflect concerns about excessive human-likeness in AI technologies and the resulting social or ethical implications. The flexibility and universality of the proposed approach are also evident in its potential application to comparative studies, for example, across different service sectors or user groups.

This is no coincidence–the human relationship with AI has ceased to be purely functional and has entered the realm of emotions, trust, and social symbolism. Models such as AIDUA, combined with IFSM, thus prove to be analytical tools and mirrors reflecting young people's contemporary attitude toward technology: hopeful yet ambivalent, enthusiastic yet cautious.

By incorporating ambiguity and indecision in respondent assessments, the proposed method provides a significant methodological contribution to technology acceptance research. It can serve as a valuable tool both in academic studies and in the practical management of AI implementation in service environments.

Appendix

Construct	Item NO	Items
Social influence	SI1	Using AI devices reflects status symbol in my social networks (e.g., friends, family and co-workers)
	SI2	People who influence my behavior would want me to utilize AI devices
	SI3	People in my social networks who would utilize AI devices have more prestige than those who don't
	SI4	People whose opinions that I value would prefer that I utilize AI devices

(continued)

(*continued*)

Construct	Item NO	Items
	SI5	People who are important to me would encourage me to utilize it
	SI6	People in my social networks who would utilize artificially intelligent have a high profile
Hedonic motivation	HM1	I have fun interacting with AI
	HM2	Interacting with AI devices is fun
	HM3	Interacting with AI devices is entertaining
	HM4	Interaction with AI is enjoyable
	HM5	The actual process of interacting would be pleasant
Anthropomorphism	A1	AI devices have a mind of their own
	A2	AI devices have consciousness
	A3	AI devices have their own free will
	A4	AI devices will experience emotions
Performance Expectancy	PE1	AI devices are more accurate than human beings
	PE2	AI devices are more accurate with less human errors
	PE3	AI devices provide more consistent service than human beings
	PE4	Information provided by AI devices are more consistent
Effort Expectancy	EE1	Using AI devices takes too much of my time
	EE2	Working with AI devices is so difficult to understand and use in services
	EE3	It takes me too long to learn how to interact with AI devices
Emotion	E1	Bored-relaxed
	E2	Malancholic-contented
	E3	Despairing-hopeful
	E4	Unsatisfied-satisfied
	E5	Annoyed-pleased
Willingness to Accept the Use of AI Devices	I1	I am willing to receive AI device services
	I2	I will feel happy to interact with AI devices

(*continued*)

(*continued*)

Construct	Item NO	Items
	I3	I am likely to interact with AI devices
Objection to the Use of AI Devices	O1	The information is processed in a less humanized manner
	O2	I prefer human contact in service transactions
	O3	People need emotion exchange during service transactions
	O4	Interaction with AI devices lacks social contact

References

Abed, S.: Acceptance and use of artificial intelligence in online tourism services by generation Z in Saudi Arabia. IEEE Access. 1–1.10 (2024)

Alsaad, A.: The dual effect of anthropomorphism on customers' decisions to use artificial intelligence devices in hotel services. J. Hosp. Market. Manag. **32**(8), 1048–1076 (2023)

Al-Sharafi, M.A., et al.: Generation Z use of artificial intelligence products and its impact on environmental sustainability: a cross-cultural comparison. Comput. Hum. Behav. **143**, 107708 (2023)

Atanassov, K.T.: Intuitionistic fuzzy sets. Fuzzy Sets Syst. **20**, 87–96 (1986)

Belanche, D., Casaló Ariño, L., Flavian, C., Schepers, J.: Service robot implementation: a theoretical framework and research agenda. Serv. Ind. J. **40**, 203–225 (2020)

Bhuiyan, K.H., Ahmed, S., Jahan, I.: Consumer attitude toward using artificial intelligence (AI) devices in hospitality services. J. Hospitality Tourism Insights (2024)

Davis, F.D., Bagozzi, R.P., Warshaw, P.R.: User acceptance of computer technology: a comparison of two theoretical models. Manage. Sci. **35**(8), 982–1003 (1989)

Fryer, L., Ainley, M., Thompson, A., Gibson, A., Sherlock, Z.: Stimulating and sustaining interest in a language course: an experimental comparison of Chatbot and Human task partners. Comput. Hum. Behav. **75**, 461–468 (2017)

Gerlich, M.: Perceptions and acceptance of artificial intelligence: a multi-dimensional study. Soc. Sci. **12**(9), 502 (2023)

Gross-Gołacka, E., Szkudlarek, E.: Strategic Employer Branding for IT Companies. Routledge, An Interdisciplinary Perspective (2024)

Gross-Gołacka, E., Kavoura, A., Pawłowski, K.: Perception of artificial intelligence by generation Z—Results of a Pilot Study in Poland. In: Kavoura, A., Briciu, VA., Briciu, A. (eds.) Strategic Innovative Marketing and Tourism. ICSIMAT 2024. Springer Proceedings in Business and Economics. Springer, Cham (2025). https://doi.org/10.1007/978-3-031-81962-9_92

Gursoy, D., Chi, O.H., Lu, L., Nunkoo, R.: Consumers acceptance of artificially intelligent (AI) device use in service delivery. Int. J. Inf. Manage. **49**, 157–169 (2019)

Ionescu, I.-M., Ardelean, A.: The Romanian consumer's perspective on the integration of artificial intelligence in e-commerce. Rom. Stat. Rev. **4**, 78–92 (2024)

Jefmański, B.: Intuitionistic fuzzy synthetic measure for ordinal data. In: Jajuga, K., Batóg, J., Walesiak, M. (eds.) Classification and Data Analysis: Theory and Applications. Studies in Classification, Data Analysis, and Knowledge Organization, pp. 53–72. Springer, Cham (2020)

Jefmański, B., Roszkowska, E., Kusterka-Jefmańska, M.: Intuitionistic fuzzy synthetic measure on the basis of survey responses and aggregated ordinal data. Entropy **23**(12) (2021)
Jeon, M.M., Lee, S., Jeongm, M.: E-Social influence and customers' behavioral intentions on a bed and breakfast website. J. Hosp. Market. Manag. **27**(3), 366–385 (2028)
Lu, L., Cai, R., Gursoy, D.: Developing and validating a service robot integration willingness scale. Int. J. Hosp. Manag. **80**, 36–51 (2019)
Ruiz-Viñals, C., Pretel-Jiménez, M., Del Olmo Arriaga, J.L., Miró Pérez, A.: The influence of artificial intelligence on generation Z's online fashion purchase intention. J. Theor. Appl. Electron. Commer. Res. **19**(4), 2813–2827 (2024)
Turnadžić, T., Peštek, A., Činjarević, M.: The role of social factors in the acceptance of artificial intelligence-based services: the example of the banking sector of Bosnia and Herzegovina. South East Eur. J. Econ. Bus. **19**(1), 145–158 (2024)
Twenge, J.M. iGen: Why Today's Super-Connected Kids Are Growing Up Less Rebellious, More Tolerant, Less Happy—and Completely Unprepared for Adulthood—and What That Means for the Rest of Us; Simon and Schuster: New York, NY, USA (2017)
Venkatesh, V., Thong, J.Y.L., Xu, X.: Consumer acceptance and use of information technology: extending the unified theory of acceptance and use of technology. MIS Q. **36**(1), 157–178 (2012)
West, A., Clifford, J., Atkinson, D.: "Alexa, build me a brand": an investigation into the impact of artificial intelligence on branding. Bus. Manage. Rev. **9**(3), 321–330 (2018)
Wirtz, J., et al.: Brave new world: service robots in the frontline. J. Serv. Manag. **29**(5), 907–931 (2018)
Xiao, B., Benbasat, I.: E-commerce product recommendation agents: use, characteristics, and impact. MIS Q. **31**(1), 137–209 (2007)

AI in Medicine and Clinical Knowledge Management

Harnessing Artificial Intelligence for Knowledge Management in Medicine

Jakub Dowejko[1(✉)], Joanna Tylkowska-Drożdż[1], Jan Petriczko[2], and Joanna Dowejko[3]

[1] University of Szczecin - Institute of Management, 22a Jana Pawla II Al., Szczecin 70-453, Poland
jakub.dowejko@usz.edu.pl
[2] Pomeranian Medical University, 1 Rybacka, Szczecin 70-204, Poland
[3] Pomeranian Medical University–Department of Genomics and Forensic Genetics, 1 Rybacka, Szczecin 70-204, Poland
https://en.usz.edu.pl/, https://www.pum.edu.pl/

Abstract. This paper explores the transformative role of artificial intelligence (AI) in managing medical knowledge, with a particular focus on plastic surgery and brief glance at biotechnology. By integrating AI into both clinical and logistical dimensions, healthcare providers can potentially improve decision-making, optimize operations, and enhance patient outcomes. The study examines the current practices of knowledge management in practical medicine, analyses key workflows in plastic surgery, assesses AI's role in augmenting decision support systems, and evaluates both the opportunities and ethical challenges associated with AI development in the medical domain.

Keywords: Artificial Intelligence · Knowledge Management · Plastic Surgery · Medical Decision Support · Healthcare Logistics · Ethics in Medical AI · Biotechnology · Genetics

1 Introduction to AI-Driven Knowledge Management

1.1 Foundations and Applications

In the era of data-driven healthcare, the ability to capture, interpret, and apply clinical knowledge efficiently has become a critical determinant of both patient outcomes and operational excellence. Knowledge management (KM), traditionally focused on the systematic organization, sharing, and utilization of medical information, is undergoing a fundamental transformation driven by artificial intelligence. Current medical knowledge is vast which is beyond the human capacity to retain. Hence, modern-day learning is not just knowledge acquisition, but the organization of knowledge in a retrievable manner [1]. From predictive analytics to real-time decision support, AI technologies are redefining how medical knowledge is generated, distributed, and applied across the continuum

Published by Springer Nature Switzerland AG 2026
M. Pondel et al. (Eds.): AI4KAM 2025, IFIP AICT 776, pp. 463–472, 2026.
https://doi.org/10.1007/978-3-032-18920-2_44

of care. The practice of medicine is characterized by decision making in which digital techniques can provide good support. In this context, AI is becoming increasingly important, yet the topic of AI and its relevance to the field of medicine is not yet sufficiently represented in medical education. It will require integration in the curriculum and performance evaluations as well as interdisciplinary and research-related teaching formats. [2].

Among the many medical specialties, plastic and reconstructive surgery presents a particularly compelling context for exploring the integration of AI and KM. This field is inherently knowledge-intensive, requiring precise anatomical understanding, individualized treatment planning, and multidisciplinary collaboration. Moreover, it operates at the intersection of technical precision and aesthetic judgment—domains where AI tools such as computer vision, natural language processing (NLP), and machine learning have shown considerable promise. NLP is a pivotal subfield of artificial intelligence (AI) that focuses on the interaction between computers and human language. Its primary objective is to enable machines to comprehend, interpret, and generate human language in a meaningful and contextually relevant manner. NLP encompasses a wide spectrum of tasks, including language translation, sentiment analysis, speech recognition, text summarization, question answering, and more, each designed to bridge the gap between human communication and computational understanding. NLP has witnessed extensive implementation in healthcare, showcasing its remarkable ability to extract and analyze valuable insights from vast amounts of unstructured clinical data, including EHRs, medical literature, and patient-generated content [3]. AI-driven innovations now support a variety of applications including surgical simulation, outcome forecasting, patient-specific 3D modeling, and postoperative complication detection.

Despite these advancements, the implementation of AI-supported KM in clinical environments remains uneven and faces a host of challenges. These include ethical considerations related to algorithmic transparency and data privacy, institutional resistance to technological change, regulatory uncertainty, and a persistent gap between technical capability and clinical relevance. Furthermore, the success of AI applications in medicine depends not only on computational power or data availability, but also on their alignment with the values and decision-making processes of healthcare professionals. As noted by Michael R. King in his editorial - artificial intelligence holds immense promise for transforming the medical field and enhancing patient outcomes. However, the author—through a reflective narrative developed in part using conversational interactions with a chatbot—underscores the importance of addressing the inherent concerns and limitations of these technologies. King emphasizes that the successful integration of AI in medicine requires not only technical advancement but also a proactive approach to mitigating ethical, legal, and practical challenges that may arise along the way [4]. Moreover, one of the biggest challenges of utilizing AI in medicine is that physicians are reluctant to trust and adopt something that they do not fully understand and regarded as a "black box" [5]. Beyond the realm of data analytics and digital decision support, the convergence of biotechnology and artificial intelligence represents a transformative frontier for knowledge management in medicine. Biotechnology—through innovations in genomics, proteomics, tissue engineering, and biomaterials—continuously generates complex, multidimensional datasets. The effective use of such data demands advanced

AI-driven models capable of identifying molecular patterns, predicting regenerative outcomes, and optimizing patient-specific treatment protocols. In plastic and reconstructive surgery, AI-enabled analysis of biotechnological data facilitates the design of customized implants, accelerates tissue regeneration research, and enhances the precision of aesthetic and reconstructive procedures. Moreover, the integration of biotechnological data into knowledge management systems allows clinicians to bridge the gap between molecular insights and clinical decision-making, transforming empirical surgical practice into a data-informed, predictive science.

The purpose of this study is to critically examine the role of AI in transforming knowledge management within the practical domain of plastic surgery and biotechnology. By mapping the current landscape of AI integration across preoperative, intraoperative, and postoperative stages, and by analyzing both the opportunities and limitations inherent in these technologies, we aim to provide a balanced and evidence-based perspective. In doing so, this paper contributes to the ongoing discourse on how intelligent systems can enhance not only technical surgical outcomes but also the epistemic foundations of modern clinical practice. Thus, the synergy between AI and biotechnology extends the scope of knowledge management beyond information organization—it enables continuous learning systems capable of translating biological complexity into actionable clinical intelligence.

2 Background: Knowledge Management in Medicine

Healthcare is considered a pillar for keeping humans alive, thus managing their health through diagnosing and treating the disease or illness is a must, in addition to preventing it. KM in healthcare domain plays a substantial role in the implementation of different processes to ensure the existence of good and outstanding healthcare systems [6]. Effective KM systems reduce medical errors and complications, impacting patient safety [7]. Nevertheless, Forbes highlights the role of integrated CDSS in improving procedures and minimizing errors [8]. Almansoori et al. define KM as "the process of creating, sharing, using and protecting knowledge in an organization", which includes: acquisition, creation, storage, sharing, application and protection.

An alternative and compelling approach is set in the first step through the automation of documentation in large language models (LLMs) exemplifies AI's ability to streamline core KM tasks historically mired in administrative burden. Generative AI systems, such as those integrated with electronic health records (EHRs), can listen to clinician–patient interactions or parse physician dictations, and seamlessly craft structured, comprehensive clinical notes. This reduces "time-sucking administrative tasks," a challenge well documented in radiology but applicable across specialties. Second, the use of AI-powered Clinical Decision Support Systems (CDSS) represents a paradigm shift in on-the-spot clinical knowledge application. These systems analyze individual patient data—such as medical history, laboratory values, comorbidities, and current medications—and match them with evidence-based guidelines, real-time risk stratification, and genomic data to recommend personalized therapies. Applications include drug interaction alerts, dose-adjusted medication regimens, and diagnostic suggestions tailored to patient-specific nuances [9]. For example, research in medication safety demonstrates that AI-enhanced

CDSS can optimize drug alert systems in hospitals, reducing alert fatigue and improving clinician engagement with best practices [10]. Structured interviews with physicians using personalized drug-dosing platforms highlight how intelligent decision support can elevate standards of care while anchoring decisions in reliable KM systems [11]. Third, with the proliferation of patient-wearable devices and Internet of Things (IoT) sensors, the volume of health data generated has exploded. AI plays a pivotal role in monitoring and analyzing these data streams in real time.

What is more, artificial intelligence is already widely used also in biotechnology to solve a variety of problems. These include, for example, drug discovery, drug safety, functional and structural genomics proteomics, metabolomics, pharmacology, pharmacogenetics, and pharmacogenomics, among many others. Future advances in this domain depend critically on the ability of biotechnology researchers to use advanced AI solutions effectively. The biotechnology industry currently relies heavily on data storage, filtering, analysis and sharing. Biotechnology companies and various healthcare organizations around the world already maintain huge databases. Drug manufacturing, chemical analysis of various compounds, sequencing of RNA and DNA, enzyme studies, and other similar biological processes all require strong support from AI software solutions to move faster and reduce manual errors [12].

Advanced AI algorithms can detect subtle deviations in vital signs like heart rate, blood pressure, oxygen saturation, or electrocardiogram signals—long before they pose clinical danger. These systems rely on early warning scores and algorithm-driven threshold alerts that bridge frontline providers and remote patients. Reports indicate ongoing deployment of IoT-AI frameworks with cloud-based analytics that transmit actionable insights to care teams instantly, significantly improving response times and reducing emergency department admissions and rehospitalizations. Market analyses confirm that the AI-enabled healthcare IoT sector is projected to grow substantially—further embedding trend analysis and predictive modeling into routine clinical surveillance [13]. Altogether, these three modalities—automated documentation, AI-enhanced CDSS, and IoT-driven monitoring—constitute a cohesive AI-empowered KM ecosystem. They are not disparate tools but interconnected components: documentation creates structured records that feed decision systems; CDSS recommendations rely on high-quality documentation and real-time monitoring input; monitoring systems provide longitudinal data that refine AI models and expand institutional knowledge bases over time. This synergy enhances not only individual patient care, but also institutional learning, as aggregated patient data enables retrospective analysis, algorithm calibration, and continuous improvement cycles in clinical pathways. Despite these benefits, their implementation brings challenges. Administrative automation must guard against the risk of AI hallucinations and inappropriate documentation—concerns that led to frameworks for evaluating LLM-generated notes, such as the CREOLA platform, which systematically classifies errors and helps assure clinical safety [13, 14]. CDSS implementation faces obstacles including alert fatigue, data integration issues, and clinician mistrust. Optimization studies emphasize responsibly tailoring alert thresholds and developing clear validation protocols before widespread deployment. IoT-AI systems raise significant data privacy and regulatory concerns, requiring encryption, secure data pipelines, and compliance with HIPAA, GDPR, and medical device regulations. Organizational resistance presents further barriers; successful integration depends on clinician engagement, training, and transformation of existing clinical workflows [15].

Looking ahead, KM strategies anchored in AI promise future directions such as AI-fueled multimodal documentation integrating speech, text, and imaging inputs; CDSS systems that adapt dynamically using federated learning architectures across provider networks; and IoT-AI models that support precision medicine through continuous risk stratification and population health analysis. For knowledge management, this translates to fluid, real-time, data-informed systems that monitor their own performance, updating institutional knowledge constructs and embedding learning loops within the care delivery cycle.

2.1 Logistical and Decision-Making Processes in Plastic Surgery

Integrating artificial intelligence into the logistical and decision-making processes of plastic surgery marks a transformative shift, extending the role of AI far beyond technical augmentation in the operating theater and into foundational aspects of practice management and patient care. Machine learning algorithms, trained on historical appointment data, surgical durations, and cancelation patterns, are capable of dynamically allocating time slots to maximize operating room utilization and minimize idle periods. Studies within outpatient care have demonstrated AI-driven systems can reduce appointment gaps by up to 40%, while predictive reminders cut no-show rates from approximately 19% to 15% [16]. In plastic surgery clinics—where elective procedures and staged reconstructions demand strict scheduling—such improvements translate into reduced revenue losses and enhanced patient access. Behind the scenes, administrative automation is reducing cognitive burden and operational overhead. AI systems now parse patient intake forms and pre-consultation chatbots gather symptom histories and triage patients before the first in-person visit, referring routine or low-risk cases directly for appointments and flagging high-risk or complex cases for surgeon review. Parallel systems streamline coding and billing processes, automatically identifying procedural charges from the clinical documentation. While peer-reviewed studies in plastic surgery specifically are limited, analogous systems in broader surgical disciplines have delivered tangible revenue-cycle enhancements by decreasing human error and accelerating claims processing. Plastic surgery clinics now use AI to segment audiences online, tailor messaging based on geographic or demographic variables, and even deploy chatbot-powered lead capture to pre-qualify inquiries 24/7. This form of logistical efficiency aligns marketing strategy with clinic capacity and prioritizes evidence-based outreach, reinforcing the socioeconomic dimensions of knowledge management. Within the operating room, AI's logistical impact is equally pronounced. Accurate surgical duration prediction enables optimized OR schedules, reducing staff idle time and minimizing cancellations. Reinforcement learning models like BEDS allocate cases evenly and respect surgeon availability, while heuristic methods such as simulated annealing have been shown to schedule over 96% of cases within planned parameters, including pediatric plastic surgery [17]. Even though many of these systems are not developed exclusively for plastic surgery, they present scalable frameworks that clinics can adapt to high-variability operative contexts. Emerging AI tools also offer real-time OR monitoring and adaptive resource allocation. While systems capable of dynamic case re-sequencing are just entering development, studies in other surgical disciplines using Markov decision frameworks have demonstrated potential for real-time resource balancing during complex sequential procedures [18]. Once refined, this approach could alleviate resource bottlenecks typical in microsurgical or multi-stage aesthetic operations. Cesar A. Gomez-Cabello et al. noted, that AI is beginning to support intraoperative decision-making LLMs such as ChatGPT-4 and Google's Gemini Pro have been evaluated against plastic surgery scenarios, including flap viability and complication management. In a controlled study, ChatGPT-4 achieved higher accuracy (mean score 3.59 vs. 3.13; $p = 0.022$) while Gemini offered faster but more concise responses. Complementing these, generative chatbots have begun automating operative notes in real-time, enabling surgeons to focus more effectively on technique while ensuring documentation fidelity.

3 Discussion

Within the context of plastic and reconstructive surgery, this effectiveness must be evaluated across a spectrum of dimensions—clinical accuracy, timeliness of recommendations, alignment with human judgment, and contribution to improved patient outcomes. While AI systems promise unprecedented efficiency and precision, their true value lies in how meaningfully they augment, rather than replace, clinician expertise. AI-driven Clinical Decision Support Systems (CDSS) have been designed to assist physicians in interpreting complex data patterns and providing evidence-based recommendations at the point of care. These tools rely on integration with electronic health records (EHRs), patient histories, lab results, and medication databases to produce individualized therapeutic suggestions. In this respect, AI functions as a highly dynamic knowledge retrieval and application mechanism, allowing clinicians to make better-informed choices based on real-time patient profiles. In plastic surgery, decision-making is uniquely complex due to the combination of technical, functional, and aesthetic considerations involved in procedures. For example, the decision to use a specific type of flap in reconstructive microsurgery involves not only anatomical precision but also aesthetic harmony and future tissue viability. AI tools capable of simulating surgical outcomes and analyzing 3D patient imagery provide tangible support in such nuanced decisions. Moreover, machine learning algorithms can process longitudinal patient data to identify subtle risk factors or predict surgical complications, offering insights that might be overlooked in conventional decision-making workflows. The integration of artificial intelligence into bioengineering is helping to transform biological design into a systematic engineering discipline, sparking rapid progress and innovation. AI-driven tools are revolutionising protein design, unlocking opportunities in therapeutics, diagnostics, and synthetic biology. For instance, tools like RFDiffusion [19] and Chroma [20] allow the creation of proteins with desired structures and properties, while DiffDock [21] significantly improves the prediction of protein-ligand binding interactions. Tools like DeepBind accurately predict protein binding to DNA and RNA, AlphaProteo [22] allows the design of novel binders, and AI tools are being used to engineer proteins with greater stability and functionality [23]. These innovations tackle long-standing challenges such as engineering enzymes for industrial applications and combating antimicrobial resistance, while also paving the way for transformative advancements in personalised medicine, biomanufacturing, and environmental sustainability. By harnessing the power of AI, bioengineering can strive to achieve levels of precision and reproducibility akin to traditional engineering disciplines, heralding a new era of scientific discovery and application.

However, effectiveness must not be equated solely with computational accuracy. A key consideration is the concordance between AI-generated recommendations and the judgments of experienced surgeons. Studies comparing the outputs of AI systems with expert decisions in plastic surgery scenarios have found varying levels of agreement, suggesting that while AI can approximate expert logic, it cannot yet fully replicate the depth of clinical reasoning grounded in tacit knowledge and years of experience [24]. The epistemological distinction between knowing that and knowing how remains critical: AI systems excel at aggregating and applying codified knowledge, but may falter in interpreting ambiguous, context-sensitive variables that frequently shape surgical choices.

Another measure of effectiveness is the influence of AI on decision efficiency. Time-critical decisions—particularly in the operating room or emergency settings—benefit from AI's capacity to deliver instant data syntheses and risk stratifications. The deployment of LLM-based assistants capable of summarizing clinical notes, surfacing key findings, or generating procedural checklists has significantly reduced cognitive load, allowing clinicians to focus on the core task of delivering care. Yet the degree to which these tools improve final outcomes versus only streamlining workflows remains underexplored. Early evidence indicates that while AI enhances operational performance, its impact on outcome quality is moderated by how actively and judiciously human clinicians engage with the system.

The notion of augmented intelligence rather than artificial intelligence is increasingly favored in medical discourse. This framing shifts the emphasis from substitution to collaboration. It recognizes that AI's most effective role is as a knowledge extender—augmenting human cognition by expanding the horizon of recall, analysis, and pattern recognition. In doing so, AI supports more holistic decision-making that accounts for both statistical probability and individualized care. However, this augmentation depends heavily on the design and transparency of AI systems. If clinicians do not understand how an algorithm reached its conclusion—a common issue with black-box models—they may either distrust its output or over-rely on it inappropriately. Both scenarios undermine effectiveness.

Furthermore, effectiveness is inextricably linked with usability. Human–AI interaction research shows that clinicians are more likely to adopt and benefit from AI tools when the user interface is intuitive, recommendations are clearly explained, and systems can adapt to the clinician's workflow rather than impose rigid protocols. When AI tools are perceived as intrusive or incompatible with established practices, they risk rejection, regardless of their technical accuracy. This highlights the sociotechnical dimension of AI effectiveness: success is not determined by algorithms alone, but by how they are embedded within the organizational and cognitive structures of healthcare delivery. Ethical considerations also factor into effectiveness, particularly when AI recommendations influence decisions with life-altering consequences. In plastic surgery, where elective procedures intersect with deeply personal notions of identity and aesthetics, the role of

AI must be tempered with empathy, cultural awareness, and respect for patient autonomy. A technically "optimal" decision from an algorithmic standpoint may not align with a patient's values or expectations. Therefore, AI should support shared decision-making processes rather than dictate clinical actions. Effectiveness, in this light, becomes relational—it reflects how well AI contributes to informed, transparent, and respectful dialogues between physicians and patients. From a systems perspective, the long-term effectiveness of AI in decision-making also relates to institutional learning and knowledge evolution. AI tools can aggregate outcomes across patient cohorts, identifying best practices and feeding them back into clinical protocols. This creates a feedback loop wherein decision-making becomes progressively refined over time. Yet such benefits require robust data governance, interoperability between systems, and institutional cultures that prioritize continuous improvement. In poorly integrated settings, AI tools may operate in isolation, failing to disseminate valuable insights or harmonize with broader organizational goals. The evaluation of effectiveness should also consider disparities in access and performance. AI systems trained on limited or biased datasets may produce recommendations that are less accurate—or even harmful—for underrepresented populations. In the context of global plastic surgery, where demographic, genetic, and cultural diversity shapes surgical indications and outcomes, the need for inclusive training data is paramount. Without it, AI effectiveness becomes a privilege of the majority, leaving minorities at risk of algorithmic neglect or harm.

Ultimately, the question of effectiveness invites a multidimensional answer. AI in decision-making is not merely about replacing human thought with computational logic; it is about enhancing human reasoning, minimizing risk, supporting learning, and enriching the quality of care. To achieve this vision, AI systems must be transparent, interpretable, ethically aligned, and clinically validated. They must support—not supplant—the judgment of clinicians who bear the final responsibility for patient outcomes. Only then can we speak meaningfully of AI's effectiveness in the real-world practice of plastic surgery and medicine more broadly.

The intersection of artificial intelligence and biotechnology delineates a new epistemological frontier in the life sciences, redefining the modalities through which biological knowledge is generated, systematized, and operationalized. In an era characterized by the exponential proliferation of genomic, proteomic, and cellular data, the capacity of traditional analytical frameworks to synthesize and interpret such complexity has reached its limits. Artificial intelligence, through its capacity for abstraction, pattern recognition, and autonomous learning, now constitutes an indispensable cognitive infrastructure for modern biotechnology.

Within this evolving paradigm, AI-based knowledge management systems transcend the passive function of information storage by facilitating dynamic, self-organizing networks of biological intelligence. Through machine learning and natural language processing, these systems not only extract and structure dispersed data but also generate inferential pathways linking molecular mechanisms to therapeutic outcomes. In doing so, they transform biotechnological research from an empirical discipline into a predictive and knowledge-driven science.

Applications across drug discovery, genomics, and tissue engineering illustrate this transformation. Deep learning models accelerate molecular screening and optimize compound selection; AI-guided genomic analytics enable precision medicine tailored to individual variability; and computational modeling of cellular interactions informs the rational design of biomaterials and regenerative scaffolds. Collectively, these developments highlight the capacity of AI to serve as both a catalyst for innovation and a meta-structure for knowledge organization within biotechnology.

However, this transformation also entails significant epistemic and ethical implications. The opacity of algorithmic decision-making, concerns over data provenance, and the risk of bias embedded in training datasets all challenge the legitimacy and interpretability of AI-assisted discoveries. Consequently, the integration of AI into biotechnology must be accompanied by rigorous frameworks ensuring transparency, accountability, and methodological validation. Only under such conditions can the convergence of artificial intelligence and biotechnology yield a truly sustainable model of scientific advancement—one that unites computational sophistication with ethical stewardship and epistemological coherence.

Funding. Co-financed by the Minister of Science under the "Regional Excellence Initiative".

Ministerstwo Nauki i Szkolnictwa Wyższego

Regionalna Inicjatywa Doskonałości

References

1. Wan, L., Hooi, S.C.: The role of artificial intelligence in knowledge management for medical students and doctors. Natl. Libr. Med. (2024). https://doi.org/10.1080/0142159X.2024.2336065
2. Lang, J., Repp, H.: Artificial intelligence in medical education and the meaning of interaction with natural intelligence-an interdisciplinary approach. Natl. Libr. Med. (2020). https://doi.org/10.3205/zma001352
3. Zhang, P., Maged, N., Kamel, B.: Generative AI in medicine and healthcare: promises, opportunities and challenges Future Internet Med. Things II (2024). https://doi.org/10.3390/fi15090286
4. King, M.R.: The future of AI in medicine: a perspective from a Chatbot. Am. J. Med. **51**, 291–295 (2023). https://doi.org/10.1016/j.amjmed.2022.12.001.Springer
5. Poon, A., Sung, J.: Opening the black box of AI-Medicine. Wiley (2021). https://doi.org/10.1111/jgh.15384
6. Almansoori, A., Alshamsi, M., Salloum, S.A., Shaalan, K.: Critical review of knowledge management in healthcare (2021). https://doi.org/10.1007/978-3-030-47411-9_6
7. https://creativesoncall.com/industries/knowledge-management-systems-in-healthcare. Last accessed 05 June 2025
8. https://www.forbes.com/consent/ketch/?toURL=https://www.forbes.com/councils/forbestechcouncil/2024/06/06/leveraging-technology-for-patient-safety-the-role-of-knowledge-management-in-healthcare/. Last accessed 07 June 2025

9. Reddy, S.: Generative AI in healthcare: an implementation science informed translational path on application, integration and governance. Implement. Sci. **19**(1), 27 (2024). https://doi.org/10.1186/s13012-024-01357-9.PMID:38491544;PMCID:PMC10941464
10. Graafsma, J., et al.: The use of artificial intelligence to optimize medication alerts generated by clinical decision support systems: a scoping review. J. Am. Med. Inform. Assoc. 31(6), 1411–1422 (2024). https://doi.org/10.1093/jamia/ocae076. PMID: 38641410; PMCID: PMC11105146
11. Vijayakumar, S., Lee, V.V., Leong, Q.Y., Hong, S.J., Blasiak, A., Ho, D.: Physicians' perspectives on AI in clinical decision support systems: interview study of the CURATE. AI Personalized Dose Optimization Platform. JMIR Hum. Factors **10**, e48476 (2023). https://doi.org/10.2196/48476. PMID: 37902825. PMCID: 10644191
12. Holzinger, A., et al.: AI for life: Trends in artificial intelligence for biotechnology. New Biotechnol. **74**, 16–24 (2023). ISSN 1871-6784
13. Report: Internet of Things in Healthcare Market, By Component (Medical Devices, System and Software, others), By Connectivity (Cellular, WIFI, Others), By Application (Telemedicine, Patient Monitoring, Others), By End-Use (Hospitals, Clinical Research Organisation, Others) and By Region Forecast to 2033
14. Asgari, E., Montaña-Brown, N., Dubois, M., et al.: A framework to assess clinical safety and hallucination rates of LLMs for medical text summarisation. npj Digit. Med. **8**, 274 (2025). https://doi.org/10.1038/s41746-025-01670-7
15. https://tortus.ai/creola/. Last accessed 09 June 2025
16. Atchinson, B.K., Fox, D.M.: The politics of the health insurance portability and accountability act. Health Affairs (1997)
17. https://www.patientprism.com/ai-is-a-game-changer-for-plastic-surgery/. Accessed 09 June 2025
18. Lopes, J., Guimarães, T., Duarte, J., Santos, M.: Enhancing surgery scheduling in health care settings with metaheuristic optimization models: algorithm validation study. JMIR Med. Inform. **13**, e57231 (2025). https://doi.org/10.2196/57231. PMID: 39935008; PMCID: PMC11840878
19. Watson, J.L., Juergens, D., Bennett, N.R., Trippe, B.L., Yim, J., Eisenach, H.E., et al.: *De novo* design of protein structure and function with RFdiffusion. Nature **620**(7976), 1089–1100 (2023). https://doi.org/10.1038/s41586-023-06415-8
20. Ingraham, J.B., Baranov, M., Costello, Z., Barber, K.W., Wang, W., Ismail, A., et al.: Illuminating protein space with a programmable generative model. Nature **623**(7989), 1070–1078 (2023). https://doi.org/10.1038/s41586-023-06728-8
21. Corso, G., Stärk, H., Jing, B., Barzilay, R., Jaakkola, T.: DiffDock: diffusion steps, twists, and turns for molecular docking (2022). arXiv [q-bio.BM]. https://doi.org/10.48550/arXiv.2210.01776
22. Zambaldi, V., La, D., Chu, A.E., Patani, H., Danson, A.E., Kwan, T.O.C., et al.: *De novo*design of high-affinity protein binders with AlphaProteo (2024). arXiv [q-bio.BM]. https://doi.org/10.48550/arXiv.2409.08022
23. Sumida, K.H., Núñez-Franco, R., Kalvet, I., Pellock, S.J., Wicky, B.I.M., Milles, L.F., et al.: Improving protein expression, stability, and function with ProteinMPNN. J. Am. Chem. Soc. **146**(3), 2054–2061 (2024). https://doi.org/10.1021/jacs.3c10941
24. Gomez-Cabello, C., Borna, S., Pressman, S.M., Ali Haider, S., Forte, A.: Large language models for intraoperative decision support in plastic surgery: a comparison between ChatGPT-4 and Gemini (2024). https://doi.org/10.3390/medicina60060957

Analysis of Medical Data Integration Issues in the Concept of Regional Digital Medicine Centers. A Case Study from the Medical University of Wrocław

Agnieszka Siennicka[1], Agnieszka Matera-Witkiewicz[1], Maciej Pondel[2](✉), and Iwona Chomiak-Orsa[2]

[1] Uniwersytet Medyczny im. Piastów Śląskich we Wrocławiu, 50-367 Wrocław, Polska
{agnieszka.siennicka,adnieszka.matera-witkiewicz}@umw.edu.pl
[2] Uniwersytet Ekonomiczny we Wrocławiu, 53-345 Wrocław, Polska
{maciej.pondel,iwona.chomiak-orsa}@ue.wroc.pl

Abstract. The dynamic development of digital technologies and artificial intelligence (AI) in medicine presents healthcare institutions with new challenges in the integration of medical data. One of the key solutions aimed at streamlining diagnostic, therapeutic, and research processes is the concept of Regional Digital Medicine Centers (RCMC), which enable the efficient collection, processing, and sharing of information resources. The aim of this article is to define the problems and barriers that hinder the creation and operation of RCMCs. The article focuses on issues related to the integration of medical data within RCMCs, analyzing a case study of the Medical University of Wrocław (UMW). It presents the main challenges encountered during the digitization and harmonization of information resources used by healthcare entities, including hospitals and academic centers. Particular attention is paid to aspects such as system interoperability, data security, standardization of information formats, and the implementation of AI tools in the analysis of large clinical datasets. The article also discusses the process of creating Regional Digital Medicine Centers and the methodologies used in their implementation. The multifaceted nature of medical data, including electronic medical records (EDM), biobank repositories, and clinical trial datasets, is highlighted. Against this backdrop, the potential benefits of using AI in integration processes are presented, along with the identification of organizational, technical, and regulatory barriers that may hinder the implementation of effective solutions. The study utilizes a triangulation of research methods, including analysis of literature, reports, and design guidelines, as well as qualitative methods such as documentation analysis, IT system assessment, expert interviews, participatory observations, and a case study of the implementation of RCMC at the Medical University of Wrocław. The article provides valuable conclusions and recommendations for institutions planning to implement Regional Digital Medicine Centers. The importance of interdisciplinary collaboration, process standardization, and the development of advanced analytical technologies are emphasized as key factors enabling effective medical data integration.

Published by Springer Nature Switzerland AG 2026
M. Pondel et al. (Eds.): AI4KAM 2025, IFIP AICT 776, pp. 473–486, 2026.
https://doi.org/10.1007/978-3-032-18920-2_45

Keywords: data integration · artificial intelligence (AI) · digital physician assistant · medical data · biobank · electronic medical records · digital medicine

1 Introduction

The purpose of this article is to present the key stages and challenges associated with integrating medical data within the Regional Digital Medicine Centers (RCMC) concept, with particular emphasis on the case study of the Wrocław Medical University (UMW). The article addresses the pragmatic implementation of artificial intelligence (AI) in healthcare, highlighting the fundamental organizational, technological, and legal challenges associated with data harmonization in the medical sector.

The literature review and the authors' experience indicate that a significant research gap is the lack of integrated IT solutions that effectively utilize AI in consortia of healthcare entities, encompassing both research units (universities), medical facilities (clinical hospitals, specialist clinics), and support units (e.g., biobanks).

The article begins with an introduction to the integration of artificial intelligence in medical practice, drawing on the authors' interdisciplinary experience gained through research in medical data science. It then proceeds to a detailed discussion of the conditions for establishing digital medicine centers, particularly in the context of financing and managing university projects, and to outline a strategic methodology for implementing digital medicine centers in academic and clinical settings. The reasons for and challenges associated with implementing digital medicine centers are also analyzed, taking into account the multifaceted nature of medical data, such as electronic medical records (EMRs), biobank repositories, and clinical trial datasets. Key challenges in standardizing, digitizing, and integrating diverse data streams, which are fundamental to the effective operation of digital medicine centers, are identified.

The introduction of Regional Digital Medicine Centers in Poland is considered a strategic initiative aimed at accelerating the adoption of AI in healthcare. This article provides practical recommendations for institutions planning AI implementation, emphasizing the importance of interdisciplinary collaboration, precise data analysis, and the development of a coherent technological strategy.

The remainder of the paper is devoted to the characteristics of medical data, including its diverse origins, formats, levels of digitization, and quality. These issues are key factors influencing the planning and development of IT infrastructure supporting AI integration in clinical and scientific areas. Strategies for organizing IT resources are also presented, enabling effective data management and utilization in diagnostic processes and scientific research.

The article concludes with a concise discussion summarizing key findings and challenges related to integrating medical data within the RCMC. It also identifies directions for future research and proposes further implementation activities that could contribute to increasing the effectiveness of AI in the Polish healthcare system. The study was conducted using qualitative research methods, including documentation analysis, IT systems assessment, interviews with project participants, participatory observations and a case study analysis of RCMC implementation at the Medical University of Wrocław [1].

This article provides valuable insights and practical recommendations for institutions beginning the process of integrating medical data, emphasizing the need for standardization, harmonization of systems, and an interdisciplinary approach to implementing technological innovations in the healthcare sector.

2 Theoretical Background

2.1 Information Technologies in Medicine

The development of modern medicine is closely linked to the development of technologies specifically designed to support diagnostic, clinical, and production processes. Communication processes and the improvement of medical data storage, flow, and analysis are also important. Patient data management is a key issue, the solution to which impacts the effectiveness and efficiency of medical processes. The computerization of healthcare systems encompasses a wide range of solutions, such as electronic medical records (EDM), hospital management systems (HIS), artificial intelligence (AI) tools, and Big Data processing technologies [2, 3].

One of the key elements of the digitalization of medicine is electronic medical records (EMR), which enable quick access to patient information, eliminating the need for traditional paper documentation. This direction of computerization is one of the most important trends in IT, dedicated to the comprehensive coherence of entire healthcare systems created and managed within the framework of national and even international health systems regulations. However, for EMR systems to contribute to improving the quality of healthcare, it is necessary to integrate patient and procedure data from various sources, such as laboratory information systems (LIS), medical imaging systems (PACS), and patient monitoring systems [2–4].

Interoperability of medical systems is one of the main challenges in the integration of information technology in healthcare. The variety of data formats and communication standards, such as HL7 (Health Level Seven) or FHIR (Fast Healthcare Interoperability Resources), requires the use of advanced integration tools that enable efficient and secure processing of information within regional or national medical data management systems supervised and managed by state administration institutions as part of the financial settlement of medical procedures [4–6].

One of the greatest challenges associated with the digitalization of medicine, which involves harmonizing and integrating data from various systems, is ensuring the security and privacy of patient data. Sensitive medical information requires advanced protection mechanisms, such as data encryption, anonymization, and access control [4, 7].

Blockchain technology is also increasingly being considered, as it can increase the security and transparency of medical data exchange. Blockchain allows for the creation of decentralized patient data registries that are resistant to manipulation and unauthorized access, making this technology a promising solution for future healthcare information management systems (HMS) [8–10].

The development of artificial intelligence (AI) and big data processing technology opens up new possibilities for precision medicine and clinical decision support systems (CDSS). Decision Support Systems). AI algorithms can be used to automatically analyze medical images, predict treatment outcomes, and personalize therapy based on individual patient characteristics [11, 12]. Big Data technologies leveraging the Internet of Things (IoT) enable real-time data collection, especially in medical procedures that require continuous monitoring of patient conditions and analysis of these parameters [12–16].

In the context of integrated medical process management systems, the use of Big Data enables the effective integration and analysis of data from electronic health records, clinical trials, biobanks, and epidemiological registries [8, 17]. These technologies enable the identification of disease patterns, monitoring treatment effectiveness, and early detection of health threats in the population.

Information technology plays a key role in transforming the healthcare sector, enabling better organization of medical data, supporting clinical decisions, and integrating various information sources [8–16]. However, its implementation poses numerous challenges, such as ensuring system interoperability, protecting patient privacy, and optimizing AI algorithms for clinical applications. Effective digitalization of medicine requires interdisciplinary collaboration and further research into the integration of new technologies in clinical practice [12, 16, 18–21].

2.2 Data Integration is a Prerequisite for Implementing Artificial Intelligence Methods in Medicine

One of the most promising directions in medical development is the use of artificial intelligence methods. The potential for AI to support diagnostic, therapeutic, and administrative processes is of particular interest [11–15]. The use of machine learning algorithms and data analysis techniques enables more precise prediction of treatment outcomes, automated medical image analysis, and personalized therapy [8]. However, the effectiveness of AI methods in medicine depends largely on the quality, availability, and interoperability of medical data [7–11, 16, 18–21].

Data integration is key to effective AI implementation, as it enables the unification and harmonization of various information sources, such as electronic medical records (EMRs), biobank data, patient registries, diagnostic images, and clinical trial results [20, 21]. Without a properly structured data infrastructure, AI models face numerous limitations, including information fragmentation, inconsistent formats, and issues related to patient data privacy and security [22].

Therefore, the challenge of developing appropriate medical computerization strategies is primarily related to resolving issues arising from the lack of data integrity from various sources and data repositories. This issue often resonates with projects and initiatives utilizing artificial intelligence methods, but above all, it is a key issue that pertains to the success of any medical computerization strategy that has a chance of being successfully implemented [21–30].

2.3 Main Problems Related to Medical Data Integration

One of the most promising directions of medical development is the use of artificial intelligence methods.

The most frequently identified problems that hinder the use of resource integration and the implementation of advanced analytical methods or process automation in the area of medical records management include:

- Data fragmentation and a lack of interoperability stem from the dispersion of information resources. Patient data is stored in various formats and structures, making it difficult to aggregate and use in AI analyses. In many cases, the IT systems used in hospitals and research centers are incompatible, leading to the loss of crucial information and limiting the ability to fully utilize its potential.
- There is a lack of standardization in data formats and protocols for exchanging information between subsystems within the healthcare ecosystem. Effective data integration requires the introduction of standards that enable uniform collection and processing of medical information. Existing standards, such as Health Level Seven (HL7) and Fast Healthcare Interoperability Resources (FHIR) and Digital Imaging and Communications in Medicine (DICOM) play a key role in ensuring interoperability between medical systems [31, 32]. However, their large-scale implementation still faces challenges due to the diversity of IT systems and the lack of uniform regulations.
- The need to exercise the utmost care in maintaining the privacy and security of medical data. Patient data protection is one of the most important aspects of their integration and processing for AI purposes. Requirements related to the General Data Protection Regulation (GDPR) and HIPAA regulations (Health Insurance Portability and Accountability Act) require medical institutions to use advanced encryption, anonymization, and access control mechanisms. Technologies such as blockchain and decentralized databases offer potential solutions for improving the security and transparency of medical information processing.
- Low quality and incompleteness of data collected in fragmented domain systems. AI models require access to high-quality data that is complete, consistent, and representative of the analyzed populations. Due to this diversity, data is stored in disparate formats (e.g., HL7, DICOM, FHIR), leading to difficulties in data integration and analysis. Issues such as documentation errors resulting from human error or incomplete records, data duplication, lack of standardized units of measurement, and incomplete entries can reduce the effectiveness of machine learning algorithms. Missing patient records, inconsistent records, human error, and diverse data formats pose significant barriers to effective data integration and the implementation of AI in diagnostics and therapy.

The problems mentioned above are only the most frequently mentioned critical determinants of the lack of integration of medical data into a unified and coherent information system that could use artificial intelligence methods or other advanced IT technologies to improve and optimize medical processes [8, 10, 22–30].

2.4 Strategies for Effective Medical Data Integration

As mentioned in previous sections of this article, effective integration of medical data is a key requirement for implementing modern information technologies, including artificial intelligence (AI). This process requires the use of appropriate strategies to enable harmonization, standardization, and exchange of information between various entities, such as hospitals, clinics, laboratories, biobanks, and research units [28].

Due to the diversity of data sources and formats, as well as numerous technical and organizational challenges, the integration of medical systems must take into account interoperability, security, data quality, and effective IT infrastructure management. The most important aspects defined in data integration strategies in healthcare systems include:

a. Unifying data exchange standards and protocols. One of the most important steps in medical data integration is the implementation of uniform interoperability standards that enable the exchange of information between different IT systems. This involves:

   - Health Level Seven (HL7) – the most commonly used standard for exchanging clinical information between hospital IT systems, laboratories and research institutions.
   - Fast Healthcare Interoperability Resources (FHIR) – a modern standard developed by HL7, which enables fast and flexible processing and exchange of medical data in the cloud and mobile systems.
   - Digital Imaging and Communications in Medicine (DICOM) – a standard designed for storing and transmitting medical images (e.g. X-ray, MRI, CT).
   - Observational Medical Outcomes Partnership (OMOP) – a unified data model used in epidemiological research and Big Data analyses.

Standardization of protocols allows for improved data quality and better accessibility for AI systems and clinical analyses.

b. Creating Central Data Repositories and Medical Data Warehouses. To effectively manage large data sets, healthcare institutions are implementing integrated databases and data warehouses that aggregate information from various sources. Key elements of this approach include:

   - Regional Digital Medicine Centers (RCMC) – systems that integrate data from various facilities within a given region, enabling faster diagnosis and better treatment coordination.
   - National patient registries – databases containing detailed information on chronic diseases, clinical trials and medical procedures, e.g. Cancer Registry or Diabetes Registry.
   - Medical data warehouses – enabling the storage of structured and unstructured medical data for advanced analytics and machine learning.

Data centralization allows us to eliminate the problem of data dispersion and fragmentation, which is one of the main barriers in AI analyses.

c. Automating data processing and harmonization. Using artificial intelligence technology to automatically normalize, cleanse, and supplement missing data can significantly improve its quality and consistency. Key methods include:

   - Natural Language Processing (NLP) – enables the extraction of relevant information from unstructured data, such as medical descriptions or test results.
   - Missing data imputation – Machine learning algorithms can predict missing values based on existing patient records.
   - Error and anomaly detection – AI systems can identify inconsistencies in documentation and suggest corrections

The use of these methods enables more efficient use of information resources and improves their value in clinical analyses.

d. Patient data security and privacy. Due to the sensitive nature of medical data, it is essential to ensure its protection from unauthorized access and privacy violations. Key strategies include:
   - Data encryption – using advanced encryption algorithms (e.g. AES, RSA) to protect patient information.
   - Blockchain technologies – enabling transparent and secure storage and sharing of medical data in decentralized systems.
   - Access control systems – implementing permission policies that allow access to data only to authorized users.

Ensuring the security of medical information is a key aspect of implementing integrated systems in accordance with legal regulations such as GDPR or HIPAA.

Interdisciplinary collaboration and open data platforms. Effective data integration requires close collaboration between medical, research, and technological institutions. To this end, open platforms are being created to enable the sharing, analysis, and development of AI algorithms in medicine. Examples of such initiatives include:

- Global Alliance for Genomics and Health (GA4GH) – an organization promoting the interoperability of genomic databases at the international level.
- OpenEHR – an initiative enabling the creation of open medical documentation standards for structured EHR systems.
- Data- sharing frameworks – systems enabling the exchange of data at the national and global level in a manner consistent with legal regulations.

This type of collaboration supports the development of innovative AI solutions and enables more comprehensive analysis of medical data.

Data integration is a key requirement for the effective implementation of artificial intelligence in medicine [5, 9–11, 27–29]. Issues related to information fragmentation, lack of interoperability, data security, and data quality require complex technological and organizational strategies. Implementing standards, developing central repositories, and leveraging AI for data harmonization are essential steps to effectively harness the potential of AI in diagnostics, therapy, and research.

3 Methodology

3.1 Concept Procedures Research

The establishment of the RCMC will involve comprehensive IT and technical assessments of resources to assess their potential for integration and interoperability. Subsequent steps will involve the gradual implementation of the technology, personnel, and infrastructure necessary to implement artificial intelligence across all available datasets, including those obtained from the university's collaborating clinical hospitals.

This article presents the research assumptions defined in the RCMC development project discussed above. The planned research process is multi-stage and incorporates the concept of triangulating research methods used in both qualitative and quantitative research.

3.2 Applied Technique Research

The research procedure presented in Fig. 1 consisted of a multi-stage study in which the authors used a range of research tools to collect both quantitative data, which will be presented in subsequent articles, and qualitative data, allowing for a detailed understanding of the expectations of users of the IT solution. Given that the article focuses on detailed results obtained from participating entities, a triangulation methodological approach was employed to ensure a comprehensive analysis of both qualitative and quantitative data. This approach increased the credibility of the research results by enabling cross-validation from multiple sources and perspectives. Research used in the RDMC project are:

- **Documentation analysis:** The detailed documentation analysis covered several different thematic areas. The first focused on guidelines for projects co-financed by the Medical Research Agency. The second area of analysis aimed to provide a detailed understanding of the existing ICT solutions. The third area involved an analysis of documentation prepared as part of the collaboration agreement between the consortium members.
- **Participatory observations:** Project team members were selected from among employees of all entities within the project consortium. This was important due to their individual experience and knowledge of all business processes and medical procedures that should be included in the solution architecture.

4 Case Study – Wroclaw Medical University

4.1 Experiences of the Wroclaw Medical University

From 2020 to 2022, the Wroclaw Medical University (WMU) collaborated with the Wroclaw University of Economics (UEW) on a three-year training initiative focused on integrating data science in medicine, with cardiology as a key example. During this program, medical researchers were introduced to the potential utility of artificial intelligence algorithms in clinical data analysis. Using historical data, several experiments were conducted, specifically demonstrating the ability of AI to detect inherent phenotypic heterogeneity among cardiac patients admitted for acute exacerbation of heart failure. Experienced cardiologists have traditionally relied on their knowledge and intuition to assess clinical severity and determine appropriate medical interventions, but these insights often lacked quantifiable metrics. Our study aimed to fill this gap by applying unsupervised machine learning techniques, specifically clustering, to analyze datasets comprising 381 patients with acute heart failure (AHF) and 63 clinical and biochemical parameters. The resulting algorithm identified six distinct clusters characterized by significant differences in etiology, clinical presentation, comorbidities, laboratory findings, and lifestyle factors [16, 17]. These identified phenotypes hold great promise for informing future study designs and personalized treatment strategies.

Our research then aimed to investigate behavioral patterns among cardiac patients, which is crucial for improving patient education and fostering self-management skills in heart failure management. Using the Multidimensional Health Locus of Control (MHLC) study, conducted among 758 patients in 11 Polish cardiology centers, we aimed to predict patient behavior based on individual health-related beliefs. This study assessed three dimensions of health beliefs: internal health control, external control by powerful others, and external control by chance. The analysis revealed nine distinct clusters, each representing different combinations of these beliefs, which influence patient engagement and adherence to treatment. Furthermore, our systematic review highlighted the potential of predictive artificial intelligence models to optimize heart failure management, outperforming conventional risk assessment methodologies and offering promising opportunities for improving clinical decision-making.

In summary, the emerging field of AI in medicine holds tremendous promise, as evidenced by global research initiatives and our own pioneering data-driven studies. The challenge now is to translate these findings into practical AI implementation strategies, particularly in clinical settings collaborating with academic institutions.

4.2 The Concept of Resource Integration in Regional Digital Medicine Centres

In 2023, the Medical Research Agency (ABM) announced an initiative to establish Regional Digital Medicine Centres (RCMCs), whose overarching objective is to create an integrated ecosystem of health data by enforcing cooperation among all entities involved in medical and therapeutic processes. The RCMC concept is based on the development of a unified information infrastructure in which data resources -originating from hospitals, medical universities, research institutions, laboratories, biobanks, and other healthcare organisations—are standardised, interoperable, and can be securely shared. This integration is systemic in nature and encompasses clinical, omics, administrative, and operational data generated during therapy, diagnostics, and patient monitoring.

According to the RCMC framework, the centres function as integration hubs responsible for designing and implementing technical and organisational solutions that enable smooth information flow between all participants in medical processes. The concept assumes that effective digital medicine requires not only data centralisation, but above all a coherent technological environment in which data are harmonised and the IT systems of various institutions are capable of communication.

As part of their remit, RCMCs develop and implement advanced digital tools, including prognostic, predictive and therapeutic algorithms, as well as clinical decision-support models that utilise integrated data resources. Key areas include, among others: optimisation of treatment processes, personalisation of therapy, patient monitoring, analysis of adverse-event risk, and the design of drug-dosing algorithms. Integrating data from multiple institutions enables the creation of models of greater precision and quality—models that would be impossible to develop within any single organisation.

Wroclaw Medical University initiated the consortium application, with Wroclaw University of Economics and Business joining as a strategic partner, contributing expert knowledge in digitalisation, data architecture, and the implementation of AI algorithms. The university also plays a crucial role in training and preparing data-science specialists essential for the operation of the RCMCs. The third partner is a renowned clinical hospital, providing clinical resources and practical knowledge of therapeutic pathways.

Thanks to the synergy of all partners' expertise and the innovative concept of medical-data integration, the project received the highest evaluation in Poland and secured substantial funding. The RCMC initiative will therefore establish the foundation for large-scale implementation of artificial intelligence and advanced medical data analytics, enabling a significant leap forward in the development of digital medicine over the next five years.

4.3 Ideas for AI Algorithms Used for Clinical Purposes Within the RCMC Development

When developing a strategy for implementing AI in medical applications, we based our approach on the use of specific algorithms. For example, based on previous research, we aimed to segment patients by simultaneously considering multiple variables, distinguishing clusters of patients with similar clinical characteristics regardless of the number of features analyzed simultaneously. This ability is facilitated by clustering algorithms such as K-means or Gaussian Mixture Model (GMM). Additionally, we envisioned using neural networks to densify the data dimensionality and perform clustering based on feature maps, thus revealing clinical patterns to identify similar patient subgroups.

In addition, our plan included the use of predictive algorithms such as logistic regression models, decision trees, neural networks, or Random Forest algorithms to support clinical reasoning by accurately predicting the risk of complications or acute events such as hospitalization or patient mortality.

Furthermore, we recognized the potential of algorithms to improve diagnostics by reducing misdiagnosis. It was hoped that classification and clustering methods, such as decision tree algorithms, neural networks, and support vector machines (SVMs), would help distinguish subpopulations of patients with similar clinical characteristics, thereby improving diagnostic accuracy.

In addition, algorithms were expected to speed up the diagnostic process by analyzing previously collected medical data to suggest necessary tests for accurate diagnosis or to speed up the detection of anomalies, facilitating the identification of cases requiring further investigation.

Finally, we envisioned using artificial intelligence to analyze imaging studies, specifically using convolutional neural networks (CNNs). This approach enables automatic feature extraction from images and the identification of abnormal structures or malignant areas in histopathological tumor images, thus increasing diagnostic accuracy and efficiency. A significant element of WMU's activities is conducting clinical trials, where algorithms can play a key role in various aspects. First, algorithms can aid in the design of the trials themselves by analyzing clinical datasets to identify relevant features for specific patient groups. Additionally, they can facilitate the selection of optimal patient randomization methods, thus increasing the precision and efficiency of trial results. Furthermore, algorithms can streamline the data collection process, leading to significant time and cost savings while also helping to identify patients most likely to respond positively to a specific treatment regimen.

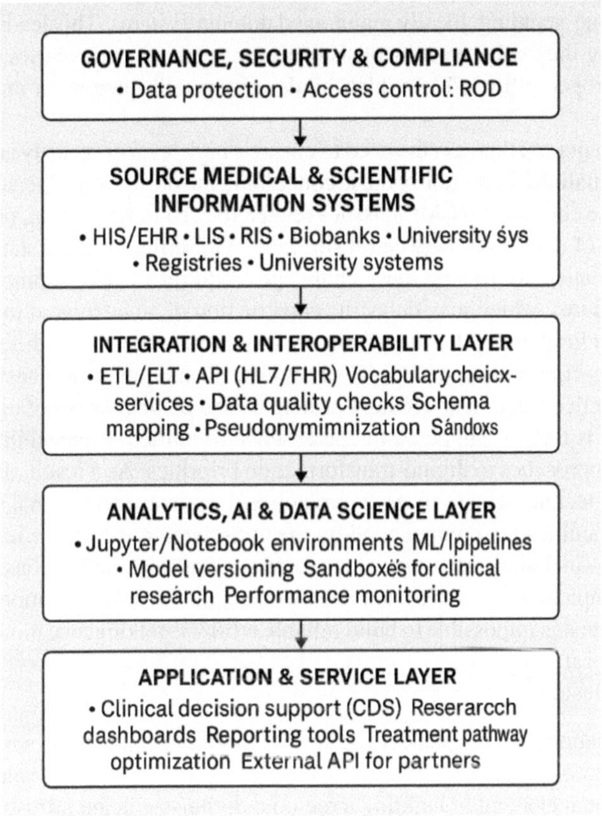

Fig. 1. Architecture Model of the Regional Digital Medicine Centre (RCMC)

Furthermore, we recognize the indispensable role of algorithms in data acquisition and preparation for subsequent analysis. This includes effectively detecting errors in medical data, such as invalid entries or missing data, as well as identifying outliers that may or may not indicate errors in the dataset. Furthermore, algorithms can help detect duplicate data entries and support the merging of records from different medical sources, facilitating data integration regardless of differences in medical terminology.

In summary, preparations for implementing AI in medical tasks encompass infrastructure, finance, organization, and conceptual aspects, culminating extensive collaboration between WMU and data scientists from UEW. The upcoming phase involves the actual implementation of the plan and its basic assumptions.

5 Discussion and Conclusions

Data integration within RCMCs reveals a range of complex technical, organisational and legal challenges arising from the dispersion of information resources across multiple healthcare entities. One of the most significant issues is the lack of unified data standards and the considerable heterogeneity of IT systems in use, ranging from modern HIS/EHR solutions to long-standing, locally maintained domain systems. This leads to difficulties in harmonising data structures, inconsistencies in medical vocabularies, and a lack of semantic interoperability, all of which hinder automated integration and data-quality verification.

Another major challenge is the need to ensure a high level of security and compliance with legal regulations, particularly those concerning the protection of personal and sensitive data. In the context of RCMCs, issues related to pseudonymisation, patient consent, and oversight of data flows between institutions with different legal statuses are especially problematic. Addressing these challenges requires the development of complex control procedures, which may delay the construction of an integrated infrastructure.

Significant limitations also stem from organisational and cultural differences among the institutions contributing to RCMC structures. Each organisation has its own documentation practices, data-management principles and distinct levels of digital maturity. Collaboration is further hampered by the absence of unified responsibility models and inconsistent approaches to digital-transformation priorities. As a result, data integration is not solely a technological process but requires deep inter-institutional coordination.

These difficulties are compounded by the considerable variability in the quality of medical data—including missing values, errors, duplicates and inconsistent records. Without the implementation of systematic mechanisms for data validation, cleaning and standardisation, it is impossible to build reliable artificial-intelligence models or conduct advanced clinical analyses.

Key Conclusions:

- Data integration requires standardisation and harmonisation at the source level.
- Developing consistent vocabularies, documentation structures and validation rules is a foundational element in building a regional digital-medicine infrastructure.
- Robust data-governance mechanisms must be implemented. An integrated centre must operate on the basis of clear rules regarding data responsibility, access control, audit processes and security policies compliant with GDPR.
- Building RCMCs is an organisational endeavour, not merely a technological one.
- Harmonising procedures, organisational cultures and levels of digital maturity is essential to achieving full interoperability.
- Data quality is fundamental to the use of artificial intelligence.
- Only after data have been standardised and cleaned is it possible to effectively train ML/AI models, create predictive tools and conduct clinical research.
- Advanced integration and monitoring tools must be embedded in the RCMC infrastructure. Automated detection of errors, duplicates and inconsistencies, as well as data-lineage systems, are essential for maintaining information reliability.
- RCMCs have the potential to play a key role in transforming the healthcare system, but only if a long-term, regionally coordinated model of cooperation is adopted-one that encompasses technology, competencies and organisational procedures.

Acknowledgements. WMU and UEW are beneficiaries of funding under the ABM/2023/2 competition related to the establishment of regional digital medicine centres (RDMC), "RCMC Disruptor", project no. 2023/ABM/02/00003-00.

Disclosure of Interests. The authors declare no competing interests relevant to the content of this article.

References

1. Siennicka, A., Matera-Witkiewicz, A., Pondel, M., Chomiak-Orsa, I.: Artificial intelligence in medical field—practical observations from the perspective of medical university. In: European Conference on Artificial Intelligence, pp. 123–135. Springer Nature Switzerland, Cham (2024)
2. Razzak, M.I., Imran, M., Xu, G.: Big data analytics for preventive medicine. Neural Comput. Appl. **32**(9), 4417–4451 (2020)
3. Nayarisseri, A., et al.: Artificial intelligence, big data and machine learning approaches in precision medicine & drug discovery. Curr. Drug Targets **22**(6), 631–655 (2021)
4. Hulsen, T., Friedecký, D., Renz, H., Melis, E., Vermeersch, P., Fernandez-Calle, P.: From big data to better patient outcomes. Clin. Chem. Lab. Med. (CCLM) **61**(4), 580–586 (2023)
5. Li, A., Bergan, R.C.: Clinical trial design: past, present, and future in the context of big data and precision medicine. Cancer **126**(22), 4838–4846 (2020)
6. Obijuru, A., Arowoogun, J.O., Onwumere, C., Odilibe, I.P., Anyanwu, E.C., Daraojimba, A.I.: Big data analytics in healthcare: a review of recent advances and potential for personalized medicine. Int. Med. Sci. Res. J. **4**(2), 170–182 (2024)
7. Pine, K.H., Bossen, C.: Good organizational reasons for better medical records: the data work of clinical documentation integrity specialists. Big Data Soc. **7**(2), 2053951720965616 (2020)
8. Ahir, S., Telavane, D., Thomas, R.: The impact of artificial intelligence, blockchain, big data and evolving technologies in coronavirus disease-2019 (COVID-19) curtailment. In: 2020 International Conference on Smart Electronics and Communication (ICOSEC), pp. 113–120. IEEE (2020)
9. Dulhare, U.N., Ahmad, K., Ahmad, K.A.B. (eds.): Machine Learning and Big Data: Concepts, Algorithms, Tools and Applications. John Wiley & Sons (2020)
10. Shafqat, F., Khan, M.N.A., Shafqat, S.: SmartHealth: IoT-enabled context-aware 5G ambient cloud platform. In: IoT in Healthcare and Ambient Assisted Living, pp. 43–67. Springer Singapore, Singapore (2021)
11. Adeghe, E.P., Okolo, C.A., Ojeyinka, O.T.: The role of big data in healthcare: a review of implications for patient outcomes and treatment personalization. World J. Biol. Pharm. Health Sci. **17**(3), 198–204 (2024)
12. Bharathi Mohan, G., Prasanna Kumar, R., Parathasarathy, S., Aravind, S., Hanish, K.B., Pavithria, G.: Text summarization for big data analytics: a comprehensive review of GPT 2 and BERT approaches. Data Analyt. Internet Things Infrastruct., 247–264 (2023)
13. Al-Quraishi, T., Al-Quraishi, N., AlNabulsi, H., AL-Qarishey, H., Ali, A.H.: Big data predictive analytics for personalized medicine: perspectives and challenges. Appl. Data Sci. Anal. **2024**, 32–38 (2024)
14. Chan, K.Y., et al.: Affective design using machine learning: a survey and its prospect of conjoining big data. Int. J. Comput. Integr. Manuf. **33**(7), 645–669 (2020)
15. Tsoukas, V., Boumpa, E., Giannakas, G., Kakarountas, A.: A review of machine learning and TinyML in healthcare. In: Proceedings of the 25th Pan-Hellenic Conference on Informatics, pp. 69–73 (2021)
16. Hassan, M., et al.: Innovations in genomics and big data analytics for personalized medicine and health care: a review. Int. J. Mol. Sci. **23**(9), 4645 (2022)
17. Islam, M.N., Inan, T.T., Rafi, S., Akter, S.S., Sarker, I.H., Islam, A.N.: A systematic review on the use of AI and ML for fighting the COVID-19 pandemic. IEEE Trans. Artif. Intell. **1**(3), 258–270 (2021)

18. Aravazhi, P.S., et al.: The integration of artificial intelligence into clinical medicine: trends, challenges, and future directions. Disease-a-Month, 101882 (2025)
19. Rehan, H.: Advancing cancer treatment with AI-Driven personalized medicine and cloud-based data integration. J. Mach. Learn. Pharm. Res. **4**(2), 1–40 (2024)
20. Gu, Y., et al.: A national survey on how to improve the integration of traditional Chinese medicine and artificial intelligence: attitudes and perceptions from medical staff. Integr. Med. Res., 101214 (2025)
21. Karalis, V.D.: The integration of artificial intelligence into clinical practice. Appl. Biosci. **3**(1), 14–44 (2024)
22. Mintser, O., Babintseva, L., Mokhnachov, S., Sukhanova, O., Hanynets, P., Sarcanich, A.: Medical data compatibility problems in the tasks of information systems integration. J. Appl. Interdisc. Res. (Special Issue), 16–20 (2024)
23. Gebreab, S., Musamih, A., Salah, K., Jayaraman, R., Boscovic, D.: Accelerating digital twin development with generative AI: a framework for 3D modeling and data integration. IEEE Access (2024)
24. Kanagarla, K.: Data fabric: a new approach to data integration (2024)
25. Badrinath, P., Deepak, G.: Challenges approaches and methodologies in data integration: a review. In: 2024 International Conference on IoT, Communication and Automation Technology (ICICAT), pp. 462–464. IEEE (2024)
26. Zhi, L., Liu, S., Dai, H., Liu, M., Wang, J.: Algorithm for data format conversion and compatibility guarantee in technical platforms for cross platform application integration. In: 2024 IEEE 4th International Conference on Data Science and Computer Application (ICDSCA), pp. 893–897. IEEE (2024)
27. Tian, D., Jiang, S., Zhang, L., Lu, X., Xu, Y.: The role of large language models in medical image processing: a narrative review. Quant. Imaging Med. Surg. **14**(1), 1108–1121 (2024)
28. Płuciennik, A., Płaczek, A., Wilk, A., Student, S., Oczko-Wojciechowska, M., Fujarewicz, K.: Data integration–possibilities of molecular and clinical data fusion on the example of thyroid cancer diagnostics. Int. J. Mol. Sci. **23**(19), 11880 (2022)
29. Kothinti, R.R.: Artificial intelligence in healthcare: revolutionizing precision medicine, predictive analytics, and ethical considerations in autonomous diagnostics. World J. Adv. Res. Rev. **19**(3), 3395–3406 (2024)
30. Yu, P., Xu, H., Hu, X., Deng, C.: Leveraging generative AI and large language models: a comprehensive roadmap for healthcare integration. In: Healthcare, vol. 11, no. 20, p. 2776. MDPI (2023)
31. Prabaharan, A.M.: Advanced data integration for smart healthcare: leveraging Blockchain and AI technologies (2024)
32. Mandl, K.D., Kohane, I.S.: Data citizenship under the 21st century cures act. N. Engl. J. Med. **382**(19), 1781–1783 (2020)

Potential of Data Science in the Analysis of Clinical and Scientific Issues in Medicine

Agnieszka Siennicka[1], Agnieszka Matera-Witkiewicz[1], Maciej Pondel[2](✉), and Iwona Chomiak-Orsa[2]

[1] Uniwersytet Medyczny im. Piastów Śląskich we Wrocławiu, 50-367 Wrocław, Polska
{agnieszka.siennicka,adnieszka.matera-witkiewicz}@umw.edu.pl
[2] Uniwersytet Ekonomiczny We Wrocławiu, 53-345 Wrocław, Polska
{maciej.pondel,iwona.chomiak-orsa}@ue.wroc.pl

Abstract.
Purpose The aim of this article is to present the possibilities and potential of data science solutions for medical procedures and clinical problems, as well as for supporting and enhancing scientific research conducted in the field of medical sciences.

Need for the Study: Medicine is becoming increasingly digitalised, with growing numbers of processes in patient care and scientific research relying on digital data. This development opens new opportunities for collaboration between clinicians and researchers and specialists in data science, including the use of artificial intelligence (AI) algorithms. However, challenges remain in implementing these solutions in clinical and research practice; therefore, it is essential to identify the areas in which data science may bring the greatest benefit.

Methodology The article is based on qualitative research methods that allow identification of the potential of data science both in improving medical procedures and in carrying out scientific research within medical sciences. The analysis draws upon the experience of Wroclaw Medical University (WMU) and Wroclaw University of Economics and Business (UEW), as well as collected examples of data science applications in clinical and medical research.

Findings: The article presents specific medical issues of both scientific and clinical relevance that may be addressed, simplified, or accomplished through data science. Particular attention is given to the early stages of implementing data analysis methods and the practical challenges encountered. The digitalisation of medical resources constitutes a foundation for implementing advanced artificial intelligence (AI) and machine learning algorithms, which can contribute to improved diagnostic efficiency, personalised therapy, and optimisation of workflows in medical institutions.

Practical Implications: The findings provide important guidance for medical and research institutions planning to implement data-science-based solutions in medicine. The study highlights the importance of interdisciplinary cooperation,

Published by Springer Nature Switzerland AG 2026
M. Pondel et al. (Eds.): AI4KAM 2025, IFIP AICT 776, pp. 487–499, 2026.
https://doi.org/10.1007/978-3-032-18920-2_46

rigorous data analysis, and a comprehensive implementation strategy—key elements for the effective use of modern technologies in clinical and scientific medical research.

Keywords: data science · artificial intelligence (AI) · medical data · electronic medical records · digital medicine · medical sciences · automation of medical processes

1 Introduction

The considerations presented in this article are introduced from two perspectives, as defined in the title.

The first section of the article is dedicated to outlining examples of scientific issues that may be supported through the use of data-science-based solutions. In this part, the authors discuss the perspectives and principal applications of data science in research procedures conducted by scholars working within the field of medical sciences.

The second section contains a description and characterisation of real examples of practical applications of data science in the domain of clinical issues. These cases are currently being implemented within the RCMC project and will continue to be developed as future research tasks commissioned for the RCMC [1].

The use of data science in medicine opens new opportunities that enhance research potential and improve medical processes [2]. Key benefits include improved efficiency in analysing clinical data, more accurate medical diagnostics, and optimisation of therapeutic procedures [3]. Moreover, data science supports scientific and research activities in medical sciences by enabling the integration of medical knowledge with advanced data-analysis methods [4].

The application of data science in medicine is commonly associated with image-based diagnostics or large-scale data analysis [5]. However, the medical field still holds many unmet needs for which data science could be transformative [6]. At the same time, it remains challenging to establish sufficiently detailed collaboration between medical and IT professionals to allow deep analysis of clinical needs [7]. Fortunately, when conducting projects aimed at digitalising medical processes, certain observations can be made that broaden the perspective on potential applications of data science in medicine [8]. When a project brings together clinical researchers and data specialists, it becomes possible to identify new, significant areas of collaboration for these two disciplines [9, 10]. This is precisely the focus of the present article.

This article was developed using qualitative research methods. The theoretical and cognitive component was prepared based on a literature review, analysis of medical and project documentation, and an examination of available IT solutions. The empirical component was developed using the action-research method, which encompassed interviews with participants of medical projects, participatory observation, analysis of the studied case, and computer simulations employing data science techniques.

2 Methodology

2.1 Concept of the Research Procedure

The defined aim of the article, together with the research problems outlined in the introduction, necessitated the development of a specific research procedure that would allow for the presentation of issues related, on the one hand, to scientific considerations focused on the potential of data science in conducting and implementing research, and, on the other hand, to the identification of practical examples in which the application of these technologies can enhance medical procedures.

Due to the limited scope of this publication, only two practical cases selected by the authors are presented in the article. This research perspective required the authors to conduct parallel research processes—both in the area of literature-based studies and in the analysis of empirical cases.

The individual stages of the research procedure are illustrated in Fig. 1

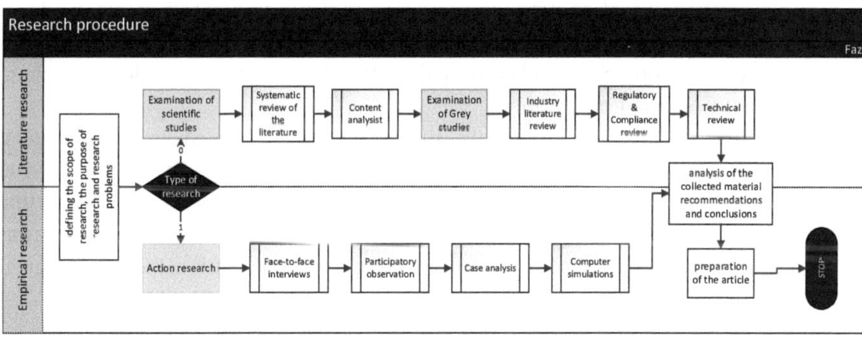

Fig. 1. Research procedure

2.2 Research Methods Used

The designed research procedure assumed the parallel implementation of studies conducted both from the perspective of literature-based research—covering scientific and industry publications—and from the perspective of empirical research, which had a qualitative character and aligned with the principles of action research. This specific type of research is dedicated to studies of a strictly qualitative nature, in which researchers may participate directly in organisational processes and influence them through the introduction of changes and improvements.

For the purposes of this article, the following research methods were applied:

A. **Within the scope of theoretical and industry publications:**

- Systematic review of the literature: Scientific publications published between 2020 and 2025 in the Scopus, Web of Science, and Google Scholar databases were analysed in terms of works containing the terms *data science*, *AI*, *medical data*, or *medical domain* in their titles, abstracts, or keywords;

- Content analysis: As a result of the systematic literature review, 25 articles were selected whose content was closely correlated with the issues presented in this article.
- Industry literaturę review: This analysis concerned publications from the field of technical and technological solutions describing trends, implementation directions, and development paths of IT systems in medical institutions.
- Regulatory & Compliance Review: This included the analysis of legal, regulatory, and normative documents defining the principles for analysing and sharing medical data that may be processed using data-science-based methods;
- Technical review: This involved analysing documentation, technical specifications, and the functional scope of IT solutions implemented at the University Clinical Hospital, Wroclaw Medical University, and the Biobank in Wrocław;

B. **Within the scope of empirical research:**

- Face-to-face interviews: Interviews were conducted with selected participants involved in specific clinical cases chosen for analysis, in which artificial intelligence methods enabling data-science-based analyses had been implemented.
- Participatory observation: The authors of this publication directly participated in all the medical processes and empirical studies discussed, which allowed them to carry out first-hand observations.
- Computer simulations: To verify the conclusions derived from theoretical research, the authors conducted experiments using available medical data from the analysed clinical cases. These experiments enabled computer simulations that made it possible to verify the research hypotheses and address research questions concerning the improvement of medical-process personalisation, the enhancement of medical procedures, and the optimisation of medical-process costs.

As a result of the research conducted, the authors collected a very extensive and comprehensive set of empirical material. The next section of the article presents selected theoretical issues and two empirical case studies.

3 Data Science in Research Problems

3.1 Verification of Data Quality and Data Preparation

A step that should precede the proper analysis of the collected data is the verification of data quality in terms of completeness and consistency [11]. Many researchers take care of their data primarily at the stage of data collection, creating forms and spreadsheets for archiving their observations and making every effort to ensure that the final dataset is complete, reliable, and accurate [12, 13]. This is, however, a laborious process and prone to unintentional errors arising mainly from human factors (fatigue, distraction).

Meanwhile, data science offers a range of methods capable of automatically detecting imperfections within a dataset [14].

One such method is the visualisation of missing-data patterns (e.g., using a heatmap), which helps identify areas with particularly high levels of missing values [15]. Applying such an approach allows researchers to determine whether the missingness is random or follows a specific pattern, which helps prevent systematic errors [16]. Another solution involves checking the value ranges (analysing the mathematical range that characterises individual variables), allowing researchers to detect atypical values that may result from data-entry errors (e.g., knowing that heart-rate measurements should not take three-digit values makes it possible to detect mistakes caused by typing errors or confusion with another measurement such as systolic blood pressure, which enables the correction of the entered value). A further issue is the assessment of consistency between variables, which can be verified using dependency rules that check whether values across different columns are logically coherent (e.g., whether a patient's age corresponds to their date of birth).

In other words, an essential element of data preparation should be the identification and interpretation of outliers, which may arise from simple errors (in which case their detection enables correction) or may represent genuine measurement results (in which case the researcher must decide whether such a value should be included in the analyses, and if not, how to justify its exclusion), as such values may significantly distort model accuracy. Classical methods such as IQR tests or standard deviation analysis perform well in such tasks. However, for more complex datasets, machine-learning algorithms such as Isolation Forest or One-Class SVM are used, as they automatically flag unusual values, indicating potential measurement errors or rare, real cases [17].

When preparing a dataset for analysis, it is also worth considering the stage of so-called data cleaning, which involves filling in missing values and/or removing duplicates. Missing data may be imputed using interpolation, linear regression, or k-NN algorithms, which estimate missing values based on similar observations. In some (though not all) analyses, such an approach is acceptable and facilitates conducting analyses despite imperfections in the dataset [18].

A particular case of preparing a dataset for analysis occurs when a researcher wishes to merge several datasets they possess. Typically, subsequent studies carried out by a researcher within a given medical or health-science discipline are similar to one another—that is, they concern similar types of participants assessed using comparable variables [19]. In such a situation—provided that proper patient consents and approvals from relevant authorities (e.g., bioethics committees) have been obtained—merging several originally independent datasets allows for maximising the scientific value of the work already completed (by increasing the sample size and reusing partly analysed information).

Manual merging of separately created datasets is tedious work burdened with a high risk of error. Here, selected AI methods may assist, depending on the types of issues present in the source data [20]. These include such groups of methods as machine learning, natural language processing, anomaly-detection algorithms, deduplication and record-matching algorithms, federated learning, and graph-based technologies [21]. The selection of an appropriate AI method should be aligned with the diagnosed problems arising from the respective data sources, and a broader description of these possibilities is presented in Table 1.

Table 1. AI Methods for Addressing Data Challenges in Multi-Source Medical Analysis.

AI Method group	Method	Challenges Addressed in the Analysis of Data from Multiple Sources
Machine Learning	Classification algorithms (Random Forest, XGBoost, SVM)	Can be used for categorising and classifying data sources and their attributes
	K-Nearest Neighbors (KNN)	May be applied to search for similar records across different datasets
	Neural Networks	Used to identify hidden dependencies and patterns across various database
Natural Language Processing (NLP)	Transformers (np. BERT, GPT-4, T5)	Enable interpretation and integration of data originating from textual documents (e.g., medical reports, scientific articles)
	Named Entity Recognition (NER)	Allows identification of key entities (e.g., patient names, medications, diagnoses) in text sources
	Word Embeddings (Word2Vec, FastText, GloVe)	Used for unifying terminology across different databases
Anomaly Detection and Data Cleaning Algorithms	Autoencoders	Used to detect errors and inconsistencies in data; particularly effective at identifying anomalies resulting from incorrect entries
	Isolation Forest	For identifying anomalies and erroneous records
	DBSCAN i k-means	For clustering and detecting unusual values
Record Deduplication and Matching Algorithms	Fuzzy Matching	For linking similar but non-identical records (e.g., misspellings in surnames)
	Levenshtein Distance/Jaccard Similarity	For measuring text similarity and eliminating duplicates
	Graph Neural Networks (GNN)	For discovering relationships between data stored in different databases

(*continued*)

Table 1. (*continued*)

AI Method group	Method	Challenges Addressed in the Analysis of Data from Multiple Sources
Federated Learning—combining data without transferring it	Federated Learning	Enables analysis of data distributed across different locations without requiring physical transfer
	Differential Privacy	Ensures privacy protection during the analysis of sensitive data
Data Integration Using Graph Technologies	Graph Databases (Neo4j, Amazon Neptune)	Represent data as connected nodes and edges, facilitating integration
	Knowledge Graphs (np. Google Knowledge Graph)	For modelling relationships between data

3.2 Use of Data Science in the Early Stages of Analysis

Many studies in the field of medical sciences (for example, most clinical trials) are based on a model in which the study protocol requires the collection of a large dataset characterising the study population at the beginning of a longer process (the so-called baseline) [3]. Such a dataset may include basic demographic data (e.g., age, sex, height, body weight), clinical data [22] (often categorical, such as diagnosis, presence of comorbidities, or procedures performed in the past), laboratory data (usually numerical values representing the concentrations of selected chemical molecules in blood or other tissues), and others [23].

During a clinical trial, the patient undergoes a controlled intervention, such as taking a medicinal product or undergoing a specific medical procedure [24]. To document and analyse the effect of the intervention on the study participant, part of the baseline measurements is repeated at regular intervals until the point defined as the end of the study [25].

The end of the study may depend on time (e.g., the protocol specifies that the experiment lasts two years) or may be determined by so-called endpoints—specific medical events defined by the investigator. These events are intended to be unambiguous, and their interpretation should facilitate the assessment of the intervention [26–28]. For example, if the purpose of the medicine is to reduce hospitalisation rates, then hospitalisation becomes one of the endpoints: if the patient does not require hospitalisation thanks to the therapy, they will continue the study until the scheduled final time point; if the patient is hospitalised, that event triggers the repetition of key assessments to enable analysis of the clinical context of the event and its potential link to the drug under investigation [29].

Ultimately, the dataset collected during such a study contains parameters measured multiple times, whose values (or changes therein) may be related to the effect of the intervention. Thus, individual parameters and their values at different stages may be assessed in terms of their predictive value—i.e., how well they might have predicted, for example, whether Patient A would require hospitalisation (due to clinical deterioration) based on their initial results. Importantly, medicine typically searches for single parameters as predictors of a patient's future clinical course.

By engaging data science, however, it is possible to search for predictors in the form of constellations of two or more variables (measured at a single time point or expressed mathematically as changes/differences between baseline and subsequent stages of the project), which may increase predictive accuracy [30]. Crucially, the accuracy of such relationships can be reliably verified because, in every clinical trial, follow-up continues until the end of the intervention [31].

Analyses using constellations of potential predictors instead of single parameters can be performed using many predictive algorithms, such as artificial neural networks, decision trees, random forests, support vector machines (SVM), k-means algorithms, and many others. Each of these methods has its specific properties that may contribute to improved predictive accuracy and the identification of patterns in the data. For example, artificial neural networks are well-suited to analysing complex non-linear patterns, while decision trees provide interpretable models that can highlight key variables influencing the outcome [32].

4 Data Science in Clinical Applications

4.1 Ensuring Personalised Medical Procedure

Contemporary medicine aims to achieve the personalisation of clinical management, meaning the tailoring of treatment and decision-making to an individual patient—based on as many well-characterised features of that person as possible. At the same time, diagnostic guidelines and classification frameworks defining diseases and subsequent management must necessarily refer to broader patient groups (e.g., "patients with a given diagnosis should be treated in the following way"). Data science can, to some extent, support at least partial personalisation of clinical actions while still respecting guidelines that apply to the general population [33].

In one of our publications, we proposed the use of clustering methods to identify clinically meaningful subgroups of patients from within an apparently homogeneous group sharing the same diagnosis [34]. This was possible because, for research purposes, extensive datasets were collected that described these patients in much greater detail than would typically be required for routine clinical care. Ultimately, having a large cohort of patients, each characterised by several dozen variables, we attempted to identify subgroups that we referred to as clinical phenotypes.

The analysis showed that the phenotypes generated by the algorithm were not only consistent with the observations of experienced clinicians but also corresponded to important medical parameters—for example, statistically significant differences in survival between the phenotypes.

Beyond its scientific value, such an approach could have practical clinical applications. For each identified cluster, one could propose more specific (i.e., personalised) clinical management—whether diagnostic (e.g., which additional tests may be warranted) or therapeutic (e.g., how to choose medications or procedures)—supplementing the standard recommendations derived from the diagnosis and common to all patients (functioning as refinements of guidelines that often leave room for choice, for instance regarding therapeutic agents).

A new patient, during initial assessment (which would include evaluation of the indicators used to derive the clusters from large patient populations), could be assigned to one of these clusters (or the algorithm could assess the degree of similarity to each phenotype). This would represent an important step towards completing the diagnostic process. It would therefore be a move towards personalisation without the need to analyse every single variable for every individual patient.

Beyond suggesting therapeutic pathways, it would also be possible to perform predictions (based on the relationships identified in earlier analyses, such as those presented in the aforementioned article) regarding not only the risk of death or complications but even the expected duration of hospitalisation or other aspects of its course (such as acute deterioration), thereby helping anticipate adverse scenarios [35].

In such a practical framework, one could also implement one of the approaches described earlier in the scientific section—namely, in predictive analyses to focus not on a single parameter but on constellations of selected measures, or on measures reflecting changes in parameters over time (e.g., change, increase, decrease of a selected biomarker within a specified period). Similarly, it would be possible to predict such changes (in single or combined indicators) and base clinical reasoning on them (e.g., for risk assessment, evaluation of disease severity, etc.).

4.2 Optimisation of Diagnostic Costs

A second important potential practical application of data science, in our view, is the possibility of suggesting a reasonable frequency for performing diagnostic tests [36, 37](with particular consideration of measurements requiring biological materials such as a patient's blood or urine).

An analysis of medical documentation shows that virtually every patient who spends several days in a hospital ward undergoes a large number of tests, some of which are repeated multiple times [38]. Apart from certain specific examinations (e.g., those related to parameters that cannot realistically change within a few days or weeks, and therefore do not need to be repeated), it is often difficult to identify a clear reason or rule determining why a given test is performed at a particular frequency.

An analysis of individual test results (i.e., the numerical values) and their variability (e.g., over time and in relation to other laboratory values and patient characteristics) could indicate the optimal testing schedule for each parameter, which would likely result in reduced laboratory diagnostic costs [39] (given that each test carries an associated financial burden).

5 Discussion and Conclusions

5.1 Discussion

The aim of this article was to demonstrate the possibilities and potential of data-science-based solutions in the medical domain. The authors chose to present this potential in the context of two key areas relevant to medicine, namely scientific research and clinical applications.

As follows from the analysis of scientific publications, the development of technologies that enhance analytical capabilities—including artificial intelligence and data science—contributes to the creation of increasingly broad research perspectives. In the literature, researchers extensively discuss the benefits arising from the use of data science in medicine and clinical studies. The most frequently cited advantages relate primarily to:

- Optimisation of research processes, through the automation of analyses and the ability to process large volumes of data.
- Application of predictive models, which can reduce the costs associated with laboratory-based experimental research.
- Reduction in the number of unsuccessful clinical trials, resulting from improved selection and recruitment of patients for research and therapeutic procedures.
- Personalisation of therapy and diagnostics, which—through genetic analyses—can more effectively identify diseases and support the selection of tailored, individualised treatment strategies.
- Optimisation of logistics and resource management, through automated adjustment of the use of laboratory equipment, hospital beds, and medical devices.
- Automation of administrative processes and the circulation of medical documentation.

The directions and perspectives for the use of data science in medicine described in the literature are highly encouraging in terms of improving the quality of medical procedures and addressing clinical challenges, as well as supporting and enhancing scientific research conducted in the field of medical sciences.

Unfortunately, the authors' experience—stemming from their direct involvement in medical processes and IT projects aimed at implementing artificial intelligence and data science within the RCMC project—does not inspire similar optimism. The scope of work, as well as numerous limitations resulting primarily from the lack of data integration between the various entities involved in medical procedures, significantly hinders and slows down the processes of implementing data-science-based solutions in medical practice. Drawing on their own research experience, the authors identify several limitations that arise in this process.

5.2 Limitations

Despite the undeniable benefits that the implementation of data science can bring to both scientific research and clinical problem-solving in medicine, the authors identify a number of barriers and limitations encountered during the execution of IT projects within medical institutions. Among the most significant challenges are:

- Problems with the quality of data collected within medical organisations, including lack of standardisation in data formats, incomplete datasets, and erroneous entries.
- Issues related to privacy and ensuring the security of storing and processing medical data.
- Difficulties in obtaining appropriately skilled personnel and IT resources required for the effective implementation and utilisation of available analytical technologies.
- Legal and regulatory constraints, which—due to the highly sensitive nature of medical data—impose very strict procedures regarding access and data processing.

These limitations constitute a substantial barrier to the implementation of data science both in the area of scientific research and in addressing clinical challenges. Nevertheless, the authors remain hopeful that these barriers will be gradually overcome, ultimately enabling the effective use of data science within the medical domain.

5.3 General Recommendations and Conclusions

The literature review and the clinical examples presented in this article indicate that the effective use of data science in the medical field requires numerous comprehensive actions, both in scientific and practical terms. In this context, the success of initiatives involving the use of data science in medicine depends on several key factors:

- Ensuring data quality through appropriate safeguarding and standardisation of data.
- Selecting appropriate artificial intelligence methods tailored to the specific nature of the medical issue.
- Improving the competencies of medical personnel, not only in terms of medical knowledge but also with regard to the analytical capabilities offered by modern technological solutions such as data science.
- Developing IT solutions that enable further growth and scalability, in line with increasing analytical needs in medical services and research.
- Creating interdisciplinary research teams, as contemporary IT technologies—such as artificial intelligence and data science—combine knowledge from multiple fields.

Implementing these recommendations will contribute to strengthening research potential and improving the quality of clinical processes.

Acknowledgements. WMU and UEW are beneficiaries of funding under the ABM/2023/2 competition related to the establishment of regional digital medicine centres (RDMC), "RCMC Disruptor", project no. 2023/ABM/02/00003-00.

Disclosure of Interests. The authors declare no competing interests relevant to the content of this article.

References

1. Siennicka, A., Matera-Witkiewicz, A., Pondel, M., Chomiak-Orsa, I.: Artificial intelligence in medical field—practical observations from the perspective of medical university. In: European Conference on Artificial Intelligence, pp. 123–135. Springer Nature Switzerland, Cham

2. Al-Antari, M.A.: Artificial intelligence for medical diagnostics—existing and future AI technology! Diagnostics **13**(4), 688 (2023)
3. Tian, M., Shen, Z., Wu, X., Wei, K., Liu, Y.: The application of artificial intelligence in medical diagnostics: a new frontier. Acad. J. Sci. Technol. **8**(2), 57–61 (2023)
4. Butryn, B., Chomiak-Orsa, I., Hauke, K., Pondel, M., Siennicka, A.: Application of machine learning in medical data analysis illustrated with an example of association rules. Procedia Comput. Sci. **192**, 3134–3143 (2021)
5. Najjar, R.: Redefining radiology: a review of artificial intelligence integration in medical imaging. Diagnostics **13**(17), 2760 (2023)
6. Gupta, P., Pandey, M.K.: Role of AI for smart health diagnosis and treatment. In: Smart Medical Imaging for Diagnosis and Treatment Planning, pp. 23–45. Chapman and Hall/CRC (2024)
7. Hammond, A., Jain, B., Celi, L.A., Stanford, F.C.: An extension to the FDA approval process is needed to achieve AI equity. Nat. Mach. Intell. **5**(2), 96–97 (2023)
8. Reverberi, C., Rigon, T., Solari, A., Hassan, C., Cherubini, P., Cherubini, A.: Experimental evidence of effective human–AI collaboration in medical decision-making. Sci. Rep. **12**(1), 14952 (2022)
9. Čartolovni, A., Tomičić, A., Mosler, E.L.: Ethical, legal, and social considerations of AI-based medical decision-support tools: a scoping review. Int. J. Med. Informatics **161**, 104738 (2022)
10. Jiang, L., et al.: Opportunities and challenges of artificial intelligence in the medical field: current application, emerging problems, and problem-solving strategies. J. Int. Med. Res. **49**(3), 03000605211000157 (2021)
11. Braun, V., Clarke, V.: Toward good practice in thematic analysis: avoiding common problems and be (com) ing a knowing researcher. Int. J. Transgender Health **24**(1), 1–6 (2023)
12. Mozumder, M.A.I., Sheeraz, M.M., Athar, A., Aich, S., Kim, H.C.: Overview: technology roadmap of the future trend of metaverse based on IoT, blockchain, AI technique, and medical domain metaverse activity. In: 2022 24th International Conference on Advanced Communication Technology (ICACT), pp. 256–261. IEEE
13. Zhang, L., Zhang, L.: Artificial intelligence for remote sensing data analysis: a review of challenges and opportunities. IEEE Geosci. Remote Sens. Mag. **10**(2), 270–294 (2022)
14. Chan, J.Y.L., et al.: Mitigating the multicollinearity problem and its machine learning approach: a review. Mathematics **10**(8), 1283 (2022)
15. Maharana, K., Mondal, S., Nemade, B.: A review: data pre-processing and data augmentation techniques. Global Transitions Proc. **3**(1), 91–99 (2022)
16. Naeem, M., et al.: Trends and future perspective challenges in big data. In: Advances in Intelligent Data Analysis and Applications: Proceeding of the Sixth Euro-China Conference on Intelligent Data Analysis and Applications, 15–18 October 2019, Arad, Romania, pp. 309–325. Springer Singapore, Singapore (2019)
17. Fenza, G., Gallo, M., Loia, V., Orciuoli, F., Herrera-Viedma, E.: Data set quality in machine learning: consistency measure based on group decision making. Appl. Soft Comput. **106**, 107366 (2021)
18. Böttcher, S., et al.: Data quality evaluation in wearable monitoring. Sci. Rep. **12**(1), 21412 (2022)
19. Chen, H., Gomez, C., Huang, C.M., Unberath, M.: Explainable medical imaging AI needs human-centered design: guidelines and evidence from a systematic review. NPJ Dig. Med. **5**(1), 156 (2022)
20. Shwedeh, F., Alzoubi, H.M.: Artificial Intelligence (AI) integration into the decision support systems of health care centers. In: International Scientific Conference Management and Engineering, pp. 165–176. Springer Nature Switzerland, Cham (2024)

21. Jussupow, E., Spohrer, K., Heinzl, A., Gawlitza, J.: Augmenting medical diagnosis decisions? An investigation into physicians' decision-making process with artificial intelligence. Inf. Syst. Res. **32**(3), 713–735 (2021)
22. Nuka, S.T.: The role of AI driven clinical research in medical device development: a data driven approach to regulatory compliance and quality assurance. Global J. Med. Case Rep. **2**(1), 1275 (2022)
23. Oyeniyi, J., Oluwaseyi, P.: Emerging trends in AI-powered medical imaging: enhancing diagnostic accuracy and treatment decisions. Int. J. Enhanced Research Sci. Technol. Eng. **13**(4), 81–94 (2024)
24. Abd-Alrazaq, A., et al.: Large language models in medical education: opportunities, challenges, and future directions. JMIR Med. Educ. **9**(1), e48291 (2023)
25. Khalifa, M., Albadawy, M.: AI in diagnostic imaging: revolutionising accuracy and efficiency. Comput. Methods Progr. Biomed. Update **5**, 100146 (2024)
26. Obuchowicz, R., Strzelecki, M., Piórkowski, A.: Clinical applications of artificial intelligence in medical imaging and image processing—a review. Cancers **16**(10), 1870 (2024)
27. Tian, D., Jiang, S., Zhang, L., Lu, X., Xu, Y.: The role of large language models in medical image processing: a narrative review. Quant. Imaging Med. Surg. **14**(1), 1108–1121 (2024)
28. Zhang, P., Kamel Boulos, M.N.: Generative AI in medicine and healthcare: promises, opportunities and challenges. Fut. Internet **15**(9), 286 (2023)
29. Esmaeilzadeh, P.: Challenges and strategies for wide-scale artificial intelligence (AI) deployment in healthcare practices: a perspective for healthcare organizations. Artif. Intell. Med. **151**, 102861 (2024)
30. Jussupow, E., Spohrer, K., Heinzl, A.: Radiologists' usage of diagnostic AI systems: the role of diagnostic self-efficacy for sensemaking from confirmation and disconfirmation. Bus. Inf. Syst. Eng. **64**(3), 293–309 (2022)
31. Fan, Z., et al.: AI hospital: benchmarking large language models in a multi-agent medical interaction simulator. In: Proceedings of the 31st International Conference on Computational Linguistics, pp. 10183–10213 (2025)
32. Maleki Varnosfaderani, S., Forouzanfar, M.: The role of AI in hospitals and clinics: transforming healthcare in the 21st century. Bioengineering **11**(4), 337 (2024)
33. Bekbolatova, M., Mayer, J., Ong, C.W., Toma, M.: Transformative potential of AI in healthcare: definitions, applications, and navigating the ethical landscape and public perspectives. In: Healthcare, vol. 12, no. 2, p. 125. MDPI (2024)
34. Nasarian, E., Alizadehsani, R., Acharya, U.R., Tsui, K.L.: Designing interpretable ML system to enhance trustworthy AI in healthcare: a systematic review of the last decade to a proposed robust framework (2023)
35. Zvuloni, E., Celi, L.A., Behar, J.A.: Generalization in medical AI: a perspective on developing scalable models. arXiv preprint: arXiv:2311.05418 (2023)
36. Wang, Y., et al.: Economic evaluation for medical artificial intelligence: accuracy vs. cost-effectiveness in a diabetic retinopathy screening case. NPJ Dig. Med. **7**(1), 43 (2024)
37. Gomez Rossi, J., Feldberg, B., Krois, J., Schwendicke, F.: Evaluation of the clinical, technical, and financial aspects of cost-effectiveness analysis of artificial intelligence in medicine: scoping review and framework of analysis. JMIR Med. Inform. **10**(8), e33703 (2022)
38. Hossain, S., Ahmed, A., Khadka, U., Sarkar, S., Khan, N.: AI-driven predictive analytics, healthcare outcomes, cost reduction, machine learning, patient monitoring. AIJMR-Adv. Int. J. Multidisc. Res. **2**(5) (2024)
39. Damoah, I.S., Ayakwah, A., Tingbani, I.: Artificial intelligence (AI)-enhanced medical drones in the healthcare supply chain (HSC) for sustainability development: a case study. J. Clean. Prod. **328**, 129598 (2021)

Graph Neural Networks (GNNs) [14] to more explicitly exploit the graph structure. The Knowledge Graph Neural Network (KGNN) [15] model was an early example that applied Graph Convolutional Network (GCN) techniques to a DDI KG [16]. KGNN aggregates features from a drug's one-hop and two-hop neighbors in the KG, thereby capturing local network structure beyond direct pairwise links. Empirically, KGNN outperformed earlier methods like Decagon that considered only immediate neighbors, confirming that incorporating higher-order connectivity yields better predictions.

Recent studies have increasingly highlighted the value of personalized knowledge graphs (KGs) in healthcare. For example, Oğuztüzün et al. [17] proposed an approach that enriches traditional KGs with polygenic risk scores, biomarker data, and clinical histories drawn from biobank resources to support individualized drug repurposing. Their findings revealed that models incorporating patient-specific features were able to uncover therapeutic opportunities overlooked by conventional population-based approaches, emphasizing the importance of personalization in biomedical prediction. Despite the considerable progress in knowledge graph-based DDI prediction, a key limitation across existing works is the lack of patient-specific factors, such as age and sex in the modelling process.

3 Methodology

In this section, we introduce utilized databases and detail the process of building our KG, including data collection, ontology design, and graph querying. The methodology encompasses data gathering from multiple sources, the definition of a unifying ontology, integration of data into a graph format, and the use of Neo4j as a storage and query engine. Figure 1 illustrates our KG schema and flowchart.

3.1 Databases

We use a hybrid approach for KG construction, where DrugBank, KEGG, and PubChem are expert-curated, while ADRs and demographics from Canada Vigilance are data-driven. Among the most comprehensive is DrugBank [18], a expert-curated database that integrates detailed information on drugs and drug targets.

KEGG (Kyoto Encyclopedia of Genes and Genomes) [19] offers a distinct approach by mapping drugs to biological pathways and disease mechanisms. The KEGG DRUG and KEGG DISEASE modules provide a framework to study DDIs in the context of enzymatic pathways and metabolic reactions.

PubChem [20], maintained by the National Center for Biotechnology Information (NCBI), is another large-scale resource offering extensive chemical and bioactivity data, including assay results and drug-target profiles.

CVAROD [9] serves as a critical source of real-world evidence on ADRs reported across Canada from 1965 to 2024. This dataset includes detailed patient-level information such as age, sex, weight, reported ADRs, suspect drugs, concomitant drugs, outcomes, and indications for use.

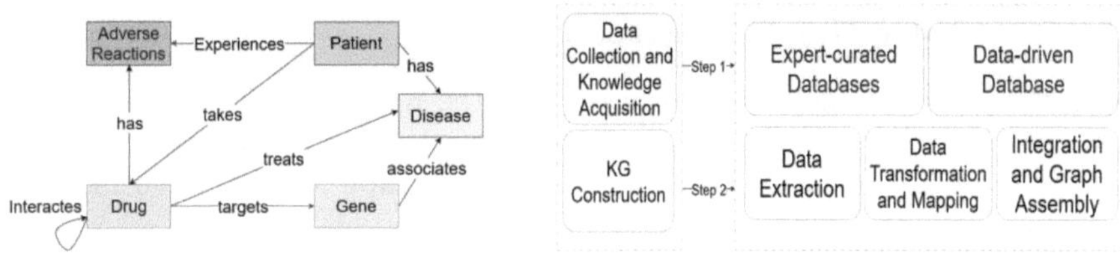

(a) Schema of KG (b) Flowchart

Fig. 1. DDI Prediction Using Patient Demographics.

3.2 Data Collection and Knowledge Acquisition

For predicting ADRs and DDIs using KGs, the initial phase centers on the meticulous collection and preprocessing of diverse datasets. This foundational step ensures the development of a robust KG-based model capable of accurate and personalized predictions. We aggregate data from a variety of reputable sources, such as DrugBank, PubChem, etc. to construct a comprehensive biomedical KG.

First we obtained detailed drug information from DrugBank, including drug identifiers, chemical properties, known targets (genes), and indicated uses (diseases/conditions the drug treats). DrugBank provided the initial list of Drug entities for our graph, along with their target Gene relationships and treatment Disease relationships as stated in its data. We used KEGG Disease database primarily for gene and disease information. We leveraged PubChem for chemical details about drugs. By linking Drug entities to their PubChem Compound IDs and attributes, we ensured consistent chemical identification. We gathered Adverse Drug Reactions (ADRs) from CVAROD. While side effects are not defined as an entity in our schema, this information was collected to augment the Drug entity. The data extracted is a series of compressed ASCII text files downloaded from Canada's Government Health website [9]. We are also going to use existing biomedical ontologies to complete our Graph. Relationships and data sources are detailed in Table 1.

Table 1. Relationships and Data Sources

Relationship	Subject Node	Object Node	Source dataset
takes	Patient	Drug	CVAROD
has	Drug	Adverse Reactions	CVAROD
experiences	Patient	Adverse Reactions	CVAROD
treats	Drug	Disease	DrugBank
targets	Drug	Gene	DrugBank
associates	Gene	Disease	KEGG Disease
has	Patient	Disease	CVAROD
interacts	Drug	Drug	DrugBank

3.3 Knowledge Graph Construction

Using the schema as a guide, we constructed the KG through a multi-step data integration pipeline. The goal was to transform the raw data from DrugBank, KEGG, PubChem, and CVAROD into a unified graph format stored in Neo4j. The key steps in the pipeline were as follows:

Data Extraction: We first extracted data from each source in a structured form. For DrugBank, this involved parsing the provided data dump (which may be in XML or TSV format) to retrieve lists of drugs along with their targets and indicated diseases. From KEGG, we downloaded the disease-gene association data and drug-related entries. Each dataset was then stored in intermediate files CSV for the next steps.

Data Transformation and Mapping: In this step, we mapped the fields from each source to the ontology schema. We standardized identifiers, for example, ensuring that the same gene appearing in both DrugBank and KEGG was represented by a single consistent Gene ID. Drug names and IDs from different sources (DrugBank vs KEGG vs PubChem) were cross-referenced, and a master list of unique drug entities was created. We also matched disease IDs between DrugBank indications and KEGG disease records to consolidate Disease entities. Where ADRs terms matched disease names or known medical terms, we noted those links as well (though these were). The transformation step included cleaning operations such as removing exact duplicate entries, resolving conflicts (e.g., when one source listed a drug-disease pair that another did not, we flagged the discrepancy but generally retained all available relationships unless they were known to be erroneous), and converting data types as needed (e.g., ensuring numeric IDs were handled uniformly).

Integration and Graph Assembly: After mapping, we integrated the data into a cohesive set of nodes and relationships. We prepared node lists for each entity type (Drug, Disease, Gene), each with a unique identifier and aggregated attributes from the original sources. For relationships, we prepared lists of triples (source node ID, relationship type, target node ID) for each relation type (e.g., treat, associated, target). At this stage, we assigned a primary key to each entity (for instance, DrugBank ID for drugs, an official gene symbol or ID for genes, and KEGG ID for diseases) to use as reference in the graph. When a drug had multiple identifiers (e.g., DrugBank, PubChem, KEGG Drug ID), we maintained those as alternate properties on the same node rather than creating separate nodes. The result of this assembly was a set of structured files ready for import-CSV files for each node type and relationship type-formatted according to Neo4j's bulk import requirements. For integration into the KG, we also encode patient features, age, sex, and weight, into a machine-readable form.

In our property graph schema, we attach age, sex, and weight information to each patient either as properties or as links to dedicated demographic nodes.

In one approach, the patient node has properties like age group, sex, and weight category set to the encoded values. Alternatively, we model demographics as first-class entities: for example, we create separate AgeGroup nodes (e.g., AgeGroup: 60–69), Sex nodes (Male, Female), and WeightRange nodes, and connect them to the patient node via relationships.

Loading Into Neo4j: The final step of construction was importing the assembled data into the Neo4j graph database. We selected Neo4j, a native property-graph database, for its intuitive schema mapping, strong community support, and expressive Cypher query language that facilitates pattern-based reasoning over biomedical entities. For example, one can query whether a 65-year-old female patient weighing 52 kg who is prescribed both Benazepril and Lanthanum Carbonate is at increased risk of experiencing a specific adverse reaction, such as nausea.

4 Conclusion and Future Work

Our work extends the KG with demographic information from the CVAROD to capture these differences. This large spontaneous reporting dataset provides real-world patient attributes alongside reported ADRs, enabling a shift towards personalized safety predictions. This study highlights the significant enhancement of the prediction of DDI achieved by integrating demographic characteristics of the patient into a biomedical knowledge graph. By embedding attributes such as age, sex, and weight within Neo4j-based KG, the proposed graph learning model effectively captures interaction patterns that vary between different subgroups of patients. This integration supports the goals of precision medicine, with the aim of tailoring therapeutic strategies to individual characteristics. KG access will be provided upon request. Here are still some challenges that need to be resolved, which are summarized as follows.

1. **Data Quality and Missing Information:** A key challenge is the missing and variable quality of demographic data in spontaneous reports. Not every adverse event report includes complete age, sex, or weight information, especially in older records.
2. **Modeling Demographic Nodes and Attributes in the KG:** Integrating demographic data into a KG as flat attributes presents both practical benefits and notable challenges. While representing age, sex, and weight directly as properties of patient or event nodes simplifies the schema and avoids graph bloat, it limits the semantic expressiveness and flexibility of the KG. Unlike node-based modeling, flat attributes reduce the ability to capture complex relationships or enable rich querying across demographic subpopulations. Another key challenge lies in preparing these attributes for machine learning tasks. Converting categorical demographic values into binary or one-hot encoded vectors is necessary for compatibility with graph learning models, but it can lead to data sparsity or dimensionality issues.

References

1. Chang, T.I., et al.:. Polypharmacy, hospitalization, and mortality risk: a nationwide cohort study. Sci. Rep., **10**(1), 18964 (2020)
2. Celebi, R., Uyar, H., Yasar, E., Gumus, O., Dikenelli, O., Dumontier, M.: Evaluation of knowledge graph embedding approaches for drug-drug interaction prediction in realistic settings. BMC Bioinform. **20**(1), 726 (2019)
3. Garg, S., Parikh, S., Garg, S.: A birds eye view of explainability with knowledge graphs, Navigating healthcare insights (2023)
4. Yu, Y., Huang, K., Zhang, C., Glass, L.M., Sun, J., Xiao, C.: Sumgnn: multi-typed drug interaction prediction via efficient knowledge graph summarization. Bioinformatics, **37**(18), 2988–2995 (2021)
5. Bhuyan, B.P., Ramdane-Cherif, A., Tomar, R., Singh, T.P.: Neuro-symbolic artificial intelligence: a survey. Neural Comput. Appl. **36**(21), 12809–12844 (2024)
6. Correia, R.B., de Araújo Kohler, L.P., Mattos, M.M., Rocha, L.M.: City-wide electronic health records reveal gender and age biases in administration of known drug-drug interactions. NPJ Digit. Med., **2**(1), 74 (2019)
7. Lule, A.P., et al.: Prevalence and factors associated with potential drug-drug interactions in prescriptions presented at private pharmacies in Mbarara city, south-western Uganda. BMC Pharmacol. Toxicol., **25**(1) (2024)
8. Shetty, V., Chowta, M.N., Nithyananda Chowta, K., Shenoy, A., Kamath, A., Kamath, P.: Evaluation of potential drug-drug interactions with medications prescribed to geriatric patients in a tertiary care hospital. J. Aging Res., **2018**, 5728957 (2018)
9. Health Canada. Government of canada, Mar (2025). https://www.canada.ca/en/health-canada/services/drugs-health-products/medeffect-canada/adverse-reaction-database.html
10. Percha, B., Altman, R.B.: Informatics confronts drug–drug interactions. Trends Pharmacol. Sci. **34**(3), 178–184 (2013)
11. Himmelstein, D.S., et al.: Hetnet connectivity search provides rapid insights into how biomedical entities are related. GigaScience, 12, December 2022. ISSN 2047-217X. https://doi.org/10.1093/gigascience/giad047
12. Zitnik, M., Agrawal, M., Leskovec, J.: Modeling polypharmacy side effects with graph convolutional networks. Bioinformatics **34**(13), i457–i466 (2018)
13. Zhang, J., et al.: A knowledge-graph-based multimodal deep learning framework for identifying drug-drug interactions. Molecules **28**(3), 1490 (2023)
14. Feng, Y.-H., Zhang, S.-W.: Prediction of drug-drug interaction using an attention-based graph neural network on drug molecular graphs. Molecules, **27**(9), 3004 (2022). ISSN 1420-3049. https://doi.org/10.3390/molecules27093004
15. Lin, X., Quan, Z., Wang, Z.-J., Ma, T., Zeng, X.: Kgnn: knowledge graph neural network for drug-drug interaction prediction. In IJCAI **380**, 2739–2745 (2020)
16. Zhong, Y., et al.: Ddi-GCN: Drug-drug interaction prediction via explainable graph convolutional networks. Artif. Intell. Med. **144**, 102640 (2023)
17. Oğuztüzün, Ç., Gao, Z., Li, H., Rong, X.: Precision drug repurposing (PDR): Patient-level modeling and prediction combining foundational knowledge graph with biobank data. J. Biomed. Inform. **163**(104786), 104786 (2025)
18. Knox, C.,et al.: DrugBank 6.0: the Drugbank Knowledgebase for (2024). Nucleic Acids Res., **52**(D1), D1265–D1275 (2024)

19. Kanehisa, M., Furumichi, M., Sato, Y., Matsuura, Y., Ishiguro-Watanabe, M.: KEGG: biological systems database as a model of the real world. Nucleic Acids Res. **53**(D1), D672–D677 (2025)
20. Kim, S.,et al.: PubChem 2025 update. Nucleic Acids Res., **53**(D1), D1516–D1525 (2025)

Author Index

Published by Springer Nature Switzerland AG 2026
M. Pondel et al. (Eds.): AI4KAM 2025, IFIP AICT 776, pp. 509–510, 2026.
https://doi.org/10.1007/978-3-032-18920-2